W0036566

CRM Series in Mathematical Physics

Springer
New York
Berlin
Heidelberg
Barcelona
Hong Kong
London
Milan
Paris
Singapore
Tokyo

CRM Series in Mathematical Physics

Semenoff and Vinet, Particles and Fields

Gordon Semenoff   Luc Vinet
Editors

Particles and Fields

With 36 illustrations

Springer

Gordon Semenoff
Department of Physics and Astronomy
University of British Columbia
Vancouver, British Columbia V6T 1Z1
Canada

Luc Vinet
Centre de Recherches Mathématiques
Université de Montréal
Montreal, Québec H3C 3J7
Canada

Library of Congress Cataloging-in-Publication Data
Particles and fields / editors, Gordon W. Semenoff, Luc Vinet
    p.   cm.— (CRM series in mathematical physics)
    ISBN 0-387-98402-X (alk. paper)
    1. Quantum field theory.   2. Gauge fields (Physics).   3. Quantum
Hall effect.   I. Semenoff, G. W.   II. Vinet, Luc.   III. Series.
QC174.45.P37   1998
530.14´3—dc21   97-46958

Printed on acid-free paper.

Production managed by Steven Pisano; manufacturing supervised by Joe Quatela.
Photocomposed pages prepared from the authors' LaTeX files.
Printed and bound by Maple-Vail Book Manufacturing Group, York, PA.
Printed in the United States of America.

9 8 7 6 5 4 3 2 1

ISBN 0-387-98402-X Springer-Verlag New York Berlin Heidelberg   SPIN 10661256

CAP-CRM summer school on particles and fields, Banff, Alberta, Canada, 1994.

Preface

The present volume has its source in the CAP-CRM summer school on "Particles and Fields" that was held in Banff in the summer of 1994. Over the years, the Division of Theoretical Physics of the Canadian Association of Physicists (CAP) has regularly sponsored such schools on various theoretical and experimental topics. In 1994, the Centre de Recherches Mathématiques (CRM) lent its support to the event. This institute, located in Montreal, is one of Canada's national research centers in the mathematical sciences. Its mandate includes the organization of scientific events across Canada and since 1994 the CRM has been holding a yearly summer school in Banff as part of its thematic program. The summer school, whose lectures are collected here, has thus become a tradition.

The focus of the school was integrable theories, matrix models, statistical systems, field theory and its applications to condensed matter physics, as well as certain aspects of algebra, geometry, and topology. This covers some of the most significant advances in modern theoretical physics. The present volume updates and expands these lectures and reflects the high pedagogical level of the school.

The first chapter by E. Corrigan describes some of the remarkable features of the integrable Toda field theories which are associated with affine Dynkin diagrams. The second chapter by J. Feldman, H. Knörrer, D. Lehmann, and E. Trubowitz covers the lectures given at the school by the first author. It describes a gas of interacting fermions in the field of a lattice of magnetic ions; interestingly, this system is shown to be a Fermi liquid in infinite volume. In the third chapter, written by D. Freed, the question of how quantum groups arise in three-dimensional topological quantum field theory is discussed. This is done using path integrals and Chern-Simons theories associated with finite gauge groups.

In Chapter 4, the reader will find the lecture notes of T. Miwa who described a method for computing correlation functions of solvable lattice models using the six-vertex model as an example. Chapter 5, written by A. Morozov, provides a review of matrix models from the point of view of integrable systems. In the four lectures that form Chapter 6, A. Niemi discusses and explains the localization of certain path integrals in the framework of equivariant cohomology. Chapter 7 contains a thorough description by S. Ruijsenaars of systems of Calogero-Moser type and records the original material that he presented at the school. The nonrelativistic models as well as the relativistic ones that Ruijsenaars himself discovered are analyzed in both the classical and quantum domains. Chapter 8 is authored by M. de Wild Propitius and F. A. Bais, and is based on lectures delivered by F. A. Bais. It covers planar gauge theories whose symmetries are broken down, via the Higgs mechanism, to a finite gauge group. These lectures, which focus on the discrete gauge theory describing the long-distance physics of such models, tie in remarkably well with the lecture of D. Freed.

A number of invited seminars were also presented at the school and complemented the program very nicely. Two are included in these proceedings: the first written by A. Capelli, C. A. Trugenberger, and G. R. Zemba describes a characterization of quantum Hall fluids based on the representation theory of the $W_{1+\infty}$ algebra; the second by P. I. Etingof addresses issues in the spectral theory of quantum vertex operators for the quantum affine algebra $u_q(\mathfrak{sl}_2)$.

Financial support for the school came from NSERC (Canada) and CRM which is funded by NSERC, FCAR (Quebec), the Université de Montréal, and private donations. We take this occasion to gratefully acknowledge their support and to thank the CRM staff for the organization of the school and the preparation of this volume.

Gordon Semenoff
Luc Vinet

Contents

List of Contributors

Andrea Cappelli Dipartimento di Fisica, I.N.F.N.-Sezione di Firenze, Largo E. Fermi 2, 50125 Firenze, Italy
cappelli@andrea.fi.infn.it

Edward Corrigan Department of Mathematical Sciences, University of Durham, South Road, Durham DH1 3LE, UK
Edward.Corrigan@durham.ac.uk

Pavel I. Etingof Department of Mathematics, Harvard University, Cambridge, MA 02138, USA
etingof@math.harvard.edu

Joel S. Feldman Department of Mathematics, University of British Columbia, Vancouver, BC V6T 1Z2, Canada
feldman@math.ubc.ca

Daniel S. Freed Department of Mathematics, University of Texas, Austin, TX 78712, USA
dafr@math.utexas.edu

Horst Knörrer Department Mathematik, Rämistrasse 101, ETH Zentrum, 8092 Zürich, Switzerland
horst.knoerrer@math.ethz.ch

Detlef Lehmann Department Mathematik, Rämistrasse 101, ETH Zentrum, 8092 Zürich, Switzerland
lehmann@math.ethz.ch

> *Present address:* Department of Mathematics, University of British Columbia, Vancouver, BC V6T 1Z2, Canada
> lehmann@math.ubc.ca

Tetsuji Miwa Research Institute for Mathematical Sciences, Kyoto University, Kyoto, 606-01, Japan
miwa@kurims.kyoto-u.ac.jp

Alexei Morozov Institute for Theoretical and Experimental Physics, B. Cheremushkinskaja, 25, 117259 Moscow, Russia
morozov@vitep5.itep.ru

Antti J. Niemi Department of Theoretical Physics, Uppsala University, P.O. Box 803, 75108 Uppsala, Sweden
niemi@tethis.teorfys.uu.se

Simon N. M. Ruijsenaars Centrum voor Wiskunde en Informatica, P.O. Box 4079, 1090 AB Amsterdam, The Netherlands
siru@wxs.nl

Eugene Trubowitz Department Mathematik, Rämistrasse 101, ETH Zentrum, 8092 Zürich, Switzerland
eugene.trubowitz@math.ethz.ch

Carlo A. Trugenberger Département de Physique Théorique, Université de Genève, 24, quai Ernest-Ansermet, 1211 Genève 4, Switzerland
cat@kalymnos.unige.ch

Mark de Wild Propitius Laboratoire de Physique Théorique et Haute Énergies, Université Pierre et Marie Curie, 4, place Jussieu, 75252 Paris cedex 05, France
mdwp@lpthe.jussieu.fr

Guillermo R. Zemba I.N.F.N. and Dipartimento di Fisica Teorica, Università degli Studi di Torino, Via Pietro Giuria, 1, 10125 Torino, Italy
zemba@to.infn.it

1

Recent Developments in Affine Toda Quantum Field Theory

E. Corrigan

1 Introduction

It is not intended to give a detailed review of all the recent activities in the area of Toda field theory, but rather to highlight some of the interesting developments, and to point out some of the currently outstanding problems. The list of references is by no means exhaustive.

Affine Toda field theory [1, 2] is a theory of r scalar fields in two-dimensional Minkowski space-time, where r is the rank of a compact semi-simple Lie algebra g. The classical field theory is determined by the Lagrangian density

$$\mathcal{L} = \frac{1}{2}\partial_\mu \phi^a \partial^\mu \phi^a - V(\phi), \tag{1.1}$$

where

$$V(\phi) = \frac{m^2}{\beta^2} \sum_0^r n_i e^{\beta \alpha_i \cdot \phi}. \tag{1.2}$$

In (1.2), m and β are real (classically unimportant) constants, α_i $i = 1,$ ..., r are the simple roots of the Lie algebra g, and $\alpha_0 = -\sum_1^r n_i \alpha_i$ is a linear combination of the simple roots; it corresponds to the extra spot on an extended Dynkin diagram for g, at least in so far as representing its inner products with the simple roots is concerned. For notational reasons, $n_0 = 1$ in (1.2), but the other integers n_i are characteristic for each type of theory. They are tabulated in many places, for example, in Kac [3]. The quantity $h = \sum_0^r n_i$ is called the Coxeter number. For most purposes, in the present context α_0 will not represent a simple root of the affine algebra $\hat{g}$.

If the term $i = 0$ is omitted from (1.2) in the Lagrangian (1.1), then the theory, both classically and after quantization, is conformal, and will be referred to as conformal Toda field theory or, simply, as Toda field theory. With the term $i = 0$, the conformal symmetry is broken but the theory remains classically integrable, in the sense that there are infinitely many independent conserved charges in involution. The recent renewal of interest in Toda field theories was stimulated by Zamolodchikov's ideas concerning perturbations of conformal field theories [4, 5]. The possible root systems which can be used in the Lagrangian (1.1), maintaining the classical integrability, are in one-to-one correspondence with the untwisted and twisted affine Dynkin-Kac diagrams [2]. However, in what follows, it is useful to distinguish those which are unchanged (apart from a possible flip) under the transformation

$$\alpha_i \rightarrow 2\frac{\alpha_i}{|\alpha_i|^2}, \qquad (1.3)$$

and those which are "dual"' pairs under this transformation. The self-dual set are $a_n^{(1)}$, $d_n^{(1)}$, $e_n^{(1)}$ whose roots are all of equal length (conventionally, the longest root satisfies $|\alpha_i|^2=2$) and $a_{2n}^{(2)}$; the dual pairs are $(b_n^{(1)}, a_{2n-1}^{(2)})$, $(c_n^{(1)}, d_{n+1}^{(2)})$, $(g_2^{(1)}, d_4^{(3)})$, $(f_4^{(1)}, e_6^{(2)})$.

Each of the members of the self-dual set are untwisted with roots of equal length, except for $a_{2n}^{(2)}$ which is twisted and contains roots of three different lengths. The affine Toda theory corresponding to the simplest case $a_1^{(1)}$ is recognized to be the sinh-Gordon theory (for real coupling), or the sine-Gordon theory (for imaginary coupling, or real coupling and imaginary fields).

Each of the non-simply laced or twisted root systems can be obtained by "folding" one of the simply laced Dynkin or affine Kac-Dynkin diagrams, respectively. A pair of examples will suffice to illustrate this. Consider the Dynkin diagram for d_4 (first diagram).

$$\qquad (1.4)$$

It has a symmetry $\alpha_1 \rightarrow \alpha_2 \rightarrow \alpha_3 \rightarrow \alpha_1$, under which $\beta_1 = \alpha_4$ and $\beta_2 = (\alpha_1 + \alpha_2 + \alpha_3)/3$ are clearly invariant. The two roots β_1, β_2 are simple roots for g_2 (that is, the second diagram, where the shorter root β_2 corresponds to the black spot). The extra root $\alpha_0 = -(\alpha_1 + \alpha_2 + \alpha_3 + 2\alpha_4)$ for d_4 is also invariant and becomes the extra root $\beta_0 = -(2\beta_1 + 3\beta_2)$ for $g_2^{(1)}$. On the other hand, the dual of $g_2^{(1)}$, $d_4^{(3)}$ is obtained using the threefold symmetry of the extended Kac-Dynkin diagram of $e_6^{(1)}$ (first diagram).

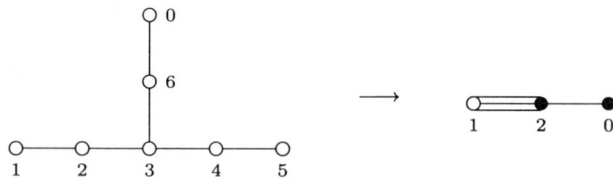

In this case, $\beta_1 = \alpha_3$, $\beta_2 = (\alpha_2 + \alpha_4 + \alpha_6)/3$ and $\beta_0 = (\alpha_0 + \alpha_1 + \alpha_5)/3 = -\beta_1 - 2\beta_2$ are the invariant combinations under the symmetry and provide the root system for $d_4^{(3)}$ (the second diagram, in which the additional root attached to the g_2 Dynkin diagram is short).

The two types of g_2 extension lead to quite different classical field theories.

2   Classical Integrability and Classical Data

Toda field theory is classically integrable and indeed conformal [6]. Affine Toda field theory is classically integrable and also, in a generalized version, conformal [7]. Consider a conformal transformation in light-cone variables:

$$x_\pm = \frac{x^0 \pm x^1}{2} \to \bar{x}_\pm(x_\pm). \tag{2.1}$$

Clearly, since the second derivative of the scalar fields transforms via

$$\partial_+\partial_-\phi \to \bar{\partial}_+\bar{\partial}_-\phi = \frac{\partial x_+}{\partial \bar{x}_+}\frac{\partial x_-}{\partial \bar{x}_-}\partial_+\partial_-\phi,$$

the equations of motion are invariant provided the potential term also scales in a suitable manner:

$$\sum_{i=1}^{r} n_i\alpha_i e^{\beta\alpha_i\cdot\phi} \to \frac{\partial x_+}{\partial \bar{x}_+}\frac{\partial x_-}{\partial \bar{x}_-}\sum_{i=1}^{r} n_i\alpha_i e^{\beta\alpha_i\cdot\phi}.$$

The latter requires the fields themselves to transform according to

$$\phi(x_+, x_-) \to \bar{\phi}(\bar{x}_+, \bar{x}_-) = \phi(x_+, x_-) + \frac{\rho}{\beta}\ln\left(\frac{\partial x_+}{\partial \bar{x}_+}\frac{\partial x_-}{\partial \bar{x}_-}\right),$$

where the vector ρ enjoys the property

$$\rho\cdot\alpha_i = 1, \quad i = 1, 2, 3, \ldots, r.$$

Since the fundamental weights satisfy

$$2\lambda_i \cdot \frac{\alpha_j}{|\alpha_j|^2} = \delta_{ij},$$

ρ may be expressed in terms of the fundamental weights:

$$\rho = \sum_{i=1}^{r} \frac{2}{|\alpha_i|^2}\lambda_i.$$

It is immediately clear, since

$$\rho \cdot \alpha_0 = -\sum_{i=1}^{r} n_i = 1 - h,$$

that adding the extra term in the Lagrangian (proportional to n_0) breaks the conformal symmetry.

On the other hand, suppose the extra affine term is included, and further suppose that the set of roots is actually taken to be the set of simple roots for the affine algebra itself. Then, the set $\hat{\alpha}_i$, $i = 0, 1, 2, \ldots, r$ are independent, lying in a Minkowski space of signature $(r + 1, 1)$ and, once again, a vector $\hat{\rho}$ can be found for which

$$\hat{\rho} \cdot \hat{\alpha}_i = 1 \quad i = 0, 1, 2, \ldots, r.$$

Using this, the argument of the last paragraph may be repeated to conclude the theory is conformal even with the affine term included [7]. The penalty being paid for this is that the scalar fields ϕ no longer take values in a Euclidean space and the energy is no longer a positive definite functional of the field components. Restricting the fields to a Euclidean space breaks the conformal invariance and, effectively, introduces a mass scale.

This situation is reminiscent of string theory which, in its most basic form, contrives to describe families of massive states starting from a conformally invariant Lagrangian whose fields take values in space-time.

Once conformal Toda field theory is quantized, it provides a coupling dependent representation of the Virasoro algebra whose central charge (*ade* series) is given by Ref. 6:

$$c(\beta) = r + 48\pi|\rho|^2 \left(\frac{\beta}{4\pi} + \frac{1}{\beta}\right)^2. \tag{2.2}$$

This central charge is clearly symmetric under the transformation $\beta \to 4\pi/\beta$, revealing that the quantum conformal field theory enjoys a weak-strong coupling symmetry not apparent in the original Lagrangian.

It will be assumed the fields take values in an r-dimensional Euclidean space, spanned by the simple roots of the Lie algebra g.

The classical integrability of the affine Toda field theories relies on the existence of a Lax pair from which the conserved quantities may be established. The details of this is a story in itself [2, 8] but from our present perspective it is enough to be aware of some of the main results. First of all, it is relatively straightforward to check the equivalence between the zero curvature property

$$F_{01} = \partial_0 A_1 - \partial_1 A_0 + [A_0, A_1] = 0,$$

and the affine Toda field equations provided the two components of the two-dimensional vector potential A_μ are given by:

$$A_0 = H \cdot \frac{\partial_1 \phi}{2} + \sum_0^r m_i \left(\lambda E_{\alpha_i} - \frac{1}{\lambda} E_{-\alpha_i} \right) e^{\alpha_i \cdot \phi / 2}$$

$$A_1 = H \cdot \frac{\partial_0 \phi}{2} + \sum_0^r m_i \left(\lambda E_{\alpha_i} + \frac{1}{\lambda} E_{-\alpha_i} \right) e^{\alpha_i \cdot \phi / 2},$$

(2.3)

where H, E_{α_i} and $E_{-\alpha_i}$ are the Cartan subalgebra and the generators corresponding to the simple roots and the extra root, respectively, of g. Thus, in particular,

$$[\mathbf{H}, E_{\alpha_i}] = \alpha_i \, E_{\alpha_i}$$

$$[E_{\alpha_i}, E_{-\alpha_j}] = \delta_{ij} \frac{2\alpha_j \cdot \mathbf{H}}{|\alpha_j|^2}.$$

The spectral parameter is λ, and the coefficients m_i are chosen to satisfy

$$m_i^2 = \frac{n_i \alpha_i^2}{8}.$$

For convenience, the classically unimportant constants m and β have been scaled away.

Since the path ordered integral of A_1,

$$T(a, b; \lambda) = P \exp \int_a^b dx^1 A_1$$

satisfies

$$\frac{d}{dt} T = T A_0(b) - A_0(a) T,$$

then

$$Q(\lambda) = \operatorname{tr} T(-\infty, \infty; \lambda)$$

is time independent when $\partial_1 \phi \to 0$ as $x^1 \to \pm\infty$ and $\phi(\infty) = \phi(-\infty) + 2\kappa$, where $\kappa \cdot \alpha_i$ is an integer.

An important fact about the Lax pair is the possibility of performing a gauge transformation after which the potentials lie in a Cartan subalgebra h_i of g, two members of which are

$$E_{\pm 1} = \sum_{i=0}^r m_i E_{\pm \alpha_i}.$$

Once this gauge transformation has been done, the potential A_1 takes the form

$$a_1 = \lambda E_1 + \sum_{s \geq 1} \lambda^{-s} h_s I_0^{(s)},$$

where the sum on the right-hand side runs over the exponents of the algebra g (another characteristic set of integers which will be met again below), modulo h, the Coxeter number. The elements of the Cartan subalgebra are conveniently labeled by the r exponents, and $h_{s+nh} = h_s$. The zero curvature condition reads

$$\partial_0 a_1 = \partial_1 a_0,$$

and therefore the integral of a_1 over the whole line is conserved. Since λ is arbitrary, there are infinitely many conserved quantities

$$Q_s = \int_{-\infty}^{\infty} dx^1 \, I_0^{(s)}.$$

Adding or subtracting the equations (2.3) reveals that λ scales under a Lorentz transformation ($\lambda \to l\lambda$) in order to guarantee the correct transformation of the light-cone components of the vector potential. Consequently, the conserved quantities Q_s must scale under the transformation by a factor l^s. (There is an alternative Abelianization for which there is a similar expression for a_1 after the gauge transformation expressed as a series of positive powers in λ. From this, a matching set of conserved quantities of the opposite spin is obtained.)

It is possible to demonstrate the involutary nature of the charges by first demonstrating the existence of a classical r-matrix for which

$$\{T(\lambda) \otimes T(\mu)\} = \left[r\left(\frac{\lambda}{\mu}\right), T(\lambda) \otimes T(\mu) \right], \quad T(\lambda) \equiv T(-\infty, \infty; \lambda),$$

follows from the canonical equal-time Poisson bracket between the fields and their conjugate momenta. Indeed, Olive and Turok [8] give r in the form:

$$r\left(\frac{\lambda}{\mu}\right) = \frac{\mu^h + \lambda^h}{\mu^h - \lambda^h} \sum_{i=1}^{r} H_i \otimes H_i$$

$$+ \frac{2}{\mu^h - \lambda^h} \sum_{\alpha > 0} \frac{|\alpha|^2}{2} (\lambda^{l(\alpha)} \mu^{h-l(\alpha)} E_\alpha \otimes E_{-\alpha} + \lambda^{h-l(\alpha)} \mu^{l(\alpha)} E_{-\alpha} \otimes E_\alpha),$$

where the sum is over all positive roots of g (i.e. all those roots expressible as combinations of simple roots with positive integer coefficients), and $l(\alpha)$ is the length of a root (i.e. the sum of the integers in its expansion in terms of simple roots).

As mentioned above, the classically conserved charges are two dimensional Lorentz tensors, labeled by their "spin" in light-cone coordinates, the possible spins being the exponents of the algebra repeated modulo its Coxeter number h. In other words, the conserved charges may be denoted Q_{s+kh}, where s is an exponent and k is an integer. The quantities

$Q_{\pm 1}$ correspond to the light-cone components of the energy-momentum. If the quantized field theory retains the integrability property, it is expected that the conserved quantities will survive as mutually commuting quantum operators whose eigenstates are the particles of the theory. Thus, for single-particle states,

$$Q_p|a\rangle = q_p^a e^{p\theta_a}|a\rangle \quad p = s + kh, \tag{2.4}$$

where θ_a is the rapidity of the particle labeled a:

$$p_a \equiv m_a(\cosh\theta_a, \sinh\theta_a),$$

and m_a is its mass.

Taking the classical Lagrangian as the starting point for the definition of a quantum field theory, the classical masses can be computed by expanding the potential (1.2) as far as the quadratic term. Thus the mass matrix is

$$(M^2)^{ab} = m^2 \sum_0^r n_i \alpha_i^a \alpha_i^b.$$

For most cases, the mass matrix was diagonalized some time ago [2]. However, more recently, it was noticed [5, 9] and then proved Lie algebraically by Freeman and others [10] that except for the twisted cases the eigenvalues of the mass matrix m_a^2 were themselves the squares of the components of the lowest eigenvalue eigenvector of the Cartan matrix corresponding to g. In other words, it is possible to choose an ordering of the masses so that $\mathbf{m} = (m_1, m_2, \ldots, m_r)$ and

$$C\mathbf{m} = 4\sin^2\frac{\pi}{2h}\mathbf{m}. \tag{2.5}$$

This is quite a remarkable result since it allows the particles to be assigned unambiguously (up to mass degeneracies), to the Dynkin diagram for g. Curiously, the gravitational ordering once this assignment is made follows the "weight" ordering in terms of the dimension of the fundamental representations also assigned to the spots on the Dynkin diagram. For example, the $a_n^{(1)}$ masses ($m_a = 2m\sin(a\pi/h)$) increase from the ends of the Dynkin diagram working in, and are doubly degenerate corresponding to the folding symmetry of the diagram; for $e_8^{(1)}$ the masses are assigned as follows:

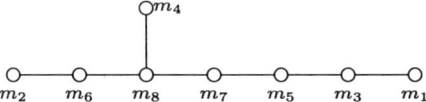

Even more remarkably, for the ade series of simply laced algebras (and for one of the twisted cases $a_{\text{even}}^{(2)}$), the classical mass ratios are preserved in perturbative field theory at least to one-loop order [11, 12], suggesting in turn that the set of eigenvalues q_1^a in (2.4) is an eigenvector of the Cartan matrix for g. In a while, a generalization of this result will be discussed. The fact that the radiative corrections to the classical masses in most of the non-simply laced cases are not universal is the first hint that these theories will be rather different as quantum field theories.

Again at the classical level, it is interesting to examine the cubic term in the expansion of (1.2) since this defines the classical three-point couplings, needed to carry out, for example, the one-loop check mentioned above. Once the mass eigenstates are known, it is possible to compute the couplings, $c^{abc} = \sum_0^r n_i \alpha_i^a \alpha_i^b \alpha_i^c$. For many triples, the coupling vanishes. However, when the coupling is not zero it is proportional always [11, 12] to the area of a triangle whose sides have lengths equal to the masses of the three participating particles a, b, c. One consequence of this is that the coupling defines a set of angles (the angles in the triangle) by, for example,[1]

$$m_c^2 = m_b^2 + m_b^2 - 2m_a m_b \cos \bar{\theta}_{ab}^c, \qquad (2.6)$$

where

$$\bar{\theta} = \pi - \theta.$$

Just which couplings are nonzero will be explained further below once some of the geometry of root systems has been explored.

It is tempting to suppose eq. (2.5) generalizes (at least for the simply laced cases) and the other conserved quantities have values constituting the components of the remaining eigenvectors of the Cartan matrix [13]:

$$C\mathbf{q}^{(s)} = 4 \sin^2 \frac{s\pi}{2h} \mathbf{q}^{(s)}. \qquad (2.7)$$

This is true in the quantum theory, in the sense that it is consistent with other known facts. Again, a fuller discussion is deferred.

2.1   Geometry Associated with the Coxeter Element

There is some very useful geometry associated with roots and weights which is less familiar than facts about representation theory. For that reason it will be reviewed briefly here—further details may be found in several books, for example, Bourbaki or Humphreys [14].

A simple Weyl reflection w_i corresponds to a linear transformation on the root lattice representing a reflection in a plane orthogonal to the simple root α_i given by

$$w_i \colon x \to x - 2\frac{\alpha_i \cdot x}{\alpha_i^2}\alpha_i.$$

---

[1]There is a convention in the literature that the outside angles of the triangle are denoted by θ_{ab}^c, etc.

A Coxeter element of the Weyl group is a product over the simple roots of the simple Weyl reflections. Clearly, once a set of simple roots have been chosen (i.e. a set of r independent roots such that any other root is either a linear combination of them with positive coefficients, or a linear combination with negative coefficients), this product could be taken with different orderings of the individual simple reflections. However, different orderings lead to conjugate Coxeter elements. Alternative choices of simple roots also lead to conjugate Coxeter elements. For our present purposes, there is a special ordering which is extremely useful and which relies on the fact that Dynkin diagrams have no closed loops. The latter fact allows the simple roots to be divided into two sets such that the roots within each set are mutually orthogonal (i.e. members of the same set are not joined by a line in the Dynkin diagram). The members of the two sets are distinguished by assigning a color to them, either black or white. Thus, for example, the e_8 diagram can be colored in this way as follows:

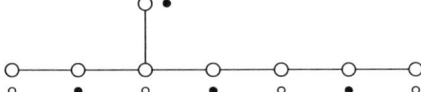

The same is true of any other Dynkin diagram, as you can easily check.[2] Obviously, the product of Weyl reflections corresponding to simple roots within one of these special sets no longer matters since the Weyl reflections commute. With this choice, the Coxeter element is only ambiguous up to the relative black-white ordering, and for definiteness, the Coxeter element will be chosen for the rest of these talks to be

$$w = w_\bullet w_\circ \equiv \prod_{k \in \bullet} w_k \prod_{k \in \circ} w_k.$$

Notice, that each of the factors $w_\bullet$ and $w_\circ$ separately squares to unity. Notice, too, that there is a close relationship between the two factors of the Coxeter element and the Cartan matrix of g:

$$(w_\bullet + w_\circ)\alpha_i = \sum_j (2\delta_{ij} - C_{ij})\alpha_j.$$

This is easily checked on the black and white roots separately. On the other hand,

$$(w_\bullet + w_\circ)^2 = 2 + w + w^{-1},$$

and therefore

$$(2 + w + w^{-1})\sum_i x_i \alpha_i = \sum_i x_i (2 - C)^2_{ij}\alpha_j, \qquad (2.8)$$

---

[2]It is not true of all extended Dynkin diagrams, however; think of $a^{(1)}_{even}$.

revealing a close relationship between the eigenvectors of the Coxeter element and the Cartan matrix. Indeed, the eigenvalues of the Cartan matrix have been mentioned already in connection with the classical data, and using them it is straightforward to deduce the eigenvalues of the Coxeter element. The eigenvalues of the Cartan matrix are given in (2.7), therefore the eigenvalues of the Coxeter element are also labeled by the spins s and are computed from (2.8) to be

$$e^{2i\pi s/h}.$$

Hence, the order of the Coxeter element is h.

To understand how the Coxeter element affects the roots, it is convenient to define certain linear combinations of the simple roots whose coefficients are the eigenvectors of the Cartan matrix. Consider, for each spin s,

$$a_\bullet^{(s)} = \sum_{i\in\bullet} q_i^{(s)}\alpha_i \qquad\qquad l_\bullet^{(s)} = \sum_{i\in\bullet} q_i^{(s)}\lambda_i$$

$$a_\circ^{(s)} = \sum_{i\in\circ} q_i^{(s)}\alpha_i \qquad\qquad l_\circ^{(s)} = \sum_{i\in\circ} q_i^{(s)}\lambda_i$$

where the λ_i are fundamental weights. Then,

$$w_\bullet a_\bullet^{(s)} = -a_\bullet^{(s)}, \quad w_\bullet a_\circ^{(s)} = a_\circ^{(s)} + 2\cos\theta_s a_\bullet^{(s)}, \quad \theta_s = \frac{s\pi}{h}. \qquad (2.9)$$

The first of (2.9) follows directly from the definition of $w_\bullet$ and the mutual orthogonality of the black roots; the second is less straightforward and requires a sequence of steps. Since the white roots have inner products with the black roots represented by the entries of the Cartan matrix,

$$w_\bullet a_\circ^{(s)} = a_\circ^{(s)} - \sum_{\substack{i\in\circ \\ j\in\bullet}} q_i^{(s)} C_{ij}\alpha_j$$

$$= a_\circ^{(s)} - \sum_{\substack{i\in\circ\cup\bullet \\ j\in\bullet}} q_i^{(s)} C_{ij}\alpha_j + \sum_{\substack{i\in\bullet \\ j\in\bullet}} q_i^{(s)} C_{ij}\alpha_j$$

$$= a_\circ^{(s)} - \lambda^{(s)} a_\bullet^{(s)} + 2a_\bullet^{(s)},$$

and the last line is the second relation in (2.9). Thus,

$$\left|w_\bullet a_\circ^{(s)}\right|^2 = \left|w_\circ a_\circ^{(s)}\right|^2 + 4\cos\theta_s a_\circ^{(s)} \cdot a_\bullet^{(s)} + 4\cos^2\theta_s \left|a_\bullet^{(s)}\right|^2 = \left|a_\circ^{(s)}\right|^2,$$

and there is a similar relation with black and white interchanged. Comparing the two leads to

$$\left|a_\circ^{(s)}\right|^2 = \left|a_\bullet^{(s)}\right|^2 \quad\text{and}\quad a_\circ^{(s)} \cdot a_\bullet^{(s)} = -\cos\theta_s \left|a_\circ^{(s)}\right|\left|a_\bullet^{(s)}\right|.$$

Using the fact relating simple roots to fundamental weights, one also has

$$a_\bullet^{(s)} = \sum_{\substack{i\in\bullet \\ j\in\bullet\cup\circ}} q_i^{(s)} C_{ij}\lambda_j = 2\left(l_\bullet^{(s)} - \cos\theta_s l_\circ^{(s)}\right),$$

with a similar expression for $a_\circ^{(s)}$. Hence,

$$l_\circ^{(s)} = \frac{a_\circ^{(s)} + \cos\theta_s a_\bullet^{(s)}}{2\sin^2\theta_s}, \quad l_\bullet^{(s)} = \frac{a_\bullet^{(s)} + \cos\theta_s a_\circ^{(s)}}{2\sin^2\theta_s},$$

from which it is easily seen that

$$l_\bullet^{(s)} \cdot a_\bullet^{(s)} = 0 = l_\circ^{(s)} \cdot a_\circ^{(s)}, \quad l_\bullet^{(s)} \cdot l_\circ^{(s)} = \frac{\cos\theta_s}{4\sin^2\theta_s}|a_\bullet^{(s)}|^2$$

$$|l_\bullet^{(s)}|^2 = |l_\circ^{(s)}|^2 = \frac{1}{4\sin^2\theta_s}|a_\bullet^{(s)}|^2.$$

Clearly, all four vectors lie in a plane, for each choice of s. The Coxeter element acts as a clockwise rotation in this plane through an angle $2\theta_s$. Notice that although it might happen that $a_\bullet^{(s)}$ and $a_\circ^{(s)}$ lie on the same Coxeter orbit, this will never be the case for $-a_\bullet^{(s)}$ and $a_\circ^{(s)}$ (nor indeed for $l_\bullet^{(s)}$ and $l_\circ^{(s)}$). The various vectors are illustrated in the diagram below.

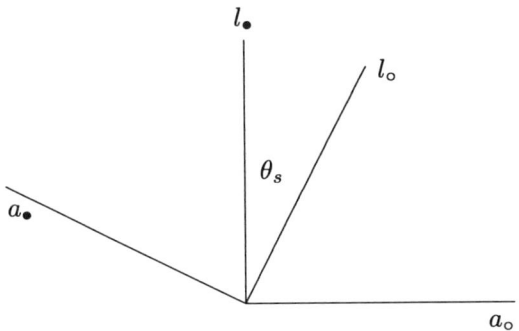

FIGURE 1. Vectors in the s plane.

The sets $w^p\alpha_i$, $i \in \circ$, and $-w^p\alpha_i$, $i \in \bullet$, for $p = 1, \ldots, h$, of images of simple roots do not intersect for distinct simple roots. They provide rank r orbits each of h elements, together providing the full set of roots for the algebra g. Since it is possible to normalize the eigenvectors of the Cartan matrix so that

$$\sum_s q_i^{(s)} q_j^{(s)} = \delta_{ij},$$

the relationship between $a_\circ^{(s)}$ and $a_\bullet^{(s)}$ can be inverted to find that each simple root has a component in the spin s plane, either along $a_\bullet^{(s)}$ or along $a_\circ^{(s)}$, according to its color. Moreover, the images of each simple root α_i under repeated application of the Coxeter element each have a component in this plane of the same magnitude, equal to $q_i^{(s)}$. In particular, for $s = 1$, and according to the earlier observation (2.5), each orbit has a classical mass associated with it and therefore the whole orbit may be assigned to a particular particle.

Consider three roots which make a triangle. The projection of this triangle onto the $s = 1$ Coxeter plane provides another triangle, the sides of which have lengths equal to the masses of the particles associated with the three orbits to which the three roots belong.

More interestingly, it is now possible to give a characterization of the three-point couplings:

> *The coupling c^{abc} between three particles a, b, and c is nonzero if and only if there are three vectors, one from each of the orbits representing the particles, which sum to zero.*

This was first proved on a case by case basis by Dorey [15] and then deduced from the classical Lagrangian by Fring, Liao, and Olive by extending the ideas of Freeman [10]. For all cases, the couplings actually correspond to a subset of the Clebsch-Gordan series in the sense that if a coupling abc is nonzero, then the tensor product of the representations assigned to the particles according to their assignment to the Dynkin diagram contains the trivial representation, i.e. $\mathbf{a} \otimes \mathbf{b} \otimes \mathbf{c} \supset \mathbf{1}$. Except for the cases corresponding to $a_n^{(1)}$, $d_4^{(1)}$ the converse is not true. The relationship between the couplings and the Clebsch-Gordan series, and other matters, has been further elucidated by Braden [16].

There is another way to label the Coxeter orbits [17] which will turn out to be useful in the next section, and which naturally incorporates the minus sign. Let the elementary Weyl reflections and the roots be labelled so that

$$w = w_\bullet w_\circ = w_1 \cdots w_b w_{b+1} \cdots w_r,$$

then set

$$\phi_k = w_r w_{r-1} \cdots w_{k+1} \alpha_k = \begin{cases} \alpha_k & \text{for } k \in \circ \\ w_\circ \alpha_k = -w^{-1} \alpha_k & \text{for } k \in \bullet \end{cases}$$
$$= (1 - w^{-1}) \lambda_k. \tag{2.10}$$

The last two lines of (2.10) follow directly from the definition in the first. One consequence of the second fact is that the set of images of distinct ϕ_k never overlap and, therefore, these vectors may be used equally well to label the orbit which has been associated with a particle. A curious property of these vectors, which turns out to have a use in the next section, is the following. The image of each of them under the inverse Coxeter element is a positive root and successive images remain positive until the middle of the orbit, after which they all change sign, remaining negative subsequently for the rest of the orbit.

As an illustration, consider d_4 labeled as before, with the outer spots colored black and the center spot white. Then

$$\phi_k = \alpha_k + \alpha_4 \text{ for } k = 1, 2, 3 \quad \phi_4 = \alpha_4,$$

and the orbits of the inverse Coxeter element are

$$
\begin{array}{ll}
1: & \alpha_1 + \alpha_4; \alpha_2 + \alpha_3 + \alpha_4; \alpha_1; -\alpha_4 - \alpha_1; \ldots \\
2: & \alpha_2 + \alpha_4; \alpha_1 + \alpha_3 + \alpha_4; \alpha_2; -\alpha_4 - \alpha_2; \ldots \\
3: & \alpha_3 + \alpha_4; \alpha_2 + \alpha_1 + \alpha_4; \alpha_3; -\alpha_4 - \alpha_3; \ldots \\
4: & \alpha_4; \alpha_1 + \alpha_2 + \alpha_3 + 2\alpha_4; \alpha_1 + \alpha_2 + \alpha_3 + \alpha_4; -\alpha_4; \ldots.
\end{array}
\tag{2.11}
$$

Clearly, these orbits provide the full set of roots as promised.

3   Aspects of the Quantum Field Theory

The intention of this discussion is to provide certain basic facts and formulae which have proved to be remarkably universal. Most of the time, the *ade* sequence of theories will be considered. The affine diagrams for these are invariant under the transformation (1.3). For background, and further references on S-matrix theory in two dimensions, the review article by Zamolodchikov and Zamolodchikov [18] is strongly recommended.

It will be supposed, as a working hypothesis, that (1) after quantization the conserved charges remain conserved and in involution—i.e. commute with one another, and (2) the particle spectrum of any one of these theories is as simple as possible—in other words, the particles are exactly r in number, stable and distinguishable, if not by their masses, then by one or other of the conserved charges.[3] With this in mind, it will be assumed that there is a set of one-particle states which are eigenstates of the quantum conserved charge operators (which have not been properly constructed yet), i.e. (2.4):

$$Q_p|a\rangle = q_a^p e^{p\theta_a}|a\rangle \quad p = s + kh,$$

where θ_a is the rapidity of particle a. In addition two-particle states are also assumed to be eigenstates of the conserved charge operators, i.e.:

$$Q_p|a, b\rangle = (q_a^p e^{p\theta_a} + q_b^p e^{p\theta_b})|a, b\rangle,$$

and so on.

---

[3]The sine-Gordon model is not as simple as this. There are two "soliton" states which are only distinguished by a zero spin charge. Such a distinction is already too relaxed for the present purposes; it permits the mixing of soliton and antisoliton in a scattering process, leading, in turn, to a greatly enriched spectrum of bound states.

There is no elementary definition of these particle states, although they may be approximated perturbatively. If it is further supposed that two-particle states, which are functions of a pair of rapidities, one for each particle, may under certain circumstances be dominated by a single-particle state, then

$$q_a^p e^{p\theta_a} + q_b^p e^{p\theta_b} = q_{\bar{c}}^p e^{p\theta_{\bar{c}}},$$

where the particle $\bar{c}$ must itself be part of the conjectured spectrum. If the particle c is to be stable, then this situation cannot occur for real rapidity difference $\Theta_{ab} = \theta_a - \theta_b$. Rather, considering the spin ± 1 charges (the light-cone components of energy-momentum, $q_k^{\pm 1} = m_k$), the situation may arise only when the rapidity difference satisfies

$$m_{\bar{c}}^2 = m_a^2 + m_b^2 + m_a m_b \cosh \Theta_{ab} = m_a^2 + m_b^2 + m_a m_b \cos U_{ab}^c,$$

and the masses m_a, m_b, and m_c are the sides of a triangle with internal angles $\bar{U}_{ab}^c$, $\bar{U}_{ac}^b$, $\bar{U}_{bc}^a$. The same triangle equally well permits a description of the energy-momentum conservation for the virtual processes $ac \to \bar{b}$ and $bc \to \bar{a}$. For these special rapidity differences, the rapidities themselves may be written conveniently as

$$\theta_a = \theta_{\bar{c}} - i\bar{U}_{ac}^b, \quad \theta_b = \theta_{\bar{c}} + i\bar{U}_{bc}^a.$$

One might expect that for a certain rapidity difference the vacuum state may dominate a particle-antiparticle state. For this, energy momentum requires $\Theta_{a\bar{a}} = i\pi$ and therefore,

$$q_a^p e^{p\theta_a} + q_{\bar{a}}^p e^{p(\theta_a - i\pi)} = 0, \quad \text{i.e. } q_{\bar{a}}^p = (-)^{p+1} q_a^p. \tag{3.1}$$

One consequence of this is immediate. Particles and antiparticles are distinguished only by even spin charges. Affine Toda theories for which the exponents are odd must contain self-conjugate particles only (this includes $e_7^{(1)}$ and $e_8^{(1)}$ which have no mass-degenerate states, and $d_{\text{even}}^{(1)}$ which has mass degenerate states corresponding to the prongs of the fork in the Dynkin diagram).

More generally, using (3.1), the $ab \to \bar{c}$ conserved charge relation may be rewritten

$$q_a^p e^{ip(U_{ac}^b + U_{bc}^a)} + q_c^p e^{ipU_{bc}^a} + q_b^p = 0, \tag{3.2}$$

which represents a series of "triangular" relations, one for each p.

At this stage, it is tempting to identify the set of triangle relations (3.2) with the projections of the root triangles which represent the classical couplings described in the last section [15]. This would require the masses of the particles in the quantum spectrum to be essentially the same (up to an overall scale) as the mass parameters in the classical Lagrangian; the coupling angles U_{ab}^c would also be the same as those for the classical mass triangles (2.6). The eigenvalues of the conserved quantities q_a^p would repeat modulo h, and the first r of them, labeled by the exponents of the algebra, would be the components of the corresponding Cartan eigenvector (2.7). These consequences of the initial hypotheses are very strong and would need to be verified by direct calculation. In fact, if one examines the members of the self-dual affine Toda theories perturbatively, all infinities may be removed by normal ordering, and a calculation of the "bubble" diagrams which contribute to mass corrections at lowest order reveals that the identification of the classical masses with the quantum masses is natural in the sense that the mass corrections are independent of particle type. This is quite definitely not the case for those theories which have a different dual partner. For them, the mass corrections are type-dependent and it would seem unnatural to insist on the quantum masses being the same as the classical ones. In the next lecture an alternative and more attractive approach to these will be presented.

Given the large number of conserved charges and the set of distinguishable particles, the two particle scattering of affine Toda particles is expected to be simple in the sense that the character of the particles is unchanged, there is no production, and the initial and final momenta are the same [18]. Indeed, the "in" and "out" states may differ only by a phase which may at most (because of Lorentz invariance) be a function of the rapidity difference of the two particles and of the coupling β^2 (or $\hbar$). That is,

$$|a, b\rangle_{\text{out}} = S_{ab}(\Theta_{ab}; \beta)|a, b\rangle_{\text{in}}. \qquad (3.3)$$

For each pair of particles there will be such a phase factor. The set of factors will be called the two-particle S-matrix although there is no real scattering going on. The phase factors regarded as functions of complex rapidity difference are far from trivial, however. Indeed, they are analytic functions of the rapidity difference,[4] with an intricate set of zeroes and poles characteristic of each specific theory. The "physical strip" consists of the region $0 < \text{Im}(\Theta) < i\pi$, the boundary $\text{Im}(\Theta) = 0$ being the physical s-channel, the boundary $\text{Im}(\Theta) = i\pi$ being the physical t-channel. The region $0 > \text{Im}(\Theta) > -i\pi$ is the second, unphysical sheet from the point of view of the Mandelstam variables. The continuation of the unitarity and crossing relations away from $\text{Im}(\Theta) = 0$ requires

$$S_{ab}^{-1}(\Theta) = S_{ab}(-\Theta) \quad \text{and} \quad S_{a\bar{b}}(\Theta) = S_{ab}(i\pi - \Theta), \qquad (3.4)$$

respectively. Taken together, these imply the S-matrix elements are $2\pi i$ periodic.

---

[4]Using the rapidity variable effectively removes the s, t threshold cuts. There are no others because there is no production.

If it is further assumed that the scattering is factorizable (one more feature which will need substantiating ultimately), then the three-particle S-matrix elements may be regarded as products of two-particle S-matrix elements. The ordering ambiguity (normally resolved by the Yang-Baxter equation) is absent here because of the special nature of the two-particle scattering (there is no reflection). Thus, the Yang-Baxter equation itself plays no role. On the other hand, a two-particle state for complex rapidity difference may share the quantum numbers of a single particle state. The signal for this is a direct channel pole in the physical strip at a purely imaginary rapidity difference. The fusing idea allows a set of "bootstrap" relations to be formulated which relates the scattering of particle d, say, with a and b, to the scattering of d with $\bar{c}$. That is, pictorially,

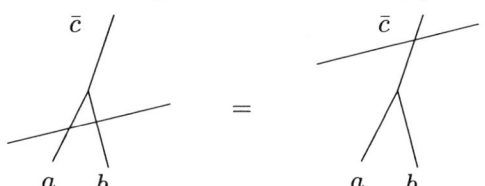

and algebraically [19]:

$$S_{d\bar{c}}(\Theta) = S_{da}(\Theta - i\bar{U}^b_{ac})S_{db}(\Theta + i\bar{U}^a_{bc}). \qquad (3.5)$$

The latter, in the case of the two particle state a, $\bar{a}$ at a relative rapidity of $i\pi$, is in agreement with the crossing relation.

Using (3.4), (3.5) can be rearranged to

$$S_{da}(\Theta + iU^b_{ac} + iU^a_{bc})S_{dc}(\Theta + iU^a_{bc})S_{db}(\Theta) = 1,$$

which is a "product" version of the sum rule (3.2).

The equation (3.5) is extremely useful but it does not fix the S-matrix elements uniquely. What it does do is supply a set of consistency conditions which must be supplemented by other data or prejudices. A natural idea, given the hypothesis concerning the masses of the particles, is to suppose that the possible fusings for which the bootstrap works are to be given precisely by the classical couplings c^{abc} and their associated angles [1, 5, 11, 12]. Before checking this, however, it is also necessary to make some remarks concerning the coupling dependence of the S-matrix elements.

Clearly, when $\beta = 0$ the S-matrix elements ought to be unity since the particles are free. When $\beta \neq 0$, the poles indicating the fusings are at fixed positions and these must be the only poles on the physical strip since it has been presumed that the classical spectrum is complete. Therefore, the S-matrix elements must contain traveling zeroes on the physical strip which coincide with the fixed poles to cancel them when $\beta = 0$. Because of unitarity, each zero has an accompanying pole which must be situated on the unphysical strip for any choice of the coupling in the range $0 \leq \beta \leq \infty$.

For the simply laced conformal Toda theories, it was pointed out (2.2) that there is a symmetry between strong and weak coupling in the sense that the Virasoro central charge is actually invariant under $\beta \to 4\pi/\beta$. There is an elegant solution to the bootstrap which also enjoys this symmetry and which neatly parametrizes the coupling dependence to ensure the other desirable properties.

It is useful to have a convenient notation for the basic ratio of functions satisfying the periodicity and unitarity relations [11]. Set

$$(x)_\Theta = \sinh\left(\frac{\Theta}{2} + \frac{i\pi x}{2h}\right) \bigg/ \sinh\left(\frac{\Theta}{2} - \frac{i\pi x}{2h}\right),$$

bearing in mind that the fusing angles are always multiples of π/h. Often, this will be referred to merely as (x). The fixed poles will be represented by terms of this kind. The coupling dependence may be incorporated by assembling blocks of this type as follows:

$$\{x\}_\Theta = \frac{(x-1)(x+1)}{(x-1+B)(x+1-B)},$$

where B is coupling dependent and, in fact, universal,

$$B(\beta) = \frac{1}{2\pi} \frac{\beta^2}{1 + \beta^2/4\pi}.$$

Clearly, for small β, $\{x\}_\Theta$ approaches unity and, because $B(\beta) = 2 - B(4\pi/\beta)$, for large β, exactly the same is true; the pole-cancelling zeroes cross over and cancel the opposite pole, as β runs from 0 to ∞. In principle, other functions of β might be acceptable under these constraints but this is the one originally suggested by Arinshtein el al. [1] for the $a_n^{(1)}$ series, on the basis of a comparison with the sin/sinh-Gordon model [18].

Rather than simply writing down the conjectures for the S-matrices, it is instructive to build one up, watching the bootstrap in action. An interesting case is $d_4^{(1)}$ labeled as in Fig. 1.4. There are four distinguished particles 1, 2, 3 with mass $\sqrt{2}m$ and 4 with mass $\sqrt{6}m$, and possible couplings

$$c^{123} \quad c^{aa4} \ (a = 1, 2, 3) \quad \text{and} \quad c^{444}.$$

In this case, the Coxeter number $h = 6$. The particles are self-conjugate and therefore the S-matrix elements are crossing symmetric.

To begin with, make the simplest compatible conjecture for S_{12}, say. It ought to have a pole at $\Theta = 2i\pi/3$ with a positive residue, and a crossed partner at $\Theta = i\pi/3$ with negative residue, and no other fixed poles. In the above notation, a reasonable guess would be

$$S_{12}(\Theta) = \{3\} \equiv \frac{(2)(4)}{(2+B)(4-B)} \sim \frac{i}{\Theta - 2\pi i/3} \frac{\pi B}{6} \quad (= S_{13} = S_{23}).$$

Using this pole together with the bootstrap (3.5), leads to

$$S_{33}(\Theta) = S_{13}\left(\Theta - \frac{i\pi}{3}\right)S_{13}\left(\Theta + \frac{i\pi}{3}\right) = \{1\}\{5\}$$

$$\equiv \frac{(0)(2)(4)(6)}{(B)(2-B)(4+B)(6-B)} \quad (= S_{11} = S_{22}),$$

which is again crossing symmetric. However, because $(6) = -1$, it is the pole at $\Theta = i\pi/3$ which has the positive residue this time, indicating the nonzero coupling c^{334} (or c^{114}, c^{224}). Using the latter with the bootstrap yields

$$S_{14}(\Theta) = S_{13}\left(\Theta - \frac{i\pi}{6}\right)S_{13}\left(\Theta + \frac{i\pi}{6}\right) = \{2\}\{4\}$$

$$\equiv \frac{(1)(3)^2(5)}{(1+B)(3-B)(3+B)(5-B)} \quad (= S_{24} = S_{34}), \quad (3.6)$$

and

$$S_{44}(\Theta) = S_{14}\left(\Theta - \frac{i\pi}{6}\right)S_{14}\left(\Theta + \frac{i\pi}{6}\right) = \{1\}\{3\}^2\{5\}$$

$$\equiv -\frac{(2)^3(4)^3}{(B)(2-B)(2+B)^2(4-B)^2(4+B)(6-B)}. \quad (3.7)$$

All other bootstrap relations are verified by direct checking.

The S-matrix elements may be represented conveniently in the following diagram.

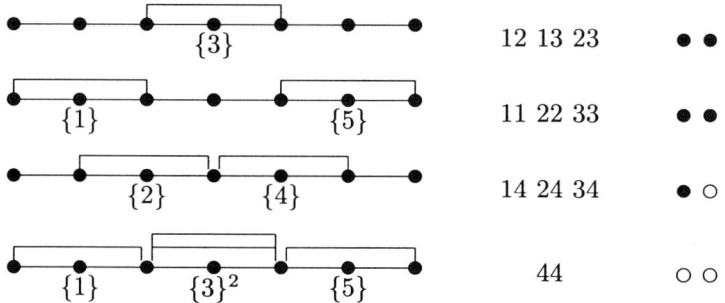

In the diagram, the physical strip is marked at intervals of $\pi/6$ and the boxes represent the basic block factors in the S-matrix elements whose labels appear on the right. The vertical lines represent the fixed pole positions on the physical strip.

Now, return to the Coxeter element orbits provided in (2.11). There is a striking correlation between the coefficient of α_i in the expressions for the vectors in the positive part of the orbit of ϕ_j (i.e. the vectors $w^{-p}\phi_j$ for

$p = 0, 1, 2$), and the boxes which appear in the diagram representing S_{ij}. Indeed, the boxes are labeled by $2p + 1 + \epsilon_{ij}$, where ϵ_{ij} depends only on the color of the pair i, j:

$$\epsilon_{\bullet\bullet} = \epsilon_{\circ\circ}, \quad \epsilon_{\circ\bullet} = -\epsilon_{\bullet\circ} = 1.$$

This observation suggests there is a formula for the S-matrix elements in terms of roots weights and Coxeter orbits [15]:

$$S_{ab}(\Theta) = \prod_{p=1}^{h} \{2p + 1 + \epsilon_{ab}\}_{+}^{\lambda_a \cdot w^{-p} \phi_b}. \tag{3.8}$$

(The + subscript indicates that because the blocks are all accounted for by traversing the positive part of the orbit of ϕ_b only, it is necessary, when extending the product over the whole Coxeter orbit, to realize that the numerators of the blocks are reconstructed by the positive part of the orbit and the denominators by the negative part.) Because the formula depends only on the roots/weights it promises to be universal. Actually, that is indeed the case. The S-matrix formula once postulated can be shown to be symmetrical under $a \leftrightarrow b$, to be unitary, to satisfy crossing, and to satisfy the bootstrap relation [15, 20, 21]. This is a beautiful result, which applies only to the ade series of cases, and it is a pity there is no direct derivation of it from the field theory.

The observant reader will have noticed that among the poles in the S-matrix elements (3.6) and (3.7) there are some of order two and three. These are clearly required by the bootstrap and are, in a sense, fortuitously useful. The point is that there is little hope of computing directly the S-matrix elements perturbatively, at least for arbitrary rapidity, but there is some hope of calculating the coefficients of higher-order poles. This is because the poles appear as Landau singularities in Feynman diagrams and there is a well-developed calculus for dealing with them. For example, there is no time to go into details, but the double poles all arise from singularities of box diagrams and it has been checked that the coefficients of the poles to order β^4 agree with the predictions of the S-matrix elements (not just for $d_4^{(1)}$, but in all cases). This is quite important because the observant reader will also have noticed that there was no attempt to check the bootstrap on the double pole in (3.6). The fact that the poles are an artifact of the perturbation expansion and there is no order β^2 tree-graph with a simple pole sharing the same pole position strongly suggests this is a correct interpretation of the bootstrap rules. On the other hand, third-order poles (and in general odd-order poles) appear as dressings of tree processes and one would expect that their existence does really signal a bound state which ought to participate in the bootstrap. It has been found that there are a number of diagrams of different types contributing to the third-order poles (all two-loop diagrams, since the leading contribution to the third-order pole is order β^6), but never more than twenty-six (!) as one ranges over the *ade* series. The sum of the contributions from these diagrams agrees exactly with the prediction from the conjectured S-matrix elements whenever the cubic poles occur in the *ade* series. The number of diagrams to be computed is prohibitively large for the fourth and higher poles (up to order twelve in the $e_8^{(1)}$ S-matrix elements), and for these a direct check is not possible. The type of checking advocated here is complicated and makes it abundantly clear how inefficient the perturbation series is from a computational point of view. These poles have been checked to order three in Ref. 22, and other perturbative matters have been investigated in Ref. 23 .

4   Dual Pairs

The theories based on non-simply laced algebras work in a very different way which will be partially explained by reference to a particular example. Further details are obtainable in the recent literature [24–29] although there is much yet to do before the final version of the story can be told.

For definiteness, consider the pair of classical theories based on the extended Dynkin diagrams $g_2^{(1)}$ and $d_4^{(3)}$:

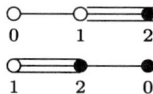

In each case, the black spots denote short roots and it is clear that to obtain one diagram from the other involves an inversion of roots (1.3).

In each case, there are two particles labeled 1 and 2, but their mass ratios are different [11, 12]. In the first case, the classical mass parameters are simply those of $d_4^{(1)}$ without the mass degeneracy, since this has been removed by the folding corresponding to the threefold symmetry of the d_4 Dynkin diagram. In the second case, the root system is obtained by applying the folding procedure to the extended Dynkin diagram $e_6^{(1)}$, which also has a threefold symmetry; hence, in this case the masses are a subset of those to be found in the $e_6^{(1)}$ theory. In summary, the two mass ratios are

$$\left.\frac{m_1}{m_2}\right|_{g_2^{(1)}} = \frac{\sin(\pi/6)}{\sin(2\pi/6)}, \quad \left.\frac{m_1}{m_2}\right|_{d_4^{(3)}} = \frac{\sin(\pi/12)}{\sin(2\pi/12)}.$$

Moreover, the nonzero three-point couplings for the two cases are:

$$g_2^{(1)} : \ 111, \ 112, \ 222, \quad d_4^{(3)} : \ 111, \ 112, \ 222, \ 221. \tag{4.1}$$

Hence, from a classical point of view these two theories are very different.

Some time ago, it was also noted that guesses for the S-matrix for g_2 were problematical if based on maintaining poles at the positions of the classical masses [11]. There were always extra singularities whose origin could not be traced in perturbation theory. It was also found that radiative mass corrections, which worked very well for the simply laced theories, did not preserve the classical mass ratios, suggesting either that cases such as g_2 were in a sense anomalous and therefore not quantum integrable or, that they were quantum integrable but that the relationship with the classical theory was much less clear-cut. The principal step in suggesting a resolution of these difficulties has been provided by Delius et al. [24]. They have noted how the bootstrap might be satisfied, even in a situation where there are particles with coupling dependent masses, in such a manner that the small coupling approximation is provided by the $g_2^{(1)}$ theory and the large coupling limit is provided by the $d_4^{(3)}$ theory. In other words, there is a quantum field theory corresponding to the pair together rather than either classical theory separately, and the transformation

$$\beta \to 4\pi/\beta,$$

effectively implements the inversion (1.3) which interchanges the two extended Dynkin diagrams. A similar mechanism is working for all the non-simply laced algebras which come in the pairs listed previously and related by (1.3). The exceptions to this are the members of the $a_{2n}^{(2)}$ sequence which are "self-dual."

The first thing to note is that the masses may be parametrized conveniently by setting

$$\left.\frac{m_1}{m_2}\right|_\beta = \frac{\sin\big(\pi/H(\beta)\big)}{\sin\big(2\pi/H(\beta)\big)} \quad \text{with } 6 \le H(\beta) \le 12 \text{ for } 0 \le \beta \le \infty,$$

where the functional dependence of H on β is really a matter of informed conjecture. A few words will be said about it at the end of the section. That the masses do depend on the coupling has been confirmed by Watts and Weston [27] who have investigated the coupling dependence in a Monte-Carlo lattice simulation of the model. Their results leave little doubt that the masses do indeed flow with the coupling although the numerical accuracy of the simulation is not yet sufficient to pin down the actual dependence on β.

The couplings (4.1) are more problematical since the two theories have different numbers of three-point couplings. However, the two self-couplings are clearly permitted whatever the masses might be and always correspond to a coupling angle of $2\pi/3$ in the notation introduced before (eq. (2.6)). Also, the 112 coupling is quite natural with a coupling angle of $2i\pi/H$, since

$$\sin^2\left(\frac{2\pi}{H}\right) \equiv 2\sin^2\left(\frac{\pi}{H}\right) + 2\sin^2\left(\frac{\pi}{H}\right)\cos\left(\frac{2\pi}{H}\right),$$

whatever the value of H might be, whereas the coupling 221 is quite un-natural. As far as an ab S-matrix element is concerned, one would expect the abc couplings to emerge as poles (or possibly multiple poles) in the physical strip with a positive coefficient (times i). That was certainly what happened in the simply laced sequences of models. However, in this and other similar cases, the mere positivity of the pole coefficient is not enough and it appears to be necessary to strengthen the requirement to *positivity over the whole range of β*. Once this is done it is found that there is a consistent set of bootstrap conditions satisfied by a subset of the classical couplings, but not all of them.

To examine the S-matrix, it is helpful to use a diagrammatic representation (see below) which displays the poles on the physical strip (solid lines) and compensating zeroes (dashed lines), as they travel from their positions at $\beta = 0$ to their positions at $\beta = \infty$. The filled circles represent points on the physical strip at intervals of π/h or $\pi/h^\vee$. Thus, the upper row represents the physical strip marked at intervals of $\pi/6$ for $g_2^{(1)}$ while the lower row represents the physical strip marked at intervals of $\pi/12$ for $d_4^{(3)}$. The dashed lines always meet solid lines at the top and bottom of the diagram indicating that the poles and zeroes precisely cancel there, as they ought because the S-matrix elements are unity at $\beta = 0$ or ∞. The first of the diagrams represents $S_{11}(\Theta)$

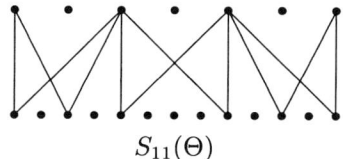

$$S_{11}(\Theta)$$

for which the algebraic expression is

$$\frac{(0)(2)}{(H/3-2)(4-H/3)} \frac{(H/3)(2H/3)}{(4)(H-4)} \frac{(H-2)(H)}{(2+2H/3)(4H/3-4)},$$

where the bracket notation has been adjusted to represent

$$(x) = \sinh\left(\frac{\Theta}{2} + \frac{xi\pi}{2H}\right) \Big/ \sinh\left(\frac{\Theta}{2} - \frac{xi\pi}{2H}\right).$$

It is clear from the diagram that there are two physical simple poles and their crossed partners—the third vertical line represents the self-coupling $11 \to 1$, and the oblique line next to it represents the $11 \to 2$ coupling. Using the bootstrap relation on the $11 \to 2$ coupling leads directly to the S-matrix element $S_{12}(\Theta)$ for which the algebraic expression is

$$\frac{(1)(2H/3-1)}{(H-5)(5-H/3)} \frac{(H/3+1)(H-1)}{(4H/3-5)(5)}$$

and which is represented diagrammatically by

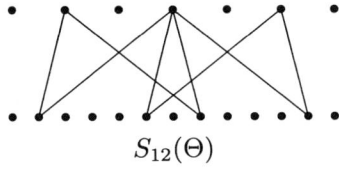

$$S_{12}(\Theta)$$

In this case, the rightmost solid line represents the physical pole for the expected bound state in the channel $12 \to 2$ and it looks as if there is another pole at $\Theta = (2/3 - 1/H)\pi i$ which meets its crossed partner at the top of the diagram but is separated from it at the bottom of the diagram. However, moving down the diagram, the coefficient of this pole has the wrong sign to be interpreted as a bound state until it is crossed by a zero represented by a dotted line. There the coefficient changes sign and remains positive up to the bottom of the diagram. A reasonable interpretation of this is that near the $d_4^{(3)}$ theory this pole looks like the one appropriate for the $12 \to 2$ coupling but, far away it does not. A reasonable hypothesis amends the bootstrap principle to include just those poles which never change sign over the whole interval. At first sight this seems strange. However, changing β is actually equivalent to adjusting $\hbar$ (remember, there is no classical coupling really), and therefore in a sense it is merely being suggested that the structure of the quantum field theory should be independent of a particular scale choice for $\hbar$. It would be difficult to check this statement in perturbative field theory because the zero in a pole coefficient is hard to find. Nevethess, Delius et al. do give preliminary perturbative arguments for the "floating" masses [24]. On the other hand, there is another argument, based on the Coleman-Thun mechanism [30], which suggests that a pole with an indefinite coefficient might be best thought of as a double pole with a compensating zero. There is no time to pursue this argument here but it is described in some detail in Ref. 28.

The bootstrap principle applied to the coupling $11 \to 2$ also yields the third S-matrix element $S_{22}(\Theta)$ whose diagram is

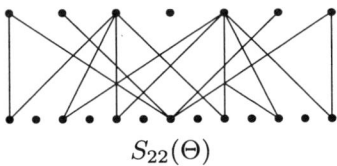

$$S_{22}(\Theta)$$

with the corresponding algebraic form

$$S_{22}(\Theta) = \frac{(0)(2H/3 - 2)}{(H-6)(4 - H/3)} \frac{(2)(2H/3)}{(H-4)(6 - H/3)}$$
$$\times \frac{(H/3)(H-2)}{(4H/3 - 6)(4)} \frac{(H/3 + 2)(H)}{(H/3 - 4)(6)}.$$

This matrix element is quite fascinating. There are a number of poles but all but two of them have coefficients which change sign. The poles (at $\Theta = 2\pi i/3$, corresponding to the self-coupling $22 \to 2$) which do not change sign fail to do so because two zeroes happen to collide. All the other poles can be accommodated within the extended Coleman-Thun scheme.

The other non-simply laced cases have all been listed elsewhere [9, 24, 28] and will not be mentioned explicitly here. In every case, the generalized bootstrap principle alluded to above is consistent and all poles not included in the bootstrap have a plausible explanation within the extended Coleman-Thun scheme.

One intriguing question in all of this is, what (if anything) replaces the beautiful structure surrounding the Coxeter element which plays such a unifying role in the simply laced cases? What replaces the formula (3.8)? Presumably, whatever the structure is, it transcends the root lattices of the pair but offers a geometrical setting for both of them (see Ref. [29]).

Finally, it is worth mentioning that there is a convenient way to parametrize the coupling angles, and the floating masses [21]. Define first, generalizing the earlier notation,

$$B_g(\beta) \equiv B(\beta; g, g^\vee) = \frac{2\beta^2}{\beta^2 + 4\pi(h/h^\vee)},$$

where g and $g^\vee$ have extended Dynkin diagrams related by (1.3), and associated Coxeter numbers h and $h^\vee$, respectively. Then, the following identity is true

$$B(\beta; g, g^\vee) = 2 - B\left(\frac{4\pi}{\beta}; g^\vee, g\right).$$

The duality of the coupling angles for the positive definite poles is then rendered transparent by setting

$$\Theta^c_{ab}(\beta) = \frac{2 - B_g}{2}\Theta^c_{ab}(0) + \frac{B_g}{2}\Theta^c_{ab}(\infty).$$

A straightforward comparison with the coupling angles for the $g_2^{(1)} - d_4^{(3)}$ case yields the consistent choice

$$\frac{1}{H(\beta)} = \frac{1}{12}\left(2 - \frac{B_{g_2^{(1)}}(\beta)}{2}\right).$$

5   A Word on Solitons

If complex solutions to the affine Toda field equations are permitted, then there is a whole extra dimension to the Toda activity. At first sight, the idea of allowing the Toda field to be complex is unattractive since the classical Hamiltonian will not be positive definite and it is not immediately clear how such solutions ought to be interpreted, or what their role in the quantum Toda theory might be. On the other hand, it has been pointed out by Hollowood [31] that the soliton solutions, although complex, actually have real energy and momentum associated with them, despite the fact that their energy-momentum density is complex. In addition, the masses associated with the solitons are closely related to the particle masses in the real theory and their couplings, in the sense of a fusing rule, are also identical to the couplings of the real particles, at least for the *ade* sequence of possibilities. The static solitons are labeled by "topological charges" corresponding to weights of the fundamental representations of the Lie algebra underlying the Toda theory. This is quite easy to check, but there is something of a mystery associated with the topological charges in the sense that complete sets of weights are only rarely found (in the $a_n^{(1)}$ sequence of theories, and even then only in the smallest dimension representations). This puzzle will be mentioned again at the end of this section.

To see that the possibility of soliton solutions exists is not difficult. It is enough to note that the Toda potential has local stationary points whenever the field is constant and taken to be

$$\phi = \frac{2i\pi\lambda}{\beta} \quad \text{with } \lambda \cdot \alpha_k = \text{integer}, \ k = 0, 1, \ldots, r.$$

At each of these values of the field, the potential vanishes. Soliton solutions to the equations of motion interpolate pairs of these "vacua" with, typically,

$$\phi(-\infty, t) = 0 \quad \phi(\infty, t) = \frac{2i\pi\lambda}{\beta}.$$

The topological charge is defined to be

$$\lambda = \frac{\beta}{2i\pi}\left(\phi(\infty, t) - \phi(-\infty, t)\right). \tag{5.1}$$

At each of these values of the field, the potential vanishes. Notice that this set up generalizes the situation to be found in the sine-Gordon theory which may be regarded as a purely imaginary version of the $a_1^{(1)}$ affine Toda theory. Note, too, that the sine-Gordon theory supplies the only example for which all the soliton solutions are effectively real.

Each of the affine Toda theories contains soliton solutions and many of the solutions have been catalogued elsewhere [32]. For definiteness and ease of computation, the $a_n^{(1)}$ types will be illustrated here using the so-called Hirota method as it was originally adopted by Hollowood. This relies on the ansatz

$$\phi = -\frac{1}{\beta}\sum_0^r \alpha_k \ln \tau_k \tag{5.2}$$

for which the (time-independent) Toda field equations reduce to:

$$\sum_0^r \alpha_k \left(\frac{\tau_k''}{\tau_k} - \frac{\tau_k'\tau_k'}{\tau_k^2} + \prod_l \tau_l^{-\alpha_k \cdot \alpha_l} \right) = 0. \tag{5.3}$$

These are the relevant equations for static solitons. Using the explicit form of the $a_r^{(1)}$ Cartan matrix (5.3) may be rewritten

$$\sum_0^r \alpha_k \left(\frac{\tau_k''}{\tau_k} - \frac{\tau_k'\tau_k'}{\tau_k^2} + \frac{\tau_{k-1}\tau_{k+1}}{\tau_k^2} \right) = 0,$$

and solved by setting

$$\frac{\tau_k''}{\tau_k} - \frac{\tau_k'\tau_k'}{\tau_k^2} + \frac{\tau_{k-1}\tau_{k+1}}{\tau_k^2} = 1 \quad \text{for } k = 0, 1, 2, \ldots, r, \tag{5.4}$$

where

$$\tau_k = 1 + \Omega_k e^{\sigma x + x_0},$$

provided

$$\sigma^2 \Omega_k - 2\Omega_k + \Omega_{k-1} + \Omega_{k+1} = 0$$
$$\Omega_{k-1}\Omega_{k+1} - \Omega_k^2 = 0 \tag{5.5}$$
$$\Omega_{k+r+1} = \Omega_k.$$

The last pair of eqs. (5.5) are solved by taking

$$\Omega_k^{(a)} = e^{2i\pi ak/r+1} = \omega^{ak} \quad \text{for each choice } a = 1, 2, \ldots, r,$$

where ω is the primitive $r + 1$st root of unity. The first of the eqs. (5.5) then imply a corresponding constraint on σ leading (for each choice of a) to

$$\sigma^{(a)} = 2\sin\left(\frac{\pi a}{r+1}\right).$$

The replacement $x \to -x$ gives another solution, and x_0 is an arbitrary constant. Assembling all these pieces, there is a solution for each a of the form (5.2):

$$\phi^{(a)} = -\frac{1}{\beta}\sum_0^r \alpha_k \ln(1 + \omega^{ak} e^{\sigma^{(a)}x + x_0^{(a)}})$$

$$= -\frac{1}{\beta}\sum_1^r \alpha_k \ln\left(\frac{1 + \omega^{ak} e^{\sigma^{(a)}x + x_0^{(a)}}}{1 + e^{\sigma^{(a)}x + x_0^{(a)}}}\right). \tag{5.6}$$

These solutions are generally complex. The same solutions may be obtained via a more sophisticated and general algebraic method given by Olive et al. based on the work of Leznov and Saveliev who pioneered a general approach to Toda wave equations some years ago [33–35]. Moreover, it appears there are no other single-soliton solutions to be found using the more general techniques. This is perhaps surprising given the special nature of the ansatz (5.2) and the particular choice of solution within it, eq. (5.4).

To calculate the energy of these solutions, it is extremely convenient to use a formula for the energy-momentum tensor established in the article by Olive et al. [35], using arguments rooted in conformal Toda field theory. They found

$$T_{\mu\nu} = (\eta_{\mu\nu}\partial^2 - \partial_\mu\partial_\nu)C,$$

where the function C for solitons is given by

$$C = \frac{-2}{\beta^2} \sum_0^r \ln \tau_k.$$

Using this, the energy of a static soliton (i.e. its mass) can be calculated

$$M^{(a)} = \int_{-\infty}^{\infty} dx\, T_{00} = \left.\frac{\partial C}{\partial x}\right|_{-\infty}^{\infty} = \frac{2}{\beta^2} \sum_0^r \left.\frac{\tau_k'}{\tau_k}\right|_{-\infty}^{\infty} = \frac{2(r+1)}{\beta^2}\sigma^{(a)}.$$

It is worthy of note that the mass is real despite the fact that the energy density is complex and, moreover, each mass is proportional to the mass of a corresponding elementary scalar particle in the real coupling Toda theory provided the label a is suitably interpreted.

To provide the interpretation, first recall from Section 1 that the scalar particles of the real coupling Toda theory are associated with the fundamental representations of the Lie algebra a_n, the lightest particles corresponding to the smallest ($n + 1$ dimensional) representation of the algebra or its conjugate, the next lightest to the representations of dimension $n(n + 1)/2$, and so on. The solitons, on the other hand, are labeled naturally by their topological charges which may be calculated from the explicit solutions using (5.1). The calculation of the topological charges is slightly tricky and must be performed with some care. First of all, note that the argument of the logarithm in eq. (5.6) must never vanish or diverge for any choice of x, otherwise the solution will be singular. This requires that the constant $x_0^{(a)}$ (which may be complex) has an imaginary part which is not entirely arbitrary; it is confined to regions in the range $[0, 2\pi]$ whose boundaries correspond to those choices of Im x_0 for which at least one of the logarithmic arguments will vanish or diverge. Hence, the number of such boundary points provides the maximum number of possibly different topological charges which might be described by the solution (5.6). Provided the boundary points are avoided, the arguments of the logarithms change continuously, the logarithm cannot jump its branch and the topological charge is defined uniquely. McGhee [36] has calculated all the topological charges that are possible given (5.6), and has confirmed that the topological charges of the solution whose mass corresponds to that of the classical particle a do indeed lie among the weights of the associated representation. However, he has also noted that the total number of possible topological charges typically falls far short of the number required to fill up the whole representation. Indeed there is a neat formula for the total number of topological charges obtainable:

$$\text{number of charges of type } a = \frac{n+1}{\gcd(a, n+1)}.$$

Indeed, the only representation with its full complement of topological weights is the representation of dimension $n + 1$, or its conjugate.

McGhee has also examined a number of other theories and has found that the Hirota solution in all other cases always fails to fill the associated weight set and often the discrepancy is huge [37]. A selection of the results are given below, the numbers below the Dynkin diagram points denoting the number of possible topological charges obtainable via the Hirota ansatz.

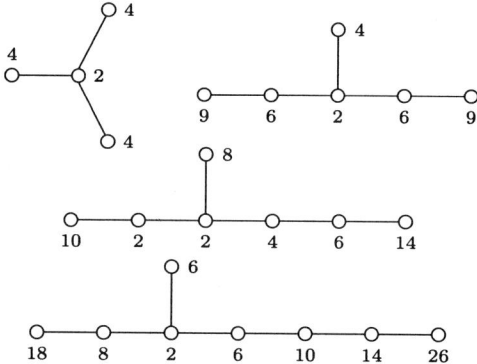

For the e_8 case the topological charges present are a small fraction of the conjectured total. If it is really the case that the approach of Olive et al. captures all the solitons but is effectively equivalent to the Hirota ansatz, then one must take seriously the "gaps' in the topological spectrum.

It has been suggested that there might be a consistent quantum field theory corresponding to the complex Toda theories (a generalization of the sine-Gordon situation), in which the particle spectrum consists of multiplets corresponding to the representations of a quantized affine Lie algebra (as is the case for the sine-Gordon theory in which the soliton and anti-soliton are a doublet of $su_q(2)$). This would appear to be natural but at the same time very mysterious without a detailed mechanism to explain the enormous gaps in the classical soliton spectrum. Presumably, the classical spectrum must be enhanced in the quantum theory. Hollowood [38], and Bernard and LeClair [39] have presented arguments suggesting that the quantum theory ought to have such an enlarged spectrum. Certainly, it is possible to obtain solutions to the Yang-Baxter equations, based on quantum group ideas. There is no time to consider these arguments here but one ought to bear in mind the intriguing behavior of the real coupling non-simply laced theories which would appear to be difficult to mirror in the complex theory, since it is not at all clear which quantum group one ought to be choosing. There is evidence that the quantum corrections to the soliton masses preserve the classical mass ratios for the *ade* cases but fail to do so for the others. Hollowood [40], using old ideas of Dashen et al., has calculated the lowest order quantum mass corrections to the soliton mass spectrum for $a_n^{(1)}$. Remarkably, the mass corrections do not spoil the mass ratios. Very recently, both Delius and Mackay and Watts [41] have performed similar calculations for non-simply laced solitons, and the classical mass ratios are not preserved. One would imagine that something akin to the duality going on in the real coupling theories should persist provided the quantum field theory of the complex theories really makes sense. It is conceivable that a truncated spectrum will be necessary in order to permit the S-matrix elements to enjoy floating bound-state poles. That some truncation of the spectrum might be needed is also indicated by a need for unitarity in a theory with a non-Hermitian Hamiltonian; an apparently serious fault which might be alleviated by removing parts of the spectrum to leave a unitary core.

It remains to be seen how this extremely interesting story will unfold.

6   Other Matters

There are several interesting developments which cannot be described here. For example, a full understanding of the quantum field theory would require much more than the S-matrix/conserved quantity considerations presented here. Indeed, there is a sizeable literature, concerning the calculation of form factors for Toda theory and related topics (for example, see Ref. 42).

What happens if affine Toda field theory is restricted to a segment of the real line, or to a half-line? The general question of integrability in the presence of boundary conditions has its own literature, but some recent articles dealing specifically with Toda theories are given in Ref. 43. Surprisingly, there are strong constraints on the possible form of the boundary condition maintaining classical integrability, but it is not clear how these will affect the quantum theory—another question to be resolved in the future.

Acknowledgments: I am grateful to the organizers of the school for the opportunity to talk about Toda field theories, and to many colleagues and students for stimulating interactions. In particular, I would like to thank Harry Braden, Patrick Dorey, Richard Hall, Tim Hollowood, Niall Mackay, William McGhee, Rachel Rietdijk, Ryu Sasaki, Gérard Watts, and Robert Weston for enjoyable discussions and collaborations. I am indebted to Patrick Dorey for various pictorial representations of the singularities of S-matrix elements some of which have been used to illustrate the dual pairs of non-simply laced models.

7   REFERENCES

1. A. E. Arinshtein, V. A. Fateev, and A. B. Zamolodchikov. Quantum S-matrix of the 1+1 dimensional Toda chain. *Phys. Lett.*, B87: 389–392, 1979.

2. A. V. Mikhailov, M. A. Olshanetsky, and A. M. Perelomov. Two-dimensional generalized Toda lattice. *Commun. Math. Phys.*, 79 (4): 473–488, 1981; G. Wilson. The modified Lax and two-dimensional Toda lattice equations associated with simple Lie algebras. *Ergod. Th. and Dynam. Sys.*, 1 (3): 361–380, 1981; D. I. Olive and N. Turok. The symmetries of Dynkin diagrams and the reduction of Toda field equations. *Nucl. Phys.*, B215 (4): 470–494, 1983.

3. V. G. Kac. *Infinite-Dimensional Lie Algebras. An Introduction*, volume 44 of *Progress in Mathematics*. Birkhäuser, Boston, 1983.

4. A. B. Zamolodchikov. Integrals of motion in scaling 3-state Potts model field theory. *Int. J. Mod. Phys.*, A3 (3): 743–750, 1988; T. J. Hollowood and P. Mansfield. Rational conformal field theories at, and away from, criticality as Toda field theories. *Phys. Lett.*, B226 (1-2): 73–79, 1989; T. Eguchi and S.-K. Yang. Deformations of conformal field theories and soliton equations. *Phys. Lett.*, B224 (4): 373–378, 1989; V. A. Fateev and A. B. Zamolodchikov. Conformal field theory and purely elastic S-matrices. *Int. J. Mod. Phys.*, A5 (6): 1025–1048, 1990; P. Christe. S-matrices of the tri-critical Ising model and Toda systems. In *Proceedings of the NATO Conference on Differential Geometric Methods in Theoretical Physics*, Lake Tahoe, U.S.A., 1989, 1990. Plenum, New York, pages 213–222; G. Mussardo. Away from criticality: some results from the S-matrix approach. In *Proceedings of the NATO Conference on Differential Geometric Methods in Theoretical Physics*, Lake Tahoe, U.S.A., 1989, 1990. Plenum, New York, pages 297–307.

5. H. W. Braden, E. Corrigan, P. E. Dorey, and R. Sasaki. Aspects of perturbed conformal field theory, affine toda field theory and exact

S-matrices. In *Proceedings of the NATO Conference on Differential Geometric Methods in Theoretical Physics*, Lake Tahoe, U.S.A., 1989, 1990. Plenum, New York, pages 169–182.

6. J.-L. Gervais and A. Neveu. New quantum treatment of Liouville field theory. *Nucl. Phys.*, B224 (2): 329–348, 1983; P. Mansfield. Light-cone quantisation of the Liouville and Toda field theories. *Nucl. Phys.*, B222 (3): 419–445, 1983; E. Braaten, T. Curtright, G. Ghandour, and C. Thorn. A class of conformally invariant quantum field theories. *Phys. Lett.*, B125: 301–304, 1983.

7. O. Babelon and L. Bonora. Sinh-Gordon model as a spontaneously broken conformal theory. *Phys. Lett.*, B267: 71–80, 1991; L. Bonora. Conformal affine Toda theories. Topological and quantum group methods in field theory and condensed matter physics. *Int. J. Mod. Phys.*, B6 (11-12): 2015–2040, 1992.

8. D. Olive and N. Turok. Local conserved densities and zero-curvature conditions for Toda lattice field theories. *Nucl. Phys.*, B257 (2): 277–301, 1985.

9. P. G. O. Freund, T. Klassen, and E. Melzer. *S*-matrices for perturbations of certain conformal field theories. *Phys. Lett.*, B229 (3): 243–247, 1989.

10. M. D. Freeman. On the mass spectrum of affine Toda field theory. *Phys. Lett.*, B261 (1-2): 57–61, 1991; A. Fring, H. C. Liao, and D. I. Olive. The mass spectrum and coupling in affine Toda theories. *Phys. Lett.*, B266 (1-2): 82–86, 1991.

11. H. W. Braden, E. Corrigan, P. E. Dorey, and R. Sasaki. Affine Toda field theory and exact *S*-matrices. *Nucl. Phys.*, B338 (3): 689–746, 1990.

12. P. Christe and G. Mussardo. Integrable systems away from criticality: the Toda field theory and *S*-matrix of the tricritical ising model. *Nucl. Phys.*, B330 (2-3): 465–487, 1990; P. Christe and G. Mussardo. Elastic *S*-matrices in $(1+1)$ dimensions and Toda field theories. *Int. J. Mod. Phys.*, A5 (24): 4581–4627, 1990.

13. T. R. Klassen and E. Melzer. Purely elastic scattering theories and their ultraviolet limits. *Nucl. Phys.*, B338 (3): 485–528, 1990.

14. N. Bourbaki. *Groupes et algèbres de Lie*, chapter IV–VI. Hermann, Paris, 1968; J. E. Humphreys. *Reflection Groups and Coxeter Groups*, volume 29 of *Cambridge Studies in Advanced Mathematics*. Cambridge University Press, Cambridge, 1990.

15. P. E. Dorey. Root systems and purely elastic S-matrices. *Nucl. Phys.*, B358 (3): 654–676, 1991; P. E. Dorey. Root systems and purely elastic S-matrices. II. *Nucl. Phys.*, B374 (3): 741–761, 1992.

16. H. W. Braden. A note on affine Toda couplings. *J. Phys.*, A25 (1): L15–L20, 1992.

17. B. Kostant. The principal three-dimensional subgroup and the Betti numbers of a complex simple Lie group. *Amer. J. Math.*, 81: 973–1032, 1959.

18. A. B. Zamolodchikov and Al. B. Zamolodchikov. Factorized S-matrices in two dimensions as the exact solutions of certain relativistic quantum field theory models. *Ann. Phys.*, 120 (2): 253–291, 1979.

19. M. Karowski. On the bound state problem in $(1+1)$-dimensional field theories. *Nucl. Phys.*, B153: 244–252, 1979.

20. A. Fring and D. I. Olive. The fussing rule and the scattering matrix of affine Toda theory. *Nucl. Phys.*, B379 (1-2): 429–447, 1992.

21. P. E. Dorey. Hidden geometrical structures in integrable models. In *Integrable Quantum Field Theories*, Como, Italy, 1992, 1993. Plenum, New York, pages 83–98.

22. H. W. Braden, E. Corrigan, P. E. Dorey, and R. Sasaki. Multiple poles and other features of affine Toda field theory. *Nucl. Phys.*, B356 (2): 469–498, 1991.

23. H. W. Braden and R. Sasaki. The S-matrix coupling dependence for a, d, and e affine Toda field theory. *Phys. Lett.*, B255 (3): 343–352, 1991; H. W. Braden and R. Sasaki. Affine Toda perturbation theory. *Nucl. Phys.*, B379 (1-2): 377–428, 1992; H. W. Braden, H. S. Cho, J. D. Kim, I. G. Koh, and R. Sasaki. Singularity analysis in A_n affine Toda theories. *Prog. Theor. Phys.*, 88: 1205–1212, 1992.

24. G. W. Delius, M. T. Grisaru, and D. Zanon. Exact S-matrices for non-simply-laced affine Toda theories. *Nucl. Phys.*, B382 (2): 365–406, 1992.

25. H. S. Cho, I. G. Koh, and J. D. Kim. Duality in the $D_4^{(3)}$ affine Toda theory. *Phys. Rev.*, D47 (6): 2625–2628, 1993.

26. H. G. Kausch and G. M. T. Watts. Duality in quantum Toda theory and W-algebras. *Nucl. Phys.*, B386 (1): 166–190, 1992.

27. G. M. T. Watts and R. A. Weston. G_2^1 affine Toda field theory: A numerical test of exact S-matrix results. *Phys. Lett.*, B289: 61–66, 1992.

28. E. Corrigan, P. E. Dorey, and R. Sasaki. On a generalised bootstrap principle. *Nucl. Phys.*, B408 (3): 579–599, 1993.

29. P. E. Dorey. A remark on the coupling dependence in affine Toda field theory. *Phys. Lett.*, B312 (3): 291–298, 1993.

30. S. Coleman and H. Thun. On the prosaic origin of the double poles in the sine-Gordon S-matrix. *Commun. Math. Phys.*, 61 (1): 31–39, 1978.

31. T. J. Hollowood. Solitons in affine Toda field theories. *Nucl. Phys.*, 384 (3): 523–540, 1992.

32. H. Aratyn, C. P. Constantinidis, L. A. Ferreira, J. F. Gomes, and A. H. Zimerman. Hirota's solitons in the affine and the conformal affine Toda models. *Nucl. Phys.*, B406 (3): 727–770, 1993; N. J. Mackay and W. A. McGhee. Affine Toda solitons and automorphisms of Dynkin diagrams. *Int. J. Mod. Phys.*, A8 (16): 2791–2807, 1993; Z. Zhu and D. G. Caldi. Multi-soliton solutions of affine Toda models. *Nucl. Phys.*, B436 (3): 659–678, 1995.

33. A. N. Leznov and M. V. Saveliev. *Group-Theoretical Methods for the Integration of Nonlinear Dynamical Systems*, volume 15 of *Progress in Physics*. Birkhäuser Verlag, Basel, 1992.

34. D. I. Olive, N. Turok, and J. W. R. Underwood. Affine Toda solitons and vertex operators. *Nucl. Phys.*, B409 (3): 509–546, 1993.

35. D. I. Olive, N. Turok, and J. W. R. Underwood. Solitons and the energy-momentum tensor for affine Toda theory. *Nucl. Phys.*, B401 (3): 663–697, 1993.

36. W. A. McGhee. The topological charges of the a_n^1 affine Toda solitons. *Int. J. Mod. Phys.*, A9 (15): 2645–2665, 1994.

37. W. A. McGhee. *On the topological charges of the affine Toda solitons.* Ph.D. thesis, University of Durham, 1994.

38. T. J. Hollowood. Quantizing sl(n) solitons and the Hecke algebra. *Int. J. Mod. Phys.*, A8 (5): 947–981, 1993.

39. D. Bernard and A. LeClair. Quantum group symmetries and nonlocal current in 2D QFT. *Commun. Math. Phys.*, 142 (1): 99–138, 1991; D. Bernard and A. LeClair. Quantum affine symmetry as generalized supersymmetry. *Nucl. Phys.*, B399 (1-2): 709–748, 1993.

40. T. J. Hollowood. Quantum soliton mass corrections in sl(n) affine Toda field theory. *Phys. Lett.*, B300 (1-2): 73–83, 1993.

41. G. M. T. Watts. Quantum mass corrections for C_2^1 affine Toda field theory solitons. *Phys. Lett.*, B338 (1): 40–46, 1994; N. J. Mackay and G. M. T. Watts. Quantum mass corrections for affine Toda solitons. *Nucl. Phys.*, 441 (1-2): 277–309, 1994; G. W. Delius and M. T. Grisaru. Toda soliton mass corrections and the particle-soliton duality conjecture. *Nucl. Phys.*, B441 (1-2): 259–276, 1995.

42. F. A. Smirnov. *Form Factors in Completely Integrable Models of Quantum Field Theory*, volume 14 of *Advanced Series in Mathematical Physics*. World Scientific, River Edge, NJ, 1992; A. Fring, G. Mussardo, and P. Simonetti. Form factors of the elementary field in the Bullough-Dodd model. *Phys. Lett.*, B307 (1-2): 83–90, 1993; G. Delfino and G. Mussardo. Two-point correlation function in integrable QFT with anti-crossing symmetry. *Phys. Lett.*, B324 (1): 40–44, 1994; A. Koubek. Form-factor bootstrap and the operator content of perturbed minimal models. *Nucl. Phys.*, B428 (3): 655–680, 1994; F. A. Smirnov. A new set of exact form factors. *Int. J. Mod. Phys.*, A9 (29): 5121–5143, 1994.

43. S. Ghoshal and A. B. Zamolodchikov. Boundary S matrix and boundary state in two-dimensional integrable quantum field theory. *Int. J. Mod. Phys.*, A9 (21): 3841–3885, 1994; S. Ghoshal. Bound state boundary S matrix of the sine-Gordon model. *Int. J. Mod. Phys.*, A9 (27): 4801–4810, 1994; A. Fring and R. Köberle. Factorized scattering in the presence of reflecting boundaries. *Nucl. Phys.*, B421 (1): 159–172, 1994; A. Fring and R. Köberle. Affine Toda field theory in the presence of reflecting boundaries. *Nucl. Phys.*, B419 (3): 647–662, 1994; R. Sasaki. Reflection bootstrap equations for Toda field theory. In *Interface Between Physics and Mathematics*, Hangzhou, China, 1993, 1994. World Scientific; E. Corrigan, P. E. Dorey, R. Rietdijk, and R. Sasaki. Affine Toda field theory on a half-line. *Phys. Lett.*, B333 (1-2): 83–91, 1994; E. Corrigan, P. E. Dorey, and R. Rietdijk. Aspects of affine Toda field theory on a half line. Technical Report DTP-94/29, Durham, 1994.

2

A Class of Fermi Liquids

J. Feldman, H. Knörrer, D. Lehmann, and E. Trubowitz

1 Introduction

In this chapter, we consider a many-body system that is somewhat unusual in that the Fermi surface survives the turning on of all sufficiently weak short-range interactions. The system consists of a gas of fermions with pre-scribed, strictly positive, density, together with a crystal lattice of *magnetic* ions. The fermions interact with each other through a two-body potential. The lattice provides periodic scalar and vector background potentials. Also, the ions oscillate, generating phonons and then the fermions interact with the phonons. At the present time our result is restricted to $d = 2$ space dimensions. But we believe that the difficulties preventing the extension to $d = 3$ are technical rather than physical and are working to overcome them.

To start, turn off the fermion-fermion and fermion-phonon interactions. Then we have a gas of independent fermions, each with Hamiltonian

$$H_0 = \frac{1}{2m}\left(i\boldsymbol{\nabla} + \mathbf{a}(\mathbf{x})\right)^2 + U(\mathbf{x}).$$

We assume that the vector and scalar potentials $\mathbf{a}$, U are periodic with respect to some lattice Γ in $\mathbb{R}^2$. By convention, boldface characters are two-component vectors. Because the Hamiltonian commutes with lattice translations it is possible to simultaneously diagonalize the Hamiltonian and the generators of lattice translations. Call the eigenvalues and eigen-vectors $\varepsilon_\nu(\mathbf{k})$ and $\phi_{\nu,\mathbf{k}}(\mathbf{x})$, respectively. They obey

$$\begin{aligned} H_0\phi_{\nu,\mathbf{k}}(\mathbf{x}) &= \varepsilon_\nu(\mathbf{k})\phi_{\nu,\mathbf{k}}(\mathbf{x}), \\ \phi_{\nu,\mathbf{k}}(\mathbf{x} + \gamma) &= e^{i\langle\mathbf{k},\gamma\rangle}\phi_{\nu,\mathbf{k}}(\mathbf{x}), \quad \forall\gamma \in \Gamma. \end{aligned} \tag{1.1}$$

The crystal momentum $\mathbf{k}$ runs over $\mathbb{R}^2/\Gamma^{\#}$ where

$$\Gamma^{\#} = \{b \in \mathbb{R}^2 \mid \langle b, \gamma\rangle \in 2\pi\mathbb{Z} \; \forall\gamma \in \Gamma\}$$

is the dual lattice to Γ. The band index $\nu \in \mathbb{N}$ just labels the eigenvalues for boundary condition $\mathbf{k}$ in increasing order.

In the grand canonical ensemble, the Hamiltonian H is replaced by $H - \mu N$ where N is the number operator and the chemical potential μ is used to control the density of the gas. At very low temperature, which is the physically interesting domain, only those pairs ν, $\mathbf{k}$ for which $\varepsilon_\nu(\mathbf{k}) \approx \mu$ are important. To keep things as simple as possible, we assume that $\varepsilon_\nu(\mathbf{k}) \approx \mu$ only for one value ν_0 of ν and we put on a fixed ultraviolet cutoff so that we consider only those crystal momenta for which $|\varepsilon_{\nu_0}(\mathbf{k}) - \mu|$ is smaller than some fixed small constant.

Precisely, we denote $e(\mathbf{k}) = \varepsilon_{\nu_0}(\mathbf{k}) - \mu$ and make the following assumptions.

Hypothesis 1. *The dispersion relation $e(\mathbf{k})$ is a real-valued, real analytic function on a compact subset B of $\mathbb{R}^d$. For all points $\mathbf{p} \in \mathrm{B}$,*

$$\nabla e(\mathbf{p}) \neq 0.$$

Hypothesis 2. *The Fermi curve*

$$\mathrm{F} = \{\mathbf{p} \in \mathrm{B} \mid e(\mathbf{p}) = 0\}$$

for e is a simple closed curve, whose curvature is bounded away from zero.

Hypothesis 3. *For all $\mathbf{q} \in \mathbb{R}^d$,*

$$-\mathrm{F} + \mathbf{q} \neq \mathrm{F}.$$

By definition,

$$-\mathrm{F} + \mathbf{q} = \{\mathbf{p} \in \mathbb{R}^2 \mid -\mathbf{p} + \mathbf{q} \in \mathrm{B} \text{ and } e(-\mathbf{p} + \mathbf{q}) = 0\}.$$

It is Hypothesis 3 that makes this class of models somewhat unusual and permits the system to remain a Fermi liquid when the interaction is turned on. If $\mathbf{a} = 0$ then, taking the complex conjugate of (1.1), we see that $\varepsilon_\nu(-\mathbf{k}) = \varepsilon_\nu(\mathbf{k})$ so that Hypothesis 3 is violated for $\mathbf{q} = 0$. Hence the presence of a nonzero vector potential is essential.

In order to make the hypotheses as simple looking as possible, we have made them much stronger than necessary. One model that violates these hypotheses, not only for technical reasons but because it exhibits different physics, is the Hubbard model at half filling. Its Fermi surface looks like Figure 1. This Fermi curve is not smooth, violating Hypothesis 1, has zero curvature almost everywhere, violating Hypothesis 2, and is reflection invariant so that $\mathrm{F} - -\mathrm{F}$, violating Hypothesis 3 with $\mathbf{q} = 0$.

The interacting models are formally characterized by the Euclidean Green's functions

$$\left\langle \prod_{i=1}^n \psi_{p_i} \bar{\psi}_{q_i} \right\rangle = \frac{\int (\prod_{i=1}^n \psi_{p_i} \bar{\psi}_{q_i}) e^{\mathcal{A}(\psi,\bar{\psi})} \prod_{k,\sigma} d\psi_{k,\sigma} d\bar{\psi}_{k,\sigma}}{\int e^{\mathcal{A}(\psi,\bar{\psi})} \prod_{k,\sigma} d\psi_{k,\sigma} d\bar{\psi}_{k,\sigma}}. \tag{1.2a}$$

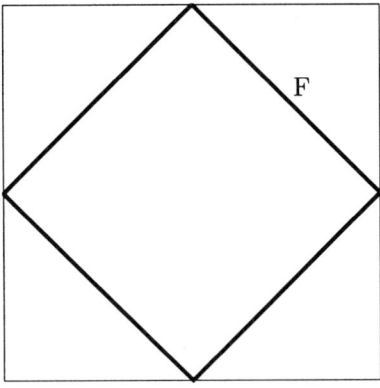

FIGURE 1.

The action

$$\mathcal{A}(\psi, \bar{\psi}) = - \int dk \big(ik_0 - e(\mathbf{k})\big) \bar{\psi}_k \psi_k$$
$$- \int dk \mathcal{E}(\lambda, \mathbf{k}) \bar{\psi}_k \psi_k - \mathcal{V}(\psi, \bar{\psi}). \quad (1.2b)$$

We now take some time to explain this formula. The fermion fields are vectors

$$\psi_k = \begin{pmatrix} \psi_{k,\uparrow} \\ \psi_{k,\downarrow} \end{pmatrix} \quad \bar{\psi}_k = \begin{pmatrix} \bar{\psi}_{k,\uparrow} & \bar{\psi}_{k,\downarrow} \end{pmatrix}$$

whose components $\psi_{k,\sigma}$, $\bar{\psi}_{k,\sigma}$, $k = (k_0, \mathbf{k}) \in \mathcal{B} = (-1, 1) \times \mathrm{B}$, $\sigma \in \{\uparrow, \downarrow\}$, are generators of an infinite-dimensional Grassmann algebra over $\mathbb{C}$. That is, the fields anticommute with each other

$$\overset{(-)}{\psi}_{k,\sigma} \overset{(-)}{\psi}_{p,\tau} = - \overset{(-)}{\psi}_{p,\tau} \overset{(-)}{\psi}_{k,\sigma}.$$

We have deliberately chosen $\bar{\psi}$ to be a row vector and ψ to be a column vector so that

$$\bar{\psi}_k \psi_p = \bar{\psi}_{k,\uparrow} \psi_{p,\uparrow} + \bar{\psi}_{k,\downarrow} \psi_{p,\downarrow} \quad \psi_k \bar{\psi}_p = \begin{pmatrix} \psi_{k,\uparrow} \bar{\psi}_{p,\uparrow} & \psi_{k,\uparrow} \bar{\psi}_{p,\downarrow} \\ \psi_{k,\downarrow} \bar{\psi}_{p,\uparrow} & \psi_{k,\downarrow} \bar{\psi}_{p,\downarrow} \end{pmatrix}.$$

In the argument $k = (k_0, \mathbf{k})$, the last d components $\mathbf{k}$ are to be thought of as a crystal momentum and the first component k_0 as the dual variable to an imaginary time. Hence the $\sqrt{-1}$ in $ik_0 - e(\mathbf{k})$. For convenience only, we have put an ultraviolet cutoff on k_0 as well as on $\mathbf{k}$. In the full model k_0 runs over $\mathbb{R}$ and $\mathbf{k}$ is replaced by $(\nu, \mathbf{k})$ with ν summed over $\mathbb{N}$ and $\mathbf{k}$ integrated over $\mathbb{R}^d / \Gamma^\#$. The relationship between the position space field $\psi(\xi)$, with $\xi = (t, \mathbf{x})$ running over (imaginary) time $\times$ space, and the momentum space field ψ_k is given, in our single-band approximation, by

$$\psi_k = \int d\xi e^{-ik_0 t}\phi_{\nu_0,\mathbf{k}}(\mathbf{x})\psi(\xi)$$

$$\psi(\xi) = \int dk e^{ik_0 t}\overline{\phi_{\nu_0,\mathbf{k}}(\mathbf{x})}\psi_k$$

(1.3)

where

$$dk = \frac{dk_0}{2\pi}\, d\mathbf{k} = \frac{d^{d+1}k}{(2\pi)^{d+1}}.$$

The general spin independent form of the interaction is

$$\mathcal{V}(\psi,\bar{\psi}) = \frac{\lambda}{2}\int \prod_{i=1}^{4} dk_i (2\pi)^{d+1}\delta(k_1 + k_2 - k_3 - k_4)\bar{\psi}_{k_1}\psi_{k_3}$$
$$\times \langle k_1, k_2|V|k_3, k_4\rangle \bar{\psi}_{k_2}\psi_{k_4}$$

Spin independence is imposed purely for notational convenience. It plays no role. The delta function δ is that for $\mathbb{R}^d/\Gamma^{\#}$ and imposes the appropriate conservation of crystal momentum for the present setting. The function $\langle k_1, k_2|V|k_3, k_4\rangle$ implements the fermion-fermion and fermion-phonon interaction. Its precise value does not concern us. We just assume

Hypothesis 4. *The interaction is short range. That is* $\langle k_1, k_2|V|k_3, k_4\rangle \in C^{\infty}$.

The net coefficient $e(\mathbf{k}) - \mathcal{E}(\lambda, \mathbf{k})$ of $\bar{\psi}_k\psi_k$ in $\mathcal{A}$ has been deliberately split into two parts, with $\mathcal{E}(\lambda, \mathbf{k})$ chosen to satisfy an explicit renormalization condition. This is called renormalization of the dispersion relation. It is done to ensure that $\langle\prod_{i=1}^{n}\psi_{p_i}\bar{\psi}_{q_i}\rangle$ is C^{∞} in λ at $\lambda = 0$. Define the proper self-energy $\Sigma(p)$ for the action $\mathcal{A}$ by the equation

$$\left(ip_0 - e(\mathbf{p}) - \Sigma(p)\right)^{-1}(2\pi)^{d+1}\delta(p-q) = \frac{\int \psi_p\bar{\psi}_q e^{\mathcal{A}(\psi)}\prod d\psi_{k,\sigma}d\bar{\psi}_{k,\sigma}}{\int e^{\mathcal{A}(\psi)}\prod d\psi_{k,\sigma}d\bar{\psi}_{k,\sigma}}.$$

The counterterm $\mathcal{E}(\lambda, \mathbf{k})$ is chosen so that

$$\Sigma(0,\mathbf{p})\big|_{\mathbf{p}\in\mathbf{F}} = 0.$$

To give a rigorous definition of (1.2) one must introduce cutoffs and then take the limit in which the cutoffs are removed. To impose an infrared cutoff in the spatial directions one may put the system in a finite periodic box $\mathbb{R}^d/L\Gamma$. To impose an infrared cutoff in the zero direction one may make the inverse temperature $\beta < \infty$. Then momenta $k = (k_0, \mathbf{k})$ are restricted to lie on the lattice

$$k_0 \in \frac{\pi}{\beta}(2\mathbb{Z}+1), \quad \mathbf{k} \in \frac{1}{L}\Gamma^{\#}.$$

The ultraviolet cutoffs further restrict $|k_0| \le 1$, $|e(\mathbf{k})| \le 1$. Then the Grassmann algebra becomes finite-dimensional and (1.2b) with the integral symbol reinterpreted as

$$\int dk\, f(k) = \frac{1}{\beta} \sum_{\substack{k_0 \in \pi(2\mathbb{Z}+1)/\beta \\ |k_0| \le 1}} \frac{1}{L^2} \sum_{\substack{\mathbf{k} \in \Gamma^\#/L \\ |e(\mathbf{k})| \le 1}} f(k)$$

is a well-defined element of that algebra.

Theorem 1.1. *Let $d = 2$ and Hypotheses 1–4 be satisfied. There is an $r > 0$ and a dispersion relation counterterm $\mathcal{E}(\lambda, \mathbf{k})$, such that the limits*

$$\lim_{\beta, L \to \infty} \frac{\int \prod_{i=1}^{n} \psi_{p_i} \bar\psi_{q_i}\, e^{\mathcal{A}(\psi, \bar\psi)} \prod d\psi_{k,\sigma} d\bar\psi_{k,\sigma}}{\int e^{\mathcal{A}(\psi, \bar\psi)} \prod d\psi_{k,\sigma} d\bar\psi_{k,\sigma}}$$

exist in the sense of distributions and are independent of the order in which the limits are taken. The counterterm and the limit are both analytic functions of the coupling constant λ for $|\lambda| < r$. Furthermore, there is a jump in the average occupation number $n_{\mathbf{k}}$ at the Fermi curve. Precisely, if

$$n_{\mathbf{k}} = \lim_{x_0 \searrow 0} \int dk_0\, e^{\imath k_0 x_0} \left(\imath k_0 - e(\mathbf{k}) - \Sigma(k_0, \mathbf{k}) \right)^{-1}$$

then

$$\lim_{\varepsilon \searrow 0} n_{\mathbf{p} - \varepsilon \nu_{\mathbf{p}}} - n_{\mathbf{p} + \varepsilon \nu_{\mathbf{p}}} = \left(1 + \imath \frac{\partial}{\partial k_0} \Sigma(0, \mathbf{p}) \right)^{-1} \ge 1 - O(\lambda)$$

for all $\mathbf{p}$ on the Fermi curve $\mathbf{F}$. Here, $\nu_{\mathbf{p}}$ is the outward pointing unit normal to $\mathbf{F}$ at $\mathbf{p}$. In other words, the infinite volume system is a Fermi liquid.

Our main goal here is to explain why this theorem is true, though the complete proof [1] is too long to include. There are two main aspects to that proof: the control of four-legged Feynman diagrams and the control of high orders of perturbation theory. The first aspect is discussed in §2 while the second is discussed in §3.

2   Four-Legged Diagrams

Spin plays no role in this section. So we suppress it. We also relax the condition $d = 2$ allowing all $d \ge 2$. Feynman diagrams in this model have lines

$$k \xrightarrow{\hspace{2cm}} = \frac{1}{\imath k_0 - e(\mathbf{k})}$$

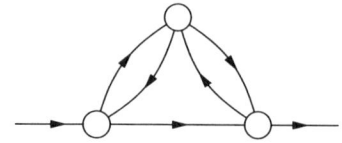

FIGURE 2.

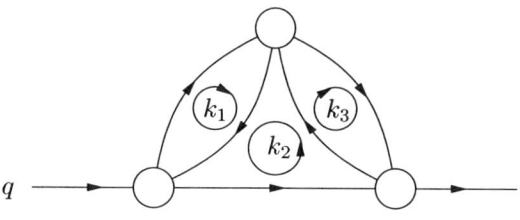

FIGURE 3.

and vertices

$$k_1 \qquad k_3$$
$$\qquad\qquad\qquad = (2\pi)^{d+1}\delta(k_1 + k_2 - k_3 - k_4)\lambda\langle k_1, k_2|V|k_3, k_4\rangle.$$
$$k_2 \qquad k_4$$

For example Figure 2 is one graph contributing to the proper self-energy. This is a three-loop graph. Choosing the loops as in Figure 3 we see that the value of this graph is

$$\int dk_1\, dk_2\, dk_3\; \frac{1}{i(k_1)_0 - e(\mathbf{k}_1)}\frac{1}{i(k_1 + k_2)_0 - e(\mathbf{k}_1 + \mathbf{k}_2)}$$
$$\times \frac{1}{i(k_2 + k_3)_0 - e(\mathbf{k}_2 + \mathbf{k}_3)}\frac{1}{i(k_3)_0 - e(\mathbf{k}_3)}\frac{1}{i(k_2 + q)_0 - e(\mathbf{k}_2 + \mathbf{q})}$$
$$\times \langle k_1 + k_2, k_3|V|k_1, k_2 + k_3\rangle\langle k_1, k_2 + q|V|q, k_1 + k_2\rangle$$
$$\times \langle k_2 + k_3, q|V|k_2 + q, k_3\rangle.$$

It is not clear that this integral converges. The domain of integration is compact, because of the ultraviolet cutoff, but the integrand is singular.

To check for convergence one first does "naive power counting" bounds. In field theory, propagator singularities occur at points. Then power counting just comes down to some simple dimensional analysis. Here there are singularities on curves, like $(k_1)_0 = 0$, $\mathbf{k}_1 \in F$. We have to have a simple yet precise way of measuring whether the integrand is large a lot. To do so we decompose the propagator

$$C(k) = \frac{1}{ik_0 - e(\mathbf{k})} = \sum_{j=-\infty}^{0} C^{(j)}$$

where

$$C^{(j)}(k) = \frac{1}{ik_0 - e(\mathbf{k})}\chi(2^j \le |ik_0 - e(\mathbf{k})| < 2^{j+1})$$

Note, the perhaps bizarre, convention that j is negative. As j tends to *minus* infinity, 2^j approaches zero and, on the support of $C^{(j)}$, $|ik_0 - e(\mathbf{k})|$ approaches zero. Naive power counting just uses

Lemma 2.1. *Let d be arbitrary and Hypothesis 1 be satisfied. Then*

(a) $\|C^{(j)}\|_\infty = \sup_k |C^{(j)}(k)| \le 2^{-j}$

(b) $\|C^{(j)}\|_1 = \int dk |C^{(j)}(k)| \le$ const 2^j

Proof. Part (a) is obvious because, by construction, $|ik_0 - e(\mathbf{k})| \ge 2^j$ on the support of $C^{(j)}(k)$.

For part (b) observe that

$$\text{vol}\{k = (k_0, \mathbf{k}) \mid C^{(j)}(k) \ne 0\} \le \text{vol}\{k_0 \mid |k_0| \le 2^{j+1}\}$$
$$\times \text{vol}\{\mathbf{k} \in B \mid |e(\mathbf{k})| \le 2^{j+1}\}$$
$$\le 2^{j+2} \text{vol}\{\mathbf{k} \in B \mid |e(\mathbf{k})| \le 2^{j+1}\}.$$

The set $\{\mathbf{k} \in B \mid |e(\mathbf{k})| \le 2^{j+1}\}$ consists of a shell of thickness $O(2^j)$ around F and hence has volume bounded by const2^j so that

$$\text{vol}\{k = (k_0, \mathbf{k}) \mid C^{(j)}(k) \ne 0\} \le \text{const } 2^{2j} \tag{2.1}$$

and

$$\|C^{(j)}\|_1 = \int dk |C^{(j)}(k)|$$
$$\le \sup_k |C^{(j)}(k)| \text{vol}\{k = (k_0, \mathbf{k}) \mid C^{(j)}(k) \ne 0\}$$
$$\le \text{const } 2^j. \qquad \square$$

We remark that the smoothness condition $\nabla e(\mathbf{k}) \ne 0$ of Hypothesis 1 was used to get the volume bound (2.1). The corresponding volume for the Hubbard model at half-filling is $|j|2^{2j}$ which leads to $\|C^{(j)}\|_1 \le$ const $|j|2^j$.

In the infrared Φ_4^4 model $C^{(j)} = \chi(2^j \le |k| < 2^{j+1})/k^2$ so that the analog of Lemma 2.1 for Φ_4^4 model is $\|C^{(j)}\|_\infty \le 2^{-2j}$, $\|C^{(j)}\|_1 \le$ const $2^{-2j}2^{4j} \le$ const 2^{2j}. The replacement $j \to 2j$ can be viewed simply as a change of units. Then it is not too surprising that Lemma 2.1 implies [2, 3] that models satisfying Hypotheses 1 and 4 obey bounds typical of strictly renormalizable models in the infrared regime. Two-legged diagrams are linearly divergent and must be renormalized. Four-legged subdiagrams are marginal and all other subdiagrams are convergent. As is normal for infrared models, the two-legged counterterm is finite and the marginality of four-legged

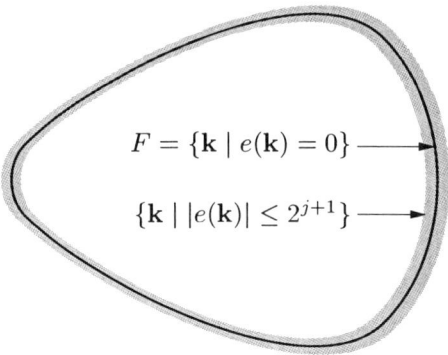

$$F = \{\mathbf{k} \mid e(\mathbf{k}) = 0\}$$

$$\{\mathbf{k} \mid |e(\mathbf{k})| \leq 2^{j+1}\}$$

FIGURE 4.

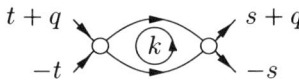

FIGURE 5.

subdiagrams does not require a counterterm. The four-legged subdiagrams are divergent only for certain exceptional momenta and then only logarithmically divergent. These logarithmic singularities are integrable and hence do not prevent diagrams from being well defined. But they can cause the values of diagrams containing many four-legged subdiagrams to be anomalously large through

$$\int dk \ln^n |k| \sim n!$$

Normally, under these circumstances one of two possibilities occur. The renormalization group flow of the four-point function is either asymptotically free or is to a nontrivial fixed point and is accompanied by some interesting physics, like mass generation or symmetry breaking. We shall now see that under Hypotheses 1–4, the bounds which give marginality of four-legged subdiagrams are not saturated. Four-legged subdiagrams are in fact convergent. The models behave more like superrenormalizable models than strictly renormalizable ones.

2.1   The Particle-Particle Bubble

As a concrete example, we'll first impose only Hypotheses 1 and 4 and do the naive power counting bound explicitly on one simple, but very important, graph—the particle-particle bubble (Figure 5). If the total momentum entering from the left is q, the value of this graph is

$$B(s,t,q) = \int dk\, C(-k+q)C(k)\langle -k+q, k|V|t+q, -t\rangle$$
$$\times \langle s+q, -s|V|-k+q, k\rangle.$$

Decomposing the two propagators into scales and then bounding the integral by the supremum of the integrand times the volume of the support of the integrand, we have

$$|B(s,t,q)|$$

$$= \left| \sum_{j_1,j_2 \leq 0} \int_B dk\, C_{j_1} C_{j_2} \langle -k+q, k|V|t+q, -t\rangle \right.$$
$$\left. \times \langle s+q, -s|V|-k+q, k\rangle \right|$$

$$\leq \sum_{j_1,j_2 \leq 0} \|V\|_\infty^2 2^{-j_1-j_2} \mathrm{vol}\{k \in \mathcal{B} \mid |ik_0 - e(k)| \leq 2^{j_1+1},$$
$$|i(-k+q)_0 - e(-\mathbf{k}+\mathbf{q})| \leq 2^{j_2+1}\}$$

$$\leq \sum_{j_1,j_2 \leq 0} \|V\|_\infty^2 2^{-j_1-j_2} 2^{\min\{j_1,j_2\}} \mathrm{vol}\{\mathbf{k} \in \mathrm{B} \mid |e(\mathbf{k})| \leq 2^{j_1+1},$$
$$|e(-\mathbf{k}+\mathbf{q})| \leq 2^{j_2+1}\} \quad (2.2)$$

since $|k_0| \leq 2^{j_1+1}$ and $|k_0 - q_0| \leq 2^{j_2+1}$. Even without using Hypotheses 2 and 3 we can bound the volume

$$\mathrm{vol}\{\mathbf{k} \in \mathrm{B} \mid |e(\mathbf{k})| \leq 2^{j_1+1}, |e(-\mathbf{k}+\mathbf{q})| \leq 2^{j_2+1}\}$$
$$\leq \min\Big\{\mathrm{vol}\{\mathbf{k} \in \mathrm{B} \mid |e(\mathbf{k})| \leq 2^{j_1+1}\},$$
$$\mathrm{vol}\{\mathbf{k} \in \mathrm{B} \mid |e(-\mathbf{k}+\mathbf{q})| \leq 2^{j_1+1}\}\Big\}$$
$$\leq \mathrm{const}\, \min\{2^{j_2}, 2^{j_2}\}$$
$$= \mathrm{const}\, 2^{\min\{j_1,j_2\}}. \quad (2.3)$$

This gives

$$\sup_{s,t,q} |B(s,t,q)| \leq \sum_{j_1,j_2 \leq 0} \mathrm{const} \|V\|_\infty^2 2^{-j_1-j_2} 2^{2\min\{j_1,j_2\}}$$
$$= \sum_{j_1,j_2 \leq 0} \mathrm{const} \|V\|_\infty^2 2^{-|j_1-j_2|}$$
$$= \sum_{j_1 \leq 0} \mathrm{const} \|V\|_\infty^2. \quad (2.4)$$

Recall that 2^j, and not j, has the units of energy. This bound allows the supremum of $|B(s,t,q)|$ to diverge logarithmically.

In the event that $e(\mathbf{k}) = e(-\mathbf{k})$, violating Hypothesis 3, and $\mathbf{q} = 0$ we have

$$\text{vol}\left\{\mathbf{k} \in B \mid e(\mathbf{k})| \le 2^{j_1+1}, |e(-\mathbf{k}+\mathbf{q})| \le 2^{j_2+1}\right\}$$
$$= \text{vol}\left\{\mathbf{k} \in B \mid |e(\mathbf{k})| \le 2^{\min\{j_1,j_2\}+1}\right\}$$
$$= O\left(2^{\min\{j_1,j_2\}+1}\right)$$

and (2.3) is saturated. In this case $q = 0$ really is an exceptional momentum for $B(q)$ which really does have a singularity at $q = 0$. We can see that the singularity is integrable by bounding

$$\sup_{s,t} \int dq|B(s,t,q)| \le \|V\|_\infty^2 \int dq\,dk|C(-k+q)C(k)|$$
$$= \|V\|_\infty^2 \int dq\,dk|C(q)C(k)|$$
$$= \sum_{j_1,j_2 \le 0} \|V\|_\infty^2 \int dq\,dk|C_{j_1}(q)C_{j_2}(k)|$$
$$= \sum_{j_1,j_2 \le 0} \text{const}\|V\|_\infty^2 \|C_{j_1}\|_1 \|C_{j_2}\|_1$$
$$\le \sum_{j_1,j_2 \le 0} \text{const}\|V\|_\infty^2 2^{j_1} 2^{j_2}$$
$$\le \text{const}\|V\|_\infty^2.$$

We now show that, when Hypotheses 2 and 3 are turned on, (2.4) is not saturated and that four-legged subgraphs are actually convergent so that the model acts superrenormalizable. By Hypotheses 3 and analyticity (or even with just Hypothesis 2 if $\mathbf{q} \ne 0$) the Fermi curve F can only meet the reflected translated Fermi curve $-F + \mathbf{q}$ transversely or with a tangency of some finite order. Hence there is an $\epsilon > 0$ such that

$$\text{vol}\left\{\mathbf{k} \in B \mid |e(\mathbf{k})| \le 2^{j_1+1}, |e(-\mathbf{k}+\mathbf{q})| \le 2^{j_2+1}\right\}$$
$$\le \text{const}\, 2^{\min\{j_1,j_2\}} 2^{\epsilon \max\{j_1,j_2\}}. \quad (2.5)$$

Here $\text{const}\, 2^{\min\{j_1,j_2\}}$ is the thickness of each component of the intersection of the two shells and $\text{const} 2^{\epsilon \max\{j_1,j_2\}}$ is a bound on the length of each component. Even though this bound is intuitively obvious, we give a complete proof in Lemma 2.2, below.

Substituting (2.5) into (2.2) gives

$$\sup_{s,t,q} |B(s,t,q)| \le \sum_{j_1,j_2 \le 0} \text{const}\|V\|_\infty^2 2^{-j_1-j_2} 2^{2\min\{j_1,j_2\}} 2^{\epsilon \max\{j_1,j_2\}}$$
$$= \sum_{j_1,j_2 \le 0} \text{const}\|V\|_\infty^2 2^{-|j_1-j_2|} 2^{\epsilon \max\{j_1,j_2\}}$$

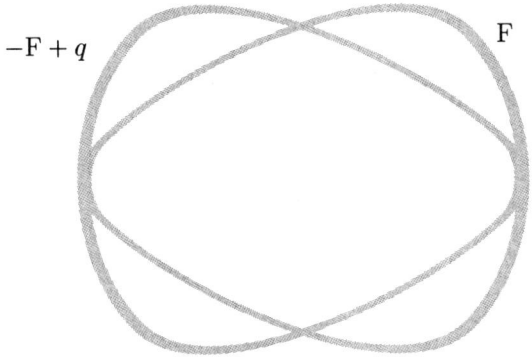

$-F + q$

F

FIGURE 6.

$$= \sum_{j \leq 0} \mathrm{const} \|V\|_\infty^2 2^{\epsilon j} < \infty$$

We conclude that, when Hypothesis 3 is turned on the particle-particle bubble becomes uniformly bounded.

Lemma 2.2. *There is a constant* const *and an* $\epsilon > 0$ *such that for all* j_1, $j_2 < 0$

$$\sup_{\mathbf{q}} \mathrm{vol}\{\mathbf{k} \mid |e(\mathbf{k})| \leq 2^{j_1+1}, |e(-\mathbf{k}+\mathbf{q})| \leq 2^{j_2+1}\}$$
$$\leq \mathrm{const}\, 2^{\min\{j_1,j_2\}} 2^{\epsilon \max\{j_1,j_2\}}.$$

Proof. We may assume without loss of generality that $j_1 \leq j_2$. Otherwise make the change of variables $\mathbf{p} = -\mathbf{k} + \mathbf{q}$. Put

$$\mathcal{M} = \{(\mathbf{q},\mathbf{k}) \in \mathbb{R}^d \times \mathbb{R}^d \mid |e(\mathbf{k})| \leq 2, |e(\mathbf{q}-\mathbf{k})| \leq 2\}.$$

Since $\mathcal{M}$ is compact, it suffices to show

Claim. *For each* $(\mathbf{q}^{(0)},\mathbf{k}^{(0)}) \in \mathcal{M}$ *there are constants* const > 0, $n \in \mathbb{N}$ *and there are neighborhoods* U *of* $\mathbf{q}^{(0)}$ *and* V *of* $\mathbf{k}^{(0)}$ *in* $\mathbb{R}^d$ *such that for all* $j_1 \leq j_2 < 0$ *and all* $\mathbf{q} \in U$

$$\mathrm{vol}\{\mathbf{k} \in V \mid |e(\mathbf{k})| \leq 2^{j_1+1}, |e(-\mathbf{k}+\mathbf{q})| \leq 2^{j_2+1}\} \leq \mathrm{const}\, 2^{j_1} 2^{j_2/n}.$$

Proof that the lemma follows from the claim. Note first that the const and the n in the statement of the claim may depend on $\mathbf{q}^{(0)}$ and $\mathbf{k}^{(0)}$. Since $\mathcal{M}$ is compact, it can be covered

$$\mathcal{M} \subset \bigcup_{i=1}^{m} U_i \times V_i$$

with U_i and V_i being the neighborhoods of the claim for some choice $(\mathbf{q}_i^{(0)}, \mathbf{k}_i^{(0)})$ of $(\mathbf{q}^{(0)}, \mathbf{k}^{(0)})$. Then the const of the lemma is bounded by $\sum_{i=1}^{m} \mathrm{const}_i$ and the ϵ of the lemma is bounded by $\min_{1 \leq i \leq m} 1/n_i$. $\square$

Proof of the claim. For maximum clarity, we first give the proof for $d = 2$. The claim is trivial if $e(\mathbf{k}^{(0)}) \neq 0$ or $e(\mathbf{q}^{(0)} - \mathbf{k}^{(0)}) \neq 0$. If $e(\mathbf{k}^{(0)}) \neq 0$, we can choose V so that $\{\mathbf{k} \in V \mid |e(\mathbf{k})| \leq 2^{j_1+1}, |e(-\mathbf{k}+\mathbf{q})| \leq 2^{j_2+1}\}$ is empty except for finitely many values of j_1 and j_2. If $e(\mathbf{k}^{(0)}) = 0$ but $e(\mathbf{q}^{(0)} - \mathbf{k}^{(0)}) \neq 0$, we can choose U, V so that $\{\mathbf{k} \in V \mid |e(\mathbf{k})| \leq 2^{j_1+1}, |e(-\mathbf{k}+\mathbf{q})| \leq 2^{j_2+1}\}$ is empty except for finitely many values of j_2 and we can bound

$$\mathrm{vol}\{\mathbf{k} \in V \mid |e(\mathbf{k})| \leq 2^{j_1+1}\} \leq \mathrm{const}\, 2^{j_1}.$$

So we assume

$$e(\mathbf{k}^{(0)}) = e(\mathbf{q}^{(0)} - \mathbf{k}^{(0)}) = 0.$$

Since $\nabla e(\mathbf{k})|_{\mathbf{k}=\mathbf{k}^{(0)}} \neq 0$ there are neighborhoods V' of $\mathbf{k}^{(0)}$ and $X \times Y$ of $(0,0)$ in $\mathbb{R}^2$ and a diffeomorphism

$$\pi \colon X \times Y \to V'$$

such that

$$e(\pi(x,y)) = y$$
$$\pi(0,0) = \mathbf{k}^{(0)}$$

and $\partial \mathbf{k}/\partial(x,y)$ is a nowhere vanishing bounded function. Define

$$E(\mathbf{q}, x) = e(\mathbf{q} - \pi(x,0)).$$

Since, for all $\mathbf{q}$ in a neighbourhood of $\mathbf{q}^{(0)}$ and all $(x,y) \in X \times Y$,

$$|E(\mathbf{q}, x)| = |e(\mathbf{q} - \pi(x,0)) - e(\mathbf{q} - \pi(x,y)) + e(\mathbf{q} - \pi(x,y))|$$
$$\leq \mathrm{const}|y| + |e(\mathbf{q} - \pi(x,y))|.$$

We have, for all $\mathbf{q}$ in a neighbourhood of $\mathbf{q}^{(0)}$

$$\mathrm{vol}\{\mathbf{k} \in V' \mid |e(\mathbf{k})| \leq 2^{j_1+1}, |e(-\mathbf{k}+\mathbf{q})| \leq 2^{j_2+1}\}$$
$$\leq \mathrm{const}\, \mathrm{vol}\{(x,y) \in X \times Y \mid |y| \leq 2^{j_1+1}, |e(\mathbf{q} - \pi(x,y))| \leq 2^{j_2+1}\}$$
$$\leq \mathrm{const}\, \mathrm{vol}\{(x,y) \in X \times Y \mid |y| \leq 2^{j_1+1}, |E(\mathbf{q}, x)| \leq \mathrm{const}\, 2^{j_2+1}\}$$
$$\leq \mathrm{const}\, 2^{j_1} \mathrm{vol}\{x \in X \mid |E(\mathbf{q}, x)| \leq \mathrm{const}\, 2^{j_2+1}\}.$$

Since $e(\mathbf{q}^{(0)} - \mathbf{k}^{(0)}) = 0$, we have that $E(\mathbf{q}^{(0)}, 0) = 0$. By analyticity, if the order of the zero of $E(\mathbf{q}^{(0)}, x)$ at $x = 0$ is not finite, Hypothesis 3 is violated. So, if X has been chosen small enough, there is a nowhere vanishing smooth function $\delta(x)$ and a natural number n such that

$$E(\mathbf{q}^{(0)}, x) = \delta(x)x^n.$$

So $E(\mathbf{q}^{(0)}, x)$ takes the normal form z^n under the diffeomorphic change of coordinates $z = \sqrt[n]{\delta(x)}x$. As we now move $\mathbf{q}$ away from $\mathbf{q}^{(0)}$, the nth order zero for $E(\mathbf{q}^{(0)}, x)$ probably splits up into a number of distinct zeros of $E(\mathbf{q}, x)$. So we cannot retain the normal form z^n for $E(\mathbf{q}, x)$. But we can have a normal form which is a polynomial of degree n.

Put, for $a = (a_1, \ldots, a_n) \in \mathbb{R}^n$

$$P(z; a) = z^n + a_1 z^{n-1} + \cdots + a_n.$$

By the theory of "universal unfoldings" there are neighborhoods $X' \subset X$, Z of 0 in $\mathbb{R}$ and U' of $\mathbf{q}^{(0)}$ in $\mathbb{R}^2$, functions $a_r(\mathbf{q})$, $0 \leq r \leq m$ on U' with

$$a_r(\mathbf{q}^{(0)}) = 0, \quad |a_r(\mathbf{q})| \leq 1$$

and a diffeomorphism

$$\begin{aligned} \Phi : X' \times U' &\to Z \times U' \\ (x, \mathbf{q}) &\mapsto (\phi_{\mathbf{q}}(x), \mathbf{q}) \end{aligned}$$

such that

(i) $\phi_{\mathbf{q}}(0) = 0$ for all $\mathbf{q} \in U'$

(ii) $E(\mathbf{q}, x) = P\Big(\phi_{\mathbf{q}}(x); \big(a_1(\mathbf{q}), \cdots, a_m(\mathbf{q})\big)\Big)$

(iii) $\big(\phi_{\mathbf{q}}^{-1}\big)^* dx = \rho(z, \mathbf{q}) \, dz$ with $|\rho(z, \mathbf{q})| \leq \text{const.}$

Therefore, for $d = 2$, Lemma 2.2 follows from Lemma 2.3 below.

For dimensions $d > 2$, $E(\mathbf{q}, x_1, \ldots, x_{d-1})$ replaces the function $E(\mathbf{q}, x)$. We can always arrange (by using a rotation in $\mathbb{R}^{d-1}$), that when $\mathbf{q} = \mathbf{q}^{(0)}$ and $x_2 = \cdots = x_{d-1} = 0$, we have $E(\mathbf{q}^{(0)}, x) = \delta(x_1) x_1^n$ with $\delta(x_1)$ bounded away from zero. By the same unfolding of singularities argument as above, for each fixed $\mathbf{q}, x_2, \ldots, x_{d-1}$, the set of allowed x_1 has volume at most const $2^{j_2/n}$. $\qquad \square$

Lemma 2.3. *There is a constant $K(m)$ such that for all $a_1, \ldots, a_m \in \mathbb{R}$ and $0 \leq \delta \leq 1$*

$$\text{vol}\big\{z \in \mathbb{R} \mid |z^m + a_1 z^{m-1} + \cdots + a_m| \leq \delta\big\} \leq K(m) \sqrt[m]{\delta}.$$

Proof. We use induction on m. The case $m = 1$ is trivial. Consider $m = 2$. We can always translate z to make $a_1 = 0$. If $a_2 \geq -\delta$

$$\big\{z \in \mathbb{R} \mid |z^2 + a_2| \leq \delta\big\} \subset \big\{z \in \mathbb{R} \mid |z^2| \leq 2\delta\big\}$$

which trivially has volume $2\sqrt{2\delta}$. If $a_2 = -a^2 < -\delta$ then

$$\big\{z \in \mathbb{R} \mid |z^2 + a_2| \leq \delta\big\} = \Big[\sqrt{a^2 - \delta}, \sqrt{a^2 + \delta}\Big] \cup \Big[-\sqrt{a^2 + \delta}, -\sqrt{a^2 - \delta}\Big]$$

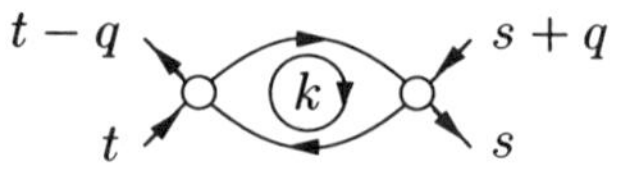

FIGURE 7.

which has volume

$$2\left(\sqrt{a^2+\delta}-\sqrt{a^2-\delta}\right)=2\frac{(a^2+\delta)-(a^2-\delta)}{\sqrt{a^2+\delta}+\sqrt{a^2-\delta}}\le 2\frac{2\delta}{\sqrt{2\delta}}$$

Now suppose $m>2$. Write

$$z^m+a_1z^{m-1}+\cdots+a_m=g(z)h(z)$$

where $g(z)$, $h(z)$ are monic polynomials of degree $m-2$ and 2 respectively. Clearly

$$\begin{aligned}&\left\{z\in\mathbb{R}\;\middle|\;|z^m+a_1z^{m-1}+\cdots+a_m|\le\delta\right\}\\&\qquad\subset\left\{z\in\mathbb{R}\;\middle|\;|g(z)|\le\delta^{(m-2)/m}\right\}\cup\left\{z\in\mathbb{R}\;\middle|\;|h(z)|\le\delta^{2/m}\right\}\end{aligned}$$

By the induction hypothesis and the case $m=2$, respectively, the volumes of the two sets is bounded by $K(m-2)\delta^{1/m}$ and $K(2)\delta^{1/m}$. □

2.2 The Particle-Hole Bubble

Of course the particle-particle bubble is just one graph. As a second example we consider the second most important graph in our class of models—the particle-hole bubble (Figure 7).

The value of this bubble is

$$B_2(s,t,q)=\int d\!\!\!{}^{-}k\,C(k+q)C(k)\langle t-q,k|\mathrm{V}|k,t\rangle\langle s,k|\mathrm{V}|k+q,s+q\rangle.$$

When Hypothesis 2 is satisfied and when q is bounded away from zero, we can apply the same argument as in the particle-particle bubble, now using the fact that shells around F and $F+q$ have small intersections

$$\begin{aligned}|B_2(s,t,q)|&\le\sum_{j_1,j_2\le 0}\|V\|_\infty^2\int_{\mathcal{B}}d\!\!\!{}^{-}k\,|C_{j_1}(k+q)C_{j_2}(k)|\\&\le\sum_{j_1,j_2\le 0}\|V\|_\infty^2 2^{\,j_1-j_2}\,\mathrm{vol}\big\{k\in\mathcal{B}\;\big|\;|ik_0-e(\mathbf{k})|\le 2^{j_1+1},\\&\qquad\qquad |i(k+q)_0-e(\mathbf{k}+\mathbf{q})|\le 2^{j_2+1}\big\}.\end{aligned}$$

In the event that $|q_0|\ge\text{const}>0$, it is only possible to satisfy the conditions $|k_0|\le 2^{j_1+1}$ and $|(k+q)_0|\le 2^{j_2+1}$ simultaneously if $\max\{j_1,j_2\}\ge$ const. In this case

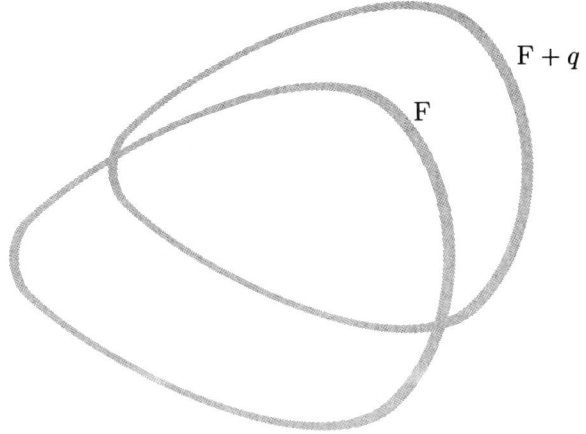

$F + q$

F

FIGURE 8.

$|B_2(s,t,q)|$

$$\leq \sum_{\substack{j_1,j_2\leq 0 \\ \max\{j_1,j_2\}\geq\text{const}}} \|V\|_\infty^2 2^{-j_1-j_2} 2^{\min\{j_1,j_2\}} \text{vol}\{k \in \mathcal{B} \mid |e(\mathbf{k})| \leq 2^{j_1+1}, \\ |e(\mathbf{k}+\mathbf{q})| \leq 2^{j_2+1}\}$$

$$\leq \sum_{\substack{j_1,j_2\leq 0 \\ \max\{j_1,j_2\}\geq\text{const}}} \text{const}\|V\|_\infty^2 2^{-j_1-j_2} 2^{2\min\{j_1,j_2\}}$$

$$= \sum_{\substack{j_1,j_2\leq 0 \\ \max\{j_1,j_2\}\geq\text{const}}} \text{const}\|V\|_\infty^2 2^{-|j_1-j_2|}$$

$$= \sum_{j=\text{const}}^{0} \text{const}\|V\|_\infty^2 < \infty.$$

On the other hand, if $|\mathbf{q}| \geq \text{const}$

$$|B_2(s,t,q)| \leq \sum_{j_1,j_2\leq 0} \|V\|_\infty^2 2^{-j_1-j_2} 2^{\min\{j_1,j_2\}} \text{vol}\{k \in \mathcal{B} \mid |e(\mathbf{k})| \leq 2^{j_1+1}, \\ |e(\mathbf{k}+\mathbf{q})| \leq 2^{j_2+1}\}$$

$$\leq \sum_{j_1,j_2\leq 0} \text{const}\|V\|_\infty^2 2^{-j_1-j_2} 2^{2\min\{j_1,j_2\}} 2^{\epsilon\max\{j_1,j_2\}} < \infty$$

Once again const $2^{\min\{j_1,j_2\}}$ is the thickness of each component of the intersection of the two shells (Figure 8) and const $2^{\epsilon\max\{j_1,j_2\}}$ is a bound on the length of each component.

So $B_2(s,t,q)$ is uniformly bounded if q is kept away from zero. As an illustration of what happens when q is small, consider $q = 0$. If we had analyticity in k_0 we could observe that

$$\int d\mathbf{k} \int dk_0 \frac{f(\mathbf{k})}{[ik_0 - e(\mathbf{k})]^2} \langle t, k|V|k, t\rangle\langle s, k|V|k, s\rangle = 0$$

simply by closing the k_0 contour in the half-plane not containing the pole $k_0 = -ie(\mathbf{k})$. However our ultraviolet cutoff destroys that analyticity so we have to work a bit harder. Changing variables to

$$x = k_0, \quad y = e(\mathbf{k})$$

and some angular variable(s) and performing the integral over the angular variable(s), we have

$$B_2(s, t, 0) = \int_B d\mathbf{k} \frac{1}{[ik_0 - e(\mathbf{k})]^2} \langle t, k|V|k, t\rangle\langle s, k|V|k, s\rangle$$

$$= \int dx\,dy \frac{1}{[ix - y]^2} I(x, y)$$

with $I(x, y)$ being some function that is C^∞ at $x = y = 0$. Making the further change of variables to polar coordinates

$$B_2(s, t, 0) = \int dr\,d\theta \frac{r}{i[re^{i\theta}]^2} I(r\cos\theta, r\sin\theta)$$

$$= \int dr\,d\theta \frac{r}{i[re^{i\theta}]^2} [I(0, 0) + O(r)].$$

The potentially logarithmically divergent term

$$\int dr\,d\theta \frac{1}{ire^{2i\theta}} I(0, 0)$$

vanishes because

$$\int_0^{2\pi} d\theta\, e^{in\theta} = 0$$

for all nonzero integers n. Hence $B_2(s, t, 0)$ is bounded. By working harder still we can bound $B_2(s, t, q)$ for all small q.

2.3 Higher-Order Diagrams

We can always Wick order the interaction. Then no tadpoles appear in Feynman diagrams and every higher-order graph falls into one of two categories. There are strings of bubbles, like Figure 9 that can be treated as above. And there are graphs which have overlapping loops, like Figure 10. In the example, the k-loop and the p-loop share a line and hence, by definition, overlap. Consequently, after the propagators have been decomposed

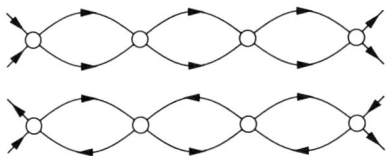

FIGURE 9.

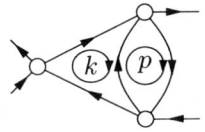

FIGURE 10.

into scales, the integrand of the value of this diagram contains all three factors $C^{(j_1)}(k)C^{(j_2)}(p)C^{(j_3)}(p-k)$. The support properties of these propagators constrains the domain of integration to

$$\{(\mathbf{k}, \mathbf{p}) \in B \times B \mid |e(\mathbf{k})| \leq 2^{j_1+1}, |e(\mathbf{p})| \leq 2^{j_2+1}, |e(\mathbf{p}-\mathbf{k})| \leq 2^{j_3+1}\}.$$

The naive bound

$$\begin{aligned}
\mathrm{vol}\{(\mathbf{k}, \mathbf{p}) \in B \times B \mid |e(\mathbf{k})| &\leq 2^{j_1+1}, |e(\mathbf{p})| \leq 2^{j_2+1}, |e(\mathbf{p}-\mathbf{k})| \leq 2^{j_3+1}\} \\
&\leq \mathrm{vol}\{(\mathbf{k}, \mathbf{p}) \in B \times B \mid |e(\mathbf{k})| \leq 2^{j_1+1}, |e(\mathbf{p})| \leq 2^{j_2+1}\} \\
&= \mathrm{vol}\{\mathbf{k} \in B \mid |e(\mathbf{k})| \leq 2^{j_1+1}\} \, \mathrm{vol}\{\mathbf{p} \in B \mid |e(\mathbf{p})| \leq 2^{j_2+1}\} \\
&\leq \mathrm{const}\ 2^{j_1} 2^{j_2}
\end{aligned}$$

allows the value of the diagram to have logarithmic singularities.

Fortunately, the third condition gives some "volume improvement" over naive power counting. To see this, make the change of variables

$$\begin{aligned}
\mathbf{k} &= \mathbf{k}(x, \theta) & x &= e(\mathbf{k}) \\
\mathbf{p} &= \mathbf{p}(y, \phi) & y &= e(\mathbf{p})
\end{aligned}$$

with θ and ϕ each being a set of $d-1$ angular variables. This change of variables has bounded Jacobean so that

$$\begin{aligned}
\mathrm{vol}\{(\mathbf{k}, \mathbf{p}) \in B \times B \mid |e(\mathbf{k})| &\leq 2^{j_1+1}, |e(\mathbf{p})| \leq 2^{j_2+1}, |e(\mathbf{p}-\mathbf{k})| \leq 2^{j_3+1}\} \\
&\leq \mathrm{const}\ \mathrm{vol}\{(x, \theta, y, \phi) \mid |x| \leq 2^{j_1+1}, |y| \leq 2^{j_2+1}, \\
&\qquad\qquad\qquad |e(\mathbf{p}(y, \phi) - \mathbf{k}(x, \theta))| \leq 2^{j_3+1}\}
\end{aligned}$$

By the Mean Value Theorem (applied twice)

$$\left| e(\mathbf{p}(y, \phi) - \mathbf{k}(x, \theta)) - e(\mathbf{p}(0, \phi) - \mathbf{k}(0, \theta)) \right| \leq \mathrm{const}\ 2^{j_1} + \mathrm{const}\ 2^{j_2}$$

so that

$$\text{vol}\{(\mathbf{k}, \mathbf{p}) \in B \times B \mid |e(\mathbf{k})| \le 2^{j_1+1}, |e(\mathbf{p})| \le 2^{j_2+1}, |e(\mathbf{p} - \mathbf{k})| \le 2^{j_3+1}\}$$
$$\le \text{const vol}\{(x, \theta, y, \phi) \mid |x| \le 2^{j_1+1}, |y| \le 2^{j_2+1},$$
$$|e(\mathbf{p}(0, \phi) - \mathbf{k}(0, \theta))| \le \text{const} 2^{\max\{j_1, j_2, j_3\}}\}$$
$$\le \text{const} 2^{j_1} 2^{j_2} \text{vol}\{(\theta, \phi) \mid |e(\mathbf{p}(0, \phi) - \mathbf{k}(0, \theta))| \le \text{const } 2^{\max\{j_1, j_2, j_3\}}\}$$
$$\le \text{const } 2^{j_1} 2^{j_2} 2^{\epsilon \max\{j_1, j_2, j_3\}}$$

for some $\epsilon > 0$. To prove the "volume improvement" bound of the last inequality, first use a compactness argument like that at the beginning of the proof of Lemma 2.1, to reduce consideration to a small ball in (ϕ, θ) space. Then recopy the last half of the proof of the claim in Lemma 2.1, in which the bound

$$\text{vol}\{x \in X \mid |E(\mathbf{q}, x)| \le \text{const } 2^{j_2+1}\} \le \text{const } 2^{j_2/n}$$

is proven. Replace 2^{j_2} by $2^{\max\{j_1, j_2, j_3\}}$, $\mathbf{q}$ by $\mathbf{p}(0, \phi)$ and x by θ. You now have a proof. Note, in particular, that for each fixed ϕ, the function $e(\mathbf{p}(0, \phi) - \mathbf{k}(0, \theta))$ cannot be identically zero in θ by Hypothesis 3. This "volume improvement" ensures that all four-legged subdiagrams have convergent, rather than marginal, power counting. See Refs. 1 and 4.

3   A Single-Slice Fermionic Cluster Expansion

In this section, we concentrate on the problem of summing high orders of perturbation theory. To do so we shall consider an artificial model which retains the essential difficulties that now interest us, but which is restricted to a single slice. The toy world consists of

- $d + 1$ dimensional Euclidean space-time

- four types of fermions, denoted $\psi_\uparrow$, $\psi_\downarrow$, $\bar{\psi}_\uparrow$, and $\bar{\psi}_\downarrow$, that play the roles of spin up and spin down electrons and positrons/holes

- momenta "morally" in the range $2^j \le |p| \le 2^{j+1}$.

This is typical of one slice of a relativistic quantum field theory. In a many-body model we would have $2^j \le |p_0| + ||\mathbf{p}| - k_F| \le 2^{j+1}$. We will discuss the implications of this set of momenta later. In a realistic model we would have to sum over j using the renormalization group, which provides a machinized version of the techniques of the previous section.

We say "morally" because momentum is never actually going to appear in the toy world. Instead we are going to mimic the assumed momentum range by two space-time properties of the model. First, because the momentum space of our toy world has volume $2^{j(d+1)}$ the Pauli exclusion principle says that there can be at most one $\psi_\uparrow$, for example, in any region of volume $2^{-j(d+1)}$ in position space. Thus we define the fields of our model to be

$$\{\psi_\uparrow(x), \psi_\downarrow(x), \bar\psi_\uparrow(x), \bar\psi_\downarrow(x) \mid x \in W := 2^{-j}\mathbb{Z}^{d+1}\}.$$

They are the generators of a Grassmann algebra, meaning

$$\overset{(-)}{\psi}_\alpha(x)\overset{(-)}{\psi}_\beta(y) = -\overset{(-)}{\psi}_\beta(y)\overset{(-)}{\psi}_\alpha(x)$$

and in particular

$$\left(\overset{(-)}{\psi}_\alpha(x)\right)^2 = 0.$$

The second concerns the propagator. That is, the free two-point Euclidean Green's function. The interacting two-point Euclidean Green's function is

$$S_2(x, x') = \frac{\int \psi_\uparrow(x)\bar\psi_\uparrow(x')e^{-\lambda V}d\mu_C}{\int e^{-\lambda V}d\mu_C},$$

where the interaction

$$V = \frac{1}{2}\sum_{y\in W} 2^{-j(d+1)}\bar\psi_\uparrow(y)\bar\psi_\downarrow(y)\psi_\downarrow(y)\psi_\uparrow(y)$$

and

$$d\mu_C = \exp\left\{\sum_{\substack{z,z'\in W \\ \sigma\in\{\uparrow,\downarrow\}}} \bar\psi_\sigma(z')C^{-1}(z',z)\psi_\sigma(z)\right\} \prod_{\substack{z\in W \\ \sigma\in\{\uparrow,\downarrow\}}} d\psi_\sigma(z)d\bar\psi_\sigma(z)$$

is the Grassmann Gaussian measure with covariance C, to be specified shortly.

Here are all of their properties of Grassmann Gaussian measures that we are going to use. The symbol $\int \cdot\, d\mu_C$ is a linear functional that assigns a complex number to every polynomial in the fields and that obeys

$$\int \psi_\sigma(x)\bar\psi_{\sigma'}(y)\, d\mu_C = \delta_{\sigma,\sigma'}C(x,y)$$

$$\int \prod_{i=1}^n \psi_{\sigma_i}(x_i)\bar\psi_{\sigma_{i'}}(y_i)\, d\mu_C = \det[\delta_{\sigma_i,\sigma'_{i'}}C(x_i,y_{i'})]_{\substack{1\le i\le n \\ 1\le i'\le n}} \tag{3.1}$$

$$\int \psi_\sigma(x)F(\psi,\bar\psi)d\mu_C = \sum_{y\in W} C(x,y)\int \frac{\delta}{\delta\bar\psi_\sigma(y)}F(\psi,\bar\psi)d\mu_C$$

$$\int \bar\psi_\sigma(y)F(\psi,\bar\psi)d\mu_C = -\sum_{x\in W} C(x,y)\int \frac{\delta}{\delta\psi_\sigma(x)}F(\psi,\bar\psi)d\mu_C. \tag{3.2}$$

Except for signs, the Grassmann derivatives behave like normal ones. They are defined by

$$
\frac{\delta}{\delta \overset{(-)}{\psi}_\alpha(x)} \prod_{j=1}^{n} \overset{(-)}{\psi}_{\sigma_j}(y_j)
$$
$$
= \sum_{k=1}^{n} (-1)^{k-1} \prod_{\substack{j=1 \\ j \neq k}}^{n} \overset{(-)}{\psi}_{\sigma_j}(y_j) \begin{cases} 1 & \text{if } \overset{(-)}{\psi}_{\sigma_k}(y_k) = \overset{(-)}{\psi}_\alpha(x); \\ 0 & \text{if } \overset{(-)}{\psi}_{\sigma_k}(y_k) \neq \overset{(-)}{\psi}_\alpha(x). \end{cases}
$$

The $(-1)^{k-1}$ in the definition of the derivative is the sign of the permutation that moves $\overset{(-)}{\psi}_{\sigma_k}(y_k)$ to the left-hand end of the product.

We assume, as the second characteristic of our momentum range, that the covariance C decays at a rate typical of a smooth function whose Fourier transform has support in a neighbourhood of $|p| = 2^j$. Precisely,

$$
|C(x,y)| \leq \kappa 2^{(d+1)j/2} e^{-2^j|x-y|}.
$$

The coefficient $2^{(d+1)j/2}$ is chosen to give power counting typical of a strictly renormalizable field theory. The position space behavior of the many-Fermion propagator is somewhat more complicated than this. More about this later.

Theorem 3.1. *Let $S_{2,n}(x,x')$ be the coefficient of λ^n in the formal power series expansion of $S_2(x,x')$ (i.e. $S_2(x,x') = \sum_{n=0}^{\infty} S_{2,n}(x,x')\lambda^n$). There exists a constant R, independent of j, x, x', such that*

$$
\sup_{x} \sum_{x'} |S_{2,n}(x,x')| \leq K_j R^n.
$$

In other words S_2 is analytic in $|\lambda| < 1/R$. That is, the sum of all connected Feynman diagrams converges for all $|\lambda| < 1/R$. Similar bounds apply to the other Euclidean Green's functions.

Logic of the proof. First, we describe the logic of the proof. Denote by $S_2(x,x';\Lambda)$ the two-point function of the model gotten by restricting the world to a finite subset Λ of W. As a preliminary step, we will show, by Hadamard's inequality, that both the numerator and denominator of $S_2(x,x';\Lambda)$ are entire functions of λ. The denominator $\int e^{-\lambda V} d\mu_C$ can have many Λ dependent zeros. But when $\lambda = 0$, the denominator is one so that $S_2(x,x';\Lambda)$ is meromorphic on all of $\mathbb{C}$ and analytic at zero. We shall develop a formal power series expansion for $S_2(x,x';\Lambda)$ with the property that for every N

$$
S_2(x,x';\Lambda) = \sum_{n=0}^{N} S_{2,n}(x,x';\Lambda)\lambda^n + O(\lambda^{N+1}). \tag{3.3a}
$$

A priori we do not claim that the tail $O(\lambda^{N+1})$ is uniform in Λ. Nevertheless, since $S_2(x, x'; \Lambda)$ is analytic at zero we must have

$$S_2(x, x'; \Lambda) = \sum_{n=0}^{\infty} \frac{\lambda^n}{n!} \frac{d^n}{d\lambda^n} S_2(x, x'; \Lambda) \Big|_{\lambda=0}$$

with

$$\frac{1}{n!} \frac{d^n}{d\lambda^n} S_2(x, x'; \Lambda) \Big|_{\lambda=0} = S_{2,n}(x, x'; \Lambda)$$

by (3.3a). Hence

$$S_2(x, x'; \Lambda) = \sum_{n=0}^{\infty} S_{2,n}(x, x'; \Lambda) \lambda^n \tag{3.3b}$$

for all λ smaller than the (possibly Λ dependent) radius convergence of the right-hand side. We remark in passing that $S_{2,n}(x, x'; \Lambda)$ must be the sum of all connected Feynman diagrams of order n with two external legs, since we have an asymptotic expansion.

The heart of the proof is to show that there exists a constant R, independent of Λ, j and a constant K_j independent of Λ such that

$$\sup_x \sum_{x'} |S_{2,n}(x, x'; \Lambda)| \leq K_j R^n. \tag{3.4}$$

As a consequence, equation (3.3b) applies for all $|\lambda| < R^{-1}$. Any zeroes of the denominator that appear in this disk must be cancelled by zeroes of the numerator. It shall also be clear from the proof of (3.4) that the limits $S_{2,n}(x, x') = \lim_{\Lambda \to W} S_{2,n}(x, x'; \Lambda)$ exist. This will prove, by the Lebesgue dominated convergence theorem, that

$$S_2(x, x') = \lim_{\Lambda \to W} S_2(x, x'; \Lambda) = \sum_{n=0}^{\infty} S_{2,n}(x, x') \lambda^n$$

for all $|\lambda| < R^{-1}$ and that the coefficients $S_{2,n}(x, x')$ obey the bound (3.4).

Analyticity in finite volume. Fix a finite subset $\Lambda \subset W$. We now show that when the interaction V is restricted to Λ, the denominator

$$Z(\Lambda) = \sum_{n=0}^{\infty} \frac{1}{n!} \left(-\frac{\lambda}{2}\right)^n 2^{-nj(d+1)} \\ \times \sum_{y_1, \ldots, y_n \in \Lambda} \int \prod_{i=1}^{n} \bar{\psi}_\uparrow(y_i) \bar{\psi}_\downarrow(y_i) \psi_\downarrow(y_i) \psi_\uparrow(y_i) \, d\mu_C$$

in the definition of $S_2(x, x'; \Lambda)$ is entire in λ. The proof that the numerator is also entire is similar.

The value of the integral is

$$\int \prod_{i=1}^{n} \bar{\psi}_\uparrow(y_i)\bar{\psi}_\downarrow(y_i)\psi_\downarrow(y_i)\psi_\uparrow(y_i\, d\mu_C$$

$$= \int \prod_{i=1}^{n} \bar{\psi}_\uparrow(y_i)\psi_\uparrow(y_i)d\mu_C \int \prod_{i=1}^{n} \bar{\psi}_\downarrow(y_i)\psi_\downarrow(y_i)d\mu_C$$

$$= \det[C(y_i, y_{i'})]_{1\leq i,i'\leq n} \det[C(y_i, y_{i'})]_{1\leq i,i'\leq n}.$$

For these determinants to be nonzero, all of the $y_{i'}$'s must be distinct. For, otherwise, the determinants have two identical columns. So, by Hadamard's inequality,

$$|\det[C(y_i, y_{i'})]_{1\leq i,i'\leq n}| \leq \prod_{i=1}^{n}\left[\sum_{i'=1}^{n}|C(y_i, y_i')|^2\right]^{1/2}$$

$$\leq \prod_{i=1}^{n}\left[\sum_{i'=1}^{n}|\kappa^2 2^{(d+1)j}e^{-2^{2j}|y_i - y_{i'}|}\right]^{1/2}$$

$$\leq \prod_{i=1}^{n}\left[\sum_{y\in\Lambda}|\kappa^2 2^{(d+1)j}e^{-2^{2j}|y_i - y|}\right]^{1/2}$$

$$\leq \tilde{\kappa}_j^n$$

and the coefficient of λ^n in the Taylor series expansion of $Z(\Lambda)$ is bounded by

$$\frac{1}{n!}\left(\frac{|\lambda|}{2}\right)^n 2^{-nj(d+1)}|\Lambda|^n\tilde{\kappa}_j^{2n}$$

which implies that $Z(\Lambda)$ is entire in λ. The bound, however, blows up badly with $|\Lambda|$. The bulk of the effort in this proof goes into deriving a Λ independent bound.

The expansion. We now describe the expansion used. To emphasize that everything is uniform in Λ, we suppress Λ. The first step is to use integration by parts (that is property 2.) to turn the $\psi_\uparrow(x)$ of the two-point function into a covariance:

$$S_2(x, x') = \frac{\int \psi_\uparrow(x)\bar{\psi}_\uparrow(x')e^{-\lambda V}d\mu_C}{\int e^{-\lambda V}d\mu_C}$$

$$= C(x, x') + \frac{\sum_y \lambda C(x, y)\int \bar{\psi}_\uparrow(x')\left(\delta V/\delta\bar{\psi}_\uparrow(y)\right)e^{-\lambda V}d\mu_C}{\int e^{-\lambda V}d\mu_C}.$$

The first term is the trivial Feynman diagram giving the free value of S_2. For the second, apply integration by parts again to turn the $\bar{\psi}_\uparrow(x')$ into another propagator.

$$S_2(x, x') = C(x, x') - \frac{\sum_{y,y'} \lambda C(x,y) C(y',x') \int \left[\frac{\delta}{\delta \bar{\psi}_\uparrow(y')} \frac{\delta}{\delta \bar{\psi}_\uparrow(y)} V \right] e^{-\lambda V} d\mu_C}{\int e^{-\lambda V} d\mu_C}$$
$$- \frac{\sum_{y,y'} \lambda^2 C(x,y) C(y',x') \int \frac{\delta V}{\delta \bar{\psi}_\uparrow(y)} \frac{\delta V}{\delta \bar{\psi}_\uparrow(y')} e^{-\lambda V} d\mu_C}{\int e^{-\lambda V} d\mu_C}.$$

In each step select any $\overset{(-)}{\psi}$ downstairs and use integration by parts to turn it into one end of a propagator. When a term has no fields downstairs, the $\int e^{-\lambda V} d\mu_C$ in the numerator exactly cancels that in the denominator, leaving a Feynman diagram. This was how the trivial diagram C arose. Leave such terms alone. Upon completion of the expansion, we have $S_2(x, x')$ expressed as the sum of all connected two point Feynman diagrams.

To illustrate the principal difficulty in bounding S_2 consider the following nth order term that arises in the midst of the expansion:

$$= \frac{\lambda^n}{2^n} 2^{-j(d+1)n} \sum_{y_1, \ldots, y_n \in W} \left(\prod_{i=1}^{n+1} C(y_{i-1}, y_i) \right)$$
$$\times \frac{\int \prod_{m=1}^{n} \bar{\psi}_\downarrow(y_m) \psi_\downarrow(y_m) e^{-\lambda V} d\mu_C}{\int e^{-\lambda V} d\mu_C}, \quad (3.5)$$

where $y_0 = x$ and $y_{n+1} = x'$. The functional integral

$$\int \prod_{m=1}^{n} \bar{\psi}_\downarrow(y_m) \psi_\downarrow(y_m) e^{-\lambda V} d\mu_C$$
$$= -\sum_{z \in W} C(z, y_1) \int \frac{\delta}{\delta \psi_\downarrow(z)} \left[\psi_\downarrow(y_1) \prod_{m=2}^{n} \bar{\psi}_\downarrow(y_m) \psi_\downarrow(y_m) e^{-\lambda V} \right] d\mu_C$$
$$= -\sum_{i=1}^{n} C(y_i, y_1) \int \psi_\downarrow(y_1) \cdots \overline{\psi_\downarrow(y_i)} \cdots \bar{\psi}_\downarrow(y_n) \psi_\downarrow(y_n) e^{-\lambda V} d\mu_C$$
$$- \frac{\lambda}{2} 2^{-j(d+1)} \sum_{y_{n+1} \in W} C(y_{n+1}, y_1) A_{n+1},$$

where

$$A_{n+1} = \int \psi_\downarrow(y_1) \cdots \psi_\downarrow(y_n) \bar{\psi}_\uparrow(y_{n+1}) \bar{\psi}_\downarrow(y_{n+1}) \psi_\uparrow(y_{n+1}) e^{-\lambda V} d\mu_C.$$

We did a single integration by parts to get rid of $\bar{\psi}_\downarrow(y_1)$ and ended up with n terms of order λ^n. The A_{n+1} terms are of order λ^{n+1}. If we perform

$n - 1$ further integrations by parts to get rid of $\bar{\psi}_\downarrow(y_2), \ldots, \bar{\psi}_\downarrow(y_n)$ we will generate $n!$ diagrams of order λ^n. Naive bounds on these $n!$ terms will fail to produce an acceptable bound on $S_{2,n}$.

Fortunately, the Pauli exclusion principle saves us. Note first that, if $y_m = y_{m'}$ for any $m \neq m'$, then $\psi_\downarrow(y_m)\psi_\downarrow(y'_m) = -\psi_\downarrow(y'_m)\psi_\downarrow(y_m)$ so that $\psi_\downarrow(y_m)\psi_\downarrow(y'_m) = 0$ and hence $\psi_\downarrow(y_1)\prod_{m=2}^n \psi_\downarrow(y_m)\psi_\downarrow(y_m) = 0$. Let

$$A_i = \int \psi_\downarrow(y_1)\cdots\cancel{\bar{\psi}_\downarrow(y_i)}\cdots\bar{\psi}_\downarrow(y_n)\psi_\downarrow(y_n)e^{-\lambda V}\,d\mu_C.$$

Then we may bound

$$\sum_{i=1}^n |C(y_i, y_1)A_i|$$

$$\leq \max_{1\leq i\leq n}|e^{2^j|y_1-y_i|/2}C(y_i,y_1)A_i|\sum_{i=1}^n e^{-2^j|y_1-y_i|/2}$$

$$\leq \max_{1\leq i\leq n}|\kappa 2^{j(d+1)/2}e^{-2^j|y_1-y_i|/2}A_i|\sum_{y\in 2^{-j}\mathbb{Z}^{d+1}} e^{-2^j|y_1-y|/2}$$

$$= \max_{1\leq i\leq n}|\kappa 2^{j(d+1)/2}e^{-2^j|y_1-y_i|/2}A_i|\sum_{x\in\mathbb{Z}^{d+1}} e^{-|x|/2}$$

$$= \mathcal{E}\max_{1\leq i\leq n}|\kappa 2^{j(d+1)/2}e^{-2^j|y_1-y_i|/2}A_i|, \tag{3.6}$$

where $\mathcal{E} = \sum_{x\in\mathbb{Z}^{d+1}}^n e^{-|x|/2} < \infty$. The crucial consequence of the Pauli exclusion principle, that the y_i's all are different, was used in going from line one to line two. Think of $2^{j(d+1)/2}e^{-2^j|y_1-y_i|/2}$ as a propagator (replacing $C(y_1, y_i)$) for a line in a graph. This propagator joins a vertex at y_1 to a vertex at y_i. The fields $\overset{(-)}{\psi}$ downstairs in the functional integral A_i are external legs for the graph.

Incorporating the A_{n+1} terms just gives

$$\left|\int\prod_{m=1}^n \bar{\psi}_\downarrow(y_m)\psi_\downarrow(y_m)e^{-\lambda V}\,d\mu_C\right|$$

$$\leq \mathcal{E}\max_{1\leq i\leq n}|\kappa 2^{j(d+1)/2}e^{-2^j|y_1-y_i|/2}A_i|$$

$$+ \frac{|\lambda|}{2}2^{-j(d+1)}\sum_{y_{n+1}\in W} C(y_{n+1},y_1)|A_{n+1}|$$

$$\leq \max\left\{\max_{1\leq i\leq n}|(\mathcal{E}+1)\kappa 2^{j(d+1)/2}e^{-2^j|y_1-y_i|/2}A_i|,\right.$$

$$\left.\frac{|\lambda|}{2}2^{-j(d+1)}\sum_{y_{n+1}\in W}(\mathcal{E}+1)\kappa 2^{j(d+1)/2}e^{-2^j|y_1-y_{n+1}|/2}|A_{n+1}|\right\}. \tag{3.7}$$

Proof of the main bound. We now develop the full bound, proceeding by induction. In each step of the induction we integrate by parts once and apply the above bounding procedure. At the end of step s of the induction procedure we will have the bound

$$|S_2(x, x')| \leq \max_{G \in \mathcal{G}_s} B(G), \tag{3.8}$$

where $\mathcal{G}_s$ is the set of all "incomplete" Feynman diagrams, like (3.5), that are formed by s or fewer integration by parts. Each $G \in \mathcal{G}_s$ has

- two one-legged vertices, labeled x and x'

- $\omega(G)$ four-legged vertices labeled $y_1, \dots, y_{\omega(G)}$

- at most s propagators with each propagator joining a pair of legs selected from the $4\omega(G) + 2$ legs of the $\omega(G) + 2$ vertices. The positions of the two vertices at the ends of line ℓ are denoted y_{i_ℓ} and y_{f_ℓ}. The set of legs of G that are not paired to form propagators is denoted $F(G)$ and consists of those fields that are downstairs in the functional integral and have not yet been either an initiator or a target of an integration by parts. If $F(G)$ is not empty the number of propagators is exactly s, because exactly s integration by parts have been performed.

The bound on G that appears in (3.8) is

$$B(G) = \frac{\lambda^{\omega(G)}}{2^{\omega(G)}} \sum_{y_1, \dots, y_{\omega(G)}} 2^{-j(d+1)\omega(G)} \prod_{\ell \in G} \left[(\mathcal{E} + 1) \kappa 2^{j(d+1)/2} e^{-2^j |y_{i_\ell} - y_{f_\ell}|/2} \right]$$
$$\times \left| \frac{\int \prod_{f \in F(G)} \overset{(-)}{\psi}_{\sigma_f}(y_f) e^{-\lambda V} d\mu_C}{\int e^{-\lambda V} d\mu_C} \right|.$$

At the very beginning of the induction $s = 0$ and $\mathcal{G}_0$ consists of precisely one graph G_0, which has two one-legged vertices, no propagators and

$$B(G_0) = \left| \frac{\int \psi_\uparrow(x) \bar{\psi}_\uparrow(x') e^{-\lambda V} d\mu_C}{\int e^{-\lambda V} d\mu_C} \right|.$$

The verification of the inductive hypothesis is virtually identical to the proof of (3.7). It suffices to replace the expression $\prod_{m=1}^{n} \bar{\psi}_\downarrow(y_m) \psi_\downarrow(y_m)$ by $\prod_{f \in F(G)} \overset{(-)}{\psi}_{\sigma_f}(y_f)$.

In order to isolate the nth order contribution to S_2, it suffices to insert the projection

$$P_{\leq n} = \left. \frac{d^n}{d\lambda^n} \right|_{\lambda=0}$$

onto nth order everywhere. Then

$$|S_{2,n}(x,x')| \le \max_{G \in \mathcal{G}_s} B_n(G) \qquad (3.9)$$

with

$$B_n(G) = \frac{1}{2^{\omega(G)}} \sum_{y_1, \dots, y_{\omega(G)}} 2^{-j(d+1)\omega(G)} \prod_{\ell \in G} \left[(\mathcal{E}+1)\kappa 2^{j(d+1)/2} e^{-2^j |y_{i_\ell} - y_{f_\ell}|/2} \right]$$

$$\left| P_{\le n - \omega(G)} \int \prod_{f \in F(G)} \overset{(-)}{\psi}_{\sigma_f}(y_f) e^{-\lambda V} d\mu_C \int e^{-\lambda V} d\mu_C \right|.$$

A graph having s propagators must be of order at least $(s-1)/2$. So (3.9) becomes independent of s for $s \ge 2n+1$ and we have

$$|S_{2,n}(x,x')| \le \frac{1}{2^n} \max_G \sum_{y_1, \dots, y_n} 2^{-j(d+1)n} \prod_{\ell \in G} \left[(\mathcal{E}+1)\kappa 2^{j(d+1)/2} e^{-2^j |y_{i_\ell} - y_{f_\ell}|/2} \right].$$

The maximum is over all connected Feynman diagrams with two one-legged vertices, labeled x, x' and n four-legged vertices labeled $y_1, \dots, y_n$. In preparation for bounding the graph G, select a spanning tree T for G. A spanning tree is a subgraph $T \subset G$ which has no loops and contains all the vertices of G. Bound all factors $e^{-2^j |y_{i_\ell} - y_{f_\ell}|/2}$ that are associated with lines $\ell \in G \setminus T$ by one. Then apply

$$\sum_{y \in W} e^{-2^j |y'-y|/2} \le \mathcal{E}$$

to each vertex of G starting with those farthest from x in the partial ordering of T. The result is

$$\sum_{x'} |S_{2,n}(x,x')|$$

$$\le \frac{1}{2^n} \max_G \sum_{y_1, \dots, y_n, x'} 2^{-j(d+1)n} \prod_{\ell \in G} \left[(\mathcal{E}+1)\kappa 2^{j(d+1)/2} e^{-2^j |y_{i_\ell} - y_{f_\ell}|/2} \right]$$

$$\le \frac{1}{2^n} 2^{-j(d+1)n} \max_G (\kappa\mathcal{E}+\kappa)^{|G|} 2^{|G|j(d+1)/2} \sum_{y_1, \dots, y_n, x'} \prod_{\ell \in T} \left[e^{-2^j |y_{i_\ell} - y_{f_\ell}|/2} \right]$$

$$\le \frac{1}{2^n} 2^{-j(d+1)n} \max_G (\kappa\mathcal{E}+\kappa)^{|G|} 2^{|G|j(d+1)/2} \mathcal{E}^{n+1}.$$

As we are currently considering an nth-order diagram contributing to the two-point function

$$|G| = \frac{2+4n}{2} = 2n+1$$

and the final bound is

$$\sum_{x'} |S_{2,n}(x,x')| \leq \frac{1}{2^n} 2^{-j(d+1)n} (\kappa\mathcal{E} + \kappa)^{2n+1} 2^{(2n+1)j(d+1)/2} \mathcal{E}^{n+1}$$

$$\leq \frac{(\mathcal{E}+1)^{3n+2}\kappa^{2n+1}}{2^n} 2^{j(d+1)/2},$$

which proves the theorem with $R = \frac{1}{2}\kappa^2(\mathcal{E}+1)^3$ and $K_j = \kappa(\mathcal{E}+1)^2 2^{j(d+1)/2}$.

$\square$

More generally, for a p-point function, $|G| = (p+4n)/2 = 2n + p/2$, the number of sums controlled by the tree decay is $n + p - 1$ and

$$\sum_{x_2,\cdots,x_p} |S_{p,n}(x_1,\cdots,x_p)|$$

$$\leq \frac{1}{2^n} 2^{-j(d+1)n} (\kappa\mathcal{E} + \kappa)^{2n+p/2} 2^{(2n+p/2)j(d+1)/2} \mathcal{E}^{n+p-1}$$

$$\leq \frac{\kappa^{2n+p/2}(\mathcal{E}+1)^{3n+3p/2-1}}{2^n} 2^{pj(d+1)/4}.$$

As we mentioned earlier, the hypothesis

$$|C(x,y)| \leq \kappa 2^{(d+1)j/2} e^{-2^j|x-y|}$$

while typical of a strictly renormalizable field theory is not satisfied by the many-Fermion propagator. The problem is that the momentum space shell $\{k \mid 2^j \leq |ik_0 - e(\mathbf{k})| < 2^{j+1}\}$ has two very different characteristic lengths: a macroscopic diameter of order 1 and a microscopic thickness of order 2^j. By decomposing the Fermi surface(Figure 11) into a union of $2^{-(d-1)j}$ "rectangles" of side 2^j, one can write the many-Fermion field at scale j as a sum $\psi^{(j)} = \sum_\alpha \psi^{(j,\alpha)}$ of $2^{-(d-1)j}$ independent fields with each "colored" field having a covariance that obeys $|C(x,y)| \leq \kappa 2^{dj} e^{-2^j|x-y|}$ One can then

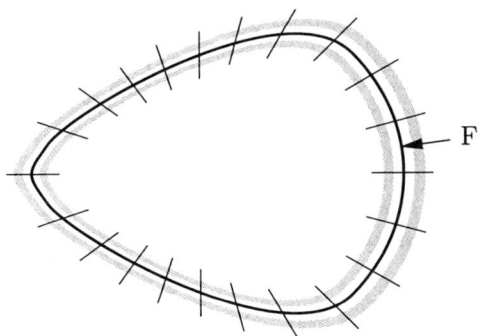

FIGURE 11.

apply the methods of Theorem 3.1. Of course one still has to control all the color sums. We have succeeded [3] in doing so for $d = 2$ and are working to extend the control to $d = 3$. This is the only place in the proof of the theorem of Section 1 that we have to restrict to $d = 2$.

Acknowledgments: Research supported in part by the Natural Sciences and Engineering Research Council of Canada.

4   REFERENCES

1. J. Feldman, H. Knörrer, D. Lehmann, and E. Trubowitz. in preparation.

2. J. Feldman and E. Trubowitz. The flow of an electron-phonon system to the superconducting state. *Helv. Phys. Acta*, 64: 213–357, 1991.

3. J. Feldman, J. Magnen, V. Rivassseau, and E. Trubowitz. An infinite volume expansion for many fermion Green's functions. *Helv. Phys. Acta*, 65: 679–721, 1992.

4. J. Feldman, M. Salmhofer, and E. Trubowitz. Perturbation theory around non-nested fermi surfaces, I: Keeping the Fermi surface fixed. *J. Stat. Phys.*, 84: 1209–1336, 1996.

3

Quantum Groups from Path Integrals

Daniel S. Freed

The goal of this discussion is to explain how quantum groups arise in three-dimensional topological quantum field theories (TQFTs). Of course, "explain how" is not the job of science, and perhaps you will find other explanations more satisfying. Let me explain!

What is a three-dimensional TQFT? At the very least it gives a topological invariant of three-dimensional manifolds. That is, to each 3-manifold X it assigns a complex number Z_X and if the invariants for X and X' are different ($Z_X \neq Z_{X'}$), then X and X' are not diffeomorphic. A three-dimensional TQFT also gives invariants of knots and links. For example, if K is a knot in ordinary 3-space, then we get a set of numerical invariants $\{I_K(\alpha)\}$ indexed by some finite set. The *Jones polynomial* of a knot, and similar polynomial invariants of knots, fit into this picture. As a historical note, Vaughan Jones [1] introduced his polynomial invariant in the mid-80s before the advent to topological quantum field theories. Those were introduced by Edward Witten in 1987 (following a suggestion of Michael Atiyah), first in 4 dimensions to give a quantum field-theoretic interpretation of Donaldson's invariants of 4-manifolds. A few years later [2] he introduced a three-dimensional TQFT which reproduces the Jones polynomial and which is our concern here. The classical action of this field theory is the *Chern-Simons invariant*, which was introduced into geometry in the early 1970s. As a mathematician I must immediately point out that Witten's methods, involving the path integral, are far from an established part of rigorous 1990s mathematics.

Shortly after Witten's paper on quantum Chern-Simons invariants, Reshetikhin and Turaev [3] showed how to start with extremely complicated algebraic data—called a *quantum group*—and again produce the Jones polynomial and its generalizations. (Their work is completely rigorous.) Subsequently, they showed [4] how to use the same data to construct invariants of 3-manifolds. The construction of a complete TQFT from this algebraic data, which involves more than invariants of 3-manifolds and knots, has been folklore ever since, and now has been described completely [5]. I remark here that instead of starting with a quantum group, one can start with certain "categorical" data instead.

The algebraic data of either a quantum group or its categorical equivalent is extremely complicated! One could hardly guess in advance that such data can produce invariants of knots and 3-manifolds. Nor can one easily construct algebraic data satisfying the necessary hypotheses. By contrast the classical Chern-Simons action is beautiful and simple! It is relatively easy to write down. One sees from the beginning that Lie groups enter the picture in a fundamental way. And if you are willing to accept the path integral (you shouldn't!), then you have a nice geometric construction of the Jones polynomial and related 3-manifold invariants. This leads us to pose the following.

Problem. Start with the Chern-Simons action and construct the quantum group which gives the same 3-manifold and knot invariants.

The goal is to explain how to do this in a simple case. As noted, the Chern-Simons theory starts with a compact Lie group G (and a piece of topological data which will be explained later). The Jones polynomial concerns the case $G = \mathrm{SU}(n)$ for variable n. In the simple case we treat, G is a *finite* group. This was first considered by Dijkgraaf and Witten [6]. The major simplification here is that the path integral is a finite sum, rather than an integral over an infinite-dimensional space, so is rigorously defined. So we immediately get a 3-manifold invariant, though it is rather simple and relatively uninteresting. The knot invariants are possibly more interesting; I don't believe that they have been investigated fully. In any case our interest is in the quantum group and our strategy is this: We exploit the fact that the path integral is well defined to introduce generalizations of the path integral. Thus one ingredient in a three-dimensional TQFT is a "quantum Hilbert space" $E(Y)$ for every surface Y. In usual quantum field theories it is constructed by canonical quantization. In our simple model we show how to get it by an exotic path integral. Something is immediately very strange—the result of an integration is a Hilbert space! Even more strange is the path integral we introduce for a 1-manifold, i.e., for a circle S. There is where we will see the quantum group emerging. In fact, the quantum groups we compute this way were written down in a paper of Dijkgraaf et al. [7]. They did not related it to the Chern-Simons invariant. It was a conjecture of Altschuler and Coste [8] that these quantum groups construct (via the Reshetikhin-Turaev prescription) the invariants of the finite group Chern-Simons theory. Our methods prove this conjecture.

This, then, is our strategy. In a d-dimensional field theory, where usually the classical action is only defined for fields in d dimensions, we will generalize the classical action to fields on manifolds of dimension less than d. We then introduce a corresponding generalization of the path integral for these exotic classical actions. Of course, one is immediately led to ask whether our constructions can be generalized, at least heuristically, in Chern-Simons theory with continuous gauge group, or possibly in other quantum field theories. At this writing the answer to this question is unknown.

The mathematics here is complicated and abstract, but not difficult. Several accounts exist. The original paper, with all of the computations, is Ref. 9. Previous joint work with Frank Quinn [10] discusses the basic theory in more detail. The brief [11] gives a heuristic account of our extension of TQFT (without mentioning path integrals or the classical theory) as well as a discussion of central extensions (which arise in Chern-Simons theory with continuous gauge groups) and invariants of "framed tangles." The conference proceedings [12] contains a heuristic explanation of our generalized path integrals as well as a brief idea of how some of this structure appears in characteristic numbers in topology. The summer school notes [13] give a more leisurely introduction to the basics of finite group Chern-Simons theory. Much of the work will be left for you in the form of guided exercises. Many of the exercises are not directly related to the topic of these lectures, but perhaps you will find them useful anyhow.

1   Classical Field Theory

We begin with a discussion of the basic ingredients in a classical field theory—spacetimes, fields, action. This is mostly to fix the ideas and notation since these same ingredients in the finite group Chern-Simons theory may otherwise appear exotic. We then discuss the Wess-Zumino-Witten action in two dimensions in some detail. Here we introduce an extension of the idea of the classical action of a field theory. Although this particular example is not the subject of our lectures, it is a familiar example and hopefully the geometry we discuss is easily accessible.

1.1   Classical Actions

A field theory has a particular dimension attached to it, which we call d. The standard examples of field theories have $d = 4$ and take place on Minkowski space. We allow more general examples. This means first of all that d is not necessarily 4.[1] Also, we allow *spacetimes* which are curved manifolds. Generally we take them to be compact, with or without boundary. Notice the terminology we will often use: A manifold is called *closed* if it is compact and has no boundary. Now Minkowski space carries a metric whose isometry group is the Poincaré group, and one usually requires that the the theory have the Poincaré group as a symmetry. We will generalize this considerably. Namely, a given field theory takes place on manifolds with some specified extra structure. This may be topological (e.g., an ori-

---

[1] In the Chern-Simons example of most interest to us $d = 3$.

entation or spin structure) or may be geometric (a conformal structure or a metric). Thus usual *relativistic* field theory should be thought of as taking place on Lorentz manifolds. We allow field theories on Riemannian manifolds and also on manifolds with less rigid structure. A field theory based on Riemannian manifolds is called a *Euclidean field theory*, one based on manifolds with a conformal structure is a *conformal field theory*, and one based on manifolds with only some topological structure is a *topological quantum field theory*. We can be even more adventurous, of course. We might allow singular manifolds, for example. For $d = 1$ there are interesting field theories defined on graphs. Or we might allow supermanifolds. Or we might take more abstract sorts of spaces.

So the first ingredients of a classical field theory are a dimension and a class of spacetimes,[2] i.e., manifolds of that dimension. The next ingredient is a space of fields $\mathcal{C}_X$ for each spacetime. This is usually a set of local functions on X if we interpret "function" liberally enough. The main object of study is the *classical action* (or simply, action) which is usually a real-valued function function on the space of fields:

$$S_X : \mathcal{C}_X \to \mathbb{R}. \tag{1.1}$$

Typically, if a field is denoted ϕ, there is a Lagrangian density $L_X(\phi)$ whose value at $x \in X$ only depends a finite number of derivatives of the field at x and on a finite number of derivatives of the geometric data (such as a metric) on X at x. Then the action is the integral of the Lagrangian density:

$$S_X(\phi) = \int_X L_X(\phi). \tag{1.2}$$

The Lagrangian density is a density(!), that is, something which can be integrated over X. If X is oriented then it can be taken to be a differential d-form.

Before reviewing the main properties of an action, let's note how some familiar examples fit into this scheme.

Exercise 1.1. Consider first classical mechanics as a field theory with $d = 1$. Look first at a particle moving in $\mathbb{R}^3$. Suppose that $X = [0, T]$. Then $\mathcal{C}_X$ is the space of paths in $\mathbb{R}^3$. If there is no potential, then what is the action? What is the Lagrangian density? What if there is now a potential function $V : \mathbb{R}^3 \to \mathbb{R}$? What if we replace $\mathbb{R}^3$ by an arbitrary Riemannian manifold M?

Exercise 1.2. A more invariant formulation of classical mechanics is as a field theory on one-dimensional Riemannian manifolds X. Can you formulate the Lagrangian density and action in this case? (The fields should be

---

[2]We use the word "spacetimes" even though in our context the words "space" and "time" may not have much significance.

paths into a fixed Riemannian manifold M.) Note that if X is diffeomorphic to an interval, then it is *isometric* to $[0, T]$ for some T. In other words, the only invariant of a Riemannian interval is its length. This is a fancy way of stating parametrization by arc length.

Exercise 1.3. Generalize the previous example to the *σ-model* in d dimensions. This is a d-dimensional field theory formulated on Riemannian d-manifolds. There is a fixed auxiliary Riemannian manifold M, and the fields on a d-dimensional Riemannian manifold X are smooth maps $\phi \colon X \to M$. The Lagrangian density is

$$L_X(\phi) = |d\phi|^2 \, d\mu_X,$$

where $d\mu_X$ is the Riemannian volume density. Write L_X in local coordinates, or in a form recognizable to you. Prove that this action is *conformally invariant* in $d = 2$ dimensions.

Exercise 1.4. Here is a simple topological example in d dimensions. Fix a manifold M (no metric!) and a d-form $\omega \in \Omega_X^d$. The spacetimes are now simply oriented d-manifolds X and the space of fields $\mathcal{C}_X$ is the space of smooth maps $\phi \colon X \to M$ (as it is for the σ-model). The Lagrangian density, or better Lagrangian form, is $L_X(\phi) = \phi^* \omega$. Write this in local coordinates. Are there any simplifications if ω is closed $(d\omega = 0)$?

Exercise 1.5. Try to formulate free field theories in this formalism. Consider for example a free scalar field or a free spinor field. What is the precise class of spacetimes considered? What is the space of fields? Do you know some interaction terms to add to these Lagrangians?

Exercise 1.6. Consider now a gauge theory. We will be doing this in more detail later, but it is a good idea for you now to think of how this fits in with our formalism. To be concrete, consider Yang-Mills in four dimensions. Now for a Riemannian 4-manifold X the space of fields is a space of connections. Can you see how to fit gauge symmetry into the picture? What is the action? Does Yang-Mills make sense in other dimensions? Show that only in $d = 4$ is the Yang-Mills Lagrangian conformally invariant. Show that in $d = 2$ it is invariant under area-preserving diffeomorphisms. (This is crucial in David Gross's lectures.)

There are two main properties of fields and classical actions: *symmetry* and *locality*. I may use the term "functoriality" for the symmetry discussed here, or perhaps "external symmetry." It is *not* like gauge symmetry (Exercise 1.6) which is a symmetry of the fields, something we might call an "internal symmetry." Rather, it refers to the symmetries of the spacetimes. Such a symmetry is a diffeomorphism $f \colon X' \to X$ which preserves all of the structure. So if we are dealing with a Euclidean field theory, the map f is required to preserve the metrics, i.e., f is an isometry. Of course, we might have $X' = X$ which is the most interesting case. In any case we require

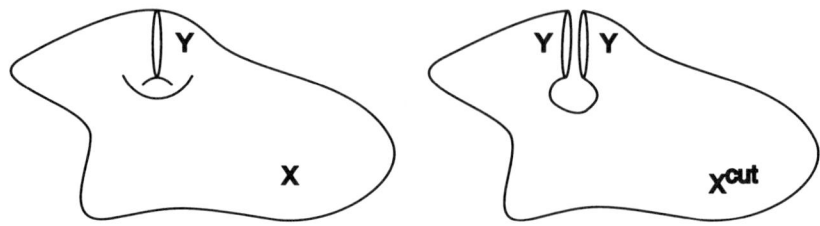

FIGURE 1. Cutting a spacetime X along Y to obtain X^{cut}.

that a symmetry induce a map on fields

$$f^* : \mathcal{C}_X \to \mathcal{C}_{X'} \tag{1.3}$$

and that the action be preserved:

$$S_{X'}(f^*\phi) = S_X(\phi), \qquad \phi \in \mathcal{C}_X. \tag{1.4}$$

If there is a Lagrangian density, then we usually have the stronger condition that the density is preserved, at least up to an exact term.

Locality is the assertion that the fields are local objects and that the action can be computed locally. There are many ways to formulate this, and for our purposes we focus on the following situation. Suppose X is a d-dimensional spacetime and $Y \hookrightarrow X$ is a closed codimension one submanifold of X. If we cut X along Y then we obtain a new spacetime called X^{cut}. Notice that we do not require that our spacetimes be connected, nor that they have connected boundaries. The usual picture is Figure 2, in which X is connected and X^{cut} has two components. But this is not necessary. Nor is it necessary that Y be connected. In any case note that X^{cut} has two new pieces in the boundary, each of which is diffeomorphic to Y. (One of them appears with the opposite orientation.) The situation is illustrated in Figure 1. Notice that there is a gluing map $g \colon X^{\text{cut}} \to X$ which identifies these two boundary components. Suppose now we have a field ϕ on X. Then there is a pullback field ϕ^{cut} on X^{cut} and we require that

$$S_{X^{\text{cut}}}(\phi^{\text{cut}}) = S_X(\phi). \tag{1.5}$$

This is trivial if the action is given by an integral as in (1.2).

Exercise 1.7. Consider a field theory formulated on Minkowski space. Show that (1.4) is the assertion that the action is invariant under the Poincaré group. How does the Poincaré group act on scalar fields? On spinor fields?

Exercise 1.8. Verify (1.4) and (1.5) for the examples considered earlier.

Exercise 1.9. If the spacetimes in a field theory have an orientation, then usually the action satisfies:

$$S_{-X}(\phi) = -S_X(\phi), \tag{1.6}$$

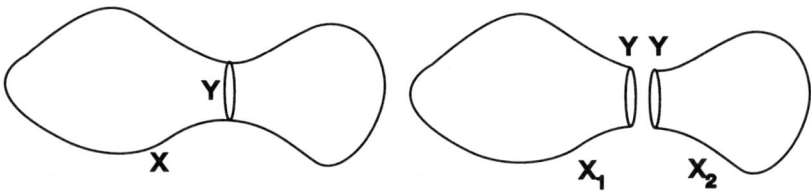

FIGURE 2. Cutting X into two pieces.

where $-X$ denotes the manifold X with the opposite orientation. Check this out in examples. On field theories formulated on Minkowski space, what does this say about Lorentz transformations not connected to the identity?

Exercise 1.10. Can you think of more general formulations of locality? What example(s) motivates your formulation?

1.2 The Wess-Zumino-Witten Action

As motivation for our extension of the notion of the classical action we consider an action which is somewhat more interesting geometrically than those indicated in the exercises above. Often this *Wess-Zumino-Witten* (WZW) action is just one term in the Lagrangian. In $d = 2$ it is added to the σ-model Lagrangian (Exercise 1.3) to define one of the fundamental conformally invariant field theories in two dimensions. From a mathematician's point of view the corresponding quantum field theory teaches us much about the representation theory of loop groups. In any case, our interest for the moment is in the geometry of the classical action.

Let $G = \mathrm{SU}(2)$ be the Lie group of 2×2 unitary matrices of determinant one. Recall that as a manifold G is diffeomorphic to the 3-sphere. Let g denote a general element of G, and then

$$\theta = g^{-1}\, dg$$

is a matrix-valued differential 1-form on G. We are interested in the (scalar) differential 3-form

$$\omega = c\, \mathrm{Tr}(\theta^3) \tag{1.7}$$

where c is chosen so that

$$\int_{\mathrm{SU}(2)} \omega = 1. \tag{1.8}$$

Exercise 1.11. Parametrize $\mathrm{SU}(2)$ by matrices $\begin{pmatrix} \alpha & \beta \\ -\bar\beta & \bar\alpha \end{pmatrix}$, where $\alpha,\ \beta \in \mathbb{C}$ and $|\alpha|^2 + |\beta|^2 = 1$. Write θ and ω in terms of α and β. Also, compute the constant c.

Exercise 1.12. More generally, let G be any compact Lie group. Define the analog of the forms θ and ω. To do this you will need to pick an invariant inner product on the Lie algebra of G. You can carry through the following discussion for this more general case if G is *simply connected*. If G is not simply connected, then you will need much fancier (but interesting) ideas.

The spacetimes in this $(1 + 1)$-dimensional field theory are oriented 2-manifolds (surfaces). Let X be a *closed* oriented surface. Then the fields we are interested in are simply maps to G:

$$\mathcal{C}_X = \mathrm{Map}(X, G).$$

Suppose that $\phi \colon X \to G$ is such a map. Since G is diffeomorphic to a 3-sphere we can always extend to a map $\Phi \colon W \to G$ where W is an oriented 3-manifold[3] whose boundary is X, and the restriction of Φ to the boundary of W is ϕ. We write $\partial \Phi = \phi$. Now define

$$S_X(\Phi) = \int_W \Phi^*(\omega). \tag{1.9}$$

Provisionally, this is the action, but we want something which only depends on ϕ, not on the extension Φ. It is straightforward to see that different choices of Φ change (1.9) by an integer (this requires (1.8), so that

$$e^{2\pi S_x(\phi)} := e^{2\pi S_X(\Phi)} \tag{1.10}$$

is independent of Φ. This is the (exponentiated) WZW action, and it takes values in the complex numbers. Note that since ω is real, it always has unit norm.

Exercise 1.13. Verify the assertion that (1.10) is independent of the extension Φ.

Exercise 1.14. Prove that the WZW action is functorial for orientation-preserving diffeomorphisms $f \colon X' \to X$, in the sense of (1.3) and (1.4).

Exercise 1.15. Prove that the exponentiated WZW action is a smooth function on $\mathcal{C}_X$. Compute its differential and find the critical points.

Notice that $S_X(\phi)$ is *not* well-defined, only the exponential $e^{2\pi i S_X(\phi)}$ is. This is fine from the path integral point of view: Typically one writes the integrand of the path integral as $e^{i/\hbar\, S_X(\phi)}$, and we can relax the normalization condition (1.8) and instead require that the "coupling constant" c in (1.7) only take a discrete set of values which makes $e^{i/\hbar\, S_X(\phi)}$ well-defined. This quantization of the coupling constant is well-known. Notice also that this action does *not* have a Lagrangian density in the sense of (1.2).

The interesting point is to define the WZW action on an oriented surface X with nontrivial boundary ∂X. Thus suppose $\phi \colon X \to G$. Since now

---

[3] Actually, it is better to say that W is an oriented 3-*chain* in the sense of homology theory.

$\partial X \neq \varnothing$ we can neither write X nor ϕ as a boundary, and so we cannot immediately find a 3-chain in G over which to integrate our 3-form ω. Thus we proceed as follows. The boundary ∂X of X is a disjoint union of circles. Let $\widehat{X} = X \cup D$ denote the closed surface obtained from X by gluing discs onto each of the boundary components. The union of those discs is denoted D.[4] Extend ϕ to a map $\phi \cup \Gamma \colon \widehat{X} \to G$, where Γ is the map on the union of the discs. Then the exponentiated action $e^{2\pi i S_{\widehat{X}}(\phi \cup \Gamma)}$ is well-defined. However, we need to investigate its dependence on the arbitrarily chosen Γ.

Exercise 1.16. Show that if Γ' is another extension, then

$$e^{2\pi i S_{\widehat{X}}(\phi \cup \Gamma')} = e^{2\pi i S_{-D \cup D}(\Gamma \cup \Gamma')} \cdot e^{2\pi i S_{\widehat{X}}(\phi \cup \Gamma)}, \tag{1.11}$$

where $-D \cup D$ is the union of 2-spheres obtained by gluing D with the opposite orientation to another copy of D. (This is usually called the *double* of D.) Now define

$$c_D(\Gamma', \Gamma) = e^{2\pi i S_{-D \cup D}(\Gamma \cup \Gamma')}. \tag{1.12}$$

Conclude that for maps $\Gamma, \Gamma', \Gamma'' \colon D \to G$ we have

$$c_D(\Gamma'', \Gamma) = c_D(\Gamma'', \Gamma') \cdot c_D(\Gamma', \Gamma).$$

Now we have a sort of exponentiated action for a field ϕ on the surface X with boundary. It is not simply a number, however. It is a (complex-valued) function depending on how we extend ϕ to $\widehat{X}$, in other words, depending on Γ. Furthermore, the dependence on Γ is fairly simple (1.11), and most importantly the factor (1.12) which arises can be computed purely in terms of Γ, Γ', i.e., it does not depend on ϕ. (This is really a manifestation of locality.) One might be content to stop here and say that this function of Γ *is* the exponentiated action. However, there is a nicer geometric way to proceed, and this is our reason for considering this example.

Here, then, is our typical mathematician's ploy. We have an example of a function which obeys a certain equation (1.11). So let's consider the set of *all* functions which satisfy the same equation. To do that we first need to identify the set of allowable Γ. Let $\gamma = \partial \phi$ be the restriction of ϕ to the boundary $\partial X = \partial D$. Then the set of Γ is the set of fields on D whose restriction to ∂D is γ:

$$\mathcal{C}_D(\gamma) = \{\Gamma \colon D \to G : \partial \Gamma = \gamma\}.$$

Now we define the set of functions which satisfy (1.11), giving it the suggestive name $L_{\partial X}(\gamma)$:

$$L_{\partial X}(\gamma) = \{\ell \colon \mathcal{C}_D(\gamma) \to \mathbb{C} : \ell(\Gamma') = c_D(\Gamma', \Gamma) \cdot \ell(\Gamma)\}. \tag{1.13}$$

---

[4]Notice that we allow ∂X to have multiple components. Recall that we do not require that any of our manifolds (for example X) to be connected.

Notice that this set depends only on γ, as is indicated by the notation. Now what does $L_{\partial X}(\gamma)$ look like? I claim that it is a one-dimensional complex vector space, also known as a *complex line*. Furthermore, I claim that it has a natural inner product, so is actually a *Hermitian line*.

Exercise 1.17. Prove these last two assertions by constructing the vector space structure and inner product and verifying that $L_{\partial X}(\gamma)$ is one dimensional.

Finally, notice that the definition of $L_{\partial X}(\gamma)$ does not use the fact that ∂X is the boundary of a surface. In other words, for any field[5] $\gamma \in \mathcal{C}_Y$ on a closed oriented 1-manifold[6] Y we use (1.13) to define a Hermitian line $L_Y(\gamma)$. (Then D is the manifold obtained by attaching a disk to each component of Y.)

So to summarize we have used the 3-form ω to define two mappings:

$$\phi \in \mathcal{C}_X \mapsto e^{2\pi i S_X(\phi)} \in L_{\partial X}(\partial \phi), \tag{1.14}$$

$$\gamma \in \mathcal{C}_Y \mapsto L_Y(\gamma). \tag{1.15}$$

Here X is a compact oriented 2-manifold and Y is a *closed* oriented 1-manifold. In other words, we allow X to have a boundary, but not Y. When X is closed the line $L_\varnothing$ is simply the "trivial" Hermitian line of complex numbers $\mathbb{C}$. The exponentiated action is then defined by (1.10). In case X has boundary (1.14) is a convenient way to look at the construction preceding Exercise 1.16).

Let me say it again: The exponentiated action on a manifold with a boundary is not a complex number, but now takes values in a Hermitian line. This is already a modification of the scheme we outlined in the previous section, but this action still has the essential locality property—or gluing law—written in (1.5), only here it is expressed in a different form— equation (1.16) of the following exercise.

Exercise 1.18. Let X be a compact oriented 2-manifold, $Y \hookrightarrow X$ a closed codimension one submanifold, and X^{cut} the manifold obtained by cutting X along Y (Figure 1). Let $\phi \in \mathcal{C}_X$ and $\phi^{\mathrm{cut}} \in \mathcal{C}_{X^{\mathrm{cut}}}$ the corresponding field on X^{cut}, as in (1.5. Prove that

$$\mathrm{Tr}_Y \, e^{2\pi i S_{X^{\mathrm{cut}}}(\phi^{\mathrm{cut}})} = e^{2\pi i S_X(\phi)}, \tag{1.16}$$

where Tr_Y is performed using the Hermitian metric in the line $L_Y(\phi|_Y)$. You will need to use the properties (1.17) and (1.18) below to make sense of (1.16).

---

[5] You should take note here that although we originally discussed fields in a d-dimensional field theory as defined on d-dimensional manifolds (spacetimes), we are now extending that notion to consider fields in a d-dimensional field theory defined on $(d-1)$-dimensional manifolds. Here it is clear that $\mathcal{C}_Y = \{\gamma\colon Y \to G\}$.

[6] This is simply a finite union of circles.

Now we make an even larger extension/modification of the scheme in the previous section, and this is our key point in this section. *We consider (1.15) as the definition of an (exponentiated) WZW action for fields on a 1-manifold.* This is not at all the usual picture. First of all, we have an action in a d-dimensional theory defined for fields in $d - 1$ dimensions. Secondly, the value of the action is not a *number*, but rather a *set* (more precisely, a Hermitian line). It is hoped that through the course of this discussion you will be convinced that this is a useful extension of the notion of a classical action. For now, here are some properties which are analogous to properties of the usual classical action. In particular, Exercise 1.20 deals with symmetry. Locality will have to wait for another time. (Think about what this would mean.)

Exercise 1.19. Construct isomorphisms

$$L_{Y_1 \sqcup Y_2}(\gamma_1 \sqcup \gamma_2) \to L_{Y_1}(\gamma_1) \otimes L_{Y_2}(\gamma_2) \tag{1.17}$$

and

$$L_{-Y}(\gamma) \to \overline{L_Y(\gamma)}. \tag{1.18}$$

Here "$\sqcup$" denotes the disjoint union (i.e., the union of two sets with empty intersection) and $\overline{L}$ is the complex conjugate vector space to L (i.e., the vector space with the same underlying addition and the complex conjugate scalar multiplication). Notice that these properties are analogous to properties of the usual classical action. (The isomorphism (1.18) is analogous to the equation (1.6). We did not write the analog of (1.17) for the usual action, but you should easily see what it is. In fact, you can view it as a special case of the gluing law where we glue along an empty manifold!) Show that these isomorphisms are actually *isometries*.

Exercise 1.20. Suppose $f \colon Y' \to Y$ is an orientation-preserving diffeomorphism of 1-manifolds. Then for any $\gamma \in C_Y$, construct an isometry

$$f^* \colon L_Y(\gamma) \to L_{Y'}(f^*\gamma). \tag{1.19}$$

Notice that (1.19) is the analog of (1.4).

The next exercise is worked out in Ref. 14. It is a geometric way to pass from the Lagrangian picture (classical action) to the Hamiltonian picture.

Exercise 1.21. Show that $L_Y(\gamma)$ depends smoothly on γ in the sense that these lines fit together into a smooth Hermitian line bundle $L_Y \to C_Y$. Define a parallel transport on paths in C_Y using the WZW action. Show that this is actually the parallel transport of a connection on L_Y. Compute the curvature of this connection.

Exercise 1.22. You can carry out the constructions in this section in much more generality. One generalization is to let G be any compact Lie group, though you will have to work harder if G is not simply connected. Another generalization is to work in d dimensions and replace $\omega \in \Omega^3(G)$ by a $(d+1)$-form on a manifold M. The normalization condition (1.8) should be replaced by the condition that the integral of this form over all $(d+1)$-cycles is an integer. You will also want to make some topological assumptions on M generalizing simple connectivity in the $d = 2$ case. So a simpler case is a $d = 1$ field theory based on an integral 2-form ω on some manifold M. In this case we can find a Hermitian line bundle $L \to M$ with a unitary connection whose curvature is $2\pi i\omega$. Show that the exponentiated classical action in this case is the parallel transport (or holonomy) of this connection, and the action (1.15) just reproduces the line bundle L. (This last assertion is not as precise as it might be—what is the precise statement?)

Suppose that $Y = S^1$ is the standard circle. Then $\mathcal{C}_{S^1} = \mathrm{Map}(S^1, G) = LG$ is the *loop group*; the multiplication of loops is defined pointwise using the multiplication in G. According to Exercise 1.21 the lines $L(\gamma) = L_{S^1}(\gamma)$ fit together to form a smooth Hermitian line bundle $L \to LG$ over the loop group. Let $\widehat{LG}$ denote the set of elements of unit norm in L; it is a principal circle bundle over LG. In the next exercise you will show that $\widehat{LG}$ is a central extension of LG. This construction of the central extension is originally due to Mickelsson [15].

Exercise 1.23. Suppose $\gamma_1, \gamma_2 \in LG$. Construct an isometry

$$L(\gamma_1) \otimes L(\gamma_2) to L(\gamma_1\gamma_2). \tag{1.20}$$

This is not trivial—it uses a formula sometimes attributed in the physics literature to Polyakov. Restrict (1.20) to the elements of unit norm to define multiplication in $\widehat{LG}$. Verify that this multiplication is associative and indeed defines a group. Construct a homomorphism $\widehat{LG} \to LG$ and show that its kernel is isomorphic to the circle group of unit complex numbers. Also, show that the kernel is *central* in $\widehat{LG}$, that is, elements in the kernel commute with every element in $\widehat{LG}$.

Exercise 1.24. Exercise 1.20 asserts that the action of Diff S^1 on LG lifts to the line bundle $L \to LG$. How does the lifted action interact with the isometry (1.20)?

For details on the material in this section, see Ref. 14.

2   Categories, Finite Groups, and Covering Spaces

We discuss in general terms what ideas are necessary to extend the notion of classical action further. The indicated extension is crucial in understanding the relationship of quantum groups to three-dimensional TQFT. Unfortunately it involves the concept of a *category*, which may be off-putting at first. Category theory has been called "the theory of abstract nonsense." Be that as it may, the notion is useful to us here. We then introduce what is surely the simplest field theory: gauge theory with finite-gauge group. The interesting structure is in the space of fields; the action we consider is trivial. In the exercises we indicate a "twisted" version of the theory which has nontrivial action. This theory exists in any dimension, though our main interest later is in the 3 case. Then it is a simple example of *Chern-Simons theory*, which can be defined for any compact gauge group. We remark that although these theories are analytically simple, they still illustrate some basic properties of gauge theory, especially the role of symmetries and reducible connections.

2.1 Going Further

What I hope you learned from the WZW example is the following. In a d-dimensional field theory we can allow actions which are not of the form (1.1) but rather of the form

$$e^{2\pi i S_X(\cdot)} \colon \mathcal{C}_X \to \mathbb{R}, \tag{2.1}$$

where $S_X(\cdot)$ may not be defined. Furthermore, I challenge you to think of any loss (from a physics point of view) in passing from (1.1) to the exponentiated form (2.1) when $S_X(\cdot)$ *is* defined. (I cannot think of any.) Here X is a *closed* d-manifold. We extend the idea of fields and action to closed $(d-1)$-manifolds Y:

$$L_Y(\cdot) \colon \mathcal{C}_Y \to \mathcal{L}, \tag{2.2}$$

where $\mathcal{L}$ is the *category* of all finite-dimensional Hilbert spaces. We will have more to say about that shortly, so please don't panic yet! Finally, if X is a d-manifold with boundary, the exponentiated action (2.1) has a generalization which we can explain using the following diagram of line bundles:

$$
\begin{array}{ccc}
r^* L_{\partial X} & \longrightarrow & L_{\partial X} \\
\downarrow & & \downarrow \\
\mathcal{C}_X & \xrightarrow{\ r\ } & \mathcal{C}_{\partial X}
\end{array}
\tag{2.3}
$$

Here r is the restriction map which restricts a field to the boundary. The line bundle $L_{\partial X} \to \mathcal{C}_{\partial X}$ is the extended action (2.2). Then the action $e^{2\pi i S_X(\cdot)}$ is a section of the line bundle $r^* L_{\partial X} \to \mathcal{C}_X$. This extended classical action has several properties, scattered in Section 1, which basically capture the idea that the action behaves like the integral over the manifold of something which depends locally on the field. The most characteristic of these properties is the *gluing law* (1.16).

Exercise 2.1. Perhaps an analogy with honest integration will help understand this idea of an extended action. The usual situation is that we have a compact oriented d-manifold X and a d-form α on X. Then the integral $\int_X \alpha$ is defined and is a single number. More generally, consider a fiber bundle $X \to \mathcal{C}$ whose typical fiber is a compact oriented d-manifold, and let $\alpha \in \Omega^d(X)$ be a d-form. Then *integration over the fibers* of $X \to Y$ produces a function $\int_{X/Y} \alpha$ on $\mathcal{C}$. This is the analog of the usual classical action. But more generally suppose that the fibers of $X \to \mathcal{C}$ are compact oriented manifolds of dimension $d - i$. Then integration along the fibers gives an i-form on the base:

$$\int_{X/Y} \alpha \in \Omega^i(\mathcal{C}). \tag{2.4}$$

So here we have a family of $(d-i)$-manifolds and we integrate something d-dimensional over the fibers to get something i-dimensional on the base. If $i = 1$ then we get a 1-form, which you might think of as analogous to a connection form on a Hermitian line bundle. This is analogous to our first extension of the classical action, which in this case is not a 1-form but rather a Hermitian line bundle.

Incidentally, the exercise here is to fill in the details if you are not already familiar with integration over the fibers. You might also want to consider the analog of Stokes's theorem in this context.

You should now raise several questions. First, how does our extended classical action work in familiar examples? Second, are there nontrivial examples other than the WZW action? Finally, can we go further and consider $(d - 1)$-manifolds with boundary, closed $(d - 2)$-manifolds, etc.? Following the old joke, we will answer these questions in the form of questions! First, the familiar examples.

Exercise 2.2. For the usual examples you considered in the previous section, show that the extended action (1.15) is trivial. However, for usual second-order Lagrangians the correct space of fields on the boundary should include a derivative. Consider classical mechanics, for example (Exercise 1.2). The space of fields attached to a point $Y = \mathrm{pt}$ should be the tangent bundle TM, not the manifold M. Then the extended action gives a trivial line bundle over TM. However the connection constructed following the idea of Exercise 1.21 is nontrivial, and its curvature is the standard symplectic form constructed from the Riemannian metric. Work out the details to the point that you recognize familiar formulas from classical mechanics.

There is another nontrivial example: the Chern-Simons action. I know this best for $d = 3$, but in principle it can be worked out in other (odd) dimensions. (The $d = 1$ case is Exercise 1.22.) Our main interest here is the Chern-Simons action for a finite-gauge group. The case of a continuous-gauge group has more geometric interest, and you can find all of the details in Ref. 14. (There is also an account in Ref. 13.) We consider the finite-gauge group case in the next section and defer to these references for the continuous-group case.

Consider again a d-dimensional field theory. We defined (in an example) an action on d-manifolds with boundary which satisfies a gluing law (1.16). Further, we asserted that (1.15) should be considered as an extension of the classical action to closed $(d-1)$-manifolds. Now we want to go further—define an action on $(d-1)$-manifolds with boundary and formulate a gluing law. Let's just see what kind of objects we should expect to run into. Since the action on a closed $(d-1)$-manifold gives a Hermitian line, we expect that on a $(d-1)$-manifold with boundary the action is some similar object. (The analogy in d dimensions is that the action on a d-manifold with boundary—an element in a complex line—is similar to the action on a closed d-manifold—a complex number.) At the very least we expect that it is a set rather than some kind of number. Then the analog of equation (1.16), the gluing law, will be an "equation" between sets. Now such equations are possible—you can say that two sets are equal—but it is also possible to say that two sets are *isomorphic* without being equal. This is an extra layer of complexity which sets have that numbers don't. We have already run into this in (1.17), (1.18), (1.19), and (1.20). Another way to put it is that sets have an "internal structure" and it is possible to have *automorphisms* of this structure. Mathematicians (notably Saunders MacLane [16]) have systematized these ideas in the notion of a *category*. It fits into the progression:

$$\text{number, \quad set, \quad category}$$

A category is a collection of *objects* and *morphisms* (maps and arrows) between objects. Two morphisms compose if the second begins where the first ends. The composition is assumed associative, and usually one assumes that there are identity morphisms as well. If we focus on the objects, then the morphisms encode the internal structure that they possess. On the other hand, it is useful to focus on the morphisms as well. For example, a category with one object is simply a set (of morphisms) with an associative composition law and an identity, also known as a *semigroup*. If a category has more than one object, then we can think of it as a "semigroup with states." The objects represent the states, and in each state there are certain morphisms which are possible. Some change the state and others (automorphisms) do not.

Exercise 2.3. As an example consider the category $\mathcal{V}$ of finite-dimensional complex vector spaces and linear maps. So an object in $\mathcal{V}$ is a vector space and a morphism $L\colon V_1 \to V_2$ between $V_1, V_2 \in \mathcal{V}$ is a linear map. Note that two vector spaces can be isomorphic but not equal. Also, every vector space of positive dimension has nontrivial automorphisms. You are familiar with the idea that we should consider isomorphic, but distinct, vector spaces as being different. Think of the tangent spaces to the standard 2-sphere. Tangent spaces at different points are isomorphic, but if we could truly think of them as equal (in a "continuous" way) we would quickly construct an everywhere nonzero vector field on the 2-sphere, which is not possible. Do you know other examples of categories? Other situations in which it is not possible to identify isomorphic objects which are not equal? What if two objects are isomorphic and have no nontrivial automorphisms? Can we safely identify them in that case?

Exercise 2.4. Next we consider maps of categories. (They are usually called *functors.*) Roughly, a functor maps the objects and morphisms of one category into the objects and morphisms of another so that it preserves compositions. As an example, consider the functor $\mathcal{V} \to \mathcal{V}$ which assigns to each vector space $V \in \mathcal{V}$ its double dual V^{**}. What does this do to morphisms?

Exercise 2.5. The extra layer of structure in a category allows us to define maps between functors, called *natural transformations.* Suppose that $\mathcal{C}_1$, $\mathcal{C}_2$ are categories and $\mathcal{F}_1$, $\mathcal{F}_2 \colon \mathcal{C}_1 \to \mathcal{C}_2$ are functors. Then a natural transformation $\theta \colon \mathcal{F}_1 \to \mathcal{F}_2$ is for each object $C \in \mathcal{C}_1$ a morphism $\theta(C) \colon \mathcal{F}_1(C) \to \mathcal{F}_2(C)$ such that it is compatible with morphisms. This means that for any morphism $C' \xrightarrow{f} C$ in $\mathcal{C}_1$ the diagram

$$
\begin{array}{ccc}
\mathcal{F}_1(C') & \xrightarrow{\mathcal{F}_1(f)} & \mathcal{F}_1(C)(C)V \\
{\scriptstyle \theta(C')}\Big\downarrow & & \Big\downarrow{\scriptstyle \theta} \\
\mathcal{F}_2(C') & \xrightarrow{\mathcal{F}_2(f)} & \mathcal{F}_2(C)
\end{array}
$$

commutes. Construct a natural transformation from the identity functor to the double dual functor of Exercise 2.4 . Show that this is in fact a natural *isomorphism*.

Now we can make an educated guess about going further. The action of a field on a closed $(d-2)$-manifold should be a category, the action of a field on a $(d-1)$-manifold with boundary should take values in the category associated to the boundary field, and there should be a gluing law analogous to (1.16) which is a morphism in a category. We can be even more precise. The type of category where the action takes its values should fit into the progression:

$$\text{complex number of unit norm,} \quad \text{Hermitian line,} \quad ?. \qquad (2.5)$$

Before describing what the "?" is, it is easier to consider the progression of *trivial* values for the action:

$$1 \in \mathbb{C}, \quad \mathbb{C}, \quad \mathcal{L}. \qquad (2.6)$$

In these lectures we only consider a trivial action, so an understanding of (2.6) will suffice. Recall that $\mathcal{L}$ is the category of all finite-dimensional Hilbert spaces. A morphism in $\mathcal{L}$ is an isometry. This category is analogous to the complex numbers as the following exercise shows.

Exercise 2.6. We construct a structure on $\mathcal{L}$ analogous to the *ring* structure on $\mathbb{C}$, that is, the addition and the multiplication. There is a further structure, which is complex conjugation. Then we can build a norm from multiplication and complex conjugation. Construct, then, an addition functor $\mathcal{L} \times \mathcal{L} \to \mathcal{L}$, a multiplication functor $\mathcal{L} \times \mathcal{L} \to \mathcal{L}$, and a complex conjugation functor $\mathcal{L} \to \mathcal{L}$. You should use direct sum, tensor product, and the conjugate linear space. Show that these functors satisfy desired properties, but be careful that where there are equalities in the complex numbers (associativity, commutativity, identity element) there are natural isomorphisms in $\mathcal{L}$. For example, the statement that $\mathbb{C} \in \mathcal{L}$ acts as a multiplicative identity is the assertion that the functor

$$\begin{array}{ccc} \mathcal{L} & \to & \mathcal{L} \\ L & \mapsto & \mathbb{C} \otimes L \end{array}$$

is naturally isomorphic to the identity functor. So part of the structure on $\mathcal{L}$ is the explicit specification of this natural isomorphism. There is a further layer of structure: *equations* among these natural isomorphisms. For example, associativity of addition is expressed by a natural isomorphism

$$\theta_{v_1, v_2, v_3} : (V_1 \otimes V_2) \otimes V_3 \to V_1 \otimes (V_2 \otimes V_3).$$

Now for four Hilbert spaces V_1, V_2, V_3, V_4 there is an equation among the various θ's. What is this equation?

In the main flow of this discussion we will not need the general case of "?" in (2.5), only the trivial case in (2.6). I don't mean to make the "?" so mysterious, and I'll give a nontrivial example in the exercises. Perhaps you have already realized that "?" fits into the analogy

complex numbers : a Hermitian line

$$= \text{category of finite-dimensional Hilbert spaces} : ? \quad (2.7)$$

We will later introduce the idea of a "2-Hilbert space" and we will see that a "?" is a one-dimensional 2-Hilbert space. In any case we now postulate that continuing the progression in (2.1) and (2.2) we have for a *closed* $(d-2)$-manifold S an assignment

$$\mathcal{L}_Y(\cdot) \colon \mathcal{C}_S \longrightarrow \text{collection of one-dimensional 2-Hilbert spaces}, \quad (2.8)$$

which satisfies properties analogous to those satisfied by (2.1) and (2.2).

Exercise 2.7. Here is a category $\mathcal{K}$ which is a nontrivial example of a one-dimensional 2-Hilbert space. Namely, let $\mathcal{K}$ be the category of all unitary representations of the unitary group $U(n)$ which are isomorphic to a direct sum of several copies of the determinant representation on $\mathbb{C}$, where the action of a unitary matrix is multiplication by the determinant. A morphism in $\mathcal{K}$ is an isomorphism of the representations (it commutes with the action of $U(n)$.) Construct a functor $\mathcal{L} \times \mathcal{K} \to \mathcal{K}$ using the tensor product. This is analogous to scalar multiplication. (Here $\mathcal{L}$ are the scalars.) Construct also the analog of vector addition, namely a functor $\mathcal{K} \times \mathcal{K} \to \mathcal{K}$, using the direct sum. Show that any irreducible (one-dimensional) representation $K_0 \in \mathcal{K}$ is a "basis" of $\mathcal{K}$. In other words, use K_0 and scalar multiplication to construct an isomorphism $\mathcal{L} \cong \mathcal{K}$. Consider how this fits with the analogy (2.7). Can you provide a rough definition for a 2-Hilbert space?

Exercise 2.8. State explicitly the properties of (2.8) which are "analogous to those satisfied by (2.1) and (2.2)." Note in particular the gluing law for (2.2):

$$L_Y(\gamma) \cong \left(L_{Y_1}(\gamma_1), L_{Y_2}(\gamma_2)\right)_{\mathcal{L}_S(\partial\gamma)}. \tag{2.9}$$

Exercise 2.9. Here is an example of a higher-dimensional 2-Hilbert space. It will be important in Section 4. Let G be a finite group and let $\mathcal{E}$ denote the category of all finite-dimensional unitary representations of G. Construct an addition $\mathcal{E} \times \mathcal{E} \to \mathcal{E}$ and a scalar multiplication $\mathcal{L} \times \mathcal{E} \to \mathcal{E}$. Then construct an inner product $\mathcal{E} \times \mathcal{E} \to \mathcal{L}$. Can you find an "orthonormal basis" of $\mathcal{E}$? What is $\dim \mathcal{E}$? This particular example of a 2-Hilbert space is also an algebra. So construct a multiplication $\mathcal{E} \times \mathcal{E} \to \mathcal{E}$.

Exercise 2.10. Here is an example of a category which is "small" by comparison with the previous examples. It should dispel any illusion you may have that categories are huge. It is also an example which will be crucial for us later. Consider a finite group G. Construct a category whose objects are elements of G. For each pair of elements $x, g \in G$ we postulate a morphism $g: x \to gxg^{-1}$. What is composition? Show that if $|G| = n$ this constructs a category with n objects and n^2 morphisms. Draw a picture for an Abelian group. For $G = S_3$.

2.2   Finite-Gauge Theory

Fix a *finite* group G. For example, G could be the cyclic group $\mathbb{Z}/n\mathbb{Z}$ of n elements, or the symmetric group S_n of $n!$ elements. There is no restriction on G. We need no other data to define our field theory, though more data is needed for the twisted theory.

Exercise 2.11. For the twisted theory we need to use the *classifying space* BG. One model is the following. Construct an embedding $G \hookrightarrow O(N)$, where $N = |G|$ is the order of G. Let EG denote the space of all N-tuples of orthonormal vectors in an infinite-dimensional dimensional real Hilbert space. Show that EG is contractible and that $O(N)$ acts freely on EG. Define BG to the the quotient EG/G, where G acts via its embedding into $O(N)$. The space BG has interesting topology, and in the twisted theory we need to fix a cohomology class[7] $\lambda \in H^d(BG; \mathbb{R}/\mathbb{Z})$. The untwisted theory in the text corresponds to $\lambda = 0$.

The theory we consider is based on "bare" manifolds. No orientation, metric, etc. is needed. Thus a spacetime is simply a compact d-manifold X with no additional structure. (The twisted theory is based on oriented manifolds, however.) Symmetries of the spacetimes are simply diffeomorphisms.

Next we define the space of fields on a spacetime X. Recall that in our extended notion of field theory we also consider fields on manifolds of dimension less than d, so we will define a space of fields for *any* manifold M. Here is the definition:

$$\mathcal{C}_M = \left\{ \begin{array}{c} P \\ \downarrow \\ M \end{array} : \begin{array}{l} P \text{ is a } \textit{principal (Galois, regular) covering} \\ \textit{space} \text{ with structure group } G. \end{array} \right\} \quad (2.10)$$

In other terms, a field is a principal bundle $P \to M$ with structure group G. Thus P is a manifold, the group G acts freely on P, and the quotient is $P/G = M$. We always take G to act on the *right*.

As an example, consider $G = \mathbb{Z}/3\mathbb{Z}$. If $M = \text{pt}$, then any bundle looks like

with G cyclically permuting the three points. If $M = S^1$ then there are three possibilities, up to isomorphism, as illustrated in Figure 3. The nontrivial coverings are pictured as pieces of helices, but the endpoints are meant to be identified. Topologically, the total space P in these covers is a circle. The total space of the trivial cover is the disjoint union of 3 circles.

You should have noticed that the space of fields (2.10) is qualitatively different than in the examples in Section 1. In usual examples the fields for a smooth manifold (usually infinite-dimensional). Here do they not even technically form a set! (The collection of all of anything is not a set—remember Russell!) Rather, they form a category.

Warning. Categories enter here in a different way than in our extended notion of classical action in Section 1. Remember that categories are a fundamental mathematical structure, just as sets are, and we should not be surprised to see them in a variety of contexts. (We shall meet them again in yet another context later.) What the idea of a category captures here is *gauge symmetry*. These are symmetries of the fields (internal symmetries),

---

[7]More precisely, we need to fix a cocycle representing this cohomology class in some model of cohomology.

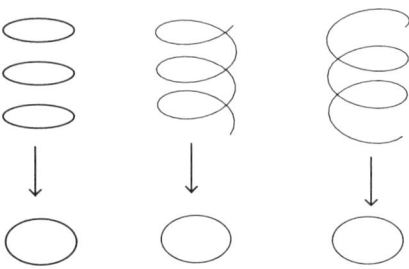

FIGURE 3. Principal $\mathbb{Z}/3\mathbb{Z}$ bundles over S^1.

not symmetries of the spacetimes (external symmetries). They are the morphisms in the category of fields $\mathcal{C}_M$. Since they are all invertible, we call them "isomorphisms."

Definition 2.1. An *isomorphism* φ from $P' \in \mathcal{C}_M$ to $P \in \mathcal{C}_M$ is a diffeomorphism $\varphi\colon P' \to P$ which commutes with the G action and such that the induced map on the quotient M is the identity.

This means that for $p' \in P'$ and $g \in G$ we have $\varphi(p' \cdot g) = \varphi(p') \cdot g$, where the first "$\cdot$" indicates the G action on P' and the second "$\cdot$" the G action on P. Any such map induces a map $\overline{\varphi}\colon M \to M$, and we restrict our isomorphism to have $\overline{\varphi} = \mathrm{id}$. More generally, we can consider diagrams

$$
\begin{array}{ccc}
P' & \xrightarrow{\ \varphi\ } & P \\
\downarrow & & \downarrow \\
M' & \xrightarrow[\ \overline{\varphi}\ ]{} & M
\end{array}
$$

where $\overline{\varphi}$ is possibly nontrivial. They enter when we consider symmetries of the manifolds as well as symmetries of the fields.

Exercise 2.12. Let $f\colon M' \to M$ be a diffeomorphism. Construct $f^*\colon \mathcal{C}_M \to \mathcal{C}_{M'}$ as required by (1.3). In this case it is a functor.

The previous exercise shows how to pull back fields under maps of spacetimes. For these to properly be considered fields we must also be able to cut and paste, as in Figure 1.

Exercise 2.13. Suppose M is a manifold and $N \hookrightarrow M$ is a closed codimension one submanifold of M. Let M^{cut} be the manifold obtained by cutting M along N. Let $g\colon M^{\mathrm{cut}} \to M$ be the gluing map. Construct a functor

$$
g^*\colon \mathcal{C}_M \to \mathcal{C}_{M^{\mathrm{cut}}}. \tag{2.11}
$$

So we get a picture of $\mathcal{C}_M$ which we schematically render in Figure 4. Each point is a field and each arrow is a symmetry of fields, that is, an

FIGURE 4. The space of fields $\mathcal{C}_M$.

isomorphism. The arrows which start and end at an object P form a *group* called Aut P, the automorphism group of P. There is a finite number of arrows from any point to any other. Aut P is also called the group of *gauge transformations* of P. We remark that this picture applies to the space of (gauge) fields in any gauge theory, except that in the general case there are continuous parameters—the space of connections on a fixed bundle—as well as discrete ones—the choice of the bundle. See Ref. 14 for a discussion.

The fields P and P' are *equivalent* ($P \cong P'$) or *isomorphic* if there is an arrow between them. Let

$$\overline{\mathcal{C}_M} = \text{set of equivalence classes of fields on } M. \qquad (2.12)$$

What makes gauge theories with finite-gauge group tractable is that $\overline{\mathcal{C}_M}$ is a finite set if M is compact.

Exercise 2.14. Determine $\overline{\mathcal{C}_M}$ for $M = pt$. For $M = S^1$. For $M = [0,1]$.

Exercise 2.15. Show that $\overline{\mathcal{C}_M}$ is a finite set for any compact manifold M.

Exercise 2.16. I claim that we can *not* make a field theory where the space of fields associated to M is $\overline{\mathcal{C}_M}$. This is because we cannot paste equivalence classes. Consider, for example, cutting the circle S^1 into an interval $[0,1]$. Show that we can cut equivalence classes of bundles, but we cannot paste them.

We can determine the space of equivalence classes of fields in terms of the *fundamental group*. Suppose M is connected. Fix *basepoints* $m \in M$ and $p \in P_m$, where P_m is the fiber of P at m. Then a field $P \to M$ determines a map

$$\{\text{loops in } M \text{ based at } m\} \longrightarrow G \qquad (2.13)$$

by taking the *holonomy* around the loop using the basepoint p. (See Figure 5.) Any loop at m lifts uniquely to a *path* in P which starts at p and ends at some p' in the fiber P_m of P over m. The holonomy is the unique $h \in G$ satisfying $p' = p \cdot h$. The holonomy only depends on the homotopy class of the loop.

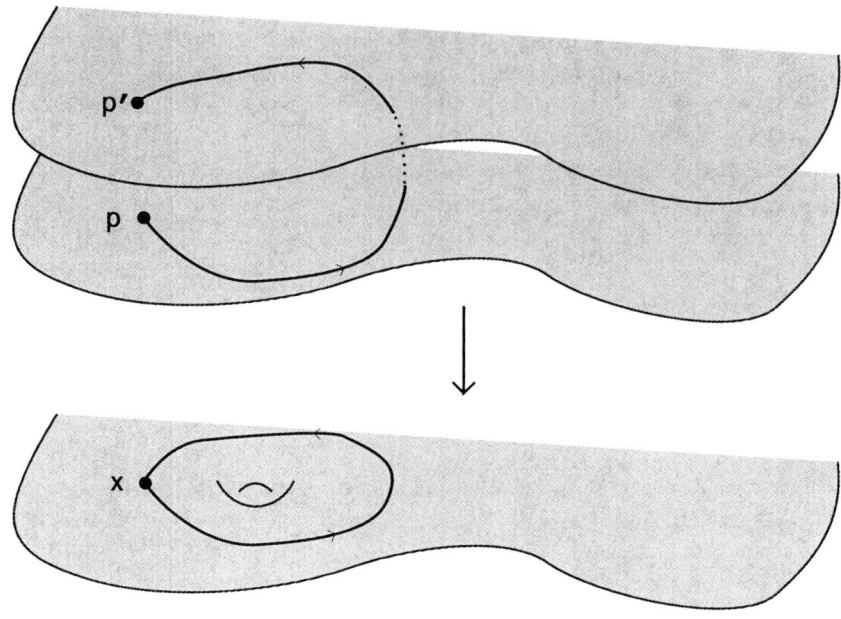

FIGURE 5. Definition of holonomy.

Exercise 2.17. (a) Check this last assertion.

(b) Show that the map $\pi_1(M, m) \to G$ defined by holonomy is a homomorphism of groups.

(c) If γ is this homomorphism, and we change the basepoint p and/or m, then the new homomorphism is $g\gamma g^{-1}$ for some $g \in G$.

(d) Suppose M is connected and $h\colon \pi_1(M, m) \to G$ is given. Construct a bundle $P \to M$ and a basepoint $p \in P_m$ such that the holonomy is h.

The assertions in Exercise 2.17 hold more generally for any *flat connection* with arbitrary gauge group. In this case they show that there is an isomorphism (of sets)

$$\overline{\mathcal{C}_M} \cong \mathrm{Hom}(\pi_1(M), G)/G \qquad (2.14)$$

if M is connected. (We omit the irrelevant basepoint in the notation for the fundamental group.)

Finally we are ready to define the action in this theory. We can summarize in one word: The action is *trivial*! Recall that we consider a d-dimensional theory, so begin with the usual (numerical) action on d-manifolds X:

$$S_X(P) = 0 \text{ for all } P \in \mathcal{C}_X. \qquad (2.15)$$

So this is an example where the usual action (1.1) is defined, not just the exponentiated action (2.1). The extended actions (2.2) and (2.8) are also trivial. So for a closed $(d-1)$-manifold Y we have

$$L_Y(Q) = \mathbb{C} \text{ for all } Q \in \mathcal{C}_Y, \tag{2.16}$$

and for a closed $(d-2)$-manifold S we have

$$\mathcal{L}_S(R) = \mathcal{L} \text{ for all } R \in \mathcal{C}_S. \tag{2.17}$$

Definitions (2.15) and (2.16) also make sense for manifolds with boundary.

Exercise 2.18. Verify the properties (discussed in Section 1) of the extended action—gluing laws, functoriality, etc. The functoriality statement is more involved since one has to worry now about internal symmetries. What is the right formulation?

Exercise 2.19. In the twisted theory the action is nontrivial [9, 10]. We give a brief indication here. Recall that the twisted theory is defined by fixing a cohomology class $\lambda \in H^d(BG; \mathbb{R}/\mathbb{Z})$, or more precisely a cocycle representing the cohomology class. Prove first that for any G bundle $P \to M$ here exists a "classifying map" $F \colon P \to EG$ which commutes with the G action, and that the map is unique up to homotopy through G-maps. So the quotient map $\overline{F} \colon M \to EG$ is unique up to homotopy. If now $P \in \mathcal{C}_X$ for X a closed *oriented* d-manifold, we define

$$e^{2\pi i S_X(P)} = e^{2\pi i \overline{F}^*(\lambda)[X]}, \tag{2.18}$$

for any classifying map F. Here $[X] \in H_d(X)$ is the fundamental class of X given by the orientation. Verify that (2.18) is well-defined.

The twisted version of (2.16) is more complicated. Try to arrive at it much as we arrived at (1.13) in our study of the WZW action. Namely consider first the expression (2.18) for a compact oriented d-manifold X with boundary. Now the orientation class is a relative homology class— $[X] \in H_d(X, \partial X)$—and the evaluation $\overline{F}^*(\lambda)[X]$ does not make sense. Instead you must choose d-cycles on X which represent $[X]$ and evaluate the chosen d-cocycle (which represents λ) on such cycles. This depends nontrivially on the choice of the d-cycle, and the dependence is encoded in a complex line which just depends on the restriction of P to ∂X. Fill in the details and you will have the twisted version of (2.16).

If you are truly ambitious you will carry this further and construct the twisted version of (2.17). This is another motivating example for our extension of the classical action.

Try to formulate (if not prove) the properties of the extended action for this twisted theory. The symmetry properties are perhaps more apparent here since the action is nontrivial. The d-dimensional action (twisted (2.15)) is invariant under gauge symmetry, but what is the analogous statement for the $(d-1)$-dimensional action (twisted (2.16))? What about for the $(d-2)$-dimensional action (twisted (2.17))?

3   Generalized Path Integrals

Now we move from the classical theory to the quantum theory. I dare say we are following Feynman, though we are so far from the original context of his path integrals that I suspect he would be either amused or more likely appalled! Our case is much simpler than usual examples in that the path integral reduces to a finite sum. Nonetheless, the formal picture—rigorous here—is the same in usual examples. We simplify our initial discussion by ignoring the gauge symmetry. The key new idea is an extension of the path integral. We interpret the usual quantum Hilbert space as the result of a path integral over fields on a $(d-1)$-manifold (in a d-dimensional theory). Naturally it involves the extended classical action, as does our extension to fields on $(d-2)$-manifolds. We work out the usual path integral and quantum Hilbert space in the finite group gauge theory. In Section 4 we will use the extended path integral to compute the quantum group relevant to those theories.

3.1   Path Integral Quantization

Imagine a d-dimensional field theory in which the space of fields C_M attached to any manifold M (of dimension $\leq d$) is a finite set. You may as well assume that this is a topological field theory. To simplify the picture even further we assume that there is no (gauge) symmetry among the fields. We put back in the symmetry later. However, here we allow a nontrivial action. The *partition function* of a closed d-manifold X is defined to be

$$Z_X = \int_{C_X} e^{2\pi i S_X(P)} \, d\mu_X(P). \tag{3.1}$$

We write a typical field as "P," keeping in mind our example. The new ingredient is a measure $d\mu_X$ on the space of fields C_X. With our assumption that C_X is a finite set, the measure is simply a positive number $d\mu_X(P)$ for each $P \in C_X$.[8] Note that the partition function Z_X is a complex number.

Exercise 3.1. Prove that Z_X is a topological invariant. That is, if $f \colon X' \to X$ is a diffeomorphism, then $Z_{X'} = Z_X$. Remember, we are assuming that this is a topological field theory.

The preceding exercise is the invariance of the partition function under symmetry. It is the quantum analog of (1.4). We now want to investigate the quantum analog of the gluing law (1.5). The following discussion is the usual argument for locality of the path integral via factorization into intermediate

---

[8] Of course, in more typical examples C_X is an infinite-dimensional space and a measure $d\mu_X$ is extremely difficult to construct and has not been constructed in many examples of interest. Nonetheless, workers in *constructive quantum field theory* have enjoyed many nontrivial successes in this pursuit.

states. In our language it goes as follows. Let X be a d-manifold, which for simplicity you may assume to be closed, and $Y \hookrightarrow X$ an embedded closed $(d-1)$-manifold. We assume that cutting X along Y splits X into two manifolds X_1 and X_2 (see Figure 2). We identify $\partial X_1 = \partial X_2 = Y$ The fields fit into the following diagram:

$$
\begin{array}{ccc}
\mathcal{C}_X & \xrightarrow{\ c\ } & \mathcal{C}_{X_1} \times \mathcal{C}_{X_2} \\
{\scriptstyle r_1}\downarrow & & \downarrow{\scriptstyle r_2} \\
\mathcal{C}_Y & \xrightarrow{\ \Delta\ } & \mathcal{C}_Y \times \mathcal{C}_Y
\end{array}
\tag{3.2}
$$

The vertical arrow r_1 is restriction to Y, the arrow r_2 is restriction to the boundaries of the X_i, the arrow Δ is the diagonal inclusion, and c is the pullback under the gluing map. ("c" stands for "cutting.") Then we propose to do the integral over $\mathcal{C}_X$ in two stages using Fubini's theorem: First integrate over the fibers of r_1 and then over $\mathcal{C}_Y$. Now the gluing law for the classical action (1.16) says that if $P \in \mathcal{C}_X$ and $\langle P_1, P_2 \rangle \in \mathcal{C}_{X_1} \times \mathcal{C}_{X_2}$ the cut field, then

$$
e^{2\pi i S_X(P)} = \left(e^{2\pi i S_{X_1}(P_1)}, e^{2\pi i S_{X_2}(P_2)} \right)_{L_Y(Q)}
\tag{3.3}
$$

where Q is the restriction of P to Y. The right-hand side is the inner product in the line $L_Y(Q)$. Carrying out the integration using the Fubini theorem we obtain:

$$
\begin{aligned}
Z_X &= \int_{\mathcal{C}_X} e^{2\pi i S_X(P)}\, d\mu_X(P) \\
&= \int_{\mathcal{C}_Y} \int_{r_1^{-1}(Q)} e^{2\pi i S_X(P)}\, d\mu_{r_1^{-1}(Q)}(P)\, d\mu_Y(Q) \\
&= \int_{\mathcal{C}_Y} \int_{r_2^{-1}(Q,Q)} e^{2\pi i S_X(P)}\, d\mu_{r_2^{-1}(Q,Q)}(P^{\mathrm{cut}})\, d\mu_Y(Q) \\
&= \int_{\mathcal{C}_Y} \left(\int_{\mathcal{C}_{X_1}(Q)} e^{2\pi i S_{X_1}(P_1)}\, d\mu_{X_1}(P_1), \right. \\
&\qquad\qquad \left. \int_{\mathcal{C}_{X_2}(Q)} e^{2\pi i S_{X_2}(P_2)}\, d\mu_{X_2}(P_2) \right)_{L_Y(Q)} d\mu_Y(Q). \tag{3.4}
\end{aligned}
$$

Here we use the definition

$$
\mathcal{C}_{X_i}(Q) = \{ P \in \mathcal{C}_{X_i} : \partial P = Q \}, \quad Q \in \mathcal{C}_Y.
\tag{3.5}
$$

Also, we make certain implicit compatibility assumptions about the measure to make this computation.

Exercise 3.2. State explicitly these compatibility assumptions.

To rewrite this last expression in a nicer form, we make the following definitions. Let X' be any d-manifold with boundary. Then the path integral is a function of a field Q on the boundary:

$$Z_{X'}(Q) = \int_{\mathcal{C}_{X'}(Q)} e^{2\pi i S_{X'}(P)} \, d\mu_{X'}(P), \quad Q \in \mathcal{C}_{\partial X'}. \tag{3.6}$$

Note that the integral is over the space of fields with fixed boundary value Q. The right-hand side of (3.6) takes values in the Hermitian line $L_{\partial X'}(Q)$, which is the extended action of the field on the boundary (cf. (2.3)). So Z_X is a section of the Hermitian line bundle $L_{\partial X} \to \mathcal{C}_{\partial X}$. Again we use the mathematician's ploy of introducing the space of *all* such sections. Hence for any closed surface Y set

$$E(Y) = \text{space of sections of the Hermitian line bundle } L_Y \to \mathcal{C}_Y. \tag{3.7}$$

Note that $E(Y)$ is a finite-dimensional complex vector space. Then (3.6) determines a *relative* invariant

$$Z_{X'} \in E(\partial X'). \tag{3.8}$$

We impose an L^2 inner product on (3.7) using a measure $d\mu_Y$ on $\mathcal{C}_Y$ and the metric on the Hermitian line bundle $L_Y \to \mathcal{C}_Y$. With these definitions we rewrite (3.4) as:

$$Z_X = (Z_{X_1}, Z_{X_2})_{E(Y)}. \tag{3.9}$$

This is the quantum gluing law—the quantum analog of (3.3)—the statement that the path integral is local.

Thus we have the standard formal ingredients of quantum field theory—the partition function Z_X on closed spacetimes and the quantum Hilbert space $E(Y)$ for closed "spaces," i.e., d-manifolds. We have indicated that symmetry and locality hold for the partition function. You should now prove that other formal properties of the path integral and quantum Hilbert space follow from the corresponding properties of the classical action. Again you will have to make some assumptions about the measures $d\mu_X$ and $d\mu_Y$, which I leave you to formulate.

Exercise 3.3 (Functoriality). Show that a diffeomorphism $f \colon Y' \to Y$ induces an isometry $f_* \colon E(Y') \to E(Y)$. In a field theory formulated on Minkowski space this is a representation of the Euclidean group of a space slice on the quantum Hilbert space; the representation of the whole Poincaré group involves the path integral as well. Show that a diffeomorphism $F \colon X' \to X$ preserves the partition function. (Consider the case where X has nontrivial boundary.)

Exercise 3.4 (Multiplicativity). What can you say about $E(Y_1 \sqcup Y_2)$? What about $Z_{X_1 \sqcup X_2}$?

Exercise 3.5. Try this point of view out in some familiar field theories, even those formulated in Minkowski space. In that case, what does the gluing law say about the usual propagation? (Here you should start by considering ordinary quantum mechanics, which is the case $d = 1$.) A good nontrivial example is the $d = 2$ theory considered in David Gross's lectures: QCD in two dimensions with a fixed gauge group. What is the Hilbert space $E(S^1)$ in this theory? Note that this is not a topological theory, but the partition function depends on the *area* of a surface. What is the correct statement of the gluing law? Of the symmetry properties?

3.2 Beyond Quantum Hilbert Spaces

You should have noticed that we have not yet considered locality—a gluing law—for the quantum Hilbert space. This is not an idea which is usually explicitly discussed in quantum field theory, though perhaps it is implicit there. For example, consider a discrete system formulated on a lattice in space, i.e., on a $(d-1)$-dimensional lattice Y. Then the Hilbert space of the theory is the tensor product over the lattice sites of a finite-dimensional Hilbert space H_y at each site:

$$E(Y) = \bigotimes_{y \in Y} H_y. \tag{3.10}$$

Then if the lattice Y is split into two pieces Y_1 and Y_2, the Hilbert spaces obviously obey the equation

$$E(Y) \cong E(Y_1) \otimes E(Y_2). \tag{3.11}$$

This is what we mean by saying that the quantum Hilbert space is local, though in general the gluing law is more complicated. Also keep in mind what we did for the classical theory. The classical counterpart to the quantum Hilbert space is the action (2.2), and the gluing law (3.11) is the quantum version of the (trivial case of the) gluing law (2.9).

The *Verlinde formulas* [17] are a nontrivial example of the locality of the quantum Hilbert space. It was originally formulated for the spaces of "conformal blocks" in two-dimensional conformal field theory. Then Witten [2] identified these spaces with the quantum Hilbert spaces of three-dimensional Chern-Simons theory, and in this context the Verlinde formula is a gluing law for the quantum Hilbert spaces. It takes the following form. Imagine that a closed surface Y is split into Y_1, Y_2 along a circle S. There is a finite number of "labels" $\lambda \in \Lambda$ and for each label a Hilbert space $E(Y_i)(\lambda)$. The Verlinde formula roughly has the form (we ignore some subtleties):

$$E(Y) \cong \bigoplus_{\lambda \in \Lambda} E(Y_1)(\lambda) \otimes E(Y_2)(\lambda). \tag{3.12}$$

Exercise 3.6. Reinterpret this formula along the following lines. Introduce the category $\mathcal{E}$ whose objects are collections of Hilbert spaces indexed by Λ. Also, introduce the "inner product" $\mathcal{E} \times \mathcal{E} \to \mathcal{E}$ defined by

$$(\{E_1(\lambda)\}, \{E_2(\lambda)\})_{\mathcal{E}} = \bigoplus_{\lambda} E_1(\lambda) \otimes E_2(\lambda). \tag{3.13}$$

Show that this makes $\mathcal{E}$ a 2-Hilbert space (cf. Exercise 2.10). Now rewrite (3.12) in terms of $\mathcal{E}$. Your formula should look like (3.9).

So how should we prove a gluing law for the quantum Hilbert space? The easiest way would be to repeat the computation (3.4). But that requires that we write the quantum Hilbert space as an integral, as in (3.1). This is what we do! It is really the crucial step in this discussion. Recall that our extended classical action (2.2) takes values in Hermitian lines. So we write exactly the same equation as (3.1), replacing the classical action $e^{2\pi i S_X(P)}$ by the classical action $L_Y(Q)$ for fields on a closed $(d-1)$-manifold:

$$E(Y) = \int_{\mathcal{C}_Y} L_Y(Q) \, d\mu_Y(Q). \tag{3.14}$$

What does this mean? It is a finite sum

$$\mu_1 \cdot L_1 + \cdots + \mu_N \cdot L_N \tag{3.15}$$

where the μ_i are positive numbers and the L_i are Hermitian lines. Now we interpret $\mu \cdot L$ as the Hermitian line with the same underlying complex vector space as L but with the inner product multiplied by μ. We interpret the sum as the orthogonal direct sum. In this way (3.14) defines a Hilbert space. In fact, it is exactly the same Hilbert space as (3.7).

Exercise 3.7. Verify this last assertion. Recall that we use the L^2 inner product on (3.7).

Now we repeat the computation (3.4) for a $(d-1)$-manifold Y split along a closed $(d-2)$-manifold S. In the course of that we will naturally introduce

$$\mathcal{E}(S) = \text{space of sections of } \mathcal{L}_S \to \mathcal{C}_S \tag{3.16}$$

which we could also write as an integral

$$\mathcal{E}(S) = \int_{\mathcal{C}_S} \mathcal{L}_S(R) \, d\mu_S(R). \tag{3.17}$$

We interpret this in the trivial case where $\mathcal{L}_S(R) = \mathcal{L}$, the category of finite dimensional Hilbert spaces. Then an element of $\mathcal{E}(S)$ is a choice of a Hilbert space W_R for each field $R \in \mathcal{C}_S$. Put differently, an element of $\mathcal{E}(S)$ is simply a Hermitian vector bundle over the space of fields $\mathcal{C}_S$. So

$$\mathcal{E}(S) = \text{Vect}(\mathcal{C}_S) \tag{3.18}$$

TABLE 1. Extended notions of classical action and path integral

dimension		classical action	path integral
d (X)		$P \mapsto e^{2\pi i S_X(P)}$ complex number of unit norm	Z_X complex number
$d-1$ (Y)		$Q \mapsto L_Y(Q)$ Hermitian line	$E(Y)$ Hilbert space
$d-2$ (S)		$R \mapsto \mathcal{L}_S(R)$ one-dimensional 2-Hilbert space	$\mathcal{E}(S)$ 2-Hilbert space

is the collection of such Hermitian vector bundles.

You can think of $\mathcal{E} = \mathcal{E}(S)$ as being a "Hilbert space over $\mathcal{L}$," analogous to the usual concept of a Hilbert space over $\mathbb{C}$. We call such an object a "2-Hilbert space." (2-vector spaces—the same object without the inner product—were introduced by Kapranov and Voevodsky [18] and Lawrence [19].) In Exercise 2.6 we already made the analogy between $\mathbb{C}$ and $\mathcal{L}$. Now go back to Exercise 2.7 and Exercise 2.10 to see what a 2-Hilbert space is. There are operations

$$+ : \mathcal{E} \times \mathcal{E} \to \mathcal{E}$$
$$\cdot : \mathcal{L} \times \mathcal{E} \to \mathcal{E} \qquad (3.19)$$
$$(\cdot, \cdot) : \mathcal{E} \times \mathcal{E} \to \mathcal{L}$$

analogous to addition, scalar multiplication, and inner product.

Exercise 3.8. Determine the operations (3.19) for the 2-Hilbert space defined in (3.18). What is a basis for this 2-Hilbert space?

Exercise 3.9. State explicitly the symmetry and gluing properties for the quantum integrals (3.1), (3.14), and (3.17). Note in particular the gluing law for the quantum Hilbert spaces:

$$E(Y) = \big(E(Y_1), E(Y_2) \big)_{\mathcal{E}(S)}. \qquad (3.20)$$

How does this fit with (3.11), (3.12), and (3.13)?

We summarize our extended notions of classical action and quantum path integral in Table 1. Here all of the manifolds are presumed closed for simplicity.

3.3   Quantum Finite-Gauge Theory

We now compute the path integral in finite-gauge theory, specializing to $d = 3$. In this section we treat the usual path integral and quantum Hilbert space. The new point is to account for the symmetry on the space of fields when carrying out the quantization. In Section 4 we will compute the quantum object $\mathcal{E}(S^1)$ associated to the circle and show how it leads to a quantum group.

You should now review the last part of Section 2 to be sure you understand the fields and classical action in this theory. (Recall that we consider
the "untwisted" case where the classical action is trivial.) We resume that
discussion taking over the notation used there. The ingredient we are missing is a measure on the space of fields, which is simply defined:

$$\mu_M(P) = \frac{1}{\# \operatorname{Aut} P}, \quad M \in \mathcal{C}_M. \tag{3.21}$$

This is the correct "counting measure" and is always how we count objects
in mathematics. Symmetries identify equivalent objects which we only want
to count once. To be consistent, then, if an object has automorphisms it
must be counted as in (3.21).

Exercise 3.10. Verify that this measure is invariant under symmetry.

The partition function is defined by (3.1), except now that we integrate
only over the space of *equivalence classes* of fields $\overline{\mathcal{C}_X}$. This makes sense
since both the classical action and the measure are invariant under sym-
metries. Writing $\overline{P}$ for a typical equivalence class we have

$$Z_X = \int_{\overline{\mathcal{C}_X}} e^{2\pi i S_X(\overline{P})} \, d\mu_X(\overline{P})$$

$$= \sum_{\overline{P}} \frac{1}{\# \operatorname{Aut} P}. \tag{3.22}$$

Suppose for simplicity that X is connected. Introduce a basepoint $x \in X$
and consider the space of bundles with a basepoint:

$$\mathcal{C}'_X = \left\{ \left\langle \begin{array}{c} P \\ \downarrow \\ M \end{array}, p \right\rangle : \begin{array}{l} P \text{ is a principal } G \text{ bundle, } p \in P_x \text{ a} \\ \text{chosen basepoint} \end{array} \right\}. \tag{3.23}$$

Here maps of bundles are required to preserve the basepoint. We need a
few facts outlined in the next exercise.

Exercise 3.11. Show that objects in $\mathcal{C}'_X$ are *rigid*, that is, have no nontrivial automorphisms. In other words, once we specify an isomorphism of
bundles on the basepoint, we know it everywhere. Show that the holonomy sets up a 1:1 correspondence between equivalence classes of elements
in $\mathcal{C}'_X$ and homomorphisms $\pi_1(X, x) \to G$. There is a G action on $\mathcal{C}'_X$: A
group element $g \in G$ simply moves the pair $\langle P, p \rangle$ to $\langle P, p \cdot g \rangle$. Show that
this action passes to the quotient $\mathcal{C}'_X$. Show that this quotient action corresponds to the action of G by conjugation on the space of homomorphisms
$\operatorname{Hom}(\pi_1(X, x), G)$.

Now the counting measure on $\mathcal{C}'_X$ weights each bundle with basepoint
with weight 1, since there is only the identity automorphism. Taking into

account the G action we find

$$Z_X = \frac{\# \operatorname{Hom}(\pi_1(X), G)}{\#G}. \tag{3.24}$$

This is obviously a topological invariant of X.

We now compute the quantum Hilbert space, defined by (3.7) or (3.14). Let Y be a closed surface. To account for the gauge symmetry we again integrate over the space of equivalence classes of fields. What this means is that we replace (3.7) by the space of *invariant* sections, that is, sections invariant under gauge transformations. Since the action is trivial (2.16), this is merely the space of invariant functions, that is, the space of functions on the quotient $\overline{C_Y} \cong \operatorname{Hom}(\pi_1(Y), G)/G$. For this last equality we assume that Y is connected. The L^2 metric is defined using the measure (3.21). Our argument with basepoints above identifies this with

$$E(Y) \cong \frac{1}{\#G} \cdot L^2(\operatorname{Hom}(\pi_1(Y, y), G), G)^G. \tag{3.25}$$

Here y is any basepoint in the connected space Y, the symbol "$(\cdot)^G$" means the invariants under the G action by conjugation, the L^2 metric weights each homomorphism with unit weight, and the prefactor $1/\#G$ multiplies this L^2 metric.

Exercise 3.12. Verify (3.25). What is the answer for $Y = S^2$? What about $Y = S^1 \times S^1$? For this example note that $\operatorname{Hom}(\pi_1(Y, y), G)$ is the set of commuting pairs of elements in G. To study the action of conjugation on this set it helps to consider the picture in Exercise 2.11 .

Exercise 3.13. By general arguments symmetries of Y are implemented as linear isometries of $E(Y)$. Do some explicit computations for $Y = S^1 \times S^1$. For example, compute the effect of the diffeomorphism defined by the matrix $T = \left(\begin{smallmatrix} 1 & 1 \\ 0 & 1 \end{smallmatrix}\right) \in \operatorname{SL}(2; \mathbb{Z})$.

Exercise 3.14. Compute the path integral on a 3-manifold with boundary. Verify the gluing law for this relative invariant.

Exercise 3.15. The $d = 2$ case of this theory is also interesting. It is the finite group version of the zero area limit (or topological limit) of two-dimensional QCD as considered in the lectures of David Gross. Now the partition function is defined for surfaces and there is a basic quantum Hilbert space E attached to the circle. Show that E can be identified as the space of *central* functions on G, that is, functions on G invariant under conjugation. What is the measure? Show that the path integral over a "pair of pants" leads to a multiplication on E. What is the multiplication? Can you diagonalize it? Use the gluing law to compute the partition function on any closed surface. Compare with (3.24) to obtain a formula which counts homomorphisms from a surface group into a finite group.

Exercise 3.16. Try to do some computations for a twisted theory (in $d = 3$) defined by a nonzero $\lambda \in H^3(BG; \mathbb{R}/\mathbb{Z})$. For example, take G to be a cyclic group of order n. Then $H^3(BG; \mathbb{R}/\mathbb{Z})$ is also cyclic of order n, and we can take λ to be the generator. Compute the result for X the projective 3-space, or more generally a lens space.

4   The Quantum Group

Finally in this discussion we produce the quantum group in finite group Chern-Simons theory. For simplicity we only consider the untwisted theory, and not the twisted theory which we indicated in the exercises of Section 2. However, we should remark that the precise computations for the twisted theory illustrate some more subtle gluing laws (along codimension two submanifolds). The interested reader should consult Ref. 9, §§8–9 for details.

Our arguments in this section are somewhat sketchy. A rigorous treatment would at the very least demand that we make precise all of the axioms for 2-Hilbert spaces, and this is already a complicated matter. The reader may refer to Ref. 9, §5 and §7 for another treatment.

4.1   The 2-Hilbert Space

We now compute the 2-Hilbert space $\mathcal{E} = \mathcal{E}(S^1)$ attached to the standard oriented circle. (We take S^1 to be the unit circle in the complex numbers with the counterclockwise orientation. We also take $1 \in S^1$ as a basepoint.) Recall from (3.17) that this is defined as a (generalized) integral over the space of fields on the circle. So our first job is to understand what that space $\mathcal{C}_{S^1}$ looks like. As in (3.23) it is best to start by rigidifying the fields by introducing a basepoint. So we consider $\mathcal{C}'_{S^1}$, the category of principal G bundles over a circle with a chosen basepoint covering the basepoint in S^1. (For brevity we call them "pointed principal bundles.") Then by starting at the basepoint p and traversing the circle in the properly oriented direction, we arrive back at a point $p \cdot x$, where $x \in G$ is the *holonomy*. (See Figure 5.) So the holonomy around S^1 defines a map

$$\mathrm{hol}\colon \mathcal{C}'_{S^1} \to G,$$

and it is easy to verify that it is an isomorphism on equivalence classes:

$$\overline{\mathcal{C}'_{S^1}} \cong G.$$

Now a change of basepoint $p \to p \cdot g$ changes the holonomy from x to $g^{-1}xg$. Thus the conjugacy class of the holonomy is independent of the basepoint, and

$$\overline{\mathcal{C}_{S^1}} \cong \text{conjugacy classes in } G, \tag{4.1}$$

which agrees with (2.14).

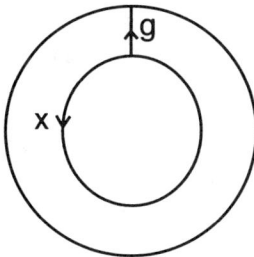

FIGURE 6. The bundle on C corresponding to $x \xrightarrow{g}$.

We know by now that it is not enough to work with equivalence classes. So as at the end of the last lecture we use basepoints to construct a good model of $\mathcal{C}_{S^1}$. Namely, introduce the category $\mathcal{C}$ whose objects are elements $x \in G$ and with a morphism $x \xrightarrow{g} gxg^{-1}$ for every pair of elements $\langle x, g \rangle$. (See Exercise 2.11.) Think of the object labeled by x as a pointed principal bundle P_x with holonomy x, and the morphism labeled by $x \xrightarrow{g} gxg^{-1}$ as the map between the pointed principal bundles $P_x \rightarrow P_{gxg^{-1}}$ which takes the basepoint in P_x to g times the basepoint in $P_{gxg^{-1}}$. These morphisms are isomorphisms of (unpointed) principal bundles, and represent all of the possible morphisms. For example, the set of arrows $x \xrightarrow{g} x$ is isomorphic to the *centralizer* C_x of x in G. This is precisely the set of automorphisms of P_x as a principal bundle. Such arguments show that $\mathcal{C}$ is a good model of $\mathcal{C}_{S^1}$.

Of course, $\mathcal{C}$ is simply a picture of the action of G on itself by conjugation.

We can also identify an arrow $x \xrightarrow{g} gxg^{-1}$ as a field on the cylinder C. We picture that in Figure 6, where we assume basepoints at the top of each circle and the group element indicates the parallel transport relative to the basepoints. (The cylinder is depicted as an annulus.)

Let $\mathcal{G}$ denote the set of arrows in $\mathcal{C}$; it is a *groupoid*. (A groupoid is a semigroup in which every arrow is invertible.) For simplicity we denote elements of $\mathcal{G}$ by $x \xrightarrow{g}$. Then the composition law in $\mathcal{G}$ is

$$(x \xrightarrow{g'}) \circ (x \xrightarrow{g}) = \begin{cases} x \xrightarrow{g'g}, & x' = gxg^{-1}; \\ \text{undefined}, & x' \neq gxg^{-1}. \end{cases} \tag{4.2}$$

We can understand this by gluing two bundles on the cylinder. The inner bundle should correspond to $x \xrightarrow{g}$ in keeping with our convention of representing parallel transport as acting on the left. (It is, after all, a bundle morphism.)

Recall from (2.17) that in this theory the classical action is trivial. (The twisted theory has a nontrivial action.) Now if there were no symmetries and the space of fields were simply $\overline{\mathcal{C}_{S^1}}$, which by (4.1) is the set of conjugacy classes in G, then by (3.18) we would identify $\mathcal{E}$ with the category of

Hermitian vector bundles over the set of conjugacy classes in G. In other words, an element of $\mathcal{E}$ would be a collection of Hermitian vector spaces indexed by the conjugacy classes of G. However, this is not correct because of the nontrivial symmetries in the fields.

When we quantize in the top dimension we deal with the symmetries by performing the path integral over the space of equivalence classes of fields (3.22). In the next dimension down, when we quantize the theory on a surface, we accommodate the symmetries by considering *invariant* sections (3.25). Now we must implement the analogous idea in codimension two: We take "invariant sections" of the trivial bundle whose fiber at each point is $\mathcal{L}$, the category of finite-dimensional Hilbert spaces. But rather than consider the large category $\mathcal{C}_{S^1}$, we use our model $\mathcal{C}$ for the space of fields on the circle. Then a section is simply a Hermitian vector bundle $W \to G$. In other words, it is a collection of Hermitian vector spaces W_x indexed by $x \in G$. What do we mean by an "invariant" section? Well, we mean that it should be invariant under the morphisms in $\mathcal{C}$, that is under the action of G on itself by conjugation. However, we do not simply mean that $W_x = W_{gxg^{-1}}$. Rather, we mean that we are given an explicit isomorphism

$$A_g : W_x \to W_{gxg^{-1}} \tag{4.3}$$

for each x, $g \in G$. This is in line with the idea that sets have an additional layer of structure: Vector spaces can be isomorphic without being equal. Of course, we presume that the isomorphisms (4.3) compose in accordance with the composition law (4.2) for the arrows. So an element of $\mathcal{E}$ is simply a vector bundle $W \to G$ together with a lift of the conjugation action of G on itself. The collection of these *equivariant vector bundles* is usually denoted $\mathrm{Vect}_G(G)$. Observe that the dimension of the fiber W_x, though constant in a conjugacy class, can vary over the group G. Also, the bundles we consider have a *Hermitian* structure.

There is one other ingredient in the definition (3.17) of $\mathcal{E}$, namely the measure on the space of fields. The correct measure to use on unpointed bundles is (3.21). However, our pointed bundles are rigid; they don't have any nontrivial automorphisms. But since we have to "divide" by the action of G (by taking invariant sections) we should use the weight $1/\#G$. (Precisely the same factor occurs when we quantize a closed surface (3.25).) So finally,

$$\mathcal{E} \cong \frac{1}{\#G} \cdot \mathrm{Vect}_G(G).$$

The standard inner product in $\mathcal{L}$ is $(V_1, V_2) = V_1 \otimes \overline{V_2}$, and that in $\mathrm{Vect}(G)$ is obtained by summing over the fibers. To account for the G action we need to take the invariants. Putting this together we see that the inner product in $\mathcal{E}$ is

$$(W_1, W_2)_{\mathcal{E}} = \frac{1}{\#G} \cdot \left(\bigoplus_x (W_1)_x \otimes \overline{(W_2)_x} \right)^G. \tag{4.4}$$

Note that the right-hand side is a finite-dimensional Hilbert space, as it should be.

Exercise 4.1. Make explicit the 2-Hilbert space structure of $\mathcal{E}$. What is the addition? Scalar multiplication?

Exercise 4.2. Rewrite (4.4) as a sum over conjugacy classes.

Exercise 4.3. Compute $\mathcal{E}$ for an Abelian group G.

Next, we observe that $\mathcal{E}$ can be identified as the category of representations of the groupoid $\mathcal{G}$. (In Exercise 2.10 we indicate how the category of representations of a finite *group* is a 2-Hilbert space. Here we replace a finite group by a finite groupoid.) Namely, associate to an element $W \in \text{Vect}_G(G)$ the finite-dimensional Hilbert space

$$W = \bigoplus_x W_x.$$

(The overloaded notation should not cause confusion.) Then the arrow $(x \xrightarrow{g})$ acts on W by $A_g|_{W_x}$; the action is trivial on $W_{x'}$ for $x' \neq x$. We can recover the fibers of the vector bundle by $W_x = (x \xrightarrow{e})(W)$, where e is the identity element of G.

We reformulate this by introducing an algebra H which we might call the "groupoid algebra," by analogy with group algebras:

$$H = \bigoplus_{x,g} \mathbb{C}\langle x, g \rangle.$$

(We use the symbol $\langle x, g \rangle$ for an element of H to distinguish it from the corresponding element $(x \xrightarrow{g})$ in $\mathcal{G}$.) So an element of H is a formal linear combination of the symbols $\langle x, g \rangle$ with complex coefficients. The multiplication in H is

$$\langle x_2, g_2 \rangle \cdot \langle x_1, g_1 \rangle = \begin{cases} \langle x_1, g_2 g_1 \rangle, & x_2 = g_1 x_1 g_1^{-1}; \\ 0, & \text{otherwise.} \end{cases} \tag{4.5}$$

The unit element is

$$1 = \sum_x \langle x, e \rangle. \tag{4.6}$$

Then $\mathcal{E}$ is the 2-Hilbert space of unitary representations of the algebra H, except that the usual inner product is divided by $\#G$.

We pass freely among these various descriptions of $\mathcal{E}$.

What is a basis for $\mathcal{E}$, thought of as a 2-Hilbert space? The geometric picture of equivariant vector bundles $W \to G$ is perhaps easiest to consider.

Note that the fiber W_x is a representation of the centralizer C_x, and that the representations $W_{x'}$ and W_x are isomorphic if x and x' are conjugate. So an indecomposable element of $\mathrm{Vect}_G(G)$ is supported on a single conjugacy class and for each x in that conjugacy class is an irreducible representation of the centralizer C_x. Let $\{W_\lambda\}$ be a set of inequivalent irreducible elements. Then it is a basis for $\mathcal{E}$ in the sense that any other element is isomorphic to $\bigoplus_\lambda (V^\lambda \otimes W_\lambda)$ for some finite-dimensional Hilbert spaces V^λ (thought of as trivial vector bundles over G).

Exercise 4.4. Write a basis of $\mathcal{E}$ for G an Abelian group. For $G = S_3$, the symmetric group on three letters.

4.2 Locality and Gluing

So far we have just used the extended notion of quantization and the structure of the space of fields on the circle. Now we want to apply the basic principles of locality and gluing to derive further structure on $\mathcal{E}$. At each stage this can be realized by introducing further structure on H. By then end we will have introduced enough structure to make H into a quantum group. We begin by considering the path integral over surfaces with boundary.

If Y is any surface with boundary, then from our general picture in Section 3 we know that the path integral $E(Y)$ is an element of $\mathcal{E}(\partial Y)$. Now ∂Y is a union of circles, and we would like to identify each of these circles with the standard circle so that $\mathcal{E}(\partial Y)$ can be identified as a tensor product of copies of $\mathcal{E} = \mathcal{E}(S^1)$. In general we cannot do this without introducing additional data (such as a parametrization of the boundary circles), but we will only need to consider surfaces $Y \subset \mathbb{C}$, that is, disks with subdisks removed. Then the boundary is made up of circles in $\mathbb{C}$, and there is a unique composition of a translation and a dilation which brings each circle to the standard circle. In this way we can identify $E(Y)$ as living in a tensor product of copies of $\mathcal{E}$. We can designate some of the boundary components as "incoming" and some as "outgoing" and use the Hermitian metric in $\mathcal{E}$ to view

$$E(Y)\colon \quad \bigotimes_{\substack{\text{incoming} \\ \text{circles}}} \mathcal{E} \to \bigotimes_{\substack{\text{outgoing} \\ \text{circles}}} \mathcal{E}.$$

This is a "linear" map of 2-Hilbert spaces, which at the simplest level is a functor between the categories indicated.

Introduce a basepoint on each boundary component of Y. Let $\mathcal{C}'_Y$ denote the category of principal G bundles over Y with chosen basepoints over the basepoints in ∂Y. Assume that Y has no closed components. Then arguments similar to those at the end of Section 3 show that we can identify

$$E(Y) \cong L^2(\overline{\mathcal{C}'_Y}). \tag{4.7}$$

Here we think of $\mathcal{E}$ as the 2-Hilbert space of representations of the groupoid $\mathcal{G}$, and the $\mathcal{G}$ action of a given boundary component is computed by gluing a cylinder to that boundary component.

Exercise 4.5. Verify (4.7) in detail.

Exercise 4.6. Suppose that $f : Y' \to Y$ is a diffeomorphism, where Y, $Y' \subset \mathbb{C}$ are surfaces as in the previous exercise. Explain how f induces an automorphism of functors $E(Y') \to E(Y)$.

Exercise 4.7. Let C denote the cylinder. Compute $E(C)$.

We will compute the path integrals over some surfaces shortly, but we begin by considering symmetries of the circle.

Exercise 4.8. Consider the group of orientation-preserving diffeomorphisms $\text{Diff}^+(S^1)$ of the circle. Inside this group is the subgroup of rigid rotations. Construct a retraction of $\text{Diff}^+(S^1)$ onto this subgroup. (Hint: Given a diffeomorphism compose with a rotation so that the composition preserves the basepoint. Now cut open the circle at the basepoint; then an orientation-preserving diffeomorphism is a monotonically increasing function $f : [0,1] \to [0,1]$, and if it preserves the basepoint then $f(0) = 0$ and $f(1) = 1$.)

Recall from Exercise 3.3 that diffeomorphisms of surfaces induce isometries of the corresponding quantum Hilbert spaces. In Exercise 3.9 you were asked to state the corresponding symmetry, or functoriality, for the 2-Hilbert space associated to a 1-manifold (or a $(d - 2)$-manifold in a d-dimensional theory). At first glance we might expect a diffeomorphism $f : S' \to S$ to induce an isometry of 2-Hilbert spaces $f_* : \mathcal{E}(S') \to \mathcal{E}(S)$, and indeed this is true. However, again there is an extra layer of structure: We expect an isotopy $F = f_t$ (path of diffeomorphisms) from f_0 to f_1 to induce a natural transformation F_* from the functor $(f_0)_*$ to the functor $(f_1)_*$. In topological theories we expect that deformations will not change this natural transformation. Applied to $f_0 = f_1 = \text{id}_S$, we obtain an action of $\pi_1(\text{Diff}^+(S))$ by automorphisms of the identity functor. Such an *automorphism of the identity* commutes with all arrows in the category $\mathcal{E}$. More explicitly, an automorphism of the identity specifies for each $W \in \mathcal{E}$ a map

$$\theta_W : W \to W$$

which commutes with all arrows in $\mathcal{E}$.

Exercise 4.9. Let $\text{Rep}(H)$ denote the 2-Hilbert space of representations of a finite group H. (See Exercise 2.10.) Determine all automorphisms of the identity of $\text{Rep}(H)$.

We apply this to the circle. From the previous exercise we know that $\text{Diff}^+(S^1)$ is homotopy equivalent to a circle, so the fundamental group is infinite cyclic. It is generated by a circle of rotations. Write an element of the circle as $e^{2\pi i x}$; then the circle of rotations is

$$f_t(e^{2\pi i x}) = e^{2\pi i(t+x)}, \quad 0 \leq t \leq 1.$$

We view this circle of rotations as a diffeomorphism τ of the cylinder C which is the identity on both boundary circles. We compute the corresponding automorphism of the identity by computing the action on $E(C)$. (See Exercise 4.6.) Now since the cylinder glued to itself is the cylinder, it follows easily that $E(C)$ is the identity map $\mathcal{E} \to \mathcal{E}$ (Exercise 4.7). We first compute the induced action on fields on the cylinder as

$$\tau^*\langle x, g \rangle = \langle x, gx \rangle = \langle x, g \rangle \cdot \langle x, x \rangle. \tag{4.8}$$

Here we use the multiplication in the algebra H (4.5). So the action on a field with holonomy x may be described by gluing on the field $\langle x, x \rangle$. Then Exercise 4.5 implies that the effect on the quantization is the operator $A_x \colon W_x \to W_x$, i.e., the automorphism of the identity is

$$\theta_W\big|_{W_x} = A_x \colon W_x \to W_x.$$

In terms of the H action we can describe it as the action of the special element

$$v = \sum_x \langle x, x \rangle. \tag{4.9}$$

This special element is the inverse of the *ribbon element* defined by Reshetikhin and Turaev [4].

Exercise 4.10. Show that (4.8) induces a diffeomorphism of the torus. In fact, it is the diffeomorphism considered in Exercise 3.13 . Compare the automorphism of the identity computed here with the action of that diffeomorphism on the vector space associated to the torus. Is there a gluing law which compares them?

Purely abstract considerations (Schur's lemma) show that an automorphism of the identity of $\mathcal{E}$ acts as multiplication by a scalar on an irreducible element W_λ. We can also see that since x is a central element of the centralizer group C_x, so acts as a scalar in any irreducible representation. These scalars are the *conformal weights* of the theory.

Exercise 4.11. Fix an element x and a representation λ of the centralizer C_x. This determines a basis element of $\mathcal{E}$. Compute the conformal weight in terms of the character of λ.

Exercise 4.12. Calculate the conformal weights explicitly for a cyclic group. Then calculate for the symmetric group S_3.

The circle also has an *orientation-reversing* diffeomorphism—a reflection —which is unique up to isotopy. We expect it to induce a map $\mathcal{E} \to \overline{\mathcal{E}}$ whose square is the identity (since a reflection squares to the identity).

Now $\overline{\mathcal{E}}$ means the conjugate 2-Hilbert space, but as a map of categories we can ignore the conjugation—that only affects the "scalar" multiplication. Again we can compute the induced action on $\mathcal{E}$ from the cylinder. Let ρ denote a fixed reflection. Then the induced map on fields is

$$\rho^* \langle x, g \rangle = \langle x^{-1}, g \rangle.$$

Note however that the incoming and outgoing circles are exchanged. This means that an element $W \in \mathcal{E}$ is taken to a new element W^* with

$$(W^*)_x = W^*_{x^{-1}}$$
$$A_g^{W^*} = (A_{g^{-1}}^W)^*.$$

In terms of the algebra H the duality is implemented by the *antipode*

$$S(\langle x, g \rangle) = \langle gx^{-1}g^{-1}, g^{-1} \rangle. \tag{4.10}$$

Exercise 4.13. In the diffeomorphism group of the circle, write the equation which relates a rotation and a reflection. In the quantization this is reflected by a relation among conformal weights. Namely, show that

$$\theta_{W^*} = \theta_W^*. \tag{4.11}$$

What does this say about conformal weights in terms of our basis?

Next, we consider the path integral over some simple surfaces $Y \subset \mathbb{C}$. For example, consider the unit disk D. Since the disk is contractible, all G bundles over D are trivial, as any two basepoints are related by an automorphism we find $\overline{C'_D}$ consists of a unique element. It follows that $E(D) \cong L^2(\overline{C'_D})$ is one-dimensional and is described by

$$E(D)_e = \begin{cases} \mathbb{C}, & x = e; \\ 0, & x \neq e, \end{cases} \tag{4.12}$$

with $C_e = \Gamma$ acting trivially on $E(D)_e$. In terms of the algebra H this particular representation is described by a *counit* $\epsilon \colon H \to \mathbb{C}$. From (4.12) we see that the counit is

$$\epsilon(\langle x, g \rangle) = \begin{cases} 1, & x = e; \\ 0, & \text{otherwise.} \end{cases} \tag{4.13}$$

More interesting is the path integral over the "pair of pants," which we realize embedded in $\mathbb{C}$ as the surface P depicted in Figure 7. We view the inside circles as incoming and the outside circle as outgoing (note the orientations), so that $E(P) \in \mathcal{E} \otimes \mathcal{E} \otimes \mathcal{E}$ induces a map

$$\odot : \mathcal{E} \otimes \mathcal{E} \to \mathcal{E}. \tag{4.14}$$

Here we denote the result of this multiplication map on W_1, W_2 as $W_1 \otimes W_2$. There is an induced map on morphisms in W_i as well: $\odot$ is a functor! (In category language it gives a "monoidal" structure to $\mathcal{E}$.)

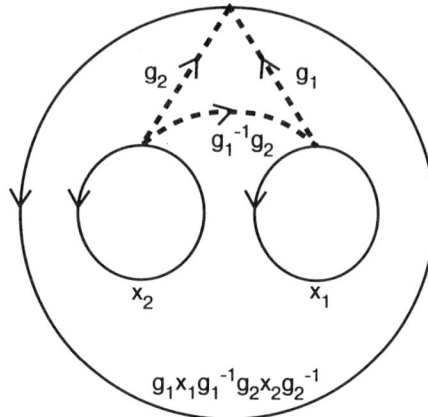

FIGURE 7. The bundle over P corresponding to $\langle x_1, g_1 \rangle \times \langle x_2, g_2 \rangle \in \mathcal{G} \times \mathcal{G}$.

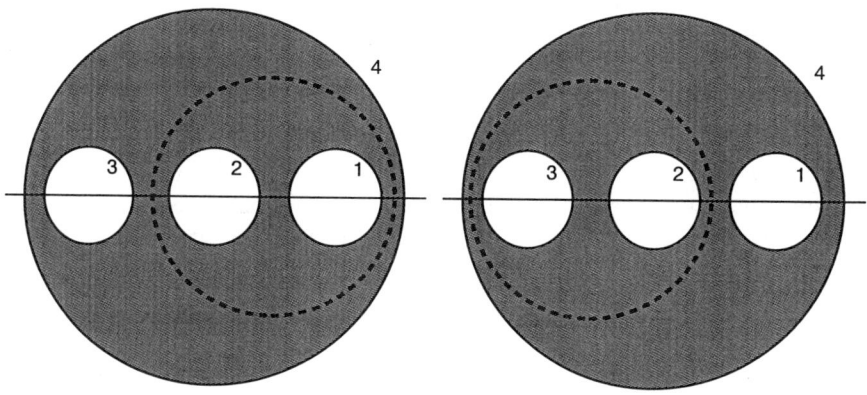

FIGURE 8. Associativity.

Exercise 4.14. Use the gluing law to describe an associativity isometry

$$\varphi_{W_1, W_2, W_3} : (W_1 \odot W_2) \odot W_3 \to W_1 \odot (W_2 \odot W_3). \qquad (4.15)$$

(See Figure 8.) In fact, you will find it to be trivial. However, in other theories (e.g., the twisted version of this theory) it is nontrivial.

Exercise 4.15. Use gluing to show that multiplication by (4.12) is isomorphic to the identity map. How is this expressed precisely?

To compute $E(P)$ we need to determine the space of fields on P. Following our procedure, we introduce basepoints on each boundary component.

Exercise 4.16. Show that there is a 1:1 correspondence between $\overline{C'_P}$ and $\mathcal{G} \times \mathcal{G}$. Hint: See Figure 7. In that figure the group elements indicate the parallel transport along the given line relative to the chosen basepoints.

Now we must compute the three different $\mathcal{G}$ actions on $E(P) \cong L^2(\overline{C_P'})$ corresponding to the three boundary circles of P. For the inner two boundary circles it is not difficult to verify that the action is by right multiplication. Namely, the action of $\langle x, g \rangle \in \mathcal{G}$ on a function $f(\cdot, \cdot)$ on $\overline{C_P'} \cong \mathcal{G} \times \mathcal{G}$ is:

$$
\begin{aligned}
f(\cdot, \cdot) &\mapsto f(\cdot \langle x, g \rangle^{-1}, \cdot), \\
f(\cdot, \cdot) &\mapsto f(\cdot, \cdot \langle x, g \rangle^{-1}).
\end{aligned}
\tag{4.16}
$$

On the other hand, the action corresponding to the outer component is

$$
\begin{aligned}
&(\langle x, g \rangle \cdot f)(\langle x_1, g_1 \rangle, \langle x_2, g_2 \rangle) \\
&= \begin{cases} f(\langle x_1, gg_1 \rangle, \langle x_2, gg_2 \rangle), & \text{if } x = g_1 x_1 g_1^{-1} g_2 x_2 g_2^{-1}; \\ 0, & \text{otherwise.} \end{cases}
\end{aligned}
\tag{4.17}
$$

Exercise 4.17. Verify (4.16) and (4.17). Recall that in each case the action is induced by gluing on a cylinder to the appropriate boundary component.

To compute $W_1 \odot W_2$, we use the inner product (4.4) in $\mathcal{E}$ to contract the first two factors (corresponding to the inner circles) in $E(P) \in \mathcal{E} \otimes \mathcal{E} \otimes \mathcal{E}$ with $W_1 \otimes W_2$. Up to a factor in the inner product, this says that $W_1 \odot W_2$ is the $\mathcal{G} \times \mathcal{G}$ invariants in $W_1 \otimes W_2 \otimes E(P)$. It is not difficult to verify that

$$
(W_1 \odot W_2)_x = \bigoplus_{x_1 x_2 = x} (W_1)_{x_1} \otimes (W_2)_{x_2}
\tag{4.18}
$$

with G action given by

$$
A_g^{W_1 \odot W_2} = A_g^{W_1} \otimes A_g^{W_2}.
\tag{4.19}
$$

Exercise 4.18. Verify (4.18)–(4.19). Also, show that the multiplication $\odot$ can be described by the following operation on $\mathrm{Vect}_G(G)$. Namely, let $\mu\colon G \times G \to G$ denote multiplication in the group G. Then given $W_1, W_2 \in \mathrm{Vect}_G(G)$, we can form their external tensor product $W_1 \boxtimes W_2 \to G \times G$. The quantum product is

$$
W_1 \odot W_2 = \mu_*(W_1 \boxtimes W_2),
\tag{4.20}
$$

where μ_* is the pushforward of vector bundles. (This is defined for finite covering maps.)

Viewing elements of $\mathcal{E}$ as representations of the algebra H, a tensor product on representations is induced by a *coproduct* $\Delta\colon H \to H \otimes H$, since then the element $h \in H$ acts in the tensor product $W_1 \otimes W_2$ by $\Delta(h)$. From (4.18) and (4.19) we see that the appropriate coproduct is

$$
\Delta(\langle x, g \rangle) = \sum_{x_1 x_2 = x} \langle x_1, g \rangle \otimes \langle x_2, g \rangle.
\tag{4.21}
$$

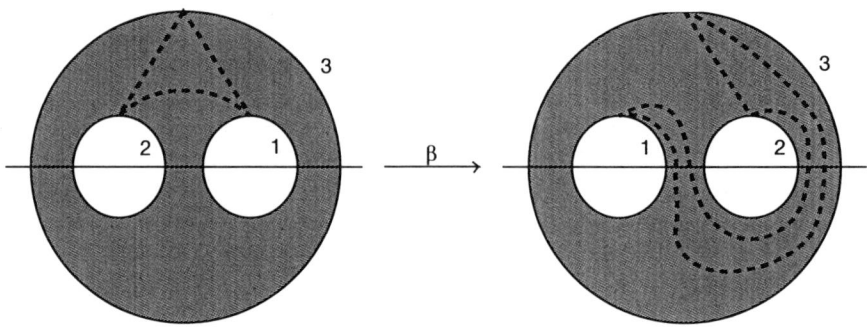

FIGURE 9. The braiding diffeomorphism β.

We content ourselves with one more computation, that of the *R-matrix*. Namely the pair of pants P has a *braiding* diffeomorphism β which acts on the quantization $E(P)$. We depict the braiding in Figure 9. The diffeomorphism β rotates circle 1 under circle 2, exchanging their positions. It is the identity on circle 3. Now by Exercise 4.6 β induces an automorphism of the multiplication $\odot$. Because two of the boundary circles are switched, this means that for each $W_1, W_2 \in \mathcal{E}$ we have a map

$$R_{W_1, W_2} : W_1 \odot W_2 \to W_2 \odot W_1 \tag{4.22}$$

which commutes with all of the arrows in $\mathcal{E}$. Before proceeding to compute it, we note one property which follows from the structure of $\mathrm{Diff}^+(P)$.

Exercise 4.19. Use pictures like Figure 9 to prove that

$$\tau_2 \tau_1 \beta = \beta^{-1} \tau_3,$$

where τ_i is a positive Dehn twist around the boundary labeled i. Show that this implies that for all $W_1, W_2 \in \mathcal{E}$ the following diagram commutes:

$$
\begin{array}{ccc}
W_1 \odot W_2 & & W_2 \odot W_1 \\
\downarrow{\scriptstyle \theta_{W_1 \odot W_2}} {\scriptstyle R_{W_1, W_2}} & & \downarrow{\scriptstyle \theta_{W_2} \odot \theta_{W_1}} \\
W_1 \odot W_2 & \xrightarrow{R^{-1}_{W_2, W_1}} & W_2 \odot W_1
\end{array}
$$

Now to the computation of R_{W_1, W_2}. As in all computations the idea is to first see the induced action on the fields. This is easily computed to be

$$\langle x_1, g_1 \rangle \times \langle x_2, g_2 \rangle \xrightarrow{\beta_*} \langle x_2, g_1 x_1 g_1^{-1} g_2 \rangle \times \langle x_1, g_1 \rangle.$$

Thus on functions $f \in L^2(\overline{C'_P}) \cong E(P)$ the induced push-forward action is:

$$(\beta_* f)(\langle x_1, g_1 \rangle, \langle x_2, g_2 \rangle) = f(\langle x_2, g_1 x_1 g_1^{-1} g_2 \rangle, \langle x_1, g_1 \rangle).$$

Now we must translate this into an action on $W_1 \odot W_2$ using the construction preceding (4.18). The result is

$$R_{W_1,W_2}: \quad (W_1)_{x_1} \otimes (W_2)_{x_2} \quad \to \quad (W_2)_{x_1 x_2 x_1^{-1}} \otimes (W_1)_{x_1}$$
$$w_1 \otimes w_2 \quad \mapsto \quad A_{x_1}^{W_2}(w_2) \otimes w_1$$

(4.23)

Exercise 4.20. Verify (4.23).

Finally, this can be implemented universally by a *quasitriangular element* $R \in H \otimes H$ which satisfies

$$R_{W_1,W_2} = \tau_{W_1,W_2} \circ (\rho_1 \otimes \rho_2)(R),$$

where $\tau_{W_1,W_2}: W_1 \otimes W_2 \to W_2 \otimes W_1$ is the transposition and ρ_i are representations of H. From (4.23) we deduce

$$R = \sum_{x_1,x_2} \langle x_1, e \rangle \otimes \langle x_2, x_1 \rangle. \tag{4.24}$$

Thus we have arrived at our goal. Namely from the quantization $\mathcal{E}$ of the circle, using symmetry and gluing laws, we have constructed an algebra H with unit (4.6), counit (4.13), antipode (4.10), comultiplication (4.21), a quasitriangular element (4.24), and a ribbon element (4.9). This is a *quasitriangular Hopf algebra* with a ribbon element. It is certainly the proper realization of a *quantum group* in this example. It was first written down in a paper of Dijkgraaf et al. [7], and it can be identified with Drinfeld's *quantum double* of the Hopf algebra of functions on G.

Exercise 4.21. Compute the Hopf algebra for G a cyclic group. For $G = S_3$.

Acknowledgments: The author is supported by NSF grant DMS-9307446, a Presidential Young Investigators award DMS-9057144, and the O'Donnell Foundation.

5 REFERENCES

1. V. F. R. Jones. A polynomial invariant of knots via von Neumann algebras. *Bull. Amer. Math. Soc.*, 12 (1): 103–111, 1985.

2. E. Witten. Quantum field theory and the Jones polynomial. *Commun. Math. Phys.*, 121 (3): 351–399, 1989.

3. N. Y. Reshetikhin and V. G. Turaev. Ribbon graphs and their invariants derived from quantum groups. *Commun. Math. Phys.*, 127 (1): 1–26, 1990.

4. N. Y. Reshetikhin and V. G. Turaev. Invariants of 3-manifolds via link polynomials and quantum groups. *Invent. Math.*, 103 (3): 547–597, 1991.

5. V. G. Turaev. *Quantum Invariants of Knots and 3-Manifolds*, volume 18 of *de Gruyter Studies in Mathematics*. Walter de Gruyter & Co., Berlin, 1994.

6. R. Dijkgraaf and E. Witten. Topological gauge theories and group cohomology. *Commun. Math. Phys.*, 129 (2): 393–429, 1990.

7. R. Dijkgraaf, V. Pasquier, and P. Roche. Quasi Hopf algebras, group cohomology and orbifold models. In *Recent Advances in Field Theory*, (Annecy-le-Vieux, 1990), volume 18B of *Nuclear Phys. B. Proc. Suppl.*, 1991. North-Holland, Amsterdam, pages 60–72.

8. D. Altschuler and A. Coste. Quasi-quantum groups, knots, 3-manifolds, and topological field theory. *Commun. Math. Phys.*, 150 (1): 83–107, 1992.

9. D. S. Freed. Higher algebraic structures and quantization. *Commun. Math. Phys.*, 159: 343–398, 1994.

10. D. S. Freed and F. Quinn. Chern-Simons theory with finite gauge group. *Commun. Math. Phys.*, 156 (3): 435–472, 1993.

11. D. S. Freed. Extended structure in topological quantum field theory. In L. H. Kauffman and R. A. Baadhio, eds., *Quantum Topology*. volume 3 of *Series on Knots and Everything*, World Scientific, River Edge, NJ, pages 162–173, 1993.

12. D. S. Freed. Characteristic numbers and generalized path integrals. In *Geometry, Topology, and Physics*. volume VI of *Conf. Proc. Lecture Notes Geom. Topology*, International Press, Cambridge, MA, pages 126–138, 1995.

13. D. S. Freed. Lectures in topological quantum field theory. In L. A. Ibort and M. A. Rodríguez, eds., *Integrable Systems, Quantum Groups, and Quantum Field Theories*. Kluwer Academic Publishers, Dordrecht, pages 95–156, 1993.

14. D. S. Freed. Classical Chern-Simons theory. I. *Ann. Math.*, 115 (2): 237–303, 1995.

15. J. Mickelsson. Kac-Moody groups and the Dirac determinant line bundle. In *Topological and Geometrical Methods in Field Theory*, (Espoo, 1986), 1986. World Scientific, Teaneck, NJ, pages 117–131.

16. S. MacLane. *Categories for the Working Mathematician*, volume 5 of *Graduate Texts in Mathematics*. Springer Verlag, New York, 1971.

17. E. Verlinde. Fusion rules and modular transformations in $2d$ conformal field theory. *Nucl. Phys.*, B300 (3): 360–376, 1988.

18. M. M. Kapranov and V. A. Voevodsky. 2-categories and Zamolodchikov tetrahedra equations. In *Algebraic Groups and their Generalizations: Quantum and Infinite-Dimensional Methods*, (University Park, PA, 1991), volume 56, Part 2 of *Proc. Sympos. Pure Math.*, 1994. Amer. Math. Soc., Providence, RI, pages 177–259.

19. R. Lawrence. Triangulations, categories and extended topological field theories. In L. H. Kauffman and R. A. Baadhio, eds., *Quantum Topology*. volume 3 of *Series on Knots and Everything*, World Scientific, Ridge River, NJ, pages 191–208, 1993.

4

Half Transfer Matrices in Solvable Lattice Models

Tetsuji Miwa

Introduction

In this discussion we explain the method for computing correlation functions of solvable lattice models [1–4]. We are going to discuss the six-vertex model on the two-dimensional lattice of infinite size.

1 The Six-Vertex Model

Consider a two-dimensional square lattice consisting of M vertical and N horizontal lines with periodic boundary condition.

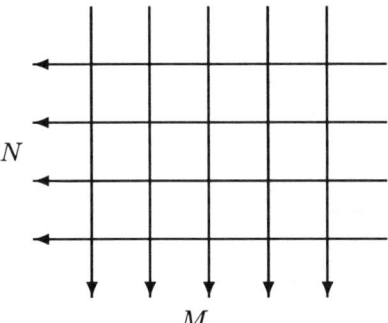

The lines are given orientation shown by the arrows. We associate a complex parameter $\zeta \in \mathbb{C} \setminus \{0\}$ to each line. Usually, we take ζ_1 for all the vertical lines, and ζ_2 for the horizontal ones. We call a crossing of two lines *a vertex*, and a line segment between two vertices *an edge*. A local variable $\sigma_j = \pm$ is associated to each edge j. A configuration C is an assignment of values of σ_j for all j. One can consider the following local configuration around a vertex

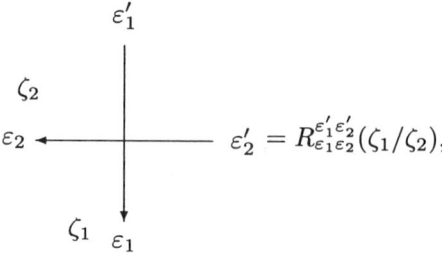

$$\varepsilon_2' = R_{\varepsilon_1\varepsilon_2}^{\varepsilon_1'\varepsilon_2'}(\zeta_1/\zeta_2),$$

to which we associate a complex number $R_{\varepsilon_1\varepsilon_2}^{\varepsilon_1'\varepsilon_2'}(\zeta_1/\zeta_2)$ depending on the local configuration of the variables ε_1, ε_1', ε_2, ε_2' and the ratio of the two parameters ζ_1 and ζ_2. It is called the Boltzmann weight. The partition function is

$$Z_{MN} = \sum_C \prod_v R_{\varepsilon_1(C,v)\varepsilon_2(C,v)}^{\varepsilon_1'(C,v)\varepsilon_2'(C,v)}(\zeta_1/\zeta_2)$$

We are to compute the partition function per site

$$\kappa(a,b,c) = \lim_{M,N\to\infty} Z_{MN}^{1/(MN)}$$

and the correlation functions

$$\langle \sigma_{j_1} \cdots \sigma_{j_n} \rangle = \lim_{M,N\to\infty} \frac{\sum_C \sigma_{j_1} \cdots \sigma_{j_n} \prod_v R_{\varepsilon_1(C,v)\varepsilon_2(C,v)}^{\varepsilon_1'(C,v)\varepsilon_2'(C,v)}(\zeta_1/\zeta_2)}{Z_{MN}}.$$

The name of the six-vertex model comes from the restriction that $R_{\varepsilon_1\varepsilon_1}^{\varepsilon_1'\varepsilon_2'}$ is nonzero only for the following six configurations:

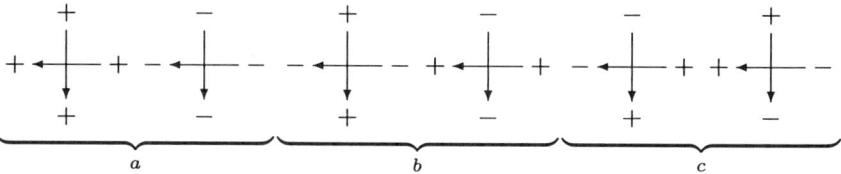

Fix a complex parameter q and consider the following parameterization of a, b, c:

$$a = \frac{1}{\kappa}, \quad b = \frac{q(1-\zeta^2)}{\kappa(1-q^2\zeta^2)}, \quad c = \frac{\zeta(1-q^2)}{\kappa(1-q^2\zeta^2)}.$$

The κ is an overall constant which seems to play no important role in the computations of the partition function and the correlation functions at a first glance. This is true as far as the results are concerned. Nevertheless we need a proper normalization later in order to go through the computations.

The reason for the above parametrization in ζ is that the following equalities called the Yang-Baxter equation is valid:

$$R_{12}(\zeta_1/\zeta_2)R_{13}(\zeta_1/\zeta_3)R_{23}(\zeta_2/\zeta_3) = R_{23}(\zeta_2/\zeta_3)R_{13}(\zeta_1/\zeta_3)R_{12}(\zeta_1/\zeta_2)$$

Let us explain the notation. We consider a two-dimensional space $V = \mathbb{C}v_+ \oplus \mathbb{C}v_-$ with the basis v_+, v_-. From the Boltzmann weights we can define a matrix $R(\zeta) \in \mathrm{End}_{\mathbb{C}}(V \otimes V)$:

$$R(\zeta)(v_{\varepsilon_1} \otimes v_{\varepsilon_2}) = \sum_{\varepsilon_1',\varepsilon_2'} v_{\varepsilon_1'} \otimes v_{\varepsilon_2'} R_{\varepsilon_1'\varepsilon_2'}^{\varepsilon_1\varepsilon_2}(\zeta).$$

Consider the tensor product $V_1 \otimes V_2 \otimes V_3$ where V_1, V_2, $V_3 \cong V$. In the Yang-Baxter equation, we denoted R_{jk} to mean the matrix R acting on the jth and the kth components.

The Yang-Baxter equation can be written graphically as

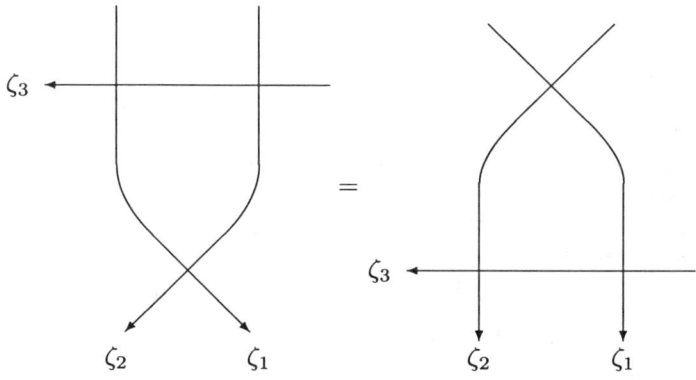

2 The Antiferromagnetic Regime

We require that the R-matrix satisfies two more relations by choosing an appropriate normalization κ:

- the unitarity

$$\left(\right) = 1, \quad R_{12}(\zeta_1/\zeta_2)R_{21}(\zeta_2/\zeta_1) = 1$$

- the crossing symmetry

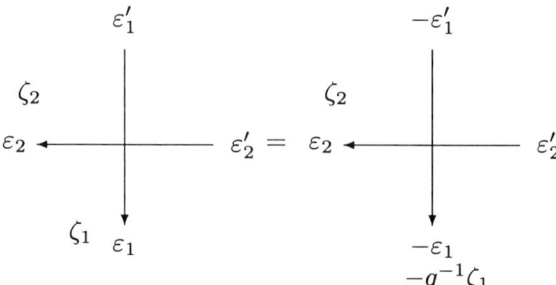

The proper choice of κ in the formula of a, b, c so that these two relations are satisfied is

$$\kappa(\zeta) = \zeta \frac{(q^4\zeta^2; q^4)_\infty (q^2\zeta^{-2}; q^4)_\infty}{(q^4\zeta^{-2}; q^4)_\infty (q^2\zeta^2; q^4)_\infty},$$

where $(z; q_1, \ldots, q_m)_\infty = \prod_{n_1, \ldots, n_m = 0}^{\infty} (1 - q_1^{n_1} \cdots q_m^{n_m} z)$. With this choice of κ and by restricting the parameters q and ζ in the region $-1 < q < 0$, and $-1 < \zeta < -q^{-1}$, we have the partition function per site $\kappa(a, b, c)$ equal to 1. In other words $\kappa(\zeta)$ is the partition function per site when we take the normalization $a = 1$.

The region $-1 < q < 0, 1 < \zeta < -q^{-1}$ corresponds to the region $c > a+b$ (and $a, b, c > 0$). In this region, among the possible configurations on the lattice, there are two which give the largest contribution to the sum in the partition function. They contain only c, i.e., $-\!\!\begin{array}{c}+\\\blacktriangleleft\!\!\mid\!\!+\\-\end{array}$ or $+\!\!\begin{array}{c}-\\\blacktriangleleft\!\!\mid\!\!-\\+\end{array}$. For example, if $M = N = 2$, they are

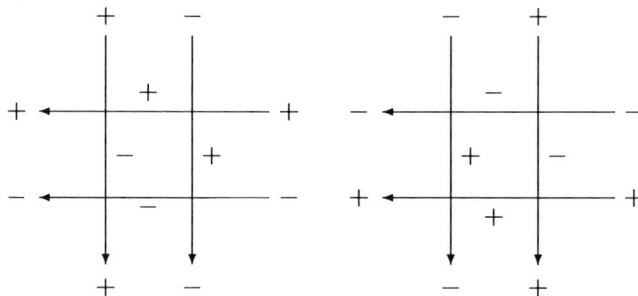

Let us call them the ground-state configurations (GSC). In this region of the parameters, which is called *the antiferromagnetic regime*, we can separate two sectors in the infinite volume limit. In the ith sector ($i = 0, 1$) we take the configuration sum over those which are different from the ith GSC at finitely many edges. Let us explain what this means in the computation of $\kappa(a, b, c)$. Choose the normalization such that $c = 1$. We have

$$Z_{NM} = 1 + NMa^2b^2 + (\text{order } 6 \text{ in } a, b).$$

Here 1 corresponds to the GSC, and NMa^2b^2 the terms that differ from the GSC at the four edges around a face. Therefore, we get

$$\kappa(a, b, 1) = 1 + a^2b^2 + (\text{order 6 in } a, b).$$

To get the full result, note that in this normalization ($c = 1$) the unitarity relation reads as

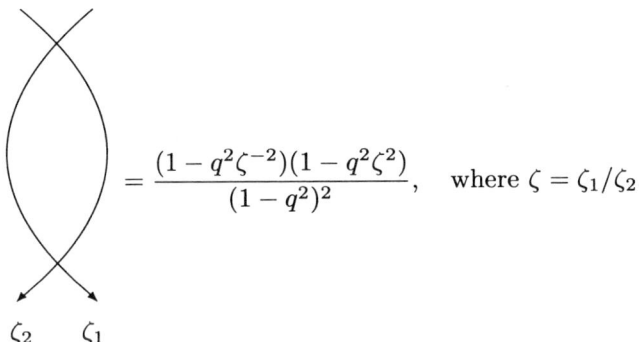

$$= \frac{(1 - q^2\zeta^{-2})(1 - q^2\zeta^2)}{(1 - q^2)^2}, \quad \text{where } \zeta = \zeta_1/\zeta_2.$$

Denote the right-hand side by $\rho(\zeta)$. Consider the following lattice.

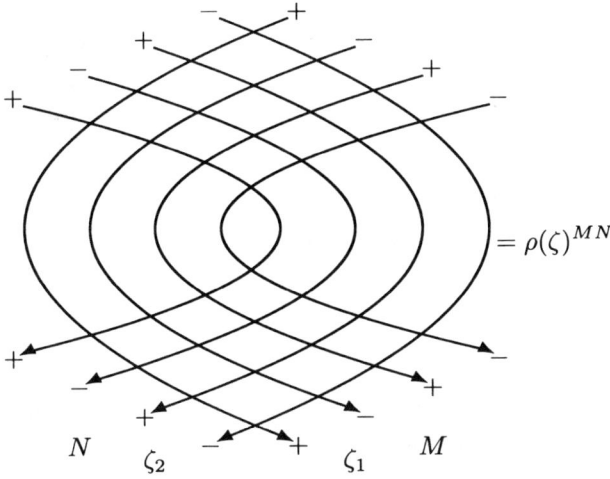

$$= \rho(\zeta)^{MN}$$

The unitarity relation implies the partition function of this lattice is equal to $\rho(\zeta)^{NM}$. On the other hand it is also equal to $Z_{MN}(\zeta) \cdot Z_{NM}(\zeta^{-1})$ approximately. In the limit $N, M \to \infty$ we can ignore the error and conclude

$$\bar\kappa(\zeta)\bar\kappa(\zeta^{-1}) = \rho(\zeta),$$

where $\bar\kappa(\zeta) = \kappa(a, b, 1)$ and $\bar\kappa(\zeta^{-1})$ is the analytic continuation of $\bar\kappa$. The crossing symmetry holds for the normalization $c = 1$. Therefore we have

$$\bar\kappa(\zeta^{-1}) = \bar\kappa(-q^{-1}\zeta).$$

We now assume that $\log \bar{\kappa}(\zeta)$ is single-valued and holomorphic in a region containing the annulus $|q|^{1/2} \leqslant |\zeta| \leqslant 1$. Then expanding it as $\log \bar{\kappa}(\zeta) = \sum_{n \in \mathbb{Z}} c_n \zeta^n$ in this annulus, and substituting this expansion to the two equations, one can uniquely determine the c_n. The result is given by

$$\bar{\kappa}(\zeta) = \kappa(\zeta)\frac{1 - q^2\zeta^2}{\zeta(1 - q^2)}$$

where $\kappa(\zeta)$ is given above.

This argument is called the inversion trick [5, 6]. In the present discussion we give more elaborate arguments of a similar kind which lead to several important equations. We can reformulate these equations in the language of the representation theory of the quantum affine algebra $U_q(\widehat{sl}_2)$. Then, the machinery in the representation theory gives answers for the correlation functions and other physical quantities. We omit, however, the details on the representation theoretical accounts. (See Ref. 7 for an exposition.)

3   Corner Transfer Matrix

Let us compute the one-point function $\langle \sigma_0 \rangle$ where the edge 0 is a vertical one. We choose a sector such that $\sigma_0 = +$ in the corresponding GSC for which $i = 1$ by definition. We introduce two kinds of operators and express $\langle \sigma_0 \rangle$ by means of the trances of certain products of these operators.

The first kind of operator, which is called *the corner transfer matrix* (CTM), was introduced by Baxter [6]. Consider the partition function of the following lattice.

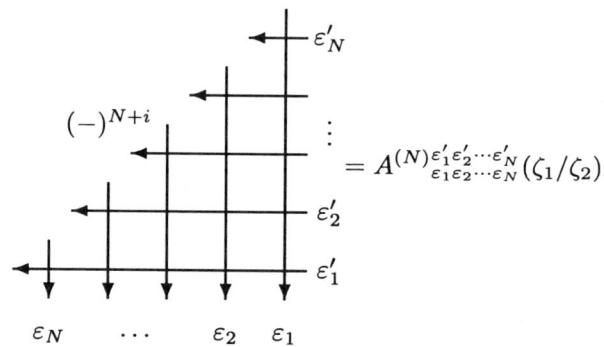

$$= A^{(N)\varepsilon'_1\varepsilon'_2\cdots\varepsilon'_N}_{\varepsilon_1\varepsilon_2\cdots\varepsilon_N}(\zeta_1/\zeta_2).$$

The boundary variables $\varepsilon_1, \varepsilon_2, \ldots, \varepsilon_N$ and $\varepsilon'_1, \varepsilon'_2, \ldots, \varepsilon'_N$ are chosen arbitrarily. The NW boundary is fixed to $(-)^{N+i}$ so that it respects the ith GSC. The vertical lines carry the parameter ζ_1 and the horizontal ones ζ_2. In this setting let us denote the partition function by $A^{(N)\varepsilon'_1\varepsilon'_2\cdots\varepsilon'_N}_{\varepsilon_1\varepsilon_2\cdots\varepsilon_N}(\zeta_1/\zeta_2)$ and consider it as a matrix element of an operator $A^{(N)}(\zeta_1/\zeta_2) \in \mathrm{End}_{\mathbb{C}}(V^{\otimes N})$.

Baxter found the following remarkable properties of CTM in the infinite volume limit $N \to \infty$:

- $\lim_{N \to \infty} A^{(N)}(\zeta) = $ (a divergent scalar) $\times \zeta^{-H_{\text{CTM}}}$.

- H_{CTM} is an operator depending only on q.

- $\text{Spec}\, H_{\text{CTM}} = \{0, 1, 2, \dots\}$.

One can introduce similar CTMs in the other three quadrants:

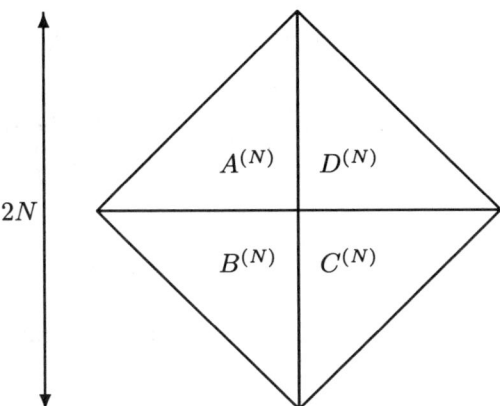

The partition function of this lattice, write it as Z_N, is given by

$$\lim_{N \to \infty} Z_N = \text{(a divergent scalar)} \times \text{tr}_{\mathcal{H}^{(i)}} DCBA,$$

where $A = \zeta^D$ and B, C, D are given below. The scalar factor will cancel out in the computation of correlation functions. We will ignore it, and write $Z^{(i)}$ for the renormalized partition function in the infinite volume limit:

$$Z^{(i)} = \text{tr}_{\mathcal{H}^{(i)}} DCBA.$$

The space $\mathcal{H}^{(i)}$ is by definition the space spanned by the eigenvectors of A. This is sensible because we have

$$B^{(N)}(\zeta) = X A^{(N)}(-\varepsilon^{-1}\zeta^{-1})$$
$$C^{(N)}(\zeta) = X A^{(N)}(\zeta) X$$
$$D^{(N)}(\zeta) = A^{(N)}(-\varepsilon^{-1}\zeta^{-1}) X$$

where X is the charge conjugation of $+$ and $-$. These equalities follow from the crossing symmetry. Therefore, we can set

$$B = (-q^{-1}\zeta^{-1})^{-H_{\text{CTM}}}, \quad C = \zeta^{-H_{\text{CTM}}}, \quad D = (-q^{-1}\zeta^{-1})^{-H_{\text{CTM}}}.$$

Note that we dropped X in the definition of B, C, D. We have the basic formula for the renormalized partition function:

$$Z^{(i)} = \mathrm{tr}_{\mathcal{H}^{(i)}} DCBA = \mathrm{tr}_{\mathcal{H}^{(i)}} q^{2H_{\mathrm{CTM}}}.$$

Because the eigenvalues of H_{CTM} are integers, we can compute $Z^{(i)}$ if we know their multiplicities. The multiplicities can be computed in the limit $q \to 0$.

It is easy to expand $A^{(N)}(\zeta)$ to the first order in u (where $\zeta = e^u$), and we get

$$\widehat{H}_{\mathrm{CTM}} = \frac{1}{1-q^2} \sum_{n=1}^{\infty} n \left\{ q(\sigma_n^x \sigma_{n+1}^x + \sigma_n^y \sigma_{n+1}^y) + \frac{1+q^2}{2} \sigma_n^z \sigma_{n+1}^z \right\}.$$

This is not renormalized (i.e., a divergent factor must be subtracted from $\widehat{H}_{\mathrm{CTM}}$ to get H_{CTM} satisfying $\mathrm{Spec}\, H_{\mathrm{CTM}} = \{0, 1, 2, \dots\}$). The renormalized one at $q = 0$ is

$$H_{\mathrm{CTM}, q=0} = \frac{1}{2} \sum_{n=1}^{\infty} n(\sigma_n^z \sigma_{n+1}^z + 1).$$

Now, the computation of the multiplicities of this Hamiltonian is a combinatorial problem. We have

$$Z = Z^{(i)} = \frac{1}{\prod_{n=0}^{\infty}(1 - x^{2n+1})}, \quad \text{where } x = q^2.$$

4   Half Transfer Matrix

We continue the computation of $\langle \sigma_0 \rangle$. The second kind of operator we will use is called *the vertex operator* in the representation theory. We will not enter the representation theory. It is good to give another name to it, *the half transfer matrix*. The full transfer matrix $T^{(N)}(\zeta) \in \mathrm{End}_{\mathbb{C}}(V^{\otimes N})$ is used to compute partition function Z_{MN}. The definition goes

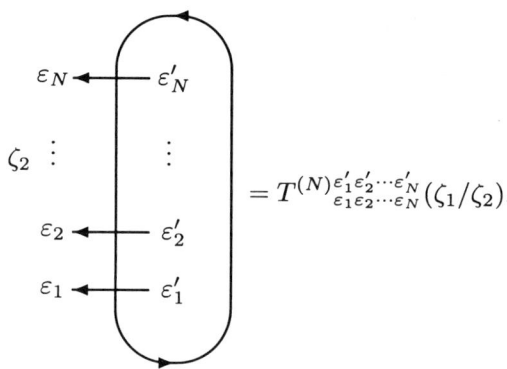

We have $Z_{MN} = \mathrm{tr}_{V^{\otimes N}}(T^{(N)}(\zeta))^M$. The diagonalization of the transfer matrix is another important subject which we will not discuss here. (See Ref. 7 for an exposition.)

Now we come to the half transfer matrix (HTM), $\Phi_\varepsilon^{(N)}(\zeta) \in \mathrm{End}_{\mathbb{C}}(V^{\otimes N})$. The definition goes

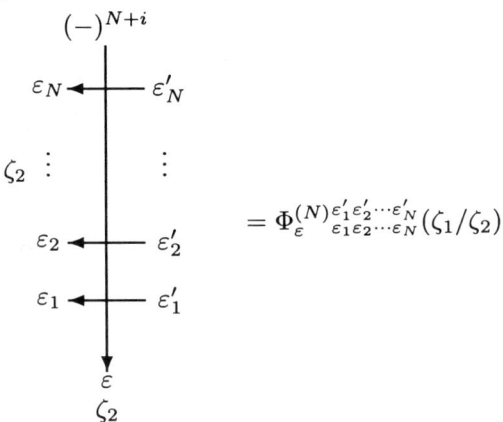

$$= \Phi_\varepsilon^{(N)}{}^{\varepsilon_1'\varepsilon_2'\cdots\varepsilon_N'}_{\varepsilon_1\varepsilon_2\cdots\varepsilon_N}(\zeta_1/\zeta_2).$$

We want to let it act from $\mathcal{H}^{(i)}$ to $\mathcal{H}^{(1-i)}$ in the infinite volume limit after an appropriate renormalization. Before going into that discussion let us rewrite $\langle\sigma_0\rangle$ by means of CTM and HTM. Consider the following lattice with the fixed value ε of the variable σ_0 at the center.

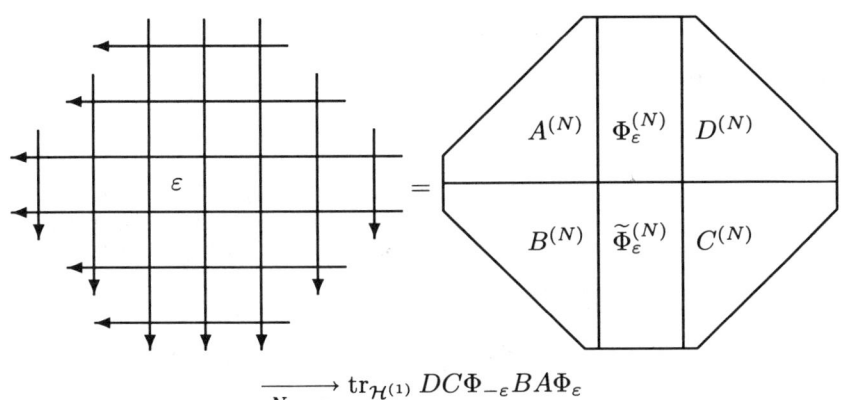

$$\xrightarrow[N\to\infty]{} \mathrm{tr}_{\mathcal{H}^{(1)}} DC\Phi_{-\varepsilon}BA\Phi_\varepsilon$$

Let us give the precise definition for the above. We are taking $2N+1$ vertical and $2N$ horizontal lines with the parameters ζ_1 and ζ_2, respectively. We assume the boundary condition in the zeroth GSC. This means the HTM Φ_ε is acting from $\mathcal{H}^{(1)}$ to $\mathcal{H}^{(0)}$. The operators A and B act from $\mathcal{H}^{(0)}$ to itself.

We then need the lower half of transfer matrix. By the crossing symmetry we see that it is equal to $\widetilde{\Phi}_\varepsilon^{(N)} = X\Phi_{-\varepsilon}^{(N)}(\zeta)X$. Canceling the X's with those attached to $B^{(N)}$ and $C^{(N)}$, we set $\Phi_{-\varepsilon}$ between C and B acting from $\mathcal{H}^{(0)}$

to $\mathcal{H}^{(1)}$. Finally we put C and D acting on $\mathcal{H}^{(1)}$, and then take the trace of the product on $\mathcal{H}^{(1)}$. We have the formula in the infinite volume limit.

$$
\begin{aligned}
\langle \sigma_0 \rangle &= \frac{\sum_\varepsilon \varepsilon \, \mathrm{tr}_{\mathcal{H}^{(1)}} \, DC\Phi_{-\varepsilon} BA\Phi_\varepsilon}{Z} \\
&= \frac{\sum_\varepsilon \varepsilon \, \mathrm{tr}_{\mathcal{H}^{(1)}} (-q)^{H_{\mathrm{CTM}}} \Phi_{-\varepsilon} (-q)^{H_{\mathrm{CTM}}} \Phi_\varepsilon}{Z}.
\end{aligned}
$$

Note that we have still not yet given a precise definition to the numerator. The precise definition must give

$$
\sum_\varepsilon \mathrm{tr}_{\mathcal{H}^{(0)}} (-q)^{H_{\mathrm{CTM}}} \Phi_{-\varepsilon} (-q)^{H_{\mathrm{CTM}}} \Phi_\varepsilon = Z.
$$

This is not at all clear from the definition. We have the following steps to solve this problem. First, we give commutation relations among $\Phi_\varepsilon(\zeta)$ and $\zeta^{H_{\mathrm{CTM}}}$. Second, we can show that the commutation relations determine $\Phi_\varepsilon(\zeta)$ uniquely up to the over all normalization. (See Ref. 8.) Lastly, we fix it by computing the two-point function of the half transfer matrices.

5   Commutation Relations

The first relation is

$$
\xi^{-H_{\mathrm{CTM}}} \Phi_\varepsilon(\zeta) = \Phi_\varepsilon(\zeta/\xi) \xi^{-H_{\mathrm{CTM}}}.
$$

This follows from the special case

$$
\zeta^{-H_{\mathrm{CTM}}} \Phi_\varepsilon(\zeta) = \Phi_\varepsilon(1) \zeta^{-H_{\mathrm{CTM}}}.
$$

Note that in the left-hand side $\Phi_\varepsilon(\zeta)$ is acting from $\mathcal{H}^{(i)}$ to $\mathcal{H}^{(1-i)}$ and $\zeta^{-H_{\mathrm{CTM}}}$ from $\mathcal{H}^{(1-i)}$ to itself. On the other hand, in the right-hand side $\zeta^{-H_{\mathrm{CTM}}}$ is acting from $\mathcal{H}^{(i)}$ to itself and $\Phi_\varepsilon(1)$ from $\mathcal{H}^{(i)}$ to $\mathcal{H}^{(1-i)}$. Graphically, the left-hand side reads as

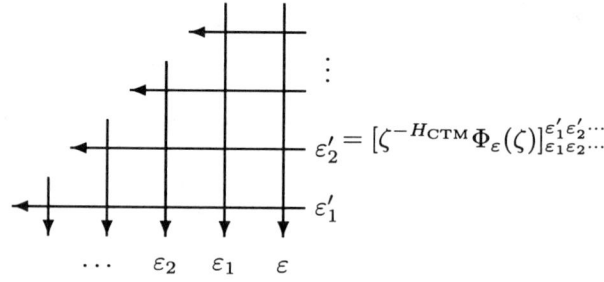

In the infinite volume limit it looks very much like $\zeta^{-H_{\mathrm{CTM}}}$ itself. The difference is that $\zeta^{-H_{\mathrm{CTM}}}$ is acting from $\mathcal{H}^{(i)}$ to itself while the one in the

discussion is acting from $\mathcal{H}^{(i)}$ to $\mathcal{H}^{(1-i)}$. We need some operator which makes the following job:

$$\mathcal{H}^{(i)} \qquad \longrightarrow \qquad \mathcal{H}^{(1-i)} \otimes V$$

$$v_{\varepsilon_2} \otimes \overset{\cap}{v}_{\varepsilon_1} \otimes v_\varepsilon \quad \longmapsto \quad (\cdots \otimes v_{\varepsilon_2} \otimes v_{\varepsilon_1}) \otimes v_\varepsilon.$$

Note that

$$R^{\varepsilon'_1 \varepsilon'_2}_{\varepsilon_1 \varepsilon_2}(1) = \delta_{\varepsilon'_1 \varepsilon_2} \delta_{\varepsilon'_2 \varepsilon_1}.$$

From this follows that

$$\Phi_{\varepsilon'}(1)(\cdots \otimes v_{\varepsilon_2} \otimes v_{\varepsilon_1} \otimes v_\varepsilon) = \delta_{\varepsilon \varepsilon'}(\cdots \otimes v_{\varepsilon_2} \otimes v_{\varepsilon_1}) \otimes v_\varepsilon.$$

Therefore, the wanted operator is $\Phi_\varepsilon(1)$.

The second relation is

$$\sum_{\varepsilon'_1, \varepsilon'_2} R^{\varepsilon'_1 \varepsilon'_2}_{\varepsilon_1 \varepsilon_2}(\zeta_1/\zeta_2) \Phi_{\varepsilon'_1}(\zeta_1) \Phi_{\varepsilon'_2}(\zeta_2) = \Phi_{\varepsilon_2}(\zeta_2) \Phi_{\varepsilon_1}(\zeta_1).$$

In this equality, the normalization of the R-matrix by $\kappa(\zeta)$ is essential; The normalization of $\Phi_\varepsilon(\zeta)$ is irrelevant, but that of R cannot be arbitrary. The "proof" goes as follows. The left-hand side reads as

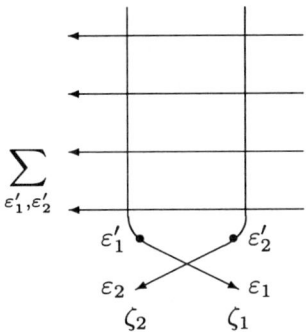

Applying the Yang-Baxter equation to this, we can deform it as follows

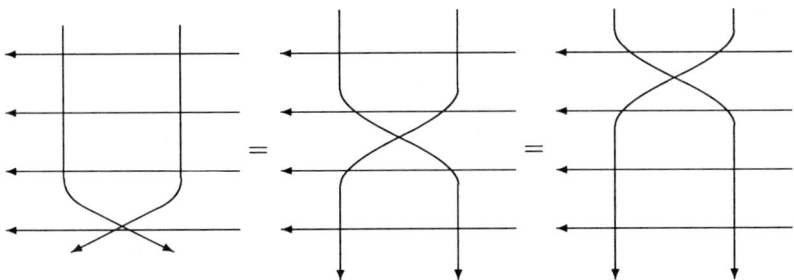

In the limit the vertex representing the insertion of R will disappear to infinity. Because the R is so normalized that the partition function per site is 1, the effect of this insertion produces no extra factor.

The third relation is

$$g \sum_\varepsilon \Phi_\varepsilon^*(\zeta) \Phi_\varepsilon(\zeta) = 1.$$

Here g is a constant depending on q but not ζ. This is not determined for the time being. It depends how we choose the normalization of $\Phi_\varepsilon(\zeta)$. In the previous discussion we used the normalization such that $g = 1$. But this is not appropriate when we make a connection to the representation theory. The operator $\Phi_\varepsilon^*(\zeta)$ is by definition given by

$$\Phi_\varepsilon^*(\zeta) = \Phi_{-\varepsilon}(-q^{-1}\zeta).$$

Graphically, it is given by

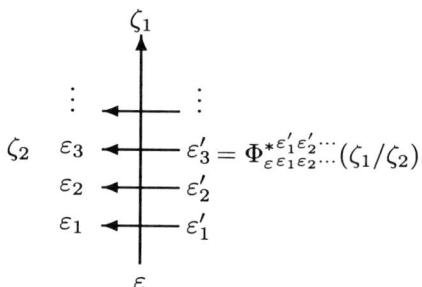

The above equality is a consequence of the crossing symmetry: The relation follows from the unitarity relation as follows.

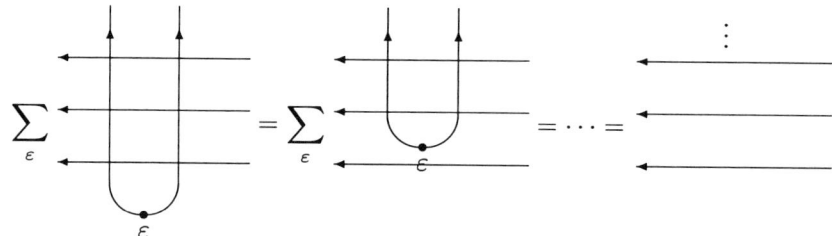

We note that the normalization of R follows from these relations: The unitarity relation follows from the second relation, and the crossing symmetry from the third.

6   Correlation Functions

Now recall the formula

$$\langle \sigma_0 \rangle = \frac{g \sum_{\varepsilon=\pm} \varepsilon \, \mathrm{tr}_{\mathcal{H}^{(1)}} (-q)^{H_{\mathrm{CTM}}} \Phi_{-\varepsilon}(\zeta) (-q)^{H_{\mathrm{CTM}}} \Phi_\varepsilon(\zeta)}{Z}.$$

We put g because we will use a normalization of $\Phi_\varepsilon(\zeta)$ different from the one used in Section 4. We will fix it in Section 7 by requiring $g\sum_\varepsilon \Phi^*_\varepsilon(\zeta)\Phi_\varepsilon(\zeta) = 1$ By using the first commutation relation in Section 5 we have

$$\langle \sigma_0 \rangle = \frac{g \sum_{\varepsilon=\pm} \varepsilon \, \mathrm{tr}_{\mathcal{H}^{(1)}} \, q^{2H_{\mathrm{CTM}}} \Phi^*_\varepsilon(1)\Phi_\varepsilon(1)}{Z}.$$

Now we fix the normalization of $\Phi_\varepsilon(\zeta)$.

Fix a mapping $\nu : \nu(\mathcal{H}^{(i)}) = \mathcal{H}^{(1-i)}$ and vectors $|i\rangle \in \mathcal{H}^{(0)}$ such that

$$H_{\mathrm{CTM}}|i\rangle = 0, \quad \nu|i\rangle = |1-i\rangle,$$
$$\nu\Phi_\varepsilon(\zeta)\nu = \Phi_{-\varepsilon}(\zeta).$$

The mapping ν is the charge conjugation, and the vectors $|i\rangle$ are the lowest eigenvectors of H_{CTM}. We normalize $\Phi_\varepsilon(\zeta)$ by requiring

$$\langle 1|\,\Phi_-(\zeta)\,|0\rangle = \langle 0|\,\Phi_+(\zeta)\,|1\rangle = 1.$$

From the parity of a, b, c, i.e.,

$$a(-\zeta) = -a(\zeta), \quad b(-\zeta) = -b(\zeta), \quad c(-\zeta) = -c(\zeta),$$

it follows that

$$\Phi_\pm(-\zeta) = \begin{cases} \mp\Phi_\pm(\zeta) \in \mathrm{End}_{\mathbb{C}}(\mathcal{H}^{(0)}, \mathcal{H}^{(1)}); \\ \pm\Phi_\pm(\zeta) \in \mathrm{End}_{\mathbb{C}}(\mathcal{H}^{(1)}, \mathcal{H}^{(0)}). \end{cases}$$

Define $\mathcal{H}_d = \{|v\rangle \in \mathcal{H}; H_{\mathrm{CTM}}|v\rangle = d|v\rangle\}$. The commutation relation $\xi^{-H_{\mathrm{CTM}}}\Phi_\varepsilon(\zeta) = \Phi_\varepsilon(\zeta/\xi)\xi^{-H_{\mathrm{CTM}}}$ implies that we can expand $\Phi_\varepsilon(\zeta)$ as

$$\Phi_\varepsilon(\zeta) = \sum_{n\in\mathbb{Z}} \Phi_{\varepsilon,n}\zeta^{-n}$$

where $\Phi_{\varepsilon,n} \in \bigoplus_k \mathrm{End}_{\mathbb{C}}(\mathcal{H}_k, \mathcal{H}_{k-n})$. In other words, the operator $\Phi_{\varepsilon,n}$ has the degree $-n$ with respect to the degree counting by H_{CTM}.

We are interested in the following two kinds of n point functions.

$$F^{(i)}_{\varepsilon_1\cdots\varepsilon_n}(\zeta_1, \ldots, \zeta_n) = \mathrm{tr}_{\mathcal{H}^{(i)}} \, q^{2H_{\mathrm{CTM}}}\Phi_{\varepsilon_1}(\zeta_1)\cdots\Phi_{\varepsilon_n}(\zeta_n) \quad (n: \text{ even}),$$
$$f^{(i)}_{\varepsilon_1\cdots\varepsilon_n}(\zeta_1, \ldots, \zeta_n) = \langle i+n|\,\Phi_{\varepsilon_1}(\zeta_1)\cdots\Phi_{\varepsilon_n}(\zeta_n)\,|i\rangle.$$

By a similar argument as in the case of $\langle \sigma_0 \rangle$ we can deduce

$$\langle \sigma_{n-1}\sigma_{n-2}\cdots\sigma_0 \rangle$$
$$= \frac{g^n}{Z} \sum_{\varepsilon_0,\ldots,\varepsilon_{n-1}=\pm} \varepsilon_0\cdots\varepsilon_{n-1}\,\mathrm{tr}_{\mathcal{H}^{(0)}}\,q^{2H_{\mathrm{CTM}}}\Phi^*_{\varepsilon_0}(1)\cdots\Phi^*_{\varepsilon_{n-1}}(1)$$
$$\times \Phi_{\varepsilon_{n-1}}(1)\cdots\Phi_{\varepsilon_0}(1),$$

if the edges corresponding to $\sigma_{n-1}, \ldots, \sigma_0$ are as follows.

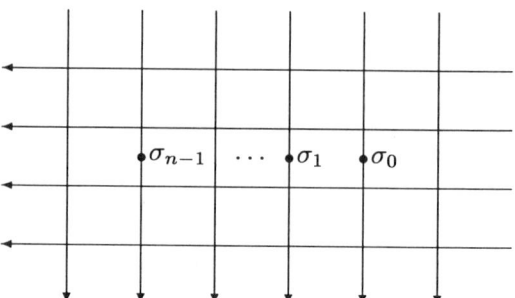

7 Two-Point Functions

Let us summarize what we have obtained so far. We have the space $\mathcal{H} = \mathcal{H}^{(0)} \oplus \mathcal{H}^{(1)}$ with the grading given by H_{CTM}. We have fixed the highest weight vectors $|i\rangle \in \mathcal{H}^{(i)}$ of degree 0. The two spaces $\mathcal{H}^{(i)}$ are connected by $\nu \in \mathrm{End}_{\mathbb{C}}(\mathcal{H})$ in such a way that $\nu(\mathcal{H}^{(i)}) = \mathcal{H}^{(1-i)}$, $\nu(|i\rangle) = |1-i\rangle$. We also use the dual space $\mathcal{H}^* = \mathcal{H}^{(0)*} \oplus \mathcal{H}^{(1)*}$, and the dual highest weight vectors $\langle i| \in \mathcal{H}^{(i)*}$ such that $\langle i|i\rangle = 1$.

We introduced the operators $\Phi_\varepsilon(\zeta) \in \mathrm{End}_{\mathbb{C}}(\mathcal{H})$. They are acting from $\mathcal{H}^{(i)}$ to $\mathcal{H}^{(1-i)}$. We have chosen the normalization $\langle 1|\Phi_-(\zeta)|0\rangle = \langle 1|\Phi_+(\zeta)|1\rangle = 1$. They satisfy $\nu\Phi_\varepsilon(\zeta)\nu = \Phi_{-\varepsilon}(\zeta)$. The following three relations are fundamental.

$$\xi^{-H_{\mathrm{CTM}}}\Phi_\varepsilon(\zeta)\xi^{H_{\mathrm{CTM}}} = \Phi_\varepsilon(\zeta/\xi),$$

$$\sum_{\varepsilon_1',\varepsilon_2'} R_{\varepsilon_1\varepsilon_2}^{\varepsilon_1'\varepsilon_2'}(\zeta_1/\zeta_2)\Phi_{\varepsilon_1'}(\zeta_1)\Phi_{\varepsilon_2'}(\zeta_2) = \Phi_{\varepsilon_2}(\zeta_2)\Phi_{\varepsilon_1}(\zeta_1),$$

$$g\sum_\varepsilon \Phi_\varepsilon^*(\zeta)\Phi_\varepsilon(\zeta) = 1.$$

Here $\Phi_\varepsilon^*(\zeta) = \Phi_{-\varepsilon}(-q^{-1}\zeta)$ and g is a constant to be determined. The constant is determined if we can compute the two-point function

$$f_{\pm\mp}^{(0)}(\zeta_1,\zeta_2) = \langle 0|\,\Phi_\pm(\zeta_1)\Phi_\mp(\zeta_2)\,|0\rangle\,.$$

We have the following equation for $f_{\varepsilon_1\varepsilon_2}^{(0)}(\zeta_1,\zeta_2)$.

$$\sum_{\varepsilon_1',\varepsilon_2'} R_{\varepsilon_1\varepsilon_2}^{\varepsilon_1'\varepsilon_2'}(\zeta_1,\zeta_2)f_{\varepsilon_1'\varepsilon_2'}^{(0)}(\zeta_1,\zeta_2) = f_{\varepsilon_2\varepsilon_1}^{(0)}(\zeta_2,\zeta_1).$$

This follows from the second commutation relation. Set $f(\zeta) = f_{+-}^{(0)}(\zeta_1,\zeta_2) + f_{-+}^{(0)}(\zeta_1,\zeta_2)$ ($\zeta = \zeta_2/\zeta_1$ unlike before). Note that the right-hand side depends only on the ratio. We have $f(-\zeta) = f_{+-}^{(0)}(\zeta_1, \zeta_2) - f_{-+}^{(0)}(\zeta_1, \zeta_2)$ because of the parity relation given in Section 6. Now the above equation is written as

$$\frac{f(\zeta)}{f(\zeta^{-1})} = \frac{1}{\kappa(\zeta)} \frac{\zeta + q}{1 + q\zeta}.$$

There are many solutions to this equation. We can multiply any function of the form $\varphi(\zeta)/\varphi(\zeta^{-1})$. Nevertheless we can determine $f(\zeta)$ by collecting a further information. Because there are no vectors of negative degrees, $f(\zeta)$ is in fact a power series in ζ. We assume that $f(\zeta)$ is holomorphic inside and on the unit circle. Then, the $f(\zeta)$ is uniquely given by

$$f(\zeta) = (1 - q\zeta) \frac{(q^6\zeta^2; q^4)_\infty}{(q^4\zeta^2; q^4)_\infty}.$$

We have

$$f_{+-}^{(0)}(\zeta) = f_{-+}^{(1)}(\zeta) = \frac{(q^6\zeta^2; q^4)_\infty}{(q^4\zeta^2; q^4)_\infty},$$

$$f_{-+}^{(0)}(\zeta) = f_{+-}^{(1)}(\zeta) = -q\zeta \frac{(q^6\zeta^2; q^4)_\infty}{(q^4\zeta^2; q^4)_\infty}.$$

Finally, from $g \sum_\varepsilon \langle 0| \Phi_{-\varepsilon}(-q^{-1}\zeta) \Phi_\varepsilon(\zeta) |0\rangle = 1$, we get

$$g = \frac{(q^4; q^4)_\infty}{(q^2; q^4)_\infty}.$$

Now, the set of commutation relations have been completed. In fact, the operators $\Phi_\varepsilon(\zeta)$ are completely determined by these relations. (See Ref. 8.)

We can also compute $F_{\varepsilon_1\varepsilon_2}^{(1)}(\zeta_1, \zeta_2) = \mathrm{tr}_{\mathcal{H}^{(1)}} q^{2H_{\mathrm{CTM}}} \Phi_{\varepsilon_1}(\zeta_1) \Phi_{\varepsilon_2}(\zeta_2)$. Set $F(\zeta) = F_{+-}^{(1)}(\zeta_1, \zeta_2) + F_{-+}^{(1)}(\zeta_1, \zeta_2)$. It satisfies the same equation

$$\frac{F(\zeta)}{F(\zeta^{-1})} = \frac{1}{\kappa(\zeta)} \frac{\zeta + q}{1 + q\zeta}.$$

Further, using the cyclic property of trace, we get

$$\mathrm{tr}_{\mathcal{H}^{(1)}} q^{2H_{\mathrm{CTM}}} \Phi_{\varepsilon_1}(\zeta_1) \Phi_{\varepsilon_2}(\zeta_2) = \mathrm{tr}_{\mathcal{H}^{(1)}} \Phi_{\varepsilon_1}(q^2\zeta_1) q^{2H_{\mathrm{CTM}}} \Phi_{\varepsilon_2}(\zeta_2)$$
$$= \mathrm{tr}_{\mathcal{H}^{(1)}} q^{2H_{\mathrm{CTM}}} \Phi_{\varepsilon_2}(\zeta_2) \Phi_{\varepsilon_1}(q^2\zeta_1).$$

This implies

$$F(\zeta) = F(q^2\zeta^{-1}).$$

Assuming the analyticity of $\log F(\zeta)$ in the neighborhood of the annulus $|q| \leq \zeta \leq 1$, we get the unique solution (up to constant multiple)

$$F(\zeta) = \frac{1}{(-q^3\zeta^{-1}; q^2)_\infty (-q\zeta; q^2)_\infty} \frac{(q^6z^{-1}; q^4, q^4)_\infty (q^4z; q^4, q^4)_\infty}{(q^4z^{-1}; q^4, q^4)_\infty (q^8z; q^4, q^4)_\infty},$$

where $z = \zeta^2$. By specializing to $\zeta = -q$, we get

$$\langle \sigma_0 \rangle = \frac{F_{-+}^{(1)}(-q^{-1}, 1) - F_{+-}^{(1)}(-q^{-1}, 1)}{F_{-+}^{(1)}(-q^{-1}, 1) + F_{+-}^{(1)}(-q^{-1}, 1)} = \frac{F(q)}{F(-q)} = \prod_{n=1}^{\infty} \left(\frac{1 - q^{2n}}{1 + q^{2n}} \right)^2.$$

This is first obtained by Baxter [9].

8   Discussion

We have shown how the method of a vertex operator is used to obtain the simplest correlation function $\langle \sigma_0 \rangle$ for the six-vertex model. We did not discuss the representation theory of $U_q(\widehat{sl}_2)$. Here is a short summary on that account. The connection of the Hopf algebra $U_q(\widehat{sl}_2)$ to the six-vertex model lies in the fact that the R-matrix for the six-vertex model is an intertwiner of two representations on $V \otimes V$ depending on two parameters ζ_1, ζ_2 [10]. This connection was the key to the discovery of the quantum groups by Drinfeld [11] and Jimbo [12].

The application of the representation theory of $U_q(\widehat{sl}_2)$ to the six-vertex model in the infinite volume limit requires the understanding of the action of $U_q(\widehat{sl}_2)$ on the infinite tensor product of V. The key statement [13] is that we can identify $\mathcal{H}^{(i)}$ ($i = 0, 1$) with the level 1 integrable irreducible highest weight representations.

In this identification, the operator H_{CTM} is identified with the scaling operator $-\rho$ ($\rho = \Lambda_0 + \Lambda_1$ in the usual notation), and the half transfer matrix $\Phi_\varepsilon(\zeta)$ with the intertwiner (or *the vertex operator*) between $\mathcal{H}^{(i)}$ and $\mathcal{H}^{(1-i)} \otimes V$. The commutation relations for $\Phi_\varepsilon(\zeta)$ given in this paper are then derived by representation theory [2]. Good news here is that the transfer matrix can be diagonalized [1, 2, 14]: The space on which the transfer matrix acts is identified with $\mathcal{H} \otimes \mathcal{H}^* = \mathrm{End}_{\mathbb{C}}(\mathcal{H})$ and the transfer matrix itself is given a nice expression

$$T(\zeta)f = g \sum_{\varepsilon} \Phi_\varepsilon(\zeta) f \Phi_{-\varepsilon}(\zeta)$$

for $f \in \mathrm{End}_{\mathbb{C}}(\mathcal{H})$. The vacuum vectors are given simply by

$$|\mathrm{vac}\rangle_i = (-q)^{H_{\mathrm{CTM}}} \in \mathrm{End}(\mathcal{H}^{(i)}).$$

The excited states are obtained by the vertex operators $\Psi_\varepsilon^*(\zeta)$ which intertwine $V \otimes \mathcal{H}^{(i)}$ with $\mathcal{H}^{(1-i)}$

$$|\xi_1, \ldots, \xi_n\rangle_{\varepsilon_1 \ldots \varepsilon_n; i} = \Psi_{\varepsilon_1}^*(\xi_1) \cdots \Psi_{\varepsilon_n}^*(\xi_n)(-q)^{H_{\mathrm{CTM}}}.$$

Correlation functions and form factors are given expressions by means of the traces of the product of the vertex operators. For the computation of these traces, we can exploit the bosonization [3, 15]: The spaces $\mathcal{H}^{(i)}$ ($i = 0$, 1) are constructed as the Fock spaces of bosons, and the operators $\Phi_\varepsilon(\zeta)$, $\Psi_\varepsilon^*(\zeta)$ are explicitly written in terms of bosons. This gives integral representations of the correlation functions and the form factors.

Many aspects of the solvable lattice models are still to be investigated. The content of this paper, that is, the part without the representation theory, has been extended to the eight-vertex model in the anti-ferromagnetic regime [4]. The corresponding representation theory is not yet developed [8, 16]. The study in the massless regime $|\Delta| < 1$ has not yet been done in this approach. There are several works for the higher-level cases [8, 17, 18] and the other models such as the ABF models [19]. But, they are not yet complete. Even in the six-vertex model in the antiferromagnetic regime, there are two major unsolved problems, that is, the completeness of the vectors $|\xi_1, \ldots, \xi_n\rangle_{\varepsilon_1 \ldots \varepsilon_n; i}$ and the complete characterization of local operators in Smirnov's sense [20]. The scaling limit of correlation functions are beyond our scope here.

Acknowledgments: I would like to express my gratitude to the organizers of the Banff summer school, 1994.

9 REFERENCES

1. M. Jimbo and T. Miwa. *Algebraic Analysis of Solvable Lattice Models*, volume 85 of *CBMS Regional Conference Series in Mathematics*. Amer. Math. Soc., Providence, RI, 1995.

2. B. Davies, O. Foda, M. Jimbo, T. Miwa, and A. Nakayashiki. Diagonalization of the XXZ Hamiltonian by vertex operators. *Commun. Math. Phys.*, 151: 89–153, 1993.

3. M. Jimbo, K. Miki, T. Miwa, and A. Nakayashiki. Correlation functions of the XXZ model for $\Delta < -1$. *Phys. Lett. A*, 168: 256–263, 1992.

4. M. Jimbo, T. Miwa, and A. Nakayashiki. Difference equations for the correlation functions of the eight-vertex model. *J. Phys.*, A26: 2199–2209, 1993.

5. Y. G. Stroganov. A new calculation method for partition functions in some lattice models. *Phys. Lett. A*, 74: 116–118, 1979.

6. R. J. Baxter. Solvable eight-vertex model on an arbitrary planar lattice. *Phil. Trans. Royal Soc. London*, 289A: 315–346, 1978.

7. T. Miwa. Quantum symmetries in solvable lattice models. In *Proceeding of Les Houches Summer School*, 1995.

8. A. H. Bougourzi and R. A. Weston. N-point correlation functions of the spin-1 XXZ model. *Nucl. Phys.*, B417: 439–462, 1993.

9. R. J. Baxter. Spontaneous staggered polarization of the F model. *J. Stat. Phys.*, 9: 145–182, 1973.

10. P. P. Kulish and N. Yu. Reshetikhin. Quantum linear problem for the sine-Gordon equation and higher representations. *J. Soviet Math.*, 23: 2435–2441, 1983.

11. V. G. Drinfeld. Quantum groups. In *Proceedings of the International Congress of Mathematicians*, (Berkeley, 1986), 1987. Amer. Math. Soc., Providence, RI, pages 798–820.

12. M. Jimbo. A q-difference analogue of $U(\mathfrak{g})$ and the Yang-Baxter equation. *Lett. Math. Phys.*, 10: 63–69, 1985.

13. O. Foda and T. Miwa. Corner transfer matrices and quantum affine algebras. In *Infinite Analysis*, (Kyoto, 1991), volume 16 of *Advanced Series in Mathematical Physics*, 1992. World Scientific Publishing, River Edge, NJ, pages 279–302.

14. K. Miki. Creation/annihilation operators and form factors of the XXZ model. *Phys. Lett. A*, 186: 217–224, 1994.

15. I. B. Frenkel and N. Jing. Vertex representations of quantum affine algebras. *Proc. Nat. Acad. Sci. USA*, 85: 9373–9377, 1988.

16. O. Foda, K. Iohara, M. Jimbo, R. Kedem, T. Miwa, and H. Yan. An elliptic quantum algebra for $\widehat{sl}_2$. *Lett. Math. Phys.*, 32: 259–268, 1994.

17. M. Idzumi. Level 2 irreducible representations of $U_q(\widehat{sl}_2)$, vertex operators, and their correlations. *Int. J. Mod. Phys.*, A9: 4449–4484, 1994.

18. M. Idzumi, T. Tokihiro, K. Iohara, M. Jimbo, T. Miwa, and T. Nakashima. Quantum affine symmetry in vertex models. *Int. J. Mod. Phys.*, A8: 1479–1511, 1993.

19. M. Jimbo, T. Miwa, and Y. Ohta. Structure of the space of states in RSOS models. *Int. J. Mod. Phys.*, A8: 1457–1477, 1993.

20. F. A. Smirnov. Counting the local fields in sine-Gordon theory. *Nucl. Phys.*, B453: 807–824, 1995.

5

Matrix Models as Integrable Systems

A. Morozov

ABSTRACT The theory of matrix models is reviewed from the point of view of its relation to integrable hierarchies. Determinantal formulas, relation to conformal field models, and the theory of generalized Kontsevich model (GKN) are discussed in some detail. Attention is also paid to the group-theoretical interpretation of τ-functions which allows us to go beyond the restricted set of the (multicomponent) KP and Toda integrable hierarchies.

1 Introduction

The purpose of these notes is to review one of the branches of modern string theory: the theory of matrix models. We put emphasis on their intrinsic integrable structure and almost ignore direct physical applications which are broadly discussed in the literature. Also most of technical details and references are omitted: they can be found in a recent review [1].

Both expressions "matrix models" and "integrability" are somewhat misleading names for the field to be discussed. They refer more to the history of the subject than to its real content. In fact the problem that is actually addressed is that of description of *nonperturbative partition functions* in quantum theory. The term "nonperturbative partition function" is now widely used to denote the generating functional of *all* the *exact* correlation functions in a given quantum model. Such a quantity is given by a functional integral where the weight in the sum over trajectories is defined by effective action, which contains either all the possible (local or nonlocal) counterterms or generic coupling to external fields, so that any correlator can be obtained as derivative with respect to appropriate coupling constants or background fields. These exact generating functionals possess new peculiar properties, resulting from the possibility to perform *arbitrary* change of integration variables in the functional integral. Such properties are never studied in the orthodox quantum field theory because there the freedom to change integration variables is severely restricted by requirements of locality and renormalizability, which lost their role as fundamental principles of physics with the creation of string theory.

Every change of integration variables can be alternatively described as some change of parameters of nonperturbative partition function (i.e., the coupling constants or background fields in effective action). Thus invariance of the integral implies certain relations (Ward identities) between partition functions at different values of parameters. Since all the fields of the model are integrated over, the set of relations is actually exhaustively large: more or less any two sets of parameters are related. Exact formulation of this property is yet unknown. The natural first step in these investigations is to look at the finite-dimensional integrals and then proceed to functional integrals by increasing the number of integrations. In turn, the natural way to do it is to make use of the well-studied "matrix models," which deal with $N \times N$ matrix integrals and their behavior in the large-N limit. It appears that nonperturbative partition functions of matrix models, at least when they can be handled with the presently available techniques, are closely related to "τ-functions," introduced originally in the study of integrable hierarchies. It looks very probable that some crucial characteristics of such partition functions are in fact not so peculiar for these simple examples, but remain true in the absolutely general setting. Extraction of such properties and construction of the adequate notion of "generalized τ-function" is the main task of further work in the theory of "matrix models" and "integrable systems."

One of the most straightforward and still promising approaches is based on interpretation of Ward identities for nonperturbative partition functions as Hirota-like equations (supplemented by a much smaller set of "string equations"), while generalized τ-functions, which are solutions to these equations, are interpreted as group-theoretical objects (generating functionals of all the matrix elements of a group element in particular representation). If successful, this approach can provide description of exact correlation functions in terms of some (originally hidden) symmetry of the given class of theories, thus raising to the new height the relation between physical theories and symmetries, which was the guiding line for development of theoretical physics during the last decades.

Analysis of nonperturbative partition functions is very important for one more reason. Construction of a generating functional is essentially "exponentiation of perturbations," i.e., it deforms original (bare) action of the model. When perturbation parameters (extra coupling constants or background fields) are noninfinitesimal, one obtains an entire *set* of models instead of just the original one. Moreover, the original model is no longer distinguished within this class: nonperturbative partition functions are associated with classes of models, not with a single model. One can easily recognize here realization of the main idea of the string program (see, e.g., Ref. 2). The study of nonperturbative partition functions even for such a simple class as ordinary matrix models can lead to much better understanding of the general idea. As usual, this can help to figure out what the adequate questions are and to develop effective technique to answer these questions.

The purpose of these notes is to briefly illustrate the general ideas with very simple examples. A lot of work is still required to obtain applications to the really interesting problems. Still, simple examples are enough to understand the ideas, and often conceptual level is no less important than that of technical effectiveness.

We begin in the following sections with consideration of the simplest matrix models, to be referred to as discrete and Kontsevich (continuous) models. Again, these names reflect more the history than the real content of the subject (continuous models were originally described nonexplicitly as specific (multiscaling) large-N limits of the discrete ones). In the context of nonperturbative partition functions the difference is that discrete models possess effective actions with all the possible counterterms added, while in the Kontsevich model an external field (source) is introduced. Analysis of these theories includes their characterization as *eigenvalue* models and derivation of related determinant representations. We also consider description of discrete models in the language of conformal field theory. It is important for connecting these matrix models to the physically relevant Liouville theory of $2d$ gravity and—more essential for our notes—to the concept of KP and Toda τ-functions.

Then we turn to the free-fermion (Grassmannian) description of the KP/Toda τ-functions and list some results from the theory of the generalized Kontsevich model [3–8]. Some other interesting models, which look *a priori* very different, are in fact just particular examples of the Kontsevich model, which actually describes a big family of theories. This example should be very instructive for the future understanding of the interplay between perturbative and nonperturbative information contained in the nonperturbative partition function.

The last topic of these notes concerns group-theoretical interpretation of τ-functions. The natural object, which arises in this way, while possessing the most important properties of conventional τ-functions, is in fact much more general. In this framework the KP/Toda τ-functions are associated with fundamental representations of SL(N) and the closely related theory of the simply laced Kac-Moody algebras of level $k = 1$. In fact the τ-function can be easily defined for *any* representation of *any* group (including also quantum groups—this can be important for the future construction of string field theory, where the idea of "third quantization" requires consideration of operator-valued τ-functions; since it takes our nonperturbative partition function as input—effective action of the full string field theory).

2   The Basic Example: Discrete 1-Matrix Model

The sample example of matrix model is that of 1-matrix integral

$$Z_N\{t\} \equiv c_N \int_{N \times N} dH e^{\sum_{k=0}^{\infty} t_k \operatorname{Tr} H^k}, \qquad (2.1)$$

where the integral is over $N \times N$ matrix H and $dH = \prod_{i,j} dH_{ij}$.[1] This measure is invariant under the conjugation $H \to UHU^\dagger$ with any unitary $N \times N$ matrix U, and the "action" $\sum_{k=0}^{\infty} t_k \operatorname{Tr} H^k$ in (2.1) is the most general one consistent with this invariance. Thus $Z_N\{t\}$ is indeed an example of the nonperturbative partition function in the sense described in the introduction of this chapter. All observables in the theory are given by algebraic combinations of $\operatorname{Tr} H^k$, and their correlation functions can be obtained by action of t_k-derivatives on $Z_N\{t\}$. Our goal now is to find some more invariant description of this quantity, not so specific as the matrix integral (2.1).

2.1   Ward Identities

Such description is provided by Ward identities. The integral is invariant under any change-of-integration matrix-variable $H \to f(H)$. It is convenient to choose the special basis in the space of such transformations:

$$\delta H = \epsilon_n H^{n+1}.$$

Here ϵ_n is some infinitesimal matrix and, of course, $n \geq -1$. Invariance of the integral implies the following identity:

$$\int_{N \times N} dH e^{\sum_{k=0}^{\infty} t_k \operatorname{Tr} H^k} = \int d(H + \epsilon_n H^{n+1}) e^{\sum_{k=0}^{\infty} t_k \operatorname{Tr}(H + \epsilon_n H^{n+1})^k},$$

i.e.,

$$\int dH e^{\sum_{k=0}^{\infty} t_k \operatorname{Tr} H^k} \left(\sum_{k=0}^{\infty} k t_k \operatorname{Tr} H^{k+n} + \operatorname{Tr} \frac{\delta H^{n+1}}{\delta H} \right) \equiv 0. \qquad (2.2)$$

In order to evaluate the Jacobian $\operatorname{Tr}(\delta H^{n+1}/\delta H)$ let us restore the matrix indices:

$$(\delta H^{n+1})_{ij} = \sum_{k=0}^{n} (H^k \delta H H^{n-k})_{ij} = \sum_{k=0}^{n} (H^k)_{il} (\delta H)_{lm} (H^{n-k})_{mj},$$

and to obtain $\operatorname{Tr}(\delta H^{n+1}/\delta H)$ put $l = i$ and $m = j$, so that

$$\operatorname{Tr} \frac{\delta H^{n+1}}{\delta H} = \sum_{k=0}^{n} \operatorname{Tr} H^k \operatorname{Tr} H^{n-k}. \qquad (2.3)$$

---

[1] This integral is often referred to as *Hermitian*. In most of our considerations we do not need to specify integration contours in matrix integrals; in particular eigenvalues of H_{ij} do not need to be real. What is indeed important is the "flatness" of the measure $dH = \prod_{i,j} dH_{ij}$. Below "Hermitian" (as opposed, for example, to "unitary") will imply just this choice of the measure and not any reality condition.

Any correlation function can be obtained as variation of the coupling constants:

$$\langle \operatorname{Tr} H^{a_1} \cdots \operatorname{Tr} H^{a_n} \rangle = \int dH e^{\sum_{k=0}^{\infty} t_k \operatorname{Tr} H^k} \operatorname{Tr} H^{a_1} \cdots \operatorname{Tr} H^{a_n}$$

$$= \frac{\partial^n}{\partial t_{a_1} \cdots \partial t_{a_n}} Z_N\{t\}.$$

This relation together with (2.3) can be used to rewrite (2.2) as:

$$L_n Z_N\{t\} = 0, \quad n \geq -1, \tag{2.4}$$

with

$$L_n \equiv \sum_{k=0}^{\infty} k t_k \frac{\partial}{\partial t_{k+n}} + \sum_{k=0}^{n} \frac{\partial^2}{\partial t_k \partial t_{n-k}}. \tag{2.5}$$

Note that according to the definition (2.1)

$$\frac{\partial}{\partial t_0} Z_N = N Z_N.$$

2.1.1   Details and Comments

Several remarks are now in order.

First of all, the expression in brackets in (2.2) represents just *all* the equations of motion for the model (2.1), and (2.4) is nothing but another way to represent the same set of equations. This example illustrates what an "exhaustively large" set of Ward identities is: it should be essentially the same as the set of all equations of motion.

Second, commutator of any two operators L_n appearing in (2.4) should also annihilate $Z_N\{t\}$. Another indication that we already have a *complete* set of constraints is that L_n's form a closed algebra:

$$[L_n, L_m] = (n - m) L_{n+m}, \quad n, m \geq -1. \tag{2.6}$$

Its particular representation (2.5) is referred to as "discrete Virasoro algebra" (to emphasize the difference with "continuous Virasoro" constraints; see eq. (3.35) below).

Third, (2.4) can be considered as invariant formulation of what is Z_N: it is a solution of this set of compatible differential equations. From this point of view eq. (2.1) is rather a particular representation of Z_N and it is sensible to look for other representations as well (we shall later discuss two of them: one in terms of CFT, another in terms of Kontsevich integrals).

Fourth, one can try to analyze the uniqueness of the solutions to (2.4). If there are not too many solutions, the set of constraints can be considered complete. A natural approach to classification of solutions to the algebra of constraints is in terms of the orbits of the corresponding group [9]. Let us consider an oversimplified example, which can still be useful to understand implications of the complete set of WI as well as to clarify the meaning of classes of universality and of integrability.

Imagine, that instead of (2.4) with L_n's defined in (2.5) we would obtain somewhat simpler equations[2]:

$$l_n Z = 0, \ n \geq 0, \quad \text{with } l_n = \sum_{k=1}^{\infty} k t_k \frac{\partial}{\partial t_{k+n}}.$$

Then operator l_1 can be interpreted as generating the shifts

$$t_2 \to t_2 + \epsilon_1 t_1,$$
$$t_3 \to t_3 + 2\epsilon_1 t_2,$$
$$\vdots$$

We can use it to shift t_2 to zero, and $l_1 Z = 0$ then implies that

$$Z(t_1, t_2, t_3, \dots) = Z(t_1, 0, \tilde{t}_3, \dots)$$

$(\tilde{t}_k = t_k - (k-1)t_2 t_{k-1}/t_1, \ k \geq 3)$.

Next, operator l_2 generates the shifts

$$t_3 \to t_3 + \epsilon_2 t_1,$$
$$t_4 \to t_4 + 2\epsilon_2 t_2,$$
$$\vdots$$

and does *not* affect t_2. We can now use $l_2 Z = 0$ to argue that

$$Z(t_1, t_2, t_3, t_4, \dots) = Z(t_1, 0, \tilde{t}_3, \tilde{t}_4, \dots) = Z(t_1, 0, 0, \tilde{\tilde{t}}_4, \dots)$$

and so on. Assuming that Z is not very much dependent on t_k with $k \to \infty$,[3] we can conclude, that

$$Z(t_1, t_2, t_3, \dots) = Z(t_1, 0, 0, \dots) = Z(1, 0, 0, \dots)$$

(at the last step we also used the equation $l_0 Z = 0$ to rescale t_1 to unity).

---

[2]One can call them "classical" approximation to (2.4), since they would arise if the variation of measure (i.e., a "quantum effect") was not taken into account in the derivation of (2.4). Though this concept is often used in physics it does not have much sense in the present context, when we are analyzing *exact* properties of functional (matrix) integrals.

[3]This, by the way, is hardly correct in this particular example, when the group has no compact orbits.

All this reasoning had sense provided $t_1 \neq 0$. Otherwise we would get $Z(0,1,0,0,\dots)$, if $t_1 = 0$, $t_2 \neq 0$, or $Z(0,0,1,0,\dots)$, if $t_1 = t_2 = 0$, $t_3 \neq 0$, etc. In other words, we obtain classes of universality (such that the value of partition function is just the same in the whole class), which in this over-simplified example are labeled just by the first nonvanishing time-variable. Analysis of the orbit structure for the actually important realizations of groups, like that connected to eq. (2.5), has never been performed in the context of matrix model theory.

In this oversimplified case the constraints actually allow one to eliminate all the dependence on the time-variables, i.e., to solve equations for Z exactly. In realistic examples one deals with less trivial representations of the constraint algebra, like (2.6). It appears that in this general framework constraints somehow imply the integrability structure of the model, what can thus be considered as a slightly more complicated version of the same solvability phenomenon.

2.2   CFT Interpretation of 1-Matrix Model

Given a complete set of the constraints on a partition function of infinitely many variables which form some closed algebra, we can now ask an inverse question: how these equations can be solved or what is the integral representation of partition function. One approach to this problem is analysis of orbits, briefly mentioned at the end of the previous subsection. Now we turn to another technique [10], which makes use of the knowledge from conformal field theory. This constructions can have some meaning from the "physical" point of view, which implies certain duality between the two-dimensional world surfaces and the spectral surfaces, associated to configuration space of the string theory. However, our present goal is more formal than discussion of this duality: we are going to use the methods of CFT for solving the constraint equations.

This is especially natural when the algebra of constraints is Virasoro algebra, as is the case with the 1-matrix model, or some other algebra known to arise as a chiral algebra in some simple conformal models. In fact the approach to be discussed is rather general and can be applied to construction of matrix models, associated with many different algebraic structures: the only requirement is existence of the (massless) free-field representation.

We begin from the set of "discrete Virasoro constraints" (2.4). The CFT formulation of interest should provide the solution to these equations in the form of some correlation function in some conformal field theory. Of course, it becomes natural if we somehow identify the operators L_n (2.5) with the harmonics of a stress-tensor T_n, which satisfy the same algebra, and manage to relate the constraint that L_n annihilate the correlator to the statement that T_n annihilate the vacuum state. Thus the procedure is naturally split into two steps. First, we should find a t-dependent operator

("Hamiltonian") $H(t)$, such that

$$L_n(t)\langle e^{H(t)}| = \langle e^{H(t)}|T_n$$

This will relate differential operators L_n to T_n's expressed through the fields of conformal model. Second, we need to enumerate the states, that are annihilated by the operators T_n with $n \geq -1$, i.e., solve equation

$$T_n|G\rangle = 0$$

for the ket-states, what is an internal problem of conformal field theory. If both ingredients $H(t)$ and $|G\rangle$ are found, solution to the problem is given by

$$\langle e^{H(t)} | G\rangle.$$

To be more explicit, for the case of the discrete Virasoro constraints we can just look for solutions in terms of the simplest possible conformal model: that of a one holomorphic scalar field

$$\phi(z) = \hat{q} + \hat{p}\log z + \sum_{k \neq 0} \frac{J_{-k}}{k}z^k$$

$$[J_n, J_m] = n\delta_{n+m,0}, \quad [\hat{q}, \hat{p}] = 1.$$

Then the procedure is as follows. Define vacuum states

$$J_k|0\rangle = 0, \quad \langle N|J_{-k} = 0, \quad k > 0,$$
$$\hat{p}|0\rangle = 0, \quad \langle N|\hat{p} = N\langle N|,$$

the stress-tensor

$$T(z) = \frac{1}{2}[\partial\phi(z)]^2 = \sum T_n z^{-n-2},$$

$$T_n = \sum_{k>0} J_{-k}J_{k+n} + \frac{1}{2}\sum_{\substack{a+b=n \\ a,b\geq 0}} J_a J_b,$$

and the Hamiltonian

$$H(t) = \frac{1}{\sqrt{2}}\sum_{k>0} t_k J_k = \frac{1}{\sqrt{2}}\oint_{C_0} U(z)J(z),$$

$$U(z) = \sum_{k>0} t_k z^k, \quad J(z) = \partial\phi(z).$$

It is easy to check that

$$L_n\langle N|e^{H(t)} = \langle N|e^{H(t)}T_n$$

and

$$T_n|0\rangle = 0, \quad n \geq -1.$$

As an immediate consequence, any correlator of the form

$$Z_N\{t \mid G\} = \langle N|e^{H(t)}G|0\rangle \tag{2.7}$$

gives a solution to (2.4) provided

$$[T_n, G] = 0, \quad n \geq -1.$$

In fact operators G that commute with the stress tensor are well known: these are just any functions of the "screening charges"[4]

$$Q_\pm = \oint J_\pm = \oint e^{\pm\sqrt{2}\phi}.$$

The correlator (2.7) will be nonvanishing only if the matching condition for zero-modes of ϕ is satisfied. If we demand the operator to depend only on Q_+, this implies that only one term of the expansion in powers of Q_+ will contribute to (2.7), so that the result is essentially independent on the choice of the function $G(Q_+)$, we can, for example, take $G(Q_+) = e^{Q_+}$ and obtain

$$Z_N\{t\} \sim \frac{1}{N!}\langle N|e^{H(t)}(Q_+)^N|0\rangle. \tag{2.8}$$

This correlator is easy to evaluate using the Wick theorem and the propagator $\phi(z)\phi(z') \sim \log(z - z')$. Finally we get

$$Z_N\{t\} = \frac{1}{N!}\langle N|:e^{(\oint_{C_0} U(z)\partial\phi(z))/\sqrt{2}}: \prod_{i=1}^N \oint_{C_i} dz_i : e^{\sqrt{2}\phi(z_i)}:|0\rangle$$

$$= \frac{1}{N!} \prod_{i=1}^N \oint_{C_i} dz_i e^{U(z_i)} \prod_{i<j}^N (z_i - z_j)^2 \tag{2.9}$$

in the form of a multiple integral. This integral does not yet look like the matrix integral (2.1). However, it is the same: (2.9) is an "eigenvalue representation" of matrix integral; see Ref. 11 and eq. (2.18) in Section 2.3.

2.2.1 Details and Comments

Thus in the simplest case we resolved the inverse problem: reconstructed an integral representation from the set of discrete Virasoro constraints.

---

[4]For notational simplicity we omit the normal ordering signs; in fact, the relevant operators are $:e^H:$ and $:e^{\pm\sqrt{2}\phi}:$.

However, the answer we got seems a little more general than (2.1) and (2.18): the r.h.s. of eq. (2.9) still depends on the contours of integration. Moreover, we can also recall that the operator G above could depend not only on Q_+, but also on Q_-. The most general formula is a little more complicated than (2.9):

$$Z_N\{t \mid C_i, C_r\} \sim \frac{1}{(N+M)!M!} \langle N | e^{H(t)} (Q_+)^{N+M} (Q_-)^M | 0 \rangle$$

$$= \frac{1}{(N+M)!M!} \prod_{i=1}^{N+M} \oint_{C_i} dz_i e^{U(z_i)} \prod_{r=1}^{M} \oint_{C_r'} dz_r' e^{U(z_r')}$$

$$\cdot \frac{\prod_{i<j}^{N+M} (z_i - z_j)^2 \prod_{r<s}^{N} (z_r' - z_s')^2}{\prod_i^{N+M} \prod_r^{M} (z_i - z_r)^2}. \qquad (2.10)$$

We refer to the papers [10] for discussion of the issue of contour-dependence. In a certain sense all these different integrals can be considered as branches of the same analytical function $Z_N\{t\}$. Dependence on M is essentially eliminated by Cauchy integration around the poles in denominator in (2.10).

The above construction can be straightforwardly applied to any other algebras of constraints, provided:

(i) The free-field representation of the algebra is known in the CFT-framework, such that the generators are *polynomials* in the fields ϕ (only in such case it is straightforward to construct a Hamiltonian H, which relates CFT-realization of the algebra to that in terms of differential operators w.r. to the t-variables; in fact under this condition H is usually linear in t's and ϕ's). There are examples (like the Frenkel-Kac representation of level $k = 1$ simply laced Kac-Moody algebras [12] or generic reductions of the WZNW model [13–17]) when generators are *exponents* of free fields; then this construction should be slightly modified.

(ii) It is easy to find a vacuum, annihilated by the relevant generators (here, e.g., is the problem with application of this approach to the case of "continuous" Virasoro and W-constraints). The resolution to this problem involves consideration of correlates on Riemann surfaces with nontrivial topologies, often of infinite genus.

(iii) The free-field representation of the "screening charges," i.e., operators that commute with the generators of the group within the conformal model, is explicitly known.

These conditions are fulfilled in many cases in CFT, including conventional **W**-algebras [18] and $\mathcal{N} = 1$[5] supersymmetric models.

---

[5]In the case of $\mathcal{N} = 2$ supersymmetry a problem arises because of the lack of reason-

For illustrative purposes we present here several formulas from the last paper of Ref. 10 for the case of the $\mathbf{W}_{r+1}$-constraints, associated with the simply laced algebras $\mathcal{G}$ of rank r.

Partition function in such "conformal multimatrix model" is a function of "time-variables" $t_k^{(\lambda)}$, $k = 0, \ldots, \infty$, $\lambda = 1, \ldots, r = \mathrm{rank}\,\mathcal{G}$, and also depends on the integer-valued r-vector $\vec{N} = \{N_1 \cdots n_r\}$. The $\mathbf{W}_{r+1}$-constraints imposed on partition function are:

$$W_n^{(a)}(t) Z_{\vec{N}}^{\mathcal{G}}\{t\} = 0, \quad n \geq 1 - a, a = 2, \ldots, r+1.$$

The form of the W-operators is somewhat complicated, for example, in the case of $r + 1 = 3$ (i.e., for $\mathcal{G} = \mathrm{SL}(3)$)

$$W_n^{(2)} = \sum_{k=0}^{\infty} \left(k t_k \frac{\partial}{\partial t_{k+n}} + k \bar{t}_k \frac{\partial}{\partial \bar{t}_{k+n}} \right) + \sum_{a+b=n} \left(\frac{\partial^2}{\partial t_a \partial t_b} + \frac{\partial^2}{\partial \bar{t}_a \partial \bar{t}_b} \right)$$

$$W_n^{(3)} = \sum_{k,l>0} \left(k t_k l t_l \frac{\partial}{\partial t_{k+n+l}} - k \bar{t}_k l \bar{t}_l \frac{\partial}{\partial t_{k+n+l}} - 2 k t_k l \bar{t}_l \frac{\partial}{\partial \bar{t}_{k+n+l}} \right)$$

$$+ 2 \sum_{k>0} \left[\sum_{a+b=n+k} \left(k t_k \frac{\partial^2}{\partial t_a \partial t_b} - k t_k \frac{\partial^2}{\partial \bar{t}_a \partial \bar{t}_b} - 2 k \bar{t}_k \frac{\partial^2}{\partial t_a \partial \bar{t}_b} \right) \right]$$

$$+ \frac{4}{3} \sum_{a+b+c=n} \left(\frac{\partial^3}{\partial t_a \partial t_b \partial t_c} - \frac{\partial^3}{\partial t_a \partial \bar{t}_b \partial \bar{t}_c} \right),$$

and two types of time-variables, denoted through t_k and $\bar{t}_k$. are associated with two *orthogonal* directions in the Cartan plane of A_2: $\mathbf{e} = \vec{\alpha}_1/\sqrt{2}$, $\bar{\mathbf{e}} = \sqrt{3}\vec{\nu}_2/\sqrt{2}$[6].

All other formulas, however, are very simple: The conformal model is usually that of the r free fields, $S \sim \int \bar{\partial}\vec{\phi}\,\partial\vec{\phi}\,d^2z$, which is used to describe representation of the level one Kac-Moody algebra, associated with $\mathcal{G}$. The Hamiltonian

$$H(t^{(1)} \cdots t^{(r+1)}) = \sum_{\lambda=1}^{r+1} \sum_{k>0} t_k^{(\lambda)} \vec{\mu}_\lambda \vec{J}_k, \tag{2.11}$$

where $\{\vec{\mu}_\lambda\}$ are associated with "fundamental weight" vectors $\vec{\nu}_\lambda$ in the Cartan hyperplane and in the simplest case of $\mathcal{G} = \mathrm{SL}(r + 1)$ satisfy

$$\vec{\mu}_\lambda \cdot \vec{\mu}_{\lambda'} = \delta_{\lambda\lambda'} - \frac{1}{r+1}, \quad \sum_{\lambda=1}^{r+1} \vec{\mu}_\lambda = 0.$$

---

able screening charges. At the most naive level the relevant operator to be integrated over superspace (over $dz d^{\mathcal{N}}\theta$) in order to produce screening charge has dimension $1 - \frac{1}{2}\mathcal{N}$, which *vanishes* when $\mathcal{N} = 2$.

[6]Such orthogonal basis is especially convenient for discussion of integrability properties of the model. These t and $\bar{t}$ are linear combinations of time-variables t_k^λ appearing in eqs. (2.11) and (2.12).

Thus only r of the time variables $t^{(1)}, \ldots, t^{(r+1)}$ are linearly independent. Relation between differential operators $W_n^{(a)}(t)$ and operators $\mathrm{W}_n^{(a)}$ in the CFT is now defined by

$$W_n^{(a)} \langle \vec{N} | e^{H(t)} = \langle \vec{N} | e^{H(t)} \mathrm{W}_i^{(a)}, \quad a = 2, \ldots, p; \ i \geq 1 - a,$$

where

$$\mathrm{W}_n^{(a)} = \oint z^{a+n-1} \mathrm{W}^{(a)}(z),$$

$$\mathrm{W}^{(a)}(z) = \sum_\lambda [\vec{\mu}_\lambda \partial \vec{\phi}(z)]^a + \cdots,$$

are spin-a generators of the $\mathbf{W}_{r+1}^{\mathcal{G}}$ algebra. The screening charges that commute with all the $\mathrm{W}^{(a)}(z)$ are given by

$$Q^{(\alpha)} = \oint J^{(\alpha)} = \oint e^{\vec{\alpha}\vec{\phi}}$$

$\{\vec{\alpha}\}$ being roots of the finite-dimensional simply laced Lie algebra $\mathcal{G}$.

Thus partition function arises in the form:

$$Z_{\vec{N}}^{\mathcal{G}}\{t\} = \langle \vec{N} | e^{H(^t} G\{Q^{(\alpha)}\} | 0 \rangle$$

where G is an exponential function of screening charges. Evaluation of the free-field correlator gives:

$$Z_{\vec{N}}^{\mathcal{G}}\{t\} \sim \int \prod_\alpha \left[\prod_{i=1}^{N_\alpha} dz_i^{(\alpha)} \exp\left(\sum_{\lambda; k > 0} t_k^{(\lambda)} (\vec{\mu}_\lambda \vec{\alpha})(z_i^{(\alpha)})^k \right) \right]$$
$$\times \prod_{(\alpha,\beta)} \prod_{i=1}^{N_\alpha} \prod_{j=1}^{N_\beta} (z_i^{(\alpha)} - z_j^{(\beta)})^{\vec{\alpha}\vec{\beta}} \quad (2.12)$$

In fact this expression can be rewritten in terms of an r-matrix integral—a "conformal multimatrix model":

$$Z_{\vec{N}}^{\mathcal{G}}\{t^{(\alpha)}\} = c_N^{p-1} \int_{N \times N} dH^{(1)} \cdots dH^{(p-1)} \prod_{\alpha=1}^{p-1} e^{\sum_{k=0}^{\infty} t_k^{(\alpha)} \mathrm{Tr}\, H_{(\alpha)}^k}$$
$$\cdot \prod_{(\alpha,\beta)} \mathrm{Det}\left(H^{(\alpha)} \otimes I - I \otimes H^{(\alpha+1)} \right)^{\vec{\alpha}\vec{\beta}} \quad (2.13)$$

In the simplest case of $\mathbf{W}_3$ algebra eq. (2.12) with insertion of only two (of the six) screenings Q_{α_1} and Q_{α_2} turns into

$$Z_{N_1,N_2}^{\mathrm{SL}(3)}(t, \bar{t}) = \frac{1}{N_1! N_2!} \langle N_1, N_2 | e^{H(t, \bar{t})} (Q^{(\alpha_1)})^{N_1} (Q^{(\alpha_2)})^{N_2} | 0 \rangle$$
$$= \frac{1}{N_1! N_2!} \prod_i \int dx_i e^{U(x_i)}$$
$$\times \prod_j \int dy_j e^{\bar{U}(y_i)} \Delta(x) \Delta(x, y) \Delta(y), \quad (2.14)$$

where $\Delta(x,y) \equiv \Delta(x)\Delta(y) \prod_{i,j}(x_i - y_j)$. This model is associated with the algebra $\mathcal{G} = \mathrm{SL}(3)$, while the original 1-matrix model (2.8)–(2.10)—with $\mathcal{G} = \mathrm{SL}(2)$.

The whole series of models (2.12)–(2.13) for $\mathcal{G} = \mathrm{SL}(r+1)$ is distinguished by its relation to the level $k = 1$ simply laced Kac-Moody algebras. In this particular situation the underlying conformal model has integer central charge $c = r = \mathrm{rank}\,\mathcal{G}$ and can be "fermionized."[7] The main feature of this formulation is that the Kac-Moody currents (which after integration turn into "screening charges" in the above construction) are quadratic in fermionic fields, while they are represented by exponents in the free-boson formulation.

In fact fermionic (spinor) model naturally possesses $\mathrm{GL}(r + 1)$ rather than $\mathrm{SL}(r + 1)$ symmetry (other simply-laced algebras can be embedded into larger GL-algebras and this provides fermionic description for them in the case of $k = 1$). The model contains $r + 1$ spin-$\frac{1}{2}$ fields ψ_i and their conjugate $\tilde{\psi}_i$ (b, c-systems);

$$S = \sum_{j=1}^{r+1} \int \tilde{\psi}_j \bar{\partial} \psi_j d^2 z,$$

central charge $c = r + 1$, and operator algebra is

$$\tilde{\psi}_j(z)\psi_k(z') = \frac{\delta_{jk}}{z - z'} + :\tilde{\psi}_j(z)\psi_k(z'):$$
$$\psi_j(z)\psi_k(z') = (z - z')\delta_{jk}:\psi_j(z)\psi_k(z'): + (1 - \delta_{jk}):\psi_j(z)\psi_k(z'):$$
$$\tilde{\psi}_j(z)\tilde{\psi}_k(z') = (z - z')\delta_{jk}:\tilde{\psi}_j(z)\tilde{\psi}_k(z'): + (1 - \delta_{jk}):\tilde{\psi}_j(z)\tilde{\psi}_k(z'):.$$

The Kac-Moody currents of the level-one $\widehat{\mathrm{GL}(r+1)}_{k=1}$ are just $J_{jk} = :\tilde{\psi}_j\psi_k:$, $j, k = 1, \ldots, r+1$, and screening charges are $Q^{(\alpha)} = iE_{jk}^{(\alpha)} \oint :\tilde{\psi}_j\psi_k:$, where $E_{jk}^{(\alpha)}$ are representatives of the roots $\vec{\alpha}$ in the matrix representation of $\mathrm{GL}(r + 1)$. The Cartan subalgebra is represented by J_{jj}, while positive and negative Borel subalgebras—by J_{jk} with $j < k$ and $j > k$, respectively. In eq. (2.10) $Q_+ = i \oint \tilde{\psi}_1\psi_2$, $Q_- = i \oint \tilde{\psi}_2\psi_1$ while in eq. (2.14) $Q^{(\alpha_1)} = i \oint \tilde{\psi}_1\psi_2$, $Q^{(\alpha_2)} = i \oint \tilde{\psi}_1\psi_3$ (and $Q^{(\alpha_3)} = i \oint \tilde{\psi}_2\psi_3$, $Q^{(\alpha_4)} = i \oint \tilde{\psi}_2\psi_1$, $Q^{(\alpha_5)} = i \oint \tilde{\psi}_3\psi_1$, $Q^{(\alpha_6)} = i \oint \tilde{\psi}_3\psi_2$). $Q^{(\alpha_6)}$ can be substituted instead of $Q^{(\alpha_2)}$ in (2.14) without changing the answer. For generic r the similar choice of "adjacent" (not simple!) roots (such that

---

[7]This is possible only for very special Kac-Moody algebras, and such formulation is important in order to deal with *conventional* formulation of integrability, which usually involves *commuting* Hamiltonian flows (not just a closed algebra of flows) and fermionic realization of the universal module space (universal Grassmannian). In fact these restrictions are quite arbitrary and can be removed (though this is not yet done in full details); see Ref. 1 and Sections 4 and 5 below for more detailed discussion.

their scalar products are $+1$ or 0) leads to selection of the following r screening operators $Q^{(1)} = i \oint \tilde{\psi}_1 \psi_2$, $Q^{(2)} = -i \oint \psi_2 \tilde{\psi}_3$, $Q^{(3)} = i \oint \tilde{\psi}_3 \psi_4$, ..., i.e., $Q^{(j)} = i \oint \tilde{\psi}_j \psi_{j+1}$ for odd j and $Q^{(j)} = -i \oint \psi_j \tilde{\psi}_{j+1}$ for even j.

2.3   1-Matrix Model in Eigenvalue Representation

In the last section we found the solution of Virasoro constraints (2.4) in the form of the multiple integral (2.9). Now we shall see that this expression is in fact equal to the original matrix integral (2.1) and arises after "auxiliary" angular variables are explicitly integrated out. These angular variables are in fact the ones associated with physical vector bosons and the possibility to solve this sector of the model explicitly is peculiar feature of the simplest class of matrix models, naturally named *eigenvalue models*. The theory of eigenvalue models is in a sense equivalent to the theory of conventional integrable hierarchies and thus is rather straightforward to work on. Inclusion of nontrivial angular integration in the general scheme is still a sophisticated task, with no universal solution found so far. Also unknown is solution of the inverse problem: how can an arbitrary eigenvalue model—with integration over eigenvalues of some matrix—be "lifted to" the full matrix model where integration goes over the entire matrix[8]; in other words, what is the way to couple vector bosons to the "topological" eigenvalue sector so that the two sectors are interacting only in the "solvable" fashion and angular integrations can be easily performed.

Let us now turn to particular example of the 1-matrix integral (2.1). First of all, this model possesses gauge symmetry, associated with the unitary (angular) rotation of matrices, $H_\alpha \to U_\alpha^\dagger H_\alpha U_\alpha$. This illustrates the general phenomenon: matrix models are usually *gauge* theories. In the case of eigenvalue models this symmetry is realized without "gauge fields" $V_{\alpha\beta}$, which would depend on pairs of indices α, β and transform like $V_{\alpha\beta} \to U_\alpha^\dagger V_{\alpha\beta} U_\beta$. In other words, eigenvalue models are gauge theories without gauge fields, i.e., they are purely topological. The case of the 1-matrix model (2.1)

$$Z_N\{t\} \equiv c_N \int_{N \times N} dH e^{\sum_{k=0}^{\infty} t_k \, \mathrm{Tr}\, H^k} \qquad (2.15)$$

is especially simple, because separation of eigenvalue and angular variables does not involve any information about unitary-matrix integrals. Take

$$H = U^\dagger D U, \qquad (2.16)$$

where U is a unitary matrix and diagonal matrix $D = \mathrm{diag}(h_1, \ldots, h_N)$ has

---

[8]See also Ref. 19.

eigenvalues of H as its entries. Then integration measure[9]

$$dH = \prod_{i,j=1}^{N} dH_{ij} = \frac{[dU]}{[dU_{\text{Cartan}}]} \prod_{i-1}^{N} dh_i \Delta^2(h),   \tag{2.17}$$

where the "Vandermonde determinant" $\Delta(h) \equiv \det_{(ij)} h_i^{j-1} = \prod_{i>j}^{N}(h_i - h_j)$ and $[dU]$ is the Haar measure of integration over unitary matrices.

It remains to note that the "action" $\operatorname{Tr} U(H) \equiv \sum_{k=0}^{\infty} t_k \operatorname{Tr} H^k$ with H substituted in the form (2.16) is independent of U:

$$\operatorname{Tr} U(H) = \sum_{i=1}^{N} U(h_i).$$

Thus

$$Z_N\{t\} = \frac{1}{N!} \prod_{i=1}^{N} \int dh_i e^{U(h_i)} \prod_{i>j}^{N} (h_i - h_j)^2$$

$$= \frac{1}{N!} \prod_{i=1}^{N} \int dh_i e^{U(h_i)} \Delta^2(h),   \tag{2.18}$$

provided c_N in (2.1) and (2.15) is chosen to be

$$c_N^{-1} = N! \frac{\operatorname{Vol}_{U(N)}}{(\operatorname{Vol}_{U(1)})^N},$$

---

[9]In order to derive eq. (2.17) one can consider the norm of infinitesimal variation

$$\|\delta H\|^2 \equiv \sum_{i,j=1}^{N} |\delta H_{ij}|^2 = \sum_{i,j=1}^{N} \delta H_{ij} \delta H_{ji} = \operatorname{Tr}(\delta H)^2$$
$$= \operatorname{Tr}(-U^\dagger \delta U U^\dagger D U + U^\dagger D \delta U + U^\dagger \delta D U)^2$$
$$= \operatorname{Tr}(\delta D)^2 + 2i \operatorname{Tr} \delta u[\delta D, D] + 2\operatorname{Tr}(-\delta u D \delta u D + (\delta u)^2 D^2),$$

where $\delta u \equiv \delta U U^\dagger / i = \delta u^\dagger$ and $\delta D = \operatorname{diag}(\delta h_1, \ldots, \delta h_N)$. The second term at the r.h.s. vanishes because both D and δD are diagonal and commute. Therefore

$$\|\delta H\|^2 = \sum_{i=1}^{N} (\delta h_i)^2 + \sum_{i,j=1}^{N} (\delta u)_{ij} (\delta u)_{ji} (h_i - h_j)^2.$$

Now it remains to recall the basic relation between the infinitesimal norm and the measure: if $\|\delta l\|^2 = G_{ab} \delta l^a \delta l^b$ then $[dl] = \sqrt{\det_{ab} G_{ab}} \prod_a dl^a$, to obtain eq. (2.17) with Haar measure $[dU] = \prod_{ij}^{N} du_{ij}$ being associated with the infinitesimal norm

$$\|\delta u\|^2 = \operatorname{Tr}(\delta u)^2 = \sum_{i,j=1}^{N} \delta u_{ij} \delta u_{ji} = \sum_{i,j=1}^{N} |\delta u_{ij}|^2$$

and $[dU_{\text{Cartan}}] \equiv \prod_{i=1}^{N} du_{ii}$.

where the volume of unitary group in Haar measure is equal to

$$\text{Vol}_{U(N)} = \frac{(2\pi)^{N(N+1)/2}}{\prod_{k=1}^{N} k!}.$$

2.4 Kontsevich-Like Representation of the 1-Matrix Model

Matrix-integral representation (2.1) is, however, not the only possible one for the given eigenvalue model (2.9). Expression (2.1) involves the "most general action," consistent with the symmetry $H \to UHU^\dagger$. As was already mentioned in the introduction, alternative representation of the partition function should instead involve the general coupling to background (source) field. In the theory of matrix models such representations are known under the name of Kontsevich models. They will be the subject of detailed discussion in the following sections of this chapter. What we need now is one simple identity, relating original (2.1) and Kontsevich-like representations of the 1-matrix theory:

$$\frac{Z_N\{t_0 = 0; t_k = -(\text{tr}\,\Lambda^{-k})/k + \delta_{k,2}/2\}}{Z_N\{t_k = \delta_{k,2}/2\}}$$

$$= \frac{\int_{N\times N} dH e^{\sum_{k=0}^{\infty} t_k \,\text{Tr}\,H^k}}{\int_{N\times N} dH e^{H^2/2}}$$

$$= \frac{e^{-\,\text{tr}\,\Lambda^2/2}}{(2\pi)^{n^2/2}(\det \Lambda)^N} \int_{n\times n} dX (\det X)^N e^{-\,\text{tr}\,X^2/2 + \text{tr}\,\Lambda X}$$

$$= \mathcal{Z}_{X^2/2}\{N,t\}, \tag{2.19}$$

where $Z_N\{t_k = \delta_{k,2}/2\} = (-2\pi)^{N^2/2} c_N$. This relation follows from another identity:

$$\frac{\int_{N\times N} dH e^{(\text{Tr}\,H^2)/2} \text{Det}(\Lambda \otimes I - I \otimes H)}{\int_{N\times N} dH e^{(\text{Tr}\,H^2)/2}}$$

$$= \frac{\int_{n\times n} dX e^{-(\text{tr}\,X^2)/2} \det^N(X + \Lambda)}{\int_{n\times n} dX e^{-(\text{tr}\,X^2)/2}},$$

which is valid for any Λ and can be proved by different methods: see Ref. 1 and references therein. Note that integrals on the right- and left-hand sides are of different sizes: $N \times N$ at the l.h.s. and $n \times n$ at the r.h.s. While N-dependence is explicit at both sides of the equation, the n-dependence at the l.h.s. enters only implicitly: through the allowed domain of variation of variables $t_k = -(\text{tr}\,\Lambda^{-k})/k + \delta_{k,2}/2$. This is the usual feature of Kontsevich integrals: explicit n-dependence disappears once the integral is expressed through the t-like variables.

Equation (2.19) can be used to perform analytical continuation in N and define what is Z_N for N, which are not positive integers. Since $c_N = 0$ for all *negative* integers (see Ref. 1), the same is true for Z_N. This property ($\tau_N = 0$ for all negative integers N) is in fact characteristic for τ-functions of *forced* hierarchies of which the partition function (2.18) is an example.

3   Generalized Kontsevich Model

Let us now proceed to investigate Kontsevich models of a rather general type. Further generalizations, leading directly to theories with physical vector bosons (see, e.g., [20]), are beyond the scope of the present notes. The basic mathematical fact, responsible for solvability (integrability) of Kontsevich models is the Duistermaat-Heckmann theorem, which allows us to evaluate explicitly the celebrated nonlinear Harish-Chandra-Itzykson-Zuber integral over unitary matrices and thus transform matrix integral into an eigenvalue model.

3.1   Kontsevich Integral. The First Step

Kontsevich integral is defined as

$$\mathcal{F}_{V,n}\{L\} = \int_{n \times n} dX e^{-\operatorname{tr} V(X) + N \operatorname{tr} \log X + \operatorname{tr} LX}. \tag{3.1}$$

In fact it depends only on *eigenvalues* of the matrix L. Indeed, substitute $X = U_X^\dagger D_X U_X$; $L = U_L^\dagger D_L U_L$ in (3.1) and denote $U \equiv U_X U_L^\dagger$. Then

$$\mathcal{F}_{V,n}\{L\} = \prod_{a=1}^{n} \int dx_a x_a^N e^{-V(x_a)} \Delta^2(x)$$
$$\times \int_{n \times n} \frac{[dU]}{[dU_{\text{Cartan}}]} \exp\left(\sum_{a,b=1}^{n} x_a l_b |U_{a\delta}|^2 \right). \tag{3.2}$$

In order to proceed further one needs to evaluate the integral over unitary matrices, which appeared at the right-hand side.

This integral can actually be represented in two different forms:

$$I_n\{X, L\} \equiv \int_{n \times n} \frac{[dU]}{[dU_{\text{Cartan}}]} e^{\operatorname{tr} XULU^\dagger} \tag{3.3}$$

$$= \int_{n \times n} \frac{[dU]}{[dU_{\text{Cartan}}]} e^{\sum_{a,b=1}^{n} x_a l_b |U_{ab}|^2}. \tag{3.4}$$

(The U's in the two integrals are related by the transformation $U \to U_X U U_L^\dagger$ and the Haar measure is both left and right invariant.) Formula

(3.3) implies that $I_n\{X, L\}$ satisfies a set of simple equations [21]:

$$\left(\mathrm{tr}\left(\frac{\partial}{\partial X_{\mathrm{tr}}}\right)^k - \mathrm{tr}\, L^k\right) I_n\{X, L\} = 0, \quad k \ge 0,$$

$$\left(\mathrm{tr}\left(\frac{\partial}{\partial L_{\mathrm{tr}}}\right)^k - \mathrm{tr}\, X^k\right) I_n\{X, L\} = 0, \quad k \ge 0,$$
(3.5)

which by themselves are not very restrictive. However, another formula, (3.4), implies that $I_n\{X, L\}$ in fact depends only on the eigenvalues of X and L, and for *such* $I_n\{X, L\} = \hat{I}\{x_a, l_b\}$ eqs. (3.5) become very restrictive[10] and allow us to determine $\hat{I}\{x_a, l_b\}$ unambiguously as a formal power series in positive powers of x_a and l_b. The final answer is

$$I_n\{X, L\} = \frac{(2\pi)^{n(n-1)/2}}{n!} \frac{\det_{ab} e^{x_a l_b}}{\Delta(x)\Delta(l)}.$$
(3.6)

Normalization constant can be defined by taking $L = 0$, when

$$I_n\{X, L = 0\} = \frac{\mathrm{Vol}_{U(n)}}{(\mathrm{Vol}_{U(1)})^n} = \frac{(2\pi)^{n(n-1)/2}}{\prod_{k=1}^{n} k!},$$

and using the fact that

$$\left.\frac{\det_{ab} f_a(l_b)}{\Delta(l)}\right|_{\{l_b = 0\}} = \left(\prod_{k=0}^{n-1} \frac{1}{k!}\right) \det_{ab} \partial^{b-1} f_a(0).$$

3.2   Itzykson-Zuber Integral and Duistermaat-Heckmann Theorem

Equation (3.6) is usually referred to as the Itzykson-Zuber formula [22]. In the mathematical literature it was earlier derived by Kharish-Chandra [23], and in fact the integral (3.3) is the basic example of the coadjoint orbit integrals [24–26], which can be exactly evaluated with the help of the Duistermaat-Heckmann theorem [27–30]. We now interrupt our discussion of the Kontsevich model for a brief illustration of this important phenomenon. The Duistermaat-Heckmann theorem claims that under certain restrictive conditions (that dynamical flow is consistent with the action of

---

[10]When acting on $\hat{I}$, which depends only on eigenvalues, matrix derivatives turn into:

$$\mathrm{tr}\,\frac{\partial}{\partial X_{\mathrm{tr}}}\hat{I} = \sum_a \frac{\partial}{\partial x_a}\hat{I};$$

$$\mathrm{tr}\,\frac{\partial^2}{\partial X_{\mathrm{tr}}^2}\hat{I} = \sum_a \frac{\partial^2}{\partial x_a^2}\hat{I} + \sum_{a \ne b} \frac{1}{x_a - x_b}\left(\frac{\partial}{\partial x_a} - \frac{\partial}{\partial x_b}\right)\hat{I};$$

and so on.

a compact group) some integrals can be expressed in a simple way through extrema of the integrand. This almost sounds like a statement that quasiclassical approximation can be exact, with two correction that *all* the extrema—not only the deepest minimum of the action—should be taken into account, and that the quantum measure should be adjusted appropriately. When applicable, the theorem states that

$$\int [d\phi] e^{-S(\phi)} \sim \sum_{\phi:\partial S/\partial\phi=0} \left(\frac{\partial^2 S}{\partial\phi^2}\right)^{-1/2} e^{-S(\phi)}. \tag{3.7}$$

The simplest example is given by the integral $\int_0^\pi [\sin\theta d\theta] e^{\mu\cos\theta} = (e^\mu - e^{-\mu})/\mu$. We shall not dwell upon the reasons why DH theorem is true in the case of the Itzykson-Zuber integral (the basic requirement: existence of the compact group action—that of unitary group—is obviously fulfilled in this case). Instead we just evaluate the r.h.s. of (3.7) provided the l.h.s. is given by $\int dU \exp(\operatorname{tr} XULU^\dagger)$. Then equation of motion for U looks like

$$[X, ULU^\dagger] = 0. \tag{3.8}$$

We assume that X and L are already diagonal matrices. Then (3.8) has an obvious solution $U = I$ (identity matrix), but it is not unique. For generic diagonal X, L the most general solution is given by arbitrary permutation matrix P: $U = P$. The "classical action" on such solution is equal to $\operatorname{tr} XULU^\dagger = \sum_a x_a l_{P(a)}$, while the preexponential factors provide Vandermonde determinants $\Delta(x)\Delta(l)$ in denominator and the sign factor $(-)^P$. Since

$$\sum_P (-)^P \exp\left(\sum_a x_a l_{P(a)}\right) = \det_{ab} e^{x_a l_b}$$

we immediately obtain the Itzykson-Zuber formula (3.6). Unfortunately the Duistermaat-Heckmann theory is not yet well developed and even if the vacuum average is exactly calculable, it does not provide immediate prescription for evaluation of correlators (or, what is essentially the same, corrections to DH formula—when it is not exactly true—are not yet described in a universal way). The very important *general* technique of *exact* evaluation of *non-Gaussian* unitary-matrix integrals is now doing its first steps (see Refs. 31–34).

3.3   Kontsevich Integral. The Second Step

Now we turn back to the eigenvalue representation of the Kontsevich integral (3.1). Substitution of (3.6) into (3.2) gives:

$$\mathcal{F}_{V,n}\{L\} = \frac{(2\pi)^{n(n-1)/2}}{\Delta(l)} \prod_{b=1}^n \int dx_b e^{-V(x_b)} \Delta(x) \frac{1}{n!} \det_{ab} e^{x_a l_b}$$

$$= \frac{(2\pi)^{n(n-1)/2}}{\Delta(l)} \prod_{b=1}^{n} \int dx_b x_b^N e^{-V(x_b) + x_b l_b} \Delta(x), \qquad (3.9)$$

where we used antisymmetry of $\Delta(x)$ under permutations of x_a's in order to change $\det_{ab} e^{x_a l_b}/n!$ for $e^{\sum_b x_b l_b}$ under the sign of the x_b integration.

We can now use the fact that $\Delta(x) = \det_{ab} x_b^{a-1}$ in order to rewrite the r.h.s. of (3.9):

$$\mathcal{F}_{V,n}\{L\} = (2\pi)^{n(n-1)/2} \frac{\det_{ab} \hat{\varphi}_{a+N}(l_b)}{\Delta(l)}, \qquad (3.10)$$

where

$$\hat{\varphi}_a(l) \equiv \int dx\, x^{a-1} e^{-V(x)+lx}, \qquad a \geq 1.$$

These functions $\hat{\varphi}(l)$ satisfy a simple recurrent relation:

$$\hat{\varphi}_a = \frac{\partial \hat{\varphi}_{a-1}}{\partial l} = \left(\frac{\partial}{\partial l}\right)^{a-1} \hat{\Phi} \qquad (3.11)$$

with

$$\hat{\Phi}(l) \equiv \hat{\varphi}_1(l) = \int dx\, e^{-V(x)+lx}.$$

This completes the transformation of Kontsevich integral to the form of eigenvalue model.

3.4   "Phases" of Kontsevich Integral. GKM as the "Quantum Piece" of $\mathcal{F}_V\{L\}$ in the Kontsevich Phase

One of the natural things to do in the study of functions, defined in the integral form, is to investigate various asymptotics when the arguments tend to various distinguished limits. In the case of Kontsevich integral the arguments are just eigenvalues l_a of the matrix L, while their distinguished values are either zero or positions of singularities of potential $V(x)$. For simplicity we assume that $V(x)$ has only a pole of the order $p+1$ at infinity, i.e., V(x) is a polynomial of degree $p+1$. Then it remains to consider separately only two different asymptotics: small l_a and large l_a. Of course, since there are many different l_a one actually has a vast variety of possibilities: some of l_a's are small and the rest are large. The two extreme cases are when all the l_a's are either small or large, and they are referred to as the "character phase," and "Kontsevich phase," respectively. The word "phase" is used instead of the more exact "asymptotics" in order to emphasize the relation of these two cases to the strong and weak coupling phases in lattice models of Yang-Mills theories. We refer to Ref. 35 for some more discussion of these phases and their properties, here only some basic things will be mentioned.

In the character phase the main observation is that the r.h.s. of the Itzykson-Zuber formula (3.6) is essentially the character of the group element $g = e^L$ of SL(n):

$$\chi_R(e^L) = \frac{\det e^{l_a x_b}}{\Delta(e^l)} \Delta(x),$$

where the set $\{x_a\}$ specifies representation R of the SL group.[11] Accordingly the integral (3.1) can be represented as a linear combination of characters with various R, with coefficients depending on the choice of potential $V(x)$ and N. This explains the reason why this limit (when everything is assumed to be expandable in positive powers of l_a or e^{l_a}) is referred to as the *character* phase. One should keep in mind, of course, that the most natural case, from this point of view, is the situation when the integral over x_a's is changed for a discrete sum over integer x_a's—this is one of the possible directions of the search for "quantum deformations" of GKM.

Let us now turn to the limit of large l_a. Then the natural expansion would be in negative powers of the arguments, but the integral (3.1) does not have such expansion, as it is. One should first extract a "quasiclassical factor" and it will be the remaining "quantum part" that will possess such an expansion. This quantum piece is the most interesting one, and it is for it that the name GKM is usually used. So, in the Kontsevich limit, integral (3.1) can be evaluated with the help of the steepest descent method, with the classical solution defined from $V'(X_0) = L$. (Note that when doing so we include the logarithmic piece in the measure, not in the action.) Let us call the solution of this equation $X_0 = \Lambda$. One could use Λ from he very beginning instead of L to parametrize the Kontsevich integral, writing $\operatorname{tr} V'(\Lambda)X$ instead of $\operatorname{tr} LX$ in the exponent in (3.1). This is a natural parameter in the Kontsevich phase, while $L = V'(\Lambda)$ plays this role in the character phase. The quasiclassical contribution to (3.1), i.e., the exponent of the classical action divided by determinant of quadratic fluctuations, is equal to

---

[11]Correction factor $\Delta(l)/\Delta(e^l)$ can be restored if one considers appropriate generalization of the Kontsevich integral (which describes not just a puncture but a hole on a surface—in terms of the naive string theory), see, for example, Ref. 20. Actually one needs to substitute $\int[dU]\exp(\operatorname{tr} XULU^\dagger)$ by the loop integral

$$\int[dU(s)]\exp\left(\int ds\operatorname{tr} X\big(U(s)\partial_s U^\dagger(s) + U(s)LU^\dagger(s)\big)\right) = \frac{\chi_R(e^L)}{d_R}$$

When this integral is evaluated for integer x_a's one essentially substitutes every item in the product $\Delta(l) = \prod_{a<b}(l_a - l_b)$ by a new infinite product over harmonics, $l_a - l_b \rightarrow \prod_k(l_a - l_b + 2\pi ik) \sim \sinh((l_a - l_b)/2) \sim e^{l_a} - e^{l_b}$.

$$\mathcal{C}_V\{\Lambda|N\} = (2\pi)^{n^2/2}\frac{\exp[\mathrm{tr}(\Lambda V'(\Lambda) - V(\Lambda))]}{\sqrt{\det V''(\Lambda)}}(\det \Lambda)^N \qquad (3.12)$$

and the partition function of GKM is by definition

$$\mathcal{Z}_{V,n}\{\Lambda \mid N\} \equiv \frac{\mathcal{F}_V\{\Lambda \mid N\}}{\mathcal{C}_V\{\Lambda \mid N\}}.$$

This function can be expanded in negative integer powers of λ_a (i.e., in *fractional* negative powers of original l_a), at least as a formal series. Moreover, there is a symmetry between all the eigenvalues λ_a, thus Z^{GKM} is in fact a function (formal series) of the "time variables" (the name is from the theory of integrable hierarchies)

$$T_k \equiv \frac{1}{k}\,\mathrm{tr}\,\Lambda^{-k}. \qquad (3.13)$$

Remarkably, if considered as a function of these T_k's $\mathcal{Z}$ becomes independent of n! We refer to Refs. 1 and 4 for more details (including exact definition of $V''(\Lambda)$ in (3.12)).

What remains to be considered here is the eigenvalue representation of $\mathcal{Z}$. If (3.10) is divided by the quasiclassical factor $\mathcal{C}_V\{\Lambda|N\}$, we get

$$\mathcal{Z}_V\{N,T\} = \frac{1}{(\det \Lambda)^N} \cdot \frac{\det_{ab} \varphi_{a+N}(\lambda_b)}{\Delta(\lambda)}. \qquad (3.14)$$

Extraction of the quasiclassical factor converts $\hat{\varphi}(l)$ into the properly normalized expansions in negative integer powers of λ:

$$\varphi_a(\lambda) = \frac{e^{-\lambda V'(\lambda)+V(\lambda)}\sqrt{V''(\lambda)}}{\sqrt{2\pi}}\hat{\varphi}_a(V'(\lambda)) = \lambda^{a-1}\big(1 + \mathcal{O}(\lambda^{-1})\big). \qquad (3.15)$$

It also changes $\Delta(l) = \Delta\big(V'(\lambda)\big)$ in the denominator of (3.10) for $\Delta(\lambda)$ in (3.14).

Note that, as a corollary of normalization condition (3.15), whenever one puts $\lambda_n = \infty$ the $n \times n$ determinant in (3.14) just turns into the $(n-1) \times (n-1)$ determinant of the same form. One can easily understand that this implies the n-independence of $\mathcal{Z}$ as a function of T-variables (3.13).

Instead of the simple recurrent relations (3.11) for $\hat{\varphi}$, the normalized functions φ satisfy

$$\varphi_a(\lambda) = \mathcal{A}\varphi_{a-1}(\lambda) = \mathcal{A}^{a-1}\Phi(\lambda), \qquad (3.16)$$

where $\Phi(\lambda) = \varphi_1(\lambda)$ and the operator

$$\mathcal{A} = \frac{1}{V''(\lambda)} \cdot \frac{\partial}{\partial \lambda} - \frac{1}{2}\frac{V'''(\lambda)}{(V''(\lambda))^2} + \lambda$$

now depends on the shape of the potential $V(x)$.

3.5   Relation Between Time- and Potential-Dependencies

Consider the vector space $\mathcal{H}$ of all formal Laurent series in some variable λ. The set of functions

$$\{\lambda^{-a}, \lambda^{a-1} \mid a = 1, 2, \ldots\}$$

is one of the many possible bases in this vector space. Let us consider a "half-space" $\mathcal{H}_+$ consisting of all the series in non-negative powers of λ. Then $\{\lambda^{a-1} \mid a = 1, 2, \ldots\}$ is a possible basis in $\mathcal{H}_+$. Every rotation of the linear subspace $\mathcal{H}_+$ within entire $\mathcal{H}$ can be represented by projection of some basis in rotated subspace onto original one. In other words, any semi-infinite set of functions $\{\phi_a(\lambda)\}$, such that

$$\phi_a(\lambda) = \lambda^{a-1}(1 + \mathcal{O}(\lambda^{-1})) = \lambda^{a-1}\left(1 + \sum_{b>0} S_{ab}\lambda^{-b}\right) \tag{3.17}$$

can be considered as describing some particular rotation of $\mathcal{H}_+$ in $\mathcal{H}$. Of course, the same rotation can be represented by different matrices S_{ab}, and in fact rotations are in one-to-one correspondence with the factor of the set $\{\phi_a\}$ modulo triangular transformations, which has the natural name of the universal Grassmannian $\mathcal{GR}$. Equation (3.14) is obviously invariant under such triangular transformations, thus $\mathcal{Z}$, as a function of its argument $V(x)$ can be considered as a function on $\mathcal{GR}$ (if the shape of potential is changed, the set $\{\phi_a\}$ is also changed).

The normalization condition (3.17) is, however, invariant not only under triangular transformations in $\{\phi_a\}$, but also under the changes $\lambda \rightarrow \lambda(1 + \mathcal{O}(\lambda^{-1}))$. Such transformations change the point of Grassmannian and they also induce a triangular linear transformation of time-variables: $T_k \rightarrow T_k + \text{lin}(T_{k+1}, T_{k+2}, \ldots)$. In other words, $\mathcal{Z}$ depends on the choice of variable λ on the "spectral curve" and on the point of $\mathcal{GR}$, i.e., it is essentially a function on the tensor product $\mathcal{GR} \times \mathcal{GR}$ of two different Grassmannians. One of them is a space of various models (related to the choice of potential in GKM); another specifies the basis in the space of observables in a given model (related to the choice of time variables). As one expects from the physical arguments and as we just saw on a more formal level these two dependencies are in fact interrelated. When consideration is restricted to the set of GKMs (from that of all the models of string theory) a much more definite statement can be made [36]. Being a priori a function of two distinct types of variables—the times $\{T_k = \text{tr } \Lambda^{-k}/k\}$ and potential $V(x)$—the GKM partition function in fact depends only on the *type* of singularities of $V(x)$ (which is a kind of "discrete" information) and on the peculiar combination of these variables. If $V(x)$ is a polynomial of degree $p + 1$ (i.e., has only finite order pole at infinity), then

$$\mathcal{Z}_V\{T_k\} \sim \tau_p\left(\frac{1}{k}\text{ tr } W(\Lambda)^{-k/p} + \frac{p}{k(p-k)}\text{ res } W(x)^{1-k/p}\, dx\right), \tag{3.18}$$

where $W(x) \equiv V'(x)$, and the shape of the function τ_p depends only on the value of p. (In order to substitute "$\sim$" by "$=$" in (3.18) one should slightly redefine the quasiclassical factor and thus $\mathcal{Z}$: one should in fact work with the variable $L^{1/p} = W(\Lambda)^{1/p}$ instead of Λ. As often happens, different variables are nice for different purposes.)

If all the arguments $\widehat{T}_k$ with $k > p$ are put equal to zero, we get a reduced τ-function

$$\tilde{\tau}_p(\widehat{T}_1, \ldots, \widehat{T}_p) \equiv \tau_p(\widehat{T}_k)|_{T_k} = 0 \text{ for } k > p$$

It appears to be a solution to "quasiclassical KP-hierarchy," which arises from pure algebraic construction and can be also identified as partition function of topological Landau-Ginsburg model with the superpotential $W(x)$.

3.6   Kac-Schwarz Problem

The function $\tau_p(T_k)$ is of course very far from arbitrary. First of all it possesses peculiar determinant representation of the type (3.14), which in turn implies some restrictive bilinear (Hirota) equations and as result τ_p appears to be a KP τ-function. Second, τ_p is further distinguished even among τ-functions by peculiar features of the functions $\varphi_a(\lambda)$. In the particular case of GKM, one of the ways to represent these properties is to use the recursive relations (3.16), $\varphi_a = \mathcal{A}^{a-1}\Phi$, supplemented by another obvious property $W(\lambda)\Phi(\lambda) = \varphi_{p+1}(\lambda)$ (it follows from invariance of the integral $\Phi(\lambda)$ under the change-of-integration x-variable). These two relations together give rise to an equation on $\varphi_1(\lambda) = \Phi(\lambda)$: $(\mathcal{A}^p - W(\lambda))\Phi(\lambda) = 0$, which is just a pth-order differential equation.

In this way one specifies nonperturbative partition function of some string model (in the case of GKM this is in fact a $(p, 1)$-minimal model coupled to $2d$-gravity) in terms of invariant points of certain operators acting on the universal module space $\mathcal{GR}$. This reformulation, though far more abstract than the original one, can be very useful for nontrivial generalizations—and also be a natural step in the search for generic configuration space of the string theory. It was first introduced by V. Kac and A. Schwarz [37], and is not yet studied as deep as it deserves, even if one deals only with the space of KP τ-functions, determinant formulas, and the ordinary universal Grassmannian. In this (actually somewhat narrow) context, the general problem is to describe common invariant points in $\mathcal{GR}$ of two operators, acting on formal Laurent series in λ: $\forall a$

$$\mathcal{A}\phi_a \in \text{Span}\{\phi_b\}, \tag{3.19}$$

$$\mathcal{K}\phi_a \in \text{Span}\{\phi_b\}. \tag{3.20}$$

In the case of GKM, $\mathcal{A}$ and $\mathcal{K}$ are differential operators of the first and zeroth order, respectively. Moreover, $\mathcal{A}$ is a "gap-one" operator (the gap is equal to g if $\mathcal{A}\phi_a = \sum_{b=1}^{a+g} \mathcal{A}_{ab}\phi_b$). See Ref. 19 for some more comments on the gap-one case. When both gaps are different from unity, the system (3.19)–(3.20) usually describes a multiparametric set of invariant points, the simplest example being associated with (p, q)-minimal models (where p and q are in fact the values of gaps for $\mathcal{K}$ and $\mathcal{A}$, respectively). In this situation Kontsevich integral describes only duality transformation of the (p, q)-model into the (q, p)-one [38]. (Note that these models do not coincide after they are coupled to $2d$ gravity.)

3.7   Ward Identities for GKM

The advantage of the Kac-Schwarz reformulation of GKM is that it is very easy to deform, since there are no real restrictions imposed on the choice of operators $\mathcal{A}$ and $\mathcal{K}$. However, this formulation does not introduce any reach structure and does not immediately provide any valuable information about the form and properties of solutions to (3.19)–(3.20), i.e., it does not explicitly reveal any nice properties which could be common for all the string models. Thus it could serve as a starting, but not the final point of analysis of nonperturbative partition functions. An alternative approach, making use of the Ward identities, provides a better description of GKM, but can appear too restrictive to allow for any interesting deformations. From the point of view of Grassmannian, the Ward identities specify some subset of points in $\mathcal{GR}$, which are invariant under certain subalgebras of $\mathcal{U}\,\mathrm{GL}(\infty)$ (the symmetry group of the entire Grassmannian). The corresponding "homogeneous spaces" are normally not discrete and contain a vast variety of points. By themselves the Ward identities are not enough to specify particular points in $\mathcal{GR}$ uniquely. They should still be supplemented by some extra conditions, like "reduction constraints" (in fact one should keep one of the Kac-Schwarz constraints, (3.20), and only another one can be usually substituted by the symmetrylike relation, coming from the Ward identities).

3.7.1   Gross-Newmann Equation

We refer to Ref. 1 for a very detailed review of the Ward identities in GKM. They are all corollaries of the simple Gross-Newmann (GN) equation [39], imposed on the Kontsevich integral (3.1) as result of its invariance under arbitrary change of the integration variable X:

$$\int dX e^{-\operatorname{tr} V(x) + N \operatorname{tr} \log X + \operatorname{tr} LX} (-V'(X) + NX^{-1} + L) = 0.$$

or

$$\left(V'\frac{\partial}{\partial L_{\mathrm{tr}}} - L - N\frac{\partial}{\partial L_{\mathrm{tr}}}^{-1}\right)\mathcal{F}_V = 0. \tag{3.21}$$

Being just an equation of motion for $\mathcal{F}_V\{L\}$, eq. (3.21) provides complete information about this function. However, this statement needs to be formulated more carefully. One of the reasons is that (3.21) does not account explicitly for a very important property of $\mathcal{F}_V\{L\}$: that it actually depends only on eigenvalues of L. This information should still be taken into account explicitly. If this is kept in mind, it becomes a tedious but straightforward work to substitute $\mathcal{F}_V = \mathcal{C}_V \mathcal{Z}_V$ and express L-derivatives through T_k-derivatives in order to derive Virasoro and W-constraints in conventional form [1]. We shall briefly sketch some pieces of this derivation and related problems in the remaining part of this section.

3.7.2 $\widetilde{W}$-Operators in Kontsevich Models

First of all, the Gross-Newmann equation (3.21) for Kontsevich models can be easily expressed in terms of the so-called $\widetilde{W}$-operators. Namely, we shall prove the following identity [40],

$$\left(\frac{\partial}{\partial \Lambda_{\mathrm{tr}}}\right)^{m+1} \mathcal{Z}\{T_k\} = (\pm)^{m+1} \sum_{l \geq 0} \Lambda^{-l-1} \widetilde{W}_{l-m}^{(m+1)}(T) \mathcal{Z}\{T_k\}, \qquad (3.22)$$

valid for *any* function $\mathcal{Z}$ that depends on $T_k = \mp \operatorname{tr} \Lambda^{-k}/k$, $k \geq 1$ and $T_0 = \pm \operatorname{tr} \log \Lambda$.

The $\widetilde{W}$-operators are defined [40] by the following construction. Consider the action of $\operatorname{Tr}(\partial^m/\partial L_{\mathrm{tr}}^m)L^n$ on $e^{\operatorname{Tr} U(L)} = e^{\sum_k t_k \operatorname{Tr} L^k}$. It gives some linear combination of terms like

$$\operatorname{tr} L^{a_1} \cdots \operatorname{tr} L^{a_l} e^{\operatorname{tr} U(L)} = \frac{\partial^l}{\partial t_{a_1} \cdots \partial t_{a_l}} e^{-\operatorname{tr} U(L)}$$

i.e., we obtain a combination of differential operators with t-derivatives, to be denoted $\widetilde{W}(t)$:

$$\widetilde{W}_{n-m}^{(m+1)}(t) e^{\operatorname{tr} U(L)} \equiv \operatorname{Tr} \frac{\partial^m}{\partial L_{\mathrm{tr}}^m} L^n e^{\operatorname{tr} U(L)}, \quad m, n \geq 0. \qquad (3.23)$$

For example,

$$\widetilde{W}_n^{(1)} = \frac{\partial}{\partial t_n}, \qquad\qquad\qquad n \geq 0;$$

$$\widetilde{W}_n^{(2)} = \sum_{k=0}^{\infty} k t_k \frac{\partial}{\partial t_{k+n}} + \sum_{k=0}^{n} \frac{\partial^2}{\partial t_k \partial t_{n-k}}, \quad n \geq -1;$$

$$\widetilde{W}_n^{(3)} = \sum_{k,l=1}^{\infty} k t_k l t_l \frac{\partial}{\partial t_{k+l+n}} + \sum_{k=1}^{\infty} k t_k \sum_{a+b=k+n} \frac{\partial^2}{\partial t_a \partial t_b}$$

$$+ \sum_{k=1}^{\infty} k t_k \sum_{a+b=n+1} \frac{\partial^2}{\partial t_a \partial t_{b+k-1}} + \sum_{a+b+c=n} \frac{\partial^3}{\partial t_a \partial t_b \partial t_c} + \binom{n-1}{2} \frac{\partial}{\partial t_n};$$

$$\cdots$$

Note, that while $\widetilde{W}_n^{(1)}$ and $\widetilde{W}_n^{(2)}$ are just the ordinary $U(1)$-Kac Moody and Virasoro operators, respectively, the higher $\widetilde{W}^{(m)}$-operators do *not* coincide with the generators of the **W**-algebras. Already

$$\widetilde{W}_n^{(3)} \neq W_n^{(3)} = \sum_{k,l=1}^{\infty} kt_k lt_l \frac{\partial}{\partial t_{k+l+n}} + 2\sum_{k=1}^{\infty} kt_k \sum_{a+b=k+n} \frac{\partial^2}{\partial t_a \partial t_b}$$

$$+ \frac{4}{3} \sum_{a+b+c=n} \frac{\partial^3}{\partial t_a \partial t_b \partial t_c}.$$

$\widetilde{W}$-operators (in variance with ordinary W-operators) satisfy the recurrent relation

$$\widetilde{W}_n^{(m+1)} = \sum_{k=1}^{\infty} kt_k \widetilde{W}_{n+k}^{(m)} + \sum_{k=0}^{m+n-1} \frac{\partial}{\partial t_k} \cdot \widetilde{W}_{n-k}^{(m)}, \quad n \geq -m.$$

Actually not too much is already known about the $\widetilde{W}$ operators and the structure of $\widetilde{\mathbf{W}}$-algebras (in particular, it remains unclear whether the negative harmonics $\widetilde{W}_n^{(m+1)}$ with $n < -m$ can be introduced in any reasonable way); see Ref. 40 for some preliminary results.

Now we can come back to the identity (3.22). Its most straightforward application is to the Gaussian Kontsevich model with potential $V(x) = x^2/2$; see the next subsection. In other cases calculations with the use of identity (3.22), accounting for the quasiclassical factor $\mathcal{C}_V\{L\}$ and the difference between $L = V'(\Lambda)$ and Λ, become somewhat more involved, though still seem sufficiently straightforward. Also, for particular potentials $V(X)$ partition function $\mathcal{Z}_V\{T\}$ is actually independent of certain (combinations of) time variables (e.g., if $V(X) = X^{p+1}/(p+1)$, it is independent of all the T_{pk}, $k \in Z_+$), and this is important for appearance of the constraints in the standard form, i.e., for certain *reduction* of $\widetilde{W}$-constraints to the ordinary W-constraints. This relation between $\widetilde{W}$- and W-operators deserves further investigation.

The proof of eq. (3.22) is provided by the following trick. Let us make a sort of Fourier transformation

$$\mathcal{Z}\{T\} = \int dH\, \mathcal{G}\{H\} e^{\sum_{k=0}^{\infty} T_k \operatorname{Tr} H^k},$$

where integral is over $N \times N$ Hermitian matrix H.[12] Then it is clear that once the identity (3.22) is established for $\mathcal{Z}\{T\}$ substituted by $e^{\operatorname{Tr} U(H)}$,

---

[12] Here it is for the first time that we encounter an important idea: matrix models— the ordinary 1-matrix model (2.1) in this case—can be considered as defining integral transformations. This view on matrix models can to large extent define their role in the future development of string theory.

$U(H) = \sum_{k=0}^{\infty} T_k \operatorname{Tr} H^k$, with any matrix H, it is valid for *any* function $\mathcal{Z}\{T\}$. The advantage of such substitution is that we can now make use of the definition (3.23) of the $\widetilde{W}$ operators in order to rewrite (3.22) in a very explicit form:

$$
\left(\frac{\partial}{\partial \Lambda_{\text{tr}}}\right)^{m+1} e^{\operatorname{Tr} U(H)}
$$

$$
= (\pm)^{m+1} \sum_{l \geq 0}^{\infty} \Lambda^{-l-1} \widetilde{W}_{l-m}^{(m+1)}(T) e^{\operatorname{Tr} U(H)}
$$

$$
= (\pm)^{m+1} \sum_{l \geq 0}^{\infty} \Lambda^{-l-1} \operatorname{Tr}\left(\frac{\partial}{\partial H_{\text{tr}}}\right)^{m} H^{l} e^{\operatorname{Tr} U(H)}
$$

$$
= (\pm)^{m+1} \operatorname{Tr}\left(\frac{\partial}{\partial H_{\text{tr}}}\right)^{m} \frac{1}{\Lambda \otimes I - I \otimes H} e^{\operatorname{Tr} U(H)}. \quad (3.24)
$$

Now an expression for T's in terms of Λ should be used. Then

$$
e^{\operatorname{Tr} U(H)} = \operatorname{Det}^{\pm 1}(\Lambda \otimes I - I \otimes H)
$$

and substituting this into (3.24) we see that (3.22) is equivalent to

$$
\left(\left(\frac{\partial}{\partial \Lambda_{\text{tr}}}\right)^{m+1} - (\pm)^{m+1} I \operatorname{Tr}\left(\frac{\partial}{\partial H_{\text{tr}}}\right)^{m} \cdot \frac{1}{\Lambda \otimes I - I \otimes H}\right)
$$
$$
\cdot \operatorname{Det}^{\pm 1}(\Lambda \otimes I - I \otimes H) = 0
$$

Here "Tr" stands for the trace in the H-space only, while $\operatorname{Det} = \operatorname{Det} \otimes \det$— for the determinant in both H and Λ spaces. After one Λ-derivative is taken explicitly, we get

$$
(I \otimes \operatorname{Tr})\left(\left(\frac{\partial}{\partial \Lambda_{\text{tr}}}\right)^{m} \otimes I - I \otimes \left(\pm \frac{\partial}{\partial H_{\text{tr}}}\right)^{m}\right)
$$
$$
\cdot \frac{\operatorname{Det}^{\pm 1}(\Lambda \otimes I - I \otimes H)}{\Lambda \otimes I - I \otimes H} = 0. \quad (3.25)
$$

This is already a matrix identity, valid for any Λ and H of the sizes $n \times n$ and $N \times N$, respectively. For example, if $m = 0$ ($\widetilde{W}^{(1)}$-case), it is obviously satisfied. If both $n = N = 1$, it is also trivially true, though for different reasons for different choice of signs: for the upper signs, the ratio at the l.h.s. is just unity and all derivatives vanish; for the lower signs we have

$$
\left(\frac{\partial}{\partial \lambda}\right)^{m} - \left(-\frac{\partial}{\partial h}\right)^{m} = \left(\sum_{\substack{a+b=m-1 \\ a,b \geq 0}} \left(\frac{\partial}{\partial \lambda}\right)^{a}\left(-\frac{\partial}{\partial h}\right)^{b}\right)\left(\frac{\partial}{\partial \lambda} + \frac{\partial}{\partial h}\right),
$$

and this obviously vanishes since $(\partial/\partial \lambda + \partial/\partial h)f(\lambda - h) \equiv 0$ for any $f(x)$.

If $m > 0$ and Λ, H are indeed *matrices*, direct evaluation becomes much more sophisticated. We present the first two nontrivial examples: $m = 1$ and $m = 2$. The following relations will be useful. Let $Q \equiv 1(\Lambda \otimes I - I \otimes H)$. Then

$$\mathrm{Det}^{\pm 1} Q \frac{\partial}{\partial \Lambda_{\mathrm{tr}}} \mathrm{Det}^{\mp 1} Q = \pm [(I \otimes \mathrm{Tr})Q];$$

$$\mathrm{Det}^{\pm 1} Q \frac{\partial}{\partial H_{\mathrm{tr}}} \mathrm{Det}^{\mp 1} Q = \mp [(\mathrm{tr} \otimes I)Q];$$

$$\left(\frac{\partial}{\partial \Lambda_{\mathrm{tr}}} \otimes I \right) Q = -[(\mathrm{tr} \otimes I)Q]Q; \tag{3.26}$$

$$\left(I \otimes \frac{\partial}{\partial H_{\mathrm{tr}}} \right) Q = [(I \otimes \mathrm{Tr})Q]Q.$$

This is already enough for the proof in the case of $m = 1$. Indeed:

$$\mathrm{Det}^{\pm 1} Q \left(\frac{\partial}{\partial \Lambda_{\mathrm{tr}}} \otimes I \mp I \otimes \frac{\partial}{\partial H_{\mathrm{tr}}} \right) Q \, \mathrm{Det}^{\mp 1} Q$$

$$= \{ -[(\mathrm{tr} \otimes I)Q]Q \pm [(I \otimes \mathrm{Tr})Q]Q \} \mp \{ [(I \otimes \mathrm{Tr})Q] \, Q \mp [(\mathrm{tr} \otimes I)Q]Q \} = 0.$$

The first two terms at the r.h.s. come from Λ, while the last two come from H-derivatives.

In the case of $m = 2$ one should take derivatives once again. This is a little more tricky, and the same compact notation is not sufficient. In addition to (3.26) we now need

$$\left(\frac{\partial}{\partial \Lambda_{\mathrm{tr}}} \otimes I \right) [(\mathrm{tr} \otimes I)Q]Q = -[(\mathrm{tr} \otimes I)Q]^2 Q - \mathcal{B}. \tag{3.27}$$

Here

$$[(\mathrm{tr} \otimes I)Q]^2 = [(\mathrm{tr} \otimes I)[(\mathrm{tr} \otimes I)Q]Q], \tag{3.28}$$

while in order to write $\mathcal{B}$ explicitly we need to restore matrix indices (Greek for the Λ-sector and Latin for the H one). The $(\alpha i, \gamma k)$-component of (3.27) looks like

$$\left(\frac{\partial}{\partial \Lambda_{\beta\alpha}} \delta^{im} \right) Q^{mj}_{\delta\delta} Q^{jk}_{\beta\gamma} = -Q^{ij}_{\delta\delta} Q^{jl}_{\beta\beta} Q^{lk}_{\alpha\gamma} - Q^{il}_{\delta\beta} Q^{lj}_{\alpha\delta} Q^{jk}_{\beta\gamma}$$

and appearance of the second term at the r.h.s. implies, that $\mathcal{B}^{ik}_{\alpha\gamma} = Q^{il}_{\delta\beta} \times Q^{lj}_{\alpha\delta} Q^{jk}_{\beta\gamma}$. Further,

$$\left(\frac{\partial}{\partial \Lambda_{\mathrm{tr}}} \otimes I \right) [(I \otimes \mathrm{Tr})Q]Q = -[(I \otimes \mathrm{Tr})[(\mathrm{tr} \otimes I)Q]Q]Q$$

$$- [(I \otimes \mathrm{Tr})[(I \otimes \mathrm{Tr})Q]Q]Q;$$

$$\left(I \otimes \frac{\partial}{\partial H_{\text{tr}}}\right)[(\text{tr} \otimes I)Q]Q = [(\text{tr} \otimes I)][(I \otimes \text{Tr})Q]Q]Q$$
$$+ [(I \otimes \text{Tr})][(\text{tr} \otimes I)Q]Q]Q;$$

$$\left(I \otimes \frac{\partial}{\partial H_{\text{tr}}}\right)[(I \otimes \text{Tr})Q]Q = [(I \otimes \text{Tr})][(I \otimes \text{Tr})Q]Q]Q + \mathcal{B}.$$

It is important that $\mathcal{B}$ that appears in the last relation in the form of $\mathcal{B}^{ik}_{\alpha\gamma} = Q^{lj}_{\alpha\delta}Q^{il}_{\delta\beta}Q^{jk}_{\beta\gamma}$ is exactly the same $\mathcal{B}$ as in eq. (3.27).

Now we can prove (3.25) for $m = 2$:

$$\text{Det}^{\pm 1} Q\left(\left(\frac{\partial}{\partial \Lambda_{\text{tr}}}\right)^2 \otimes I - I \otimes \left(\frac{\partial}{\partial H_{\text{tr}}}\right)^2\right)Q \, \text{Det}^{\mp 1} Q$$

$$= \{\pm [(I \otimes \text{Tr})Q](-[(\text{tr} \otimes I)Q]Q \pm [(I \otimes \text{Tr})Q]Q)$$
$$- (-[(\text{tr} \otimes I)][(\text{tr} \otimes I)Q]Q]Q - \mathcal{B})$$
$$\pm (-[(I \otimes \text{Tr})][(\text{tr} \otimes I)Q]Q]Q - [(\text{tr} \otimes I)][(I \otimes \text{Tr})Q]Q]Q)\}$$
$$- \{\mp [(\text{tr} \otimes I)Q]([(I \otimes \text{Tr})Q]Q \mp [(\text{tr} \otimes I)Q]Q)$$
$$+ ([(I \otimes \text{Tr})][(I \otimes \text{Tr})Q]Q]Q + \mathcal{B})$$
$$\mp ([(\text{tr} \otimes I)][(I \otimes \text{Tr})Q]Q]Q + [(I \otimes \text{Tr})][(\text{tr} \otimes I)Q]Q]Q)\}$$

where the terms 1, 2, 3, 4, 5, and 6 in the first braces cancel the terms 1, 3, 2, 4, 6, and 5 in the second braces and identity (3.28) and its counterpart with $(\text{tr} \otimes I) \to (I \otimes \text{Tr})$ is used.

Explicit proof of eq. (3.25) for generic m is unknown.

3.7.3   Discrete Virasoro Constraints for the Gaussian Kontsevich Model

As a simplest illustration we derive now the constraints for the Gaussian Kontsevich model [41] with potential $V(X) = X^2/2$:

$$\mathcal{Z}_{X^2/2}\{N, T\} = \frac{e^{-\text{tr}\, L^2/2}}{(\det L)^N} \int dX (\det X)^N e^{-\text{tr}\, X^2/2 + LX}. \qquad (3.29)$$

In this case $L = V'(\Lambda) = \Lambda$, and the time variables are just

$$T_k = \frac{1}{k} \text{tr}\, \Lambda^{-k} = \frac{1}{k} \text{tr}\, L^{-k}.$$

The model is nontrivial because of the presence of the "zerotime" variable N [42]. The Gross-Newmann equation (3.21) looks like

$$\frac{e^{-\text{tr}\, L^2/2}}{(\det L)^N}\left(\frac{\partial}{\partial L_{\text{tr}}}\right)^{n+1} \cdot \left(\frac{\partial}{\partial L_{\text{tr}}} - N\left(\frac{\partial}{\partial L_{\text{tr}}}\right)^{-1} - L\right)$$
$$\cdot (\det L)^N e^{+\text{tr}\, L^2/2} \mathcal{Z}_{X^2/2}\{N, T\} = 0. \qquad (3.30)$$

In order to get rid of the integral operator $(\partial/\partial L)^{-1}$ one should take here $n \geq 0$ rather than $n \geq -1$. In fact all the equations with $n > 0$ follow from the one with $n = 0$, and we restrict our consideration to the last one. For $n = 0$ we obtain from (3.30)

$$\left(\left(\frac{\partial}{\partial L_{\mathrm{tr}}} + \frac{N}{L} + L \right)^2 - 2N - L \left(\frac{\partial}{\partial L_{\mathrm{tr}}} + \frac{N}{L} + L \right) \right) \mathcal{Z} = 0$$

or

$$\left(\left(\frac{\partial}{\partial L_{\mathrm{tr}}} \right)^2 + \left(L + \frac{2N}{L} \right) \frac{\partial}{\partial L_{\mathrm{tr}}} + \frac{N^2}{L^2} - \frac{N}{L} \operatorname{tr} \frac{1}{L} \right) \mathcal{Z} = 0.$$

One can now use eq. (3.22) to obtain

$$\sum_{m=-1}^{\infty} \frac{1}{L^{m+2}} \left(\sum_{k=1+\delta_{m,-1}}^{\infty} \left(\operatorname{tr} \frac{1}{L^k} \right) \frac{\partial}{\partial T_{k+m}} + \sum_{k=1}^{m-1} \frac{\partial^2}{\partial T_k \partial T_{m+k}} \right.$$

$$\left. - \frac{\partial}{\partial T_{m+2}} - 2N \frac{\partial}{\partial T_m} + N^2 \delta_{m,0} - N \left(\operatorname{tr} \frac{1}{L} \right) \delta_{m,-1} \right) \mathcal{Z}$$

$$= \sum_{m=-1}^{\infty} \frac{1}{L^{m+2}} e^{NT_0} L_m(T+r) e^{-NT_0} \mathcal{Z} = 0. \quad (3.31)$$

Here $L_m(t) = \widetilde{W}_m^{(2)}(t)$ are just the generators (2.5) of "discrete" Virasoro algebra (2.4)

$$e^{Nt_0} L_m(t) e^{-Nt_0} = e^{Nt_0} \left(\sum_{k=1}^{\infty} k t_k \frac{\partial}{\partial t_{k+m}} + \sum_{k=0}^{m} \frac{\partial^2}{\partial t_k \partial t_{m-k}} \right) e^{-Nt_0}.$$

and at the r.h.s. of (3.31) $r_k = -\delta_{k,2}/2$.[13]

Thus we found that the Ward identities for the Gaussian Kontsevich model (3.29) coincide with those for the ordinary 1-matrix model (2.1), moreover the size of the matrix N in the latter model is associated with the "zero time" in the former one. This result [41] of course implies that the two models are identical:

---

[13] This small correction is manifestation of a very general phenomenon which was already mentioned in Section 3.5 above: from the point of view of symmetries (Ward identities) it is more natural to consider Z_V not as a function of T-variables, but of some more complicated combination $\hat{T}_k + r_k$, depending on the shape of potential V. If V is a polynomial of degree $p+1$, $\hat{T}_k = \operatorname{tr}(V'(\lambda))^{-k/p}/k$, while $r_k = p/(k(p-k)) \times \operatorname{Res}(V'(\mu))^{1-k/p} d\mu$. For monomial potentials these expressions become very simple: $\hat{T}_k = T_k$ and $r_k = -p\delta_{k,p+1}/(p+1)$. See Ref. 36 for more details. In most places in these notes we prefer to use invariant potential-independent times T_k, instead of $\hat{T}_k$, but then Ward identities acquire some extra terms with r_k.

$$e^{-NT_0} \mathcal{Z}_{X^2/2}\{N, T_1, T_2, \dots\} \sim Z_N\{T_0, T_1, T_2, \dots\}.$$

See Refs. 1 and 42 for explicit proof of this identity.

3.7.4   Continuous Virasoro Constraints for the $V = X^3/3$ Kontsevich Model

This example is a little more complicated than that in the previous subsection, and we do not present calculations in full details (see Refs. 4 and 43). Our goal is to demonstrate that the constraints that arise in this model, though they still form (the Borel subalgebra of) some Virasoro algebra, are *different* from (2.4). From the point of view of the CFT-formulation the relevant model is that of the *twisted* (in this particular case, antiperiodic) free fields. These so-called "continuous Virasoro constraints" give the simplest illustration of the difference between discrete and continuous matrix models: this is essentially the difference between "homogeneous" (Kac-Frenkel) and "principal" (soliton vertex operator) representation of the level $k = 1$ Kac-Moody algebra. From the point of view of integrable hierarchies this is the difference between Toda-chain-like and KP-like hierarchies.

Another (historically first) aspect of the same relation also deserves mentioning, since it also illustrates the interrelation between different models. The discrete 1-matrix model arises naturally in description of quantum $2d$ gravity as sum over 2-geometries in the formalism of random equilateral triangulations. The model, however, describes only lattice approximation to $2d$ gravity and (double-scaling) continuum limit should be taken in order to obtain the real (continuous) theory of $2d$ gravity. This limit was originally formulated [44] in terms of the constraint algebra (equations of motion or "loop" or "Schwinger-Dyson" equations—terminology is taste-dependent), leaving open the problem of what is the form of partition function $\mathcal{Z}^{\mathrm{cont}}\{T\}$ of continuous theory. Since the relevant algebra appeared to be just the set of Ward identities for the Kontsevich model (with $V(X) = X^3/3$), this proves that the latter one is exactly the continuous theory of pure $2d$ gravity. At the same time, the Kontsevich model itself can be naturally introduced as a theory of *topological* gravity (in fact this is how the model was originally discovered in Ref. 3). From this point of view the constraint algebra, to be discussed below in this subsection, plays the central role in the proof of equivalence between pure $2d$ quantum gravity and pure topological gravity (in both cases "pure" means that "matter" fields are not included).

After these introductory remarks we proceed to calculations. Actually they just repeat those from the previous subsection for the Gaussian model, but formulas get somewhat more complicated. This time we do not include zero-time N and use eq. (3.21) with $V(X) = X^3/3$. Also, this time it is much more tricky (though possible) to work in matrix notations (because fractional powers of L will be involved) and we rewrite everything in terms of the eigenvalues of L.

Substitute

$$
C_{X^3/3} = \frac{\prod_b e^{2\lambda_b^{3/2}/3}}{\sqrt{\prod_{a,b}(\sqrt{\lambda_b} + \sqrt{\lambda_a})}},
$$

$$
\left(\frac{\partial^2}{\partial L_{\mathrm{tr}}^2}\right)_{aa} = \frac{\partial^2}{\partial \lambda_a^2} + \sum_{b \neq a} \frac{1}{\lambda_a - \lambda_b} \left(\frac{\partial}{\partial \lambda_a} - \frac{\partial}{\partial \lambda_b}\right)
$$

and introduce a special notation for

$$
\frac{\mathcal{D}}{\mathcal{D}\lambda_a} \equiv C_{X^3/3}^{-1} \frac{\partial}{\partial \lambda_a} C_{X^3/3} = \frac{\partial}{\partial \lambda_a} + \sqrt{\lambda_a} - \frac{1}{4\lambda_a} - \frac{1}{2} \sum_{b \neq a} \frac{1}{\sqrt{\lambda_a}(\sqrt{\lambda_b} + \sqrt{\lambda_a})}.
$$

Then (3.21) turns into

$$
\left(\left(\frac{\mathcal{D}}{\mathcal{D}\lambda_a}\right)^2 + \sum_{b \neq a} \frac{1}{\lambda_a - \lambda_b} \left(\frac{\mathcal{D}}{\mathcal{D}\lambda_a} - \frac{\mathcal{D}}{\mathcal{D}\lambda_b}\right)\right) \mathcal{Z}_{X^3/3}\{T\} = 0. \qquad (3.32)
$$

Now we need an explicit expression for T,

$$
T_k = \frac{1}{k} L^{-k}, \qquad (3.33)
$$

and—as we already know from the previous subsection—we also need

$$
r_k = -\frac{2}{3}\delta_{k,3}.
$$

The expression $\mathcal{Z}_{X^3/3}\{T\}$ is in fact independent of all the time variables with *even* numbers (subscripts); see Refs. 1, 4 for the explanation. Therefore we can take only $k = 2l + 1$ in (3.33),

$$
T_{2l+1} = \frac{1}{2l+1} \sum_b \lambda_b^{-l-1/2},
$$

$$
r_{2l+1} = -\frac{2}{3}\delta_{l,1},
$$

and

$$
\frac{\partial}{\partial \lambda_a} \mathcal{Z}_{X^3/3}\{T\} = \sum_{l=0}^{\infty} \frac{\partial T_{2l+1}}{\partial \lambda_a} \frac{\partial \mathcal{Z}}{\partial T_{2l+1}} = -\frac{1}{2} \sum_{a=0}^{\infty} \lambda_a^{-l-3/2} \frac{\partial \mathcal{Z}}{\partial T_{2l+1}};
$$

$$
\frac{\partial^2}{\partial \lambda_a^2} \mathcal{Z}_{X^3/3}\{T\} = \frac{1}{4} \sum_{l,m=0}^{\infty} \lambda_a^{-l-m-3} \frac{\partial \mathcal{Z}}{\partial T_{2l+1}\partial T_{2m+1}}
$$

$$
+ \frac{1}{2} \sum_{l=0}^{\infty} \left(l + \frac{3}{2}\right) \lambda_a^{-l-5/2} \frac{\partial \mathcal{Z}}{\partial T_{2l+1}}.
$$

These expressions should be now substituted into (3.32) and we obtain

$$
\frac{1}{4}\sum_{l,m=0}^{\infty}\lambda_a^{-l-m-3}\frac{\partial \mathcal{Z}}{\partial T_{2l+1}\partial T_{2m+1}}
$$

$$
+\sum_{l=0}^{\infty}\left[\frac{1}{2}\sum_{a=0}^{\infty}\left(l+\frac{3}{2}\right)\lambda_a^{-l-5/2}-\frac{1}{2}\sum_{b\neq a}\frac{1}{\lambda_a-\lambda_b}(\lambda_a^{-l-3/2}-\lambda_b^{-l-3/2})\right.
$$

$$
\left.-\sum_{a=0}^{\infty}\left(\sqrt{\lambda_a}-\frac{1}{4\lambda_a}-\frac{1}{2}\sum_{b\neq a}\frac{1}{\sqrt{\lambda_a}(\sqrt{\lambda_b}+\sqrt{\lambda_a})}\right)\lambda_a^{-l-3/2}\right]\frac{\partial \mathcal{Z}}{\partial T_{2l+1}}
$$

$$
+[\dots]\mathcal{Z}=\sum_{n=-1}^{\infty}\frac{1}{\lambda_a^{n+2}}\mathcal{L}_n\mathcal{Z}\quad(3.34)
$$

with

$$
\mathcal{L}_{2n}=\sum_{l=0}^{\infty}\left(l+\frac{1}{2}\right)(T_{2l+1}+r_{2l+1})\frac{\partial}{\partial T_{2l+2n+1}}
$$

$$
+\frac{1}{4}\sum_{\substack{l+m=n-1\\l,m\geq 0}}\frac{\partial^2}{\partial T_{2l+1}\partial T_{2m+1}}+\frac{1}{16}\delta_{n,0}+\frac{1}{4}T_1^2\delta_{n,-1}
$$

$$
=\frac{1}{2}\sum_{\text{odd } k=1}^{\infty}k(T_k+r_k)\frac{\partial}{\partial T_{k+2n}}
$$

$$
+\frac{1}{4}\sum_{\text{odd } k=1}^{2n-1}\frac{\partial^2}{\partial T_k\partial T_{2n-k}}+\frac{1}{16}\delta_{n,0}+\frac{1}{4}T_1^2\delta_{n,-1}.\quad(3.35)
$$

Factor $\frac{1}{2}$ in front of the first term at the r.h.s. in (3.35) is important for $\mathcal{L}_{2n}$ to satisfy the properly normalized Virasoro algebra[14]:

$$
[\mathcal{L}_{2n},\mathcal{L}_{2m}]=(n-m)\mathcal{L}_{2n+2m}.
$$

Coefficient $\frac{1}{4}$ in front of the second term can be eliminated by rescaling of time variables: $T\to\frac{1}{2}T$; then the last term turns into $T_1^2\delta_{n,-1}/16$.

We shall not actually discuss evaluation of the coefficient in front of $\mathcal{Z}$ (with no derivatives), which is denoted by $[\dots]$ in (3.34) (see Refs. 43 and 4). In fact almost all the terms in original complicated expression cancel, giving finally

$$
[\dots]=\frac{1}{16\lambda_a^2}+\frac{T_1^2}{4\lambda_a},
$$

---

[14]Therefore it could be reasonable to use a different notation, $\mathcal{L}_n$ instead of $\mathcal{L}_{2n}$. We prefer $\mathcal{L}_{2n}$, because it emphasizes the property of the model to be a 2-reduction of the KP hierarchy (to KdV).

and this is represented by the terms with $\delta_{n,0}$ and $\delta_{n,-1}$ in expressions (3.35) for the Virasoro generators $\mathcal{L}_{2n}$.

The term with the double T-derivative in (3.34) is already of the necessary form. Of intermediate complexity is evaluation of the coefficient in front of $\partial \mathcal{Z}/\partial T_{2l+1}$ in (3.34), which we shall briefly describe now. First of all, rewrite this coefficient, reordering the items:

$$\frac{1}{2}\left[\left(l+\frac{3}{2}\right)\lambda_a^{-l-5/2} - \sum_{b\neq a}\frac{1}{\lambda_a - \lambda_b}(\lambda_a^{-l-3/2} - \lambda_b^{-l-3/2})\right]$$
$$+ \left[\frac{1}{4}\lambda_a^{-l-5/2} + \frac{1}{2}\sum_{b\neq a}\frac{\lambda_a^{-l-2}}{\sqrt{\lambda_b} + \sqrt{\lambda_a}}\right] - \lambda_a^{-l-1}. \quad (3.36)$$

The first two terms together are equal to the sum over *all* b (including $b = a$):

$$-\frac{1}{2}\sum_b \frac{1}{\lambda_a - \lambda_b}(\lambda_a^{-l-3/2} - \lambda_b^{-l-3/2})$$
$$= \frac{1}{2}\sum_b \frac{\lambda_a^{l+3/2} - \lambda_b^{l+3/2}}{\lambda_a - \lambda_b} \cdot \frac{1}{\lambda_a^{l+3/2}\lambda_b^{l+3/2}}$$
$$= \frac{1}{2\lambda_a^{l+2}}\sum_b \frac{\lambda_a^{l+2} - \lambda_a^{1/2}\lambda_b^{l+3/2}}{\lambda_a - \lambda_b} \cdot \frac{1}{\lambda_b^{l+3/2}}.$$

Similarly, the next two terms can be rewritten as

$$\frac{1}{2}\sum_b \frac{\lambda_a^{-l-2}}{\sqrt{\lambda_a} + \sqrt{\lambda_b}} = \frac{1}{2\lambda_a^{l+2}}\sum_b \frac{\sqrt{\lambda_a} - \sqrt{\lambda_b}}{\lambda_a - \lambda_b}$$
$$= \frac{1}{2\lambda_a^{l+2}}\sum_b \frac{\lambda_a^{1/2}\lambda_b^{l+3/2} - \lambda_b^{l+2}}{\lambda_a - \lambda_b} \cdot \frac{1}{\lambda_b^{l+3/2}}.$$

The sum of these two expressions is equal to

$$\frac{1}{2\lambda_a^{l+2}}\sum_b \frac{\lambda_a^{l+2} - \lambda_b^{l+2}}{\lambda_a - \lambda_b} \cdot \frac{1}{\lambda_b^{l+3/2}}.$$

Note that powers $l + 2$ are already integer and the remaining ratio can be represented as a sum of $l + 2$ terms. Adding also the last term from the l.h.s. of (3.36), we finally obtain:

$$-\frac{1}{\lambda_a^{l+1}} + \frac{1}{2}\sum_{n=-1}^{a}\frac{1}{\lambda_a^{n+2}}\sum_b \frac{1}{\lambda_b^{l-n+1/2}}$$
$$= \frac{1}{2}\sum_{n=-1}^{a}\frac{1}{\lambda_a^{n+2}}(2a - 2n + 1)(T + r)_{2l-2n+1}$$

in accordance with (3.34) and (3.35).

Calculations can be repeated for every particular monomial potential $V(x) = x^{p+1}/(p+1)$, but they become far more tedious and no general derivation of $W^{(p)}$-constraints [44] is yet found on these lines. See Ref. 45 for detailed examination of the $W^{(3)}$-constraints in the $X^4/4$-Kontsevich model.

4   Kp/Toda τ-Function in Terms of Free Fermions

There are several different definitions of τ-functions, but all of them are particular realizations of the following idea: τ-function is a generating functional of all the matrix elements of some group element in a particular representation. Since methods of geometrical quantization allow us to express all the group-theoretical objects in terms of the quantum theory of free fields, generic τ-functions can be also considered as nonperturbative partition functions of such models. The basic property of τ-function, which can be practically derived in such a general context, is that it always satisfy certain bilinear equations, of which Hirota equation for the conventional KP τ-function is the simplest example.

KP/Toda τ-functions are associated with the free particles of a peculiar type: free fermions in $1 + 1$ dimensions [46]. Existence of fermionization is a very rare property of free field theory (in variance with bosonization which is always available). If existing it leads to dramatic simplification of the formalism and to especially simple determinant formulas (instead of sophisticated and often somewhat abstract objects like chiral determinants $\det \bar{\partial}$ in the generic case). In the case of Kac-Moody algebras the corresponding τ-function is nothing but the nonperturbative partition function of the corresponding Wess-Zumino-Novikov-Witten model. Among simply laced algebras only $\widehat{GL(N)}_{k=1}$ is straightforwardly fermionized, and the formalism is much simpler in this case than for the generic Wess-Zumino-Witten model with arbitrary level k. For $N = 1$ we obtain KP/Toda τ-functions, while $N \neq 1$ are related to the "N-component KP/Toda systems." Level-one Kac-Moody algebras $\widehat{SL(N)}_{k=1}$ are distinguished because their universal enveloping is essentially the same as that of their Cartan subalgebras. This allows us to define generation functions with the help of sets of mutually commuting generators and makes evolution, described by commuting Hamiltonian flows, complete (acting transitively on the orbits of the group). This is why such systems are distinguished from the point of view of Hamiltonian integrability—and why they are the usual personages in the theory of integrable hierarchies. In the general case ($k \neq 1$) one naturally deals with the set of flows that form closed but non-Abelian algebra. In the language of matrix models, restriction to $k = 1$ and free *fermions* are essentially equivalent to restriction to *eigenvalue* models. Serious consideration of non-eigenvalue models, aimed at revealing their integrable (solvable) structure will certainly involve the theory of generic τ-functions.

4.1 Explicit Definition

Let us introduce two fields (a spin-$\frac{1}{2}$ b, c-system) $\tilde{\psi}(z)$ and $\psi(z)$ satisfying the canonical commutation relation

$$[\tilde{\psi}(\tilde{z}), \psi(z)]_+ = \delta(\tilde{z} - z)d\tilde{z}^{1/2}dz^{1/2}.$$

Then

$$\tau\{A\} \sim \langle 0| \exp\left(\oint_{d\tilde{z}} \oint_{dz} A(z, \tilde{z})\psi(z)\tilde{\psi}(\tilde{z})\right)|0\rangle. \tag{4.1}$$

Now it is usual to expand in Laurent series around $z = 0$,

$$\psi(z) = \sum_{n \in Z} \psi_n z^n dz^{1/2}, \quad \tilde{\psi}(z) = \sum_{n \in Z} \tilde{\psi}_n z^{-n-1} dz^{1/2},$$

$$[\tilde{\psi}_m, \psi_n]_+ = \delta_{m,n},$$

$$\psi_m|0\rangle = 0 \quad \text{for } m < 0, \quad \tilde{\psi}_m|0\rangle = 0 \quad \text{for } m \geq 0,$$

$$A(z, \tilde{z}) = \sum_{m,n \in \mathbb{Z}} z^{-m-1}\tilde{z}^n A_{mn} dz^{1/2}d\tilde{z}^{1/2},$$

so that

$$\oint_{d\tilde{z}} \oint_{dz} A(z, \tilde{z})\psi(z)\tilde{\psi}(\tilde{z}) = \sum_{m,n \in Z} A_{mn}\psi_m\tilde{\psi}_n.$$

In fact this expansion could be around *any* pair of points z_0, z_∞ and on a 2-surface of any topology: topological effects can be easily included as specific shifts of the functional $A(z, \tilde{z})$—by combinations of the "handle-gluing operators." Analogous shifts can imitate the change of basic functions z^n for $z^{n+\alpha}$ and more complicated expressions (holomorphic $\frac{1}{2}$-differentials with various boundary conditions on surfaces of various topologies).

One can now wonder, whether *local* functionals $A(z, \tilde{z}) = U(z)\delta(\tilde{z} - z)dz^{1/2}d\tilde{z}^{1/2}$ play any special role. The corresponding contribution to the Hamiltonian looks like

$$H_{\text{Cartan}} = \oint_{dz} U(z)\psi(z)\tilde{\psi}(z) = \oint_{dz} U(z)J(z),$$

where

$$J(z) = \psi(z)\tilde{\psi}(z) = \sum_{n \in \mathbb{Z}} J_n z^{-n-1} dz$$

is the $U(1)_{k=1}$ Kac-Moody current;

$$J_n = \sum_{m \in \mathbb{Z}} \psi_m \tilde{\psi}_{m+n}; \quad [J_m, J_n] = m \delta_{m+n,0}.$$

If scalar function (potential) $U(z)$ is expanded as $U(z) = \sum_{k \in \mathbb{Z}} t_k z^k$, then

$$H_{\text{Cartan}} = \sum_{n \in \mathbb{Z}} t_k J_k.$$

This contribution to the whole Hamiltonian can be considered distinguished for the following reason. Let us return to the original expression (4.1) and try to consider it as a generating functional for all the correlation functions of $\tilde{\psi}$ and ψ. Naively, variation w.r. to $A(z, \tilde{z})$ should produce bilinear combination $\psi(z)\tilde{\psi}(\tilde{z})$ and this would solve the problem. However, things are not just so trivial, because operators involved do not commute (and in particular, the exponential operator in (4.1) should still be defined less symbolically; see next subsection). Things would be much simpler, if we can consider *commuting* set of operators: this is where Abelian $\widehat{U(1)}_{k=1}$ subgroup of the entire $\text{GL}(\infty)_{k=1}$ (and even its purely commuting Borel subalgebra) enters the game. Remarkably, it is sufficient to deal with this Abelian subgroup in order to reproduce all the correlation functions.[15] The crucial point is the identity for free fermions (generalizable to any b, c-systems)

$$:\psi(\lambda)\tilde{\psi}(\tilde{\lambda}): = :\exp\left(\int_\lambda^{\tilde{\lambda}} J\right): \tag{4.2}$$

which is widely known in the form of bosonization formulas[16]: if $J(z) = \partial\phi(z)$,

$$\tilde{\psi}(\tilde{\lambda}) \sim :e^{\phi(\tilde{\lambda})}: \qquad (:\psi(\infty)\tilde{\psi}(\tilde{\lambda}): = :e^{(\phi(\tilde{\lambda})-\phi(\infty))}:);$$

$$\psi(\lambda) \sim :e^{-\phi(\lambda)}: \qquad (:\psi(\lambda)\tilde{\psi}(\infty): = :e^{(\phi(\infty)-\phi(\lambda))}:).$$

---

[15] We once again emphasize that this trick is specific for the free fermions and for the level $k = 1$ Kac-Moody algebras, which can be expressed entirely in terms of free fields, associated with Cartan generators (modulo some unpleasant details, related to "cocycle factors" in the Frenkel-Kac representations [12], which are in fact reminiscent of free fields associated with the non-Cartan generators (parafermions) [47]—but can, however, be put under the carpet or/and taken into account "by hands" as unpleasant but nonessential(?) sophistications).

[16] Formulas in brackets are indeed correct; they are preceded by the usual symbolic relations. Using these formulas we get

$$:\psi(\lambda)\tilde{\psi}(\tilde{\lambda}): = :e^{\phi(\tilde{\lambda})-\phi(\lambda)}: = :e^{\int_\lambda^{\tilde{\lambda}} \partial\phi}: = :e^{\int_\lambda^{\tilde{\lambda}} J}:.$$

This identity can be of course obtained within fermionic theory, one should only take into account that ψ-operators are nilpotent, so that exponent of a single ψ-operator would be just a sum of two terms (polynomial) and carefully follow the normal ordering prescription.

This identity implies that one can generate any bilinear combinations of ψ-operators by variation of potential $U(z)$ only, moreover this variation should be of specific form:

$$\Delta \oint U J = \Delta \left(\sum_{k \in \mathbb{Z}} t_k J_k \right) = \int_z^{\bar{z}} J$$

$$= \sum_{k \in \mathbb{Z}} \int_z^{\bar{z}} z^{-k-1} dz = \sum_{k \in \mathbb{Z}} \frac{1}{k} J_k \left(\frac{1}{z^k} - \frac{1}{\bar{z}^k} \right),$$

i.e.,

$$\Delta t_k = \frac{1}{k} \left(\frac{1}{z^k} - \frac{1}{\bar{z}^k} \right)$$

Note that this is *not* an infinitesimal variation and that it has exactly the form, consistent with Miwa parametrization.

Since any bilinear combination can be generated in this way from $U(z)$, it is clear that the entire Hamiltonian $\sum A_{mn} \psi_m \psi_n$ can be also considered as resulting from some transformation of V (i.e., of "time variables" t_k). In other words,

$$\tau\{A\} = \mathcal{O}_A[t] \tau\{A = U\}.$$

These operators $\mathcal{O}_A$ are naturally interpreted as elements of the group GL(∞), acting on the universal Grassmannian GR [48–50], parametrized by the matrices A_{mn} modulo changes of coordinates $z \rightarrow f(z)$. This representation for $\tau\{A\}$ is, however, not very convenient, and usually one considers *infinitesimal* version of the transformation, which just shifts A

$$\tau\{t \mid A + \delta A\} = \hat{\mathcal{O}}_{\delta A}[t] \tau\{t \mid A\}, \tag{4.3}$$

note that this transformation clearly distinguishes between the dependencies of τ on t and on all other components of A. The possibility of such representation with the privileged role of Cartan generators is the origin of all simplifications, arising in the case of free-fermion τ-functions. Relation (4.3) is the basis of the orbit interpretation of τ-functions [49].

4.2 Basic Determinant Formula for the Free-Fermion Correlator

Let us consider the following matrix element:

$$\tau_N\{t, \bar{t} \mid G\} = \langle N | e^H G e^{\overline{H}} | N \rangle \tag{4.4}$$

where

$$\psi(z) = \sum_{n \in \mathbb{Z}} \psi_n z^n dz^{1/2}, \quad \tilde{\psi}(z) = \sum_{n \in \mathbb{Z}} \tilde{\psi}_n z^{-n-1} dz^{1/2},$$

$$G = \exp\left(\sum_{m,n \in \mathbb{Z}} A_{mn} \psi_m \tilde{\psi}_n \right),$$

$$H = \sum_{k>0} t_k J_k, \quad \bar{H} = \sum_{k>0} \bar{t}_k J_{-k},$$

$$J(z) = \psi(z)\tilde{\psi}(z) = \sum_{n \in \mathbb{Z}} J_n z^{-n-1} dz, \quad J_n = \sum_k \psi_k \tilde{\psi}_{k+n}, \tag{4.5}$$

$$[\tilde{\psi}_m, \psi_n]_+ = \delta_{m,n}, \quad [J_m, J_n] = m\delta_{m+n,0},$$

$$\psi_m|N\rangle = 0, m < N, \quad \langle N|\psi_m = 0, m \geq N,$$

$$\tilde{\psi}_m|N\rangle = 0, m \geq N, \quad \langle N \mid \tilde{\psi}_m = 0, m < N,$$

$$J_m|N\rangle = 0, m > 0, \quad \langle N|J_m = 0, m < 0.$$

The "Nth vacuum" $|N\rangle$ is defined as the Dirac sea, filled up to the level N,

$$|N\rangle = \prod_{i=N}^{\infty} \tilde{\psi}_i |\infty\rangle = \prod_{i=-\infty}^{N-1} \psi_i \mid -\infty\rangle,$$

$$\langle N| = \langle\infty| \prod_{i=N}^{\infty} \psi_i = \langle-\infty| \prod_{i=-\infty}^{N-1} \tilde{\psi}_i,$$

where the "empty" (bare) and "completely filled" vacua are defined so that

$$\tilde{\psi}_m|-\infty\rangle = 0, \qquad\qquad \langle-\infty|\psi_m = 0,$$

$$\psi_m|\infty\rangle = 0, \qquad\qquad \langle\infty|\tilde{\psi}_m = 0,$$

for *any* $m \in \mathbb{Z}$. For the only reason that operators J, H, $\bar{H}$ and G are defined so that they always have $\tilde{\psi}$ at the very right and ψ at the very left, we get also:

$$J_m|-\infty\rangle = 0, \qquad\qquad \langle-\infty|J_m = 0,$$

$$G^{\pm 1}|-\infty\rangle = |-\infty\rangle, \qquad\qquad \langle-\infty|G^{\pm 1} = \langle-\infty|,$$

$$e^{\pm\bar{H}}|-\infty\rangle = |-\infty\rangle, \qquad\qquad \langle-\infty|e^{\pm H} = \langle-\infty|.$$

Now we can use all these formulas to rewrite our original correlator (4.4) as

$$\langle N|e^H G e^{\bar{H}}|N\rangle = \langle-\infty|\left(\prod_{i=-\infty}^{N-1} \tilde{\psi}_i \right) e^H G e^{\bar{H}} \left(\prod_{i=-\infty}^{N-1} \psi_i \right)|-\infty\rangle$$

$$= \langle -\infty | e^{-H} \Big(\prod_{i=-\infty}^{N-1} \tilde{\psi}_i \Big) e^{H} G e^{\bar{H}} \Big(\prod_{i=-\infty}^{N-1} \psi_i \Big) e^{-\bar{H}} | -\infty \rangle$$

$$= \langle -\infty | \prod_{i=-\infty}^{N-1} \tilde{\Psi}_i[t] \prod_{j=-\infty}^{N-1} \Psi_j^G[\bar{t}] | -\infty \rangle$$

$$= \mathrm{Det}_{-\infty < i,j < N} \langle -\infty | \tilde{\Psi}_i[t] \Psi_j^G[\bar{t}] | -\infty \rangle$$

$$= \mathrm{Det}_{i,j<0} \, \mathcal{H}_{i+N,j+N}. \qquad (4.6)$$

The last two steps here were introduction of "GL(∞)-rotated" fermions,

$$\tilde{\Psi}_i[t] \equiv e^{-H} \psi_i e^{H}, \quad \Psi_j[\bar{t}] \equiv e^{\bar{H}} \psi_j e^{-\bar{H}}, \quad \Psi_j^G[\bar{t}] \equiv G \Psi_j[\bar{t}] G^{-1}, \quad (4.7)$$

and application of the Wick theorem to express multifermion correlation function through pair correlators

$$\mathcal{H}_{ij}(t,\bar{t}) \equiv \langle -\infty | \tilde{\Psi}_i[t] \Psi_j^G[\bar{t}] | -\infty \rangle = \langle -\infty | \tilde{\Psi}_i[t] G \Psi_j[\bar{t}] | -\infty \rangle. \qquad (4.8)$$

(Once again the fact that $G^{-1} | -\infty \rangle = | -\infty \rangle$ was used.) The only nontrivial dynamical information entered through applicability of the Wick theorem, and for that it was crucial that all the operators e^H, $e^{\bar{H}}$, and G be *quadratic* exponents, i.e., can only modify the shape of the propagator, but do not destroy the quadratic form of the action (fields remain *free*). This is exactly equivalent to the statement that "Heisenberg" operators $\Psi[t]$ are just "rotations" of ψ, i.e., that transformations (4.7) are *linear*.

We shall now describe these transformations in a little more explicit form. Namely, their entire time dependence can be encoded in terms of the ordinary Schur polynomials $P_n(t)$. These are defined to have a very simple generating function (which we already encountered many times in the theory of matrix models)

$$\sum_{n \geq 0} P_n(t) z^n = \exp \Big(\sum_{k=1}^{\infty} t_k z^k \Big)$$

(i.e., $P_0 = 1$, $P_1 = t_1$, $P_2 = t_1^2/2 + t_2$, etc.), and satisfy the relation

$$\frac{\partial P_n}{\partial t_k} = P_{n-k}.$$

Since

$$\exp \Big(\sum_{k=1}^{\infty} t_k z^k \Big) = \prod_{k>0} \Big(\sum_{n_k \geq 0} \frac{1}{n_k!} t_k^{n_k} z^{kn_k} \Big),$$

Schur polynomials can be also represented as

$$P_n(t) = \sum_{\substack{\{n_k\} \\ \sum_{k>0} kn_k = n}} \Big(\prod_{k>0} \frac{1}{n_k!} t_k^{n_k} \Big). \qquad (4.9)$$

Now, since

$$e^{-B} A e^{B} = A + [A, B] + \frac{1}{2!}[[A, B], B] + \frac{1}{3!}\left[[[A, B], B], B\right] + \cdots$$

and

$$[\tilde{\psi}_i, J_k] = \tilde{\psi}_{i+k}, \quad [[\tilde{\psi}_i, J_{k_1}], J_{k_2}] = \tilde{\psi}_{i+k_1+k_2}, \ldots,$$

we have for every fixed k,

$$e^{-t_k J_k} \tilde{\psi}_i e^{t_k J_k} = \sum_{n_k \geq 0} \frac{t_k^{n_k}}{n_k!} \tilde{\psi}_{i+kn_k}.$$

It remains to note that all the harmonics of J in $H = \sum_{k>0} t_k J_k$ commute with each other, to obtain

$$\tilde{\Psi}_i(t) = e^{-H} \tilde{\psi}_i e^{H} = \left(\prod_{k>0} e^{-t_k J_k}\right) \tilde{\psi}_i \left(\prod_{k>0} e^{t_k J_k}\right)$$

$$= \sum_{n \geq 0} \tilde{\psi}_{i+n} \left(\sum_{\substack{\{n_k\} \\ \sum_{k>0} k n_k = n}} \left(\prod_{k>0} \frac{1}{n_k!} t_k^{n_k}\right)\right)$$

$$\overset{(4.9)}{=} \sum_{n \geq 0} \tilde{\psi}_{i+n} P_n(t) = \sum_{l \geq i} \tilde{\psi}_l P_{l-i}(t).$$

Similarly, relation $[J_k, \psi_j] = \psi_{k+j}$ implies that

$$\Psi_j(\bar{t}) = e^{\overline{H}} \psi_j e^{-\overline{H}} = \sum_{n \geq 0} \psi_{j+n} P_n(\bar{t}) = \sum_{m \geq j} \psi_m P_{m-j}(\bar{t})$$

and finally

$$\mathcal{H}_{ij} = \sum_{\substack{l \geq i \\ m \geq j}} \langle -\infty | \tilde{\psi}_l G \psi_m | -\infty \rangle P_{l-i}(t) P_{m-j}(\bar{t})$$

$$= \sum_{\substack{l \geq i \\ m \geq j}} T_{lm} P_{l-i}(t) P_{m-j}(\bar{t}), \qquad (4.10)$$

which implies also that

$$\frac{\partial \mathcal{H}_{ij}}{\partial t_k} = \mathcal{H}_{i+k,j},$$

$$\frac{\partial \mathcal{H}_{ij}}{\partial \bar{t}_k} = \mathcal{H}_{i,j+k}. \qquad (4.11)$$

The matrix

$$T_{lm} \equiv \langle -\infty | \tilde{\psi}_l G \psi_m | -\infty \rangle \qquad (4.12)$$

is the one that defines fermion rotations under the action of GL(∞)-group element G

$$G\psi_m G^{-1} = \sum_{l\in\mathbb{Z}} \psi_l T_{lm},$$

$$G^{-1}\tilde{\psi}_l G = \sum_{m\in\mathbb{Z}} T_{lm}\tilde{\psi}_m, \quad \text{or} \quad G\tilde{\psi}_l G^{-1} = \sum_{m\in\mathbb{Z}} (T^{-1})_{lm}\tilde{\psi}_m.$$

If $G = 1$, $T_{lm} = \delta_{lm}$. If all $t_k = \bar{t}_k = 0$, $\mathcal{H}_{ij} = T_{ij}$.

4.3 KP Hierarchy and Other Reductions

In the previous subsection a formula

$$\tau_N\{t,\bar{t}\mid G\} = \text{Det}_{i,j<0}\,\mathcal{H}_{i+N,j+N} \tag{4.13}$$

was derived for the basic correlator, which defines "Toda-lattice τ-function." For obvious reasons the variables $\bar{t}$ are often referred to as "negative-times." The τ-function can be normalized by division over the same quantity with all the time variables vanishing, but this is not always convenient. Equation (4.13) has generalizations—when similar matrix elements in a multifermion system is considered—this leads to "multicomponent Toda" (or AKNS) τ-functions. Generalizations to arbitrary conformal models should be considered as well. It has also particular "reductions," of which the most important are: KP (Kadomtsev-Petviashvili), forced (semi-infinite), and Toda-chain τ-functions. This is the subject to be discussed in this subsection.

The idea of linear reduction is that the form of operator G, or, what is the same, of the matrix T_{lm} in eq. (4.10), can be adjusted in such a way, that $\tau_N\{t,\bar{t}\mid G\}$ becomes independent of some variables, i.e., equation(s)

$$\left(\sum_k \alpha\frac{\partial}{\partial t_k} + \sum_k \bar{\alpha}\frac{\partial}{\partial\bar{t}_k} + \sum_k \beta_k D_N(k) + \gamma\right)\tau_N\{t,\bar{t}\mid G\} = 0 \tag{4.14}$$

can be solved as equations for G for all the values of t, $\bar{t}$, and N at once. (In (4.14) $D_N(k)f_N \equiv f_{N+k} - f_N$.) In this case the system of integrable equations (hierarchy), arising from Hirota equation for τ, gets reduced and one usually speaks about "reduced hierarchy." Usually equation (4.14) is applied directly to matrix $\mathcal{H}_{ij}$; then , of course, (4.14) is just a corollary.

We shall refer to the situation when (4.14) is fulfilled for *any* t, $\bar{t}$, N as "strong reduction." It is often reasonable to consider also "weak reductions," when (4.14) is satisfied on particular infinite-dimensional hyperplanes in the space of time variables. Weak reduction is usually a property of the entire τ-function as well, but not expressible in the form of a local linear equation, satisfied identical for *all* values of t, $\bar{t}$, N. Now we proceed to concrete examples.

4.3.1   Toda-Chain Hierarchy

This is a *strong* reduction. The corresponding constraint (4.14) is just

$$\frac{\partial \mathcal{H}_{ij}}{\partial t_k} = \frac{\partial \mathcal{H}_{ij}}{\partial \bar{t}_k},$$

or, because of (4.11), $\mathcal{H}_{i+k,j} = \mathcal{H}_{i,j+k}$. It has an obvious solution:

$$\mathcal{H}_{i,j} = \widehat{\mathcal{H}}_{i+j},$$

i.e., $\mathcal{H}_{ij}$ is expressed in terms of a one-index quantity $\widehat{\mathcal{H}}_i$. It is, however, not enough to say what are the restrictions on $\mathcal{H}_{ij}$—they should be fulfilled for all t and $\bar{t}$ at once, i.e., they should be resolvable as equations for T_{lm}. In the case under consideration this is simple: T_{lm} should be such that

$$T_{lm} = \widehat{T}_{l+m}.$$

Indeed, then

$$\begin{aligned}
\mathcal{H}_{ij} &= \sum_{l,m} T_{lm} P_{l-i}(t) P_{m-j}(\bar{t}) \\
&= \sum_{l,m} \widehat{T}_{l+m} P_{l-i}(t) P_{m-j}(\bar{t}) \\
&= \sum_{n \geq 0} \widehat{T}_{n+i+j} \left(\sum_{k=0}^{n} P_k(t) P_{n-k}(\bar{t}) \right),
\end{aligned}$$

and

$$\widehat{\mathcal{H}}_i = \sum_{n \geq 0} \widehat{T}_{n+i} \left(\sum_{k=0}^{n} P_k(t) P_{n-k}(\bar{t}) \right). \tag{4.15}$$

4.3.2   Volterra Hierarchy

Toda-chain τ-function can be further *weakly* reduced to satisfy the identity

$$\left. \frac{\partial \tau_{2N}}{\partial t_{2k+1}} \right|_{\{t_{2l+1}=0\}} = 0, \quad \text{for all } k, \tag{4.16}$$

i.e., τ_{2N} is requested to be even function of all odd-times t_{2l+1} (this is an example of "global characterization" of the weak reduction). Note that (4.16) is imposed only on *Toda-chain* τ-function with *even* values of zero-time. Then (4.16) will hold whenever $\widehat{\mathcal{H}}_i$ in (4.15) are even (odd) functions of t_{odd} for even (odd) values of i. Since Schur polynomials $P_k(t)$ are even (odd) functions of odd-times for even (odd) k, it is enough that the sum in (4.15) goes over even (odd) n when i is even (odd). In other words, the restriction on T_{lm} is that

$$T_{lm} = \widehat{T}_{l+m} \quad \text{and} \quad \widehat{T}_{2k+1} = 0 \text{ for all } k.$$

4.3.3   Forced Hierarchies

This is another important example of strong reduction. It also provides an example of *singular* τ-functions, arising when $G = \exp(\sum A_{mn}\psi_m\tilde\psi_n)$ blows up and normal ordered operators should be used to define regularized τ-functions. Forced hierarchy appears when G can be represented in the form [51] $G = G_0 P_+$, where projection operator P_+ is such that

$$P_+|N\rangle = \begin{cases} |N\rangle & \text{for } N \geq N_0, \\ 0 & \text{for } N < N_0. \end{cases} \tag{4.17}$$

Explicit expression for this operator is[17]

$$P_+ = \ :\exp\left(-\sum_{l<N_0} \tilde\psi_l\psi_l\right): \ = \prod_{l<N_0}(1 - \tilde\psi_l\psi_l) = \prod_{l<N_0}\psi_l\tilde\psi_l.$$

Because of (4.17), $P_+|-\infty\rangle = 0$, and the identity $G|-\infty\rangle = |-\infty\rangle$, which was essentially used in the derivation in (4.9), can be satisfied only if G_0 is singular and $T_{lm} = \infty$. In order to avoid this problem one usually introduces in the vicinity of such singular points in the universal module space a sort of normalized (forced) τ-function $\tau_N^f \equiv \tau_N\tau_{N_0}$. One can check that now $T_{lm}^f = \infty$ for all $l, m < N_0$, and τ^f can be represented as determinant of a final-dimensional matrix [51, 52]:

$$\tau_N^f = \begin{cases} \text{Det}_{N_0 \leq i,j < N}\, \mathcal{H}_{ij}^f & \text{for } N > N_0, \\ 1 & \text{for } N = N_0, \\ 0 & \text{for } N < N_0. \end{cases}$$

For $N > N_0$ we now have a determinant of a *finite*-dimensional $(N - N_0) \times (N - N_0)$ matrix. The choice of N_0 is not really essential, therefore it is better to put $N_0 = 0$ in order to simplify formulas, phrasing, and relation with the discrete matrix models (N_0 is easily restored if everywhere N is substituted by $N - N_0$). For forced hierarchies one can also represent $\hat\tau$ as

$$\tau_N^f = \text{Det}_{0 \leq i,j < N}\, \partial_1^i \bar\partial_1^j \mathcal{H}^f,$$

where $\mathcal{H}^f = \mathcal{H}_{00}^f$ and $\partial_1 = \partial/\partial t_1$, $\bar\partial_1 = \partial/\partial \bar t_1$. For *forced Toda-chain* hierarchy this turns into even simpler expression

$$\tau_N^f = \text{Det}_{0 \leq i,j < N}\, \partial_1^{i+j} \widehat{\mathcal{H}}^f,$$

---

[17]Normal ordering sign : : means that all operators $\tilde\psi$ stand to the *left* of all operators ψ. The product at the r.h.s. obviously implies both the property (4.17) and projection property $P_+^2 = P_+$.

while for the *forced Volterra* case we get a product of two Toda-chain τ-functions with twice as small value of N [53]

$$\tau_{2N}^f = \left(\mathrm{Det}_{0 \le i,j < N}\, \partial_2^{i+j}\widehat{\mathcal{H}}^f\right) \cdot \left(\mathrm{Det}_{0 \le i,j < N}\, \partial_2^{i+j}(\partial_2 \widehat{\mathcal{H}}^f)\right)$$
$$= \tau_N^f[\widehat{\mathcal{H}}^f] \cdot \tau_N^f[\partial_2 \widehat{\mathcal{H}}^f].$$

Forced τ_N^f can be *always* represented in the form of a scalar-product matrix model. Indeed,

$$\mathcal{H}_{ij} = \sum T_{lm} P_{l-i}(t) P_{m-j}(\bar{t}) = \oint\oint e^{U(h)+\overline{U}(\bar{h})} h^i \bar{h}^j T(h,\bar{h})\, dh\, d\bar{h},$$

where $T(h,\bar{h}) \equiv \sum_{lm} T_{lm} h^{-l-1} \bar{h}^{-m-1}$, and $e^{U(h)} = e^{\sum_{k>0} t_k h^k} = \sum_{l \ge 0} h^l P_l(t)$. Then, since $\mathrm{Det}_{0 \le i,j < N}\, h^i = \Delta_N(h)$—this is where it is essential that the hierarchy is forced—

$$\mathrm{Det}_{0 \le i,j < N}\, \mathcal{H}_{ij} = \prod_i \oint\oint e^{U(h_i)+\overline{U}(\bar{h}_i)} T(h_i,\bar{h}_i)\, dh_i\, d\bar{h}_i \cdot \Delta_N(h)\Delta_N(\bar{h}),$$

i.e., we obtain a scalar-product model with

$$d\mu_{h,\bar{h}} = e^{U(h)+\overline{U}(\bar{h})} T(h,\bar{h})\, dh\, d\bar{h}.$$

The inverse is also true: the partition function of every scalar-product model is a forced Toda-lattice τ-function.

4.3.4 Kp Hierarchy

In this case we just ignore the dependence of the τ-function on times $\bar{t}$. Every Toda-lattice τ-function can be considered also as KP τ-function: just operator $G^{KP} \equiv G e^{\overline{H}}$ (a point of Grassmannian) becomes $\bar{t}$-dependent. Usually N-dependence is also eliminated. This can be considered as a little more sophisticated change of G. When N is fixed, extra changes of field variables are allowed, including transformation from a Ramond to a Neveu-Schwarz sector, etc. Often the KP hierarchy is from the very beginning formulated in terms of Neveu-Schwarz (antiperiodic) fermionic fields (associated with principal representations of Kac-Moody algebras), i.e., expansions in the first line of (4.5) are in semi-integer powers of z: $\psi_{NS}(z) = \sum_{n \in \mathbb{Z}} \psi_n z^{n-1/2} dz^{1/2}$.

Given a KP τ-function one can usually construct a Toda-lattice with the *same* G, by introducing in an appropriate way dependencies on $\bar{t}$ and N. For this purpose τ^{KP} should be represented in the form of (4.13),

$$\tau^{KP}\{t \mid G\} = \mathrm{Det}_{i,j<0}\, \mathcal{H}_{ij}^{KP}, \tag{4.18}$$

where $\mathcal{H}_{ij}^{KP} = \sum_l T_{lj} P_{l-i}(t)$. Since T_{lm} is a function of G only, it does not change when we build up a Toda-lattice τ-function:

$$\tau_N\{t, \bar{t} \mid G\} = \mathrm{Det}_{i,j<0}\, \mathcal{H}_{i+N,j+N},$$

$$\mathcal{H}_{ij} = \sum_{l,m} T_{lm} P_{l-i}(t) P_{m-j}(\bar{t}) = \sum_m \mathcal{H}_{im}^{KP} P_{m-j}(\bar{t}).$$

Then

$$\tau^{KP}\{t \mid G\} = \tau_0\{t, 0 \mid G\}.$$

If we go in the opposite direction, when Toda-lattice τ-function is considered as KP τ-function,

$$\tau_0\{t, \bar{t} \mid G\} = \tau^{KP}\{t \mid \widetilde{G}(\bar{t})\},$$

$$\widetilde{\mathcal{H}}_{ij}^{KP} = \sum_m \mathcal{H}_{im} P_{m-j}(\bar{t}), \qquad (4.19)$$

$$\widetilde{T}_{lj}\{\widetilde{G}(\bar{t})\} = \sum_m T_{lm}\{G\} P_{m-j}(\bar{t}).$$

KP reduction in its turn has many further weak reductions (KdV and Boussinesq being the simplest examples).

4.4 Fermion Correlator in Miwa Coordinates

Let us now return to original correlator (4.4) and discuss in a little more detail the implications of bosonization identity (4.2). In order not to write down integrals of J, we introduce scalar field[18]:

$$\phi(z) = \sum_{\substack{k \neq 0 \\ k \in \mathbb{Z}\backslash 0}} \frac{J_{-k}}{k} z^k + \phi_0 + J_0 \log z,$$

such that $\partial\phi(z) = J(z)$. Then (4.2) states that

$$:\!\psi(\lambda)\tilde{\psi}(\tilde{\lambda})\!: \; = \; :\!e^{\phi(\tilde{\lambda})-\phi(\lambda)}\!:. \qquad (4.20)$$

"Normal ordering" here means nothing more but the requirement to neglect all mutual contractions (or correlators) of operators in between : : when the Wick theorem is applied to evaluate correlation functions. One can also get rid of the normal ordering sign at the l.h.s. of (4.20) then

$$\psi(\lambda)\tilde{\psi}(\tilde{\lambda}) = \; :\!e^{\phi(\tilde{\lambda})}\!::\!e^{-\phi(\lambda)}\!:. \qquad (4.21)$$

---

[18]One can consider ϕ as introduced for simplicity of notation, but it should be kept in mind that the scalar-field representation is in fact more fundamental for *generic* τ-functions, not related to the level $k = 1$ Kac-Moody algebras. (This phenomenon is well known in conformal filed theory; see Ref. 13 for more details.)

In distinguished coordinates on a sphere, when the free-field propagator is just $\log(z - \tilde{z})$, one also has

$$\psi(z)\tilde{\psi}(\tilde{z}) = \frac{1}{z - \tilde{z}} :\psi(z)\tilde{\psi}(\tilde{z}):.$$

Our task now is to express operators e^H and $e^{\bar{H}}$ through the field ϕ. This is simple:

$$H = \oint_0 U(z)J(z) = \oint_0 U(z)\partial\phi(z) = -\oint_0 \phi(z)\partial U(z).$$

Here as usual $U(z) = \sum_{k>0} t_k z^k$ and the integral is around $z = 0$. This is very similar to the generic linear functional of $\phi_-(\lambda) \equiv -\sum_{k>0} J_k \lambda^{-k}/k$,

$$H = \int \phi_-(\lambda)f(\lambda)\,d\lambda, \tag{4.22}$$

and one should only require that[19]

$$\partial U(z) = \oint \frac{f(\lambda)}{z - \lambda}\,d\lambda,$$

i.e.,

$$U(z) = \oint \log\left(1 - \frac{z}{\lambda}\right) f(\lambda)\,d\lambda.$$

In terms of time variables this means that

$$t_k = -\frac{1}{k}\int \lambda^{-k} f(\lambda)\,d\lambda. \tag{4.23}$$

Here we required that $U(z = 0) = 0$. Sometimes it can be more natural to introduce also

$$t_0 = \int \log \lambda\, f(\lambda)\,d\lambda.$$

This change from the time variables to "time density" $f(\lambda)$ is known as the Miwa transformation. In order to establish relation with fermionic representation and also with matrix models we shall need it in "discretized" form:

$$t_k = \frac{\xi}{k}\left(\sum_a \lambda_a^{-k} - \sum_a \tilde{\lambda}_a^{-k}\right),$$

$$t_0 = -\xi\left(\sum_a \log \lambda_a - \sum_a \log \tilde{\lambda}_a\right). \tag{4.24}$$

---

[19] The factor $2\pi i$ is included in the definition of the contour integral $\oint$.

We changed integral over λ for a discrete sum (i.e., the density function $f(\lambda)$ is a combination of δ-functions, picked at some points λ_a, $\tilde{\lambda}_a$. This is of course just another basis in the space of the linear functionals, but the change from one basis to another one is highly nontrivial. The thing is that we selected the basis where amplitudes of different δ-functions are the *same*: parameter ξ in (4.24) is *independent* of a. Thus the real parameters are just positions of the points λ_a, $\tilde{\lambda}_a$, while the amplitude is defined by the density of these points in the integration (summation) domain. This domain does not need to be a priori specified: it can be a real line, any other contour, or—better—some Riemann surface.) Parameter ξ is also unnecessary to introduce because bases with different ξ are essentially equivalent. We shall soon put it equal to *one*, but not before Miwa transformation will be discussed in a little more detail.

Our next steps will be as follows. Substitution of (4.22) into (4.24), gives

$$H = -\xi \sum_a \phi_-(\lambda_a) + \xi \sum_a \phi_-(\tilde{\lambda}_a).$$

In fact, what we need is not the operator H itself, but the state which is created by e^H from the vacuum state $\langle N |$. Then, since $\langle N \mid J_m = 0$ for $m < 0$, $\langle N | e^{-\xi \phi_-(\lambda)}$ is essentially equivalent to $\langle N \mid e^{-\xi \phi(\lambda)}$ with $\phi_-(\lambda)$ substituted by entire $\phi(\lambda)$. If $\xi = 1$, $e^{-\phi(\lambda)}$ can be further changed for $\psi(\lambda)$ and we obtain an expression for the correlator (4.4) where e^H is substituted by a product of operators $\psi(\lambda_a)$. The same is of course true for $e^{\overline{H}}$. Then Wick theorem can be applied and a new type of determinant formulas arises like, for example,

$$\tau \sim \frac{\Delta(\lambda, \tilde{\lambda})}{\Delta^2(\lambda)\Delta^2(\tilde{\lambda})} \det{}_{ab}\langle N|\psi(\lambda_a)\tilde{\psi}(\tilde{\lambda}_b)G|N\rangle.$$

It can also be obtained directly from (4.6), (4.8), and (4.10) by Miwa transformation. The rest of this subsection describes this derivation in somewhat more detail.

The first task is to substitute ϕ_- by ϕ. For this purpose we introduce the operator

$$\sum_{k=-\infty}^{\infty} t_k J_k = H_+ + H_-,$$

where $H_+ = \sum_{k>0} t_k J_k$ is just our old H, $H_- = \sum_{k\geq 0} t_{-k} J_k$, and "negative times" t_{-k} are defined by "analytical continuation" of the same formulas (4.23) and (4.24):

$$t_{-k} = \frac{1}{k} \int \lambda^k f(\lambda) d\lambda = -\frac{\xi}{k}\left(\sum_a \lambda_a^k - \sum_a \tilde{\lambda}_a^k\right).$$

Then

$$\sum_{k=-\infty}^{\infty} t_k J_k = H_+ + H_- = -\xi \left(\sum_a \phi(\lambda_a) - \sum_a \phi(\tilde{\lambda}_a) \right). \qquad (4.25)$$

Further,

$$e^{H_+ + H_-} = e^{-s(t)/2} e^{H_+} e^{H_-} = e^{s(t)/2} e^{H_-} e^{H_+},$$

where

$$s(t) \equiv \sum_{k>0} k t_k t_{-k} = -\xi^2 \sum_{k>0} \frac{1}{k} \left(\sum_a (\lambda_a^{-k} - \tilde{\lambda}_a^{-k}) \sum_b (\lambda_b^k - \tilde{\lambda}_b^k) \right)$$

$$= \xi^2 \log \left(\prod_{a,b}' \frac{(1 - \lambda_b/\lambda_a)(1 - \tilde{\lambda}_b/\tilde{\lambda}_a)}{(1 - \tilde{\lambda}_b/\lambda_a)(1 - \lambda_b/\tilde{\lambda}_a)} \right) + \text{const},$$

where prime means that the terms with $a = b$ are excluded from the product in the numerator and accounted for in the infinite "constant," added at the r.h.s. In other words,

$$e^{s(t)/2} = \text{const} \cdot \left(\frac{\prod_{a>b}(\lambda_a - \lambda_b)(\tilde{\lambda}_a - \tilde{\lambda}_b)}{\prod_a \prod_b (\lambda_a - \tilde{\lambda}_b)} \right)^{\xi^2}$$

$$= \text{const} \cdot \left(\frac{\Delta^2(\lambda) \Delta^2(\tilde{\lambda})}{\Delta(\lambda, \tilde{\lambda})} \right)^{\xi^2}. \qquad (4.26)$$

Since $\langle N | J_m = 0$ for all $m < 0$, we have $\langle N | e^{H_-} = \langle N |$, and therefore

$$\langle N | e^H \equiv \langle N | e^{H_+} = \langle N | e^{H_-} e^{H_+} = e^{-s(t)/2} \langle N | e^{H_+ + H_-}.$$

From eq. (4.25),

$$e^{H_+ + H_-} = \text{const} \cdot \prod_a :e^{-\xi\phi(\lambda_a)}: :e^{\xi\phi(\tilde{\lambda}_a)}:$$

where "const" is exactly the same as in (4.26). If $\xi = 1$, eq. (4.21) can be used to write

$$\langle N | e^H = \frac{\Delta(\lambda, \tilde{\lambda})}{\Delta^2(\lambda) \Delta^2(\tilde{\lambda})} \langle N | \prod_a \psi(\lambda_a) \prod_a \tilde{\psi}(\tilde{\lambda}_a).$$

Similarly,

$$e^{\bar{H}} | N \rangle = \prod_b \psi(\bar{\lambda}_b) \prod_b \tilde{\psi}(\bar{\tilde{\lambda}}_b) | N \rangle \frac{\Delta(\bar{\lambda}, \bar{\tilde{\lambda}})}{\Delta^2(\bar{\lambda}) \Delta^2(\bar{\tilde{\lambda}})},$$

where

$$\bar{t}_k = -\frac{1}{k}\sum_b (\bar{\lambda}_b^k - \tilde{\bar{\lambda}}_b^k)$$

and we used the fact that $J_m|N\rangle = 0$ for all $m > 0$. Finally,

$$\tau_N\{t, \bar{t} \mid G\} = \langle N|e^H G e^{\bar{H}}|N\rangle$$

$$= \frac{\Delta(\lambda, \tilde{\lambda})}{\Delta^2(\lambda)\Delta^2(\tilde{\lambda})} \frac{\Delta(\bar{\lambda}, \tilde{\bar{\lambda}})}{\Delta^2(\bar{\lambda})\Delta^2(\tilde{\bar{\lambda}})}$$

$$\cdot \langle N| \prod_a \psi(\lambda_a) \prod_a \tilde{\psi}(\tilde{\lambda}_a) G \prod_b \psi(\bar{\lambda}_b) \prod_b \tilde{\psi}(\tilde{\bar{\lambda}}_b)|N\rangle. \quad (4.27)$$

Singularities at the coinciding points are completely eliminated from this expression, since poles and zeroes of the correlator are cancelled by those coming from the Vandermonde determinants.

Let us now put $N = 0$ and define normalized τ-function

$$\hat{\tau}_0\{t, \bar{t} \mid G\} \equiv \frac{\tau_0\{t, \bar{t} \mid G\}}{\tau_0\{0, 0 \mid G\}},$$

i.e., divide r.h.s. of (4.27) by $\langle 0|G|0\rangle$. Wick theorem now allows us to rewrite the correlator at the r.h.s. as a determinant of the block matrix:

$$\det \begin{pmatrix} \frac{\langle 0|\psi(\lambda_a)\tilde{\psi}(\tilde{\lambda}_b)G|0\rangle}{\langle 0|G|0\rangle} & \frac{\langle 0|\psi(\lambda_a)G\tilde{\psi}(\tilde{\bar{\lambda}}_b)|0\rangle}{\langle 0|G|0\rangle} \\ -\frac{\langle 0|\tilde{\psi}(\bar{\lambda}_b)G\psi(\bar{\lambda}_a)|0\rangle}{\langle 0|G|0\rangle} & \frac{\langle 0|G \ \psi(\bar{\lambda}_a)\tilde{\psi}(\tilde{\bar{\lambda}}_b)|0\rangle}{\langle 0|G|0\rangle} \end{pmatrix} \quad (4.28)$$

Special choices of points $\lambda_a, \ldots, \tilde{\bar{\lambda}}_b$ can lead to simpler formulas. If $\tilde{\bar{\lambda}}_a \to \bar{\lambda}_a$, so that $\bar{t}_k \to 0$, the matrix elements at the right lower block in (4.28) blow up, so that the off-diagonal blocks can be neglected. Then

$$\tau_0\{t, \bar{t} \mid G\} \to \tau^{KP}\{t \mid G\} = \frac{\langle 0|e^H G|0\rangle}{\langle 0|G|0\rangle}$$

$$= \frac{\Delta(\lambda, \tilde{\lambda})}{\Delta^2(\lambda)\Delta^2(\tilde{\lambda})} \times \det_{ab} \frac{\langle 0|\psi(\lambda_a)\tilde{\psi}(\tilde{\lambda}_b) \ G \ |0\rangle}{\langle 0|G|0\rangle}. \quad (4.29)$$

This function no longer depends on $\bar{t}$-times and is just a KP τ-function.

The matrix element

$$\varphi(\lambda, \tilde{\lambda}) = \frac{\langle 0|\psi(\lambda)\tilde{\psi}(\tilde{\lambda})G|0\rangle}{\langle 0|G|0\rangle}$$

is singular, when $\lambda \to \tilde{\lambda}$: $\varphi(\lambda, \tilde{\lambda}) \to 1/(\lambda - \tilde{\lambda})$. If now in (4.29) all $\tilde{\lambda} \to \infty$,

$$\tau^{KP}\{t \mid G\} = \frac{\det_{ab} \varphi_b(\lambda_a)}{\Delta(\lambda)}, \tag{4.30}$$

where

$$\varphi_b(\lambda) \equiv \langle 0|\psi(\lambda)\left(\partial^{b-1}\tilde{\psi}\right)(\infty)G|0\rangle \sim \lambda^{b-1}\left(1 + \mathcal{O}\left(\frac{1}{\lambda}\right)\right).$$

This is the main determinant representation of KP τ-function in Miwa parametrization.

Starting from representation (4.30) one can restore the corresponding matrix $\mathcal{H}_{ij}^{KP}$ in eq. (4.18) [42]:

$$\mathcal{H}_{ij}^{KP}\{t\} = \oint z^i \varphi_{-j}(z) e^{\sum_k t_k z^k} \, dz, \tag{4.31}$$

i.e.,

$$T_{lj}^{KP} = \oint z^l \varphi_{-j}(z).$$

Then obviously $\partial \mathcal{H}_{ij}^{KP}/\partial t_k = \mathcal{H}_{i+k,j}^{KP}$. Now we need to prove that the τ-function is given at once by $\det \varphi_a(\lambda_\delta)/\Delta(\lambda)$ and $\mathrm{Det}\, \mathcal{H}_{ij}^{KP}\{t\}$. In order to compare these two expressions one should take $t_k = \sum_a^n \lambda_a^{-k}/k$, so that

$$\exp\left(\sum_{k>0} t_k z^k\right) = \prod_{a=1}^n \frac{\lambda_a}{\lambda_a - z} = \left(\prod_a^n \lambda_a\right) \sum_a \frac{(-)^a}{z - \lambda_a} \frac{\Delta_a(\lambda)}{\Delta(\lambda)},$$

where

$$\Delta_a(\lambda) = \prod_{\substack{\alpha > \beta \\ \alpha, \beta \neq a}} (\lambda_\alpha - \lambda_\beta) = \frac{\Delta(\lambda)}{\prod_{\alpha \neq a}(\lambda_\alpha - \lambda_a)},$$

and

$$\mathcal{H}_{ij}^{KP}\Big|_{t_k = \sum_a^n \lambda_a^{-k}/k} = \left(\prod_a^n \lambda_a\right) \sum_a \frac{(-)^{a+1}\Delta_a(\lambda)}{\Delta(\lambda)} \lambda_a^i \varphi_{-j}(\lambda_a). \tag{4.32}$$

As far as n is kept finite, determinant of the infinite-size matrix (4.32), $\mathrm{Det}_{i,j<0}\, \mathcal{H}_{ij}^{KP}\big|_{t_k = \sum_a^n \lambda_a^{-k}/k} = 0$ since it is obvious from (4.32) that the rank of the matrix is equal to n. Therefore let us consider the maximal non-vanishing determinant,

$$\mathrm{Det}_{-n \leq i,j<0}\, \mathcal{H}_{ij}^{KP}\Big|_{t_k = \sum_a^n \lambda_a^{-k}/k}$$

$$= \left(\prod_a^n \lambda_a\right)^n \det_{ia}\left(\frac{(-)^{a+1}\Delta_a(\lambda)}{\lambda_a^i \Delta(\lambda)}\right) \cdot \det_{aj} \varphi_j(\lambda_a)$$

$$= \frac{\det_{aj} \varphi_j(\lambda_a)}{\Delta(\lambda)}. \tag{4.33}$$

We used here the fact that determinant of a matrix is a product of determinants and reversed the signs of i and j. Also used were some simple relations:

$$\prod_{a=1}^{n} \frac{\Delta_a(\lambda)}{\Delta(\lambda)} = \frac{1}{\Delta^2(\lambda)},$$

$$\det_{ia} \frac{1}{\lambda_a^i} = \left(\prod_a^n \lambda_a\right)^{-1} \Delta(1/\lambda),$$

$$\Delta(1/\lambda) = \prod_{\alpha>\beta}\left(\frac{1}{\lambda_\alpha} - \frac{1}{\lambda_\beta}\right) = (-)^{n(n-1)/2}\Delta(\lambda)\left(\prod_a^n \lambda_a\right)^{-(n-1)},$$

thus

$$\left(\prod_a^n \lambda_a\right)(-)^{n(n-1)/2}\prod_{a=1}^{n} \frac{\Delta_a(\lambda)}{\Delta(\lambda)}\det_{ia}\frac{1}{\lambda_a^i} = \frac{1}{\Delta(\lambda)}.$$

Since (4.33) is true for any n, one can claim that in the limit $n \to \infty$ we recover the statement, that $\tau^{KP}\{t\} = \mathrm{Det}_{i,j<0}\,\mathcal{H}_{ij}^{KP}$ with $\mathcal{H}_{ij}^{KP}$ given by eq. (4.32) (that formula does not refer directly to Miwa parametrization and is defined for any t and any $j < 0$ and i). This relation between φ_a's and $\mathcal{H}_{ij}^{KP}$ can now be used to introduce negative times $\bar{t}_k$ according to the rule (4.19). Especially simple is the prescription for zero-time: $\mathcal{H}_{ij} \to \mathcal{H}_{i+N,j+N}$, when expressed in terms of φ just implies that

$$\frac{\det \varphi_a(\lambda_b)}{\Delta(\lambda)} \to \frac{\det \varphi_{a+N}(\lambda_b)}{(\det \Lambda)^N \Delta(\lambda)}.$$

Generalizations of (4.31), like

$$\mathcal{H}_{ij}\{t, \bar{t}\} = \oint\oint z^i \bar{z}^j \langle 0|\psi(z)G\tilde{\psi}(\bar{z})|0\rangle e^{\sum_k (t_k z^k + \bar{t}_k \bar{z}^k)}\, dz\, d\bar{z},$$

can be also considered.

4.5 1-Matrix Model versus Toda-Chain Hierarchy

At the end of this section we use an explicit example of the discrete 1-matrix model [54] to demonstrate how a more familiar Lax description of integrable hierarchies arises from determinant formulas. Lax representation appears usually after some coordinate system is chosen in the Grassmannian. In the example which we are now considering this system is introduced by the use of orthogonal polynomials.

Formalism of orthogonal polynomials was intensively used at the early days of the theory of matrix models. It is applicable to scalar-product eigenvalue models (see Ref. 1 for details about this notion) and allows to further transform (diagonalize) the remaining determinants into products. In variance with both reduction from original N^2-fold matrix integrals to the eigenvalue problem, which (when possible) reflects a physical phenomenon —decoupling of angular (unitary-matrix) degrees of freedom (associated with d-dimensional gauge bosons)—and with occurrence of determinant formulas which reflects integrability of the model, orthogonal polynomials appear more as a technical device. Essentially orthogonal polynomials are necessary if one wants to explicitly separate dependence on the the size N of the matrix in the matrix integral ("zero time") from dependencies on all other time variables and to explicitly construct variables that satisfy Toda-like equations. However, modern description of integrable hierarchies in terms of τ-functions does not require explicit separation of the zero time and treats it more or less on the equal footing with all other variables, thus making the use of orthogonal polynomials unnecessary. Still this technique remains in the arsenal of the matrix model theory[20] and we now briefly explain what it is about.

In the context of the theory of scalar-product matrix models, orthogonal polynomials naturally arise when one notes that after partition functions appear in a simple determinantal form,

$$
Z_N = \frac{1}{N!} \prod_{k=1}^{N} \int d\mu_{h_k, \bar{h}_k} \, \text{Det}_{ik} \, h_k^{i-1} \, \text{Det}_{jk} \, \bar{h}_k^{j-1}
$$

$$
= \text{Det}_{ij} \int d\mu_{h, \bar{h}} h^{i-1} \bar{h}^{j-1} = \text{Det}_{ij} \langle h^{i-1} \mid \bar{h}^{j-1} \rangle, \qquad (4.34)
$$

(of which eq. (2.18) is a simple example), any linear change of bases $h^i \to Q_i(h) = \sum_k A_{ik} h_k$, $\bar{h}^j \to \overline{Q}_j(\bar{h}) = \sum_l B_{jl} \bar{h}^l$ can be easily performed and $Z \to Z \cdot \det A \cdot \det B$. In particular, if A and B are triangular with units at diagonals, their determinants are just unities and Z does not change at all. This freedom is, however, enough, to diagonalize the scalar product and choose polynomials Q_i and $\overline{Q}_j$ so that

$$
\langle Q_i(h) \mid \overline{Q}_j(\bar{h}) \rangle = e^{\phi_i} \delta_{ij}. \qquad (4.35)
$$

The values Q_i and $\overline{Q}_j$ defined in this way up to normalization are called orthogonal polynomials. (Note that $\overline{Q}$ does not need to be a *complex* conjugate of Q: "bar" does not mean complex conjugation.) Because of the above restriction on the form of matrices A and B these polynomials are normalized so that

$$
Q_i(h) = h^i + \cdots ; \quad \overline{Q}_j(\bar{h}) = \bar{h}^j + \cdots ,
$$

---

[20]Of course, one can also use this link just with the aim of putting the rich and beautiful mathematical theory of orthogonal polynomials into the general context of string theory. Among interesting problems here is the matrix-model description of q-orthogonal polynomials.

i.e., the leading power enters with the *unit* coefficient. From (4.34) and (4.35) it follows that

$$Z_N = \text{Det}_{0 < i, j \leq N} \langle h^i \mid \bar{h}^j \rangle = \prod_{i=1}^{N} e^{\phi_{i-1}}. \qquad (4.36)$$

This formula is essentially the main outcome of orthogonal polynomials theory for matrix models: it provides complete separation of the N-dependence of Z (on the size of the matrix) from that on all other parameters (which specify the shape of potential, i.e., the measure $d\mu_{h,\bar{h}}$). This information is encoded in a rather complicated fashion in ϕ_i. As was already mentioned, any feature of matrix model can be examined already at the level of eq. (4.34), which does not refer to orthogonal polynomials and thus they are not really relevant for the subject.

Consider now the case of the *local* measure, $d\mu_{h,\bar{h}} = d\mu_h \delta(h, \bar{h})$, when $\overline{Q}_i = Q_i$. The local measure is distinguished by the property that multiplication by (any function of) h is Hermitian operator:

$$\langle h f(h) \mid g(\bar{h}) \rangle = \langle f(h) \mid \bar{h} g(\bar{h}) \rangle, \quad \text{if } d\mu_{h,\bar{h}} \sim \delta(h - \bar{h}).$$

This implies further that the coefficients c_{ij} in the recurrent relation

$$h Q_i(h) = Q_{i+1}(h) + \sum_{j=0}^{i} c_{ij} Q_j(h)$$

are almost all vanishing. Indeed: for $j < i$

$$c_{ij} = \frac{\langle h Q_i(h) \mid Q_j(\bar{h}) \rangle}{\langle Q_j(h) \mid Q_j(\bar{h}) \rangle} = \frac{\langle Q_i(h) \mid \bar{h} Q_j(\bar{h}) \rangle}{\langle Q_j(h) \mid Q_j(\bar{h}) \rangle}$$
$$= \delta_{i,j+1} \frac{\langle Q_i(h) \mid Q_i(\bar{h}) \rangle}{\langle Q_j(h) \mid Q_j(\bar{h}) \rangle} = \delta_{j,i-1} e^{\phi_i - \phi_{i-1}}.$$

In other words, polynomials, orthogonal w.r. to a local measure are obliged to satisfy the "3-term recurrent relation":

$$h Q_n(h) = Q_{n+1}(h) + c_n Q_n(h) + R_n Q_{n-1}(h)$$

(the coefficient in front of Q_{n+1} can of course be changed by the change of normalization). Parameter c_n vanishes if the measure is even (symmetric under the change $h \to -h$); then polynomials are split into two orthogonal subsets: even and odd in h. Partition function (4.36) of the *one*-component model can be expressed through parameters $R_i = e^{\phi_i - \phi_{i-1}}$ of the 3-term relation,

$$Z_N = Z_1 \prod_{i=1}^{N-1} R_i^{N-i},$$

thus defining a one-component matrix model (i.e., particular shape of potential), associated with any system of orthogonal polynomials.

Coming back to the 1-matrix model (2.18), one can say that all the information is contained in the determinant formula (4.13) together with the rule (4.11), which defines time-dependence of $\mathcal{H}_{ij}^f = \langle h^i \mid h^j \rangle = \hat{\mathcal{H}}_{i+j}^f$:

$$\frac{\partial \mathcal{H}_{ij}^f}{\partial t_k} = \mathcal{H}_{i+k,j}^f = \mathcal{H}_{i,j+k}^f \quad \text{or} \quad \frac{\partial \hat{\mathcal{H}}_i^f}{\partial t_k} = \hat{\mathcal{H}}_{i+k}^f.$$

The possibility to express everything in terms of $\mathcal{H}_i^f$ with a single matrix index i is the feature of Toda-chain reduction of generic Toda-lattice hierarchy.

However, in order to reveal the standard Lax representation we need to go into somewhat more involved considerations. Namely, we consider representation of two operators in the basis of orthogonal polynomials. First,

$$h^k Q_n(h) = \sum_{m=0}^{n+k} \frac{\langle n | h^k | m \rangle}{\langle m \mid m \rangle} Q_m(h) = \sum_{m=0}^{n+k} \gamma_{nm}^{(k)} Q_m(h)$$

(here the simplified notation is introduced for $\langle n | f(h) | m \rangle \equiv \langle Q_n | f(h) | Q_m \rangle$ and $\gamma_{nm}^{(k)} \equiv \langle n | h^k | m \rangle / \langle m \mid m \rangle$.) Second,

$$\frac{\partial Q_n(h)}{\partial t_k} = -\sum_{m=0}^{n-1} \frac{\langle n | h^k | m \rangle}{\langle m \mid m \rangle} Q_m(h) = -\sum_{m=0}^{n-1} \gamma_{nm}^{(k)} Q_m(h),$$

$$\frac{\partial \phi_n}{\partial t_k} = \frac{\langle n | h^k | n \rangle}{\langle n \mid n \rangle} = \gamma_{nn}^{(k)}.$$

(These last relations arise from differentiation of the orthogonality condition (4.35):

$$e^{\phi_n} \frac{\partial \phi_n}{\partial t_k} \delta_{nm} = \frac{\partial \langle Q_n \mid Q_m \rangle}{\partial t_k}$$

$$= \left\langle \frac{\partial Q_n}{\partial t_k} \middle| Q_m \right\rangle + \left\langle Q_n \middle| \frac{\partial Q_m}{\partial t_k} \right\rangle + \langle Q_n | h^k | Q_m \rangle$$

by looking at the cases of $m < n$ and $m = n$, respectively.)

From these relations one immediately derives the Lax-like formula:

$$\frac{\partial \gamma_{nm}^{(k)}}{\partial t_q} = -\sum_{l=m-k}^{n-1} \gamma_{nl}^{(q)} \gamma_{lm}^{(k)} + \sum_{l=m+1}^{n+k} \gamma_{nl}^{(k)} \gamma_{lm}^{(q)} \qquad (4.37)$$

or, in a matrix form,

$$\frac{\partial \gamma^{(k)}}{\partial t_q} = \left[R\gamma^{(q)}, \gamma^{(k)} \right],$$

where

$$R\gamma_{mn}^{(k)} \equiv \begin{cases} -\gamma_{mn}^{(k)} & \text{if } m > n, \\ \gamma_{mn}^{(k)} & \text{if } m < n. \end{cases}$$

(Remember that usually R-matrix acts on a function $f(h) = \sum_{n=-\infty}^{+\infty} f_n h^n$ according to the rule: $Rf(h) = \sum_{n \geq l} f_n h^n - \sum_{n < l} f_n h^n$ with some "level" l.) These $\gamma^{(k)}$ are not symmetric matrices, but one can also rewrite all the formulas above in terms of symmetric ones:

$$\mathcal{L}_{mn}^{(k)} \equiv e^{(\phi_n - \phi_m)/2} \gamma_{mn}^{(k)} = \frac{\langle m|h^k|n \rangle}{\sqrt{\langle m \mid m \rangle \langle n \mid n \rangle}}.$$

From eqs. (4.37) one can easily deduce Toda equations for ϕ_n,

$$\frac{\partial^2 \phi_n}{\partial t_k \partial t_l} = \frac{\partial}{\partial t_k} \frac{\langle n|h^l|n \rangle}{\langle n \mid n \rangle} = \left(\sum_{m>n} - \sum_{m<n} \right) \frac{\langle n|h^k|m \rangle \langle m|h^l|n \rangle}{\langle m \mid m \rangle \langle n \mid n \rangle},$$

where the r.h.s. can be expressed in terms of $R_m = e^{\phi_m - \phi_{m-1}}$. In particular,

$$\frac{\partial^2 \phi_n}{\partial t_1 \partial t_1} = R_{n+1} - R_n = e^{\phi_{n+1} - \phi_n} - e^{\phi_n - \phi_{n-1}}.$$

Let us also mention that in this formalism the Ward identities (Virasoro constraints) follow essentially from the relation

$$\left(\frac{\partial}{\partial h} \right)^\dagger = -\frac{\partial}{\partial h} - \sum_{k>0} k t_k h^{k-1},$$

where Hermitian conjugation is w.r. to the scalar product $\langle \mid \rangle$. For example, this relation implies, that

$$\left\langle Q_n \, \middle| \, \frac{\partial Q_n}{\partial h} \right\rangle = -\left\langle \frac{\partial Q_n}{\partial h} \, \middle| \, Q_n \right\rangle - \sum_{k>0} k t_k \langle Q_n|h^{k-1}|Q_n \rangle.$$

Now we note that $\partial Q_n / \partial h$ is a polynomial of degree $n - 1$, thus $\langle Q_n \mid \partial Q_n / \partial h \rangle = 0$. In fact

$$\frac{\partial Q_n}{\partial h} = -\sum_{k>0} k t_k \left(\sum_{m=0}^{n-1} \gamma_{nm}^{(k-1)} Q_m \right) = -\sum_{k>0} k t_k \frac{\partial Q_n}{\partial t_{k-1}}.$$

Also we recall that $\langle Q_n|h^{k-1}|Q_n \rangle = \langle Q_n \mid Q_n \rangle \partial \phi_n / \partial t_{k-1}$, and obtain

$$\sum_{k>0} k t_k \frac{\partial \phi_n}{\partial t_{k-1}} = 0$$

for any n. This should be supplemented by relation $\partial \phi_n / \partial t_0 = \phi_n$. In order to get the lowest Virasoro constraint (string equation), $L_{-1} Z_N = 0$ or $L_{-1} \log Z_N = 0$ it is enough just to sum over n from 0 to $N - 1$.

For more details about 1-matrix model, Toda-chain hierarchy and application of the formalism of orthogonal polynomials in this context, see Ref. 54.

5 τ-Function as a Group-Theoretical Quantity

This section contains some remarks about the general notion of τ-function on the lines suggested in Ref. 1. Examples below are taken from Refs. [55] and 56.

As mentioned in the beginning of the previous section we define the (generalized) τ-function as the generating functional of all the matrix elements of a given group element $g \in G$ in a given representation R:

$$\tau_R(t, \bar{t} \mid g) \equiv \sum_{\{m, \overline{m}\} \in R} s^R_{m, \overline{m}}(t, \bar{t}) \langle m|g|\overline{m}\rangle. \qquad (5.1)$$

The choice of functions $s^R_{m, \overline{m}}(t, \bar{t})$ is the main ambiguity in the definition of the τ-function and needs to be fixed in some clever way, not yet known in full generality. The only a priori requirement is that it is indeed a generating functional, i.e., there should be some (g-independent) operators $\mathcal{M}$, acting on $t, \bar{t}$-variables, which allow us to extract all particular matrix elements once $\tau_R(t, \bar{t})$ is known:

$$\langle m|g|\overline{m}\rangle = \langle\!\langle \mathcal{M}_{R, m, \overline{m}}(t, \bar{t}) \mid \tau_R(t, \bar{t} \mid g)\rangle\!\rangle.$$

The ambiguity in the choice of $s^R_{m, \overline{m}}(t, \bar{t})$ can be partly fixed (at least in the case of the highest weight representations R) by the requirement that

$$\tau_R(t, \bar{t} \mid g) = \langle \mathrm{vac}_R|U(t)g\overline{U}(\bar{t})|\mathrm{vac}_R\rangle \qquad (5.2)$$

where operators U and $\overline{U}$ do not depend on R. In order to be even more specific one can further request that evolution operators are group elements, i.e.,

$$\Delta U(t) = U(t) \otimes U(t) = (U(t) \otimes I)(I \otimes U(t)), \qquad (5.3)$$

where Δ denotes group comultiplication law. In the case of Lie algebras $\Delta(T_{\vec{a}}) = T_{\vec{a}} \otimes I + I \otimes T_{\vec{a}}$, and (5.3) is true at least for the evolution operators in KP/Toda systems. Later we shall see that in the case of quantum groups it can be natural to slightly modify the condition (5.3).

Remarkably, the τ-function defined in (5.1) always satisfies a family of nonlinear equations [55], relating τ_R with different R's, which reflects the fact that matrix elements of the *same* group element in different representations are not independent. The conventional bilinear Hirota equation for KP/Toda τ-functions is nothing but the particular case of this generic construction,[21] which has two (essentially identical) interpretations: in terms of fundamental representations of GL(∞) and in terms of the level $k = 1$ Kac-Moody $\widehat{U(1)}$ algebra.

5.1   From Intertwining Operators to Bilinear Equations

The following construction [55] in terms of intertwining operators is the general source of bilinear equations for the τ-function (5.1). One can easily recognize the standard free-fermion derivation of Hirota equations for KP/Toda τ-functions as a particular example (with G being the level $k = 1$ Kac-Moody algebras $\widehat{G}_{k=1}$, V a fundamental representation, and W the simplest fundamental representation corresponding to the very left root of the Dynkin diagram). The construction below involves a lot of arbitrariness. In order to make the consideration more transparent, we formulate our construction explicitly for finite-dimensional Lie algebras and their q-counterparts.

Bilinear equations which we are going to derive are relating τ-functions (5.1) for four different Verma modules V, $\widehat{V}$, V', $\widehat{V}'$. Given V, V', every allowed choice of $\widehat{V}$, $\widehat{V}'$ provides a separate set of bilinear identities. Of course, not all of these sets are actually independent and can be parametrized by source modules V and V' and by a weight of finite-dimensional representation. Also different choices of positive root systems and their ordering in (5.2) provides equations in somewhat different forms. A more invariant description of the minimal set of bilinear equations for given G would be clearly interesting to find.

1. Our starting point is embedding of Verma module $\widehat{V}$ into the tensor product $V \otimes W$, where W is some irreducible finite-dimensional representation of G (in the case of Kac-Moody algebra evaluation representation should be used). Once V and W are specified, there is only a finite number of choices for $\widehat{V}$.

   Now we define right vertex operator of the W-type as homomorphism of G-modules:

   $$E_R \colon \widehat{V} \to V \otimes W.$$

   This intertwining operator can be explicitly continued to the whole rep-

---

[21] Note that it is somewhat different from the approach advocated by V. Kac [49] (see also Ref. 1), which makes use of Casimir operators and is less universal than the one to be described below (using intertwining operators).

resentation once it is constructed for the vacuum (highest-weight) state,

$$\widehat{V} = \left\{ |\mathbf{n}_\alpha\rangle_{\widehat{V}} = \prod_{\alpha>0} \Delta(T_{-\alpha})^{n_\alpha} |0\rangle_{\widehat{V}} \right\},$$

where comultiplication Δ provides the action of G on the tensor product of representations, and

$$|0\rangle_{\widehat{V}} = \left(\sum_{\{p_\alpha, i_\alpha\}} A\{p_\alpha, i_\alpha\} \left(\prod_{\alpha>0} (T_{-\alpha})^{p_\alpha} \otimes (T_{-\alpha})^{i_\alpha} \right) \right) |0\rangle_V \otimes |0\rangle_W.$$

For finite-dimensional W's, this gives every $|\mathbf{n}_\alpha\rangle_{\widehat{V}}$ in a form of *finite* sums of states $|\mathbf{m}_\alpha\rangle_V$ with coefficients, taking values in elements of W.

2. The next step is to take another triple, defining a left vertex operator,

$$\bar{E}'_L : \widehat{V}' \to W^* \otimes V',$$

Note the change of ordering at the r.h.s.; this is different from $V' \otimes W^*$ in the case of quantum groups. The product $W \otimes W^*$ of the module W and its conjugate contains *unit* representation of $\mathcal{G}$. The projection to this unit representation

$$\pi : W \otimes W^* \to I$$

is explicitly provided by multiplication of any element of $W \otimes W^*$ by

$$\pi = {}_W\langle 0| \otimes {}_{W^*}\langle 0| \left(\sum_{\{i_\alpha, i'_\alpha\}} \pi\{i_\alpha, i'_\alpha\} \left(\prod_{\alpha>0} (T_{+\alpha})^{i_\alpha} \otimes (T_{+\alpha})^{i'_\alpha} \right) \right)$$

Using this projection, if it is not occasionally orthogonal to the image of $E \otimes E'$, one can build a new intertwining operator

$$\Gamma : \widehat{V} \otimes \widehat{V}' \xrightarrow{E \otimes E'} V \otimes W \otimes W^* \otimes V' \xrightarrow{I \otimes \pi \otimes I} V \otimes V',$$

which possesses the property

$$\Gamma(g \otimes g) = (g \otimes g)\Gamma \qquad (5.4)$$

for any group element g such that

$$\Delta(g) = g \otimes g.$$

3. It now remains to take a matrix element of (5.4) between four states,

$$_{V'}\langle k'|_V\langle k|(g \otimes g)\Gamma|n\rangle_{\widehat{V}}|n'\rangle_{\widehat{V}'} = {}_{V'}\langle k'|_V\langle k|\Gamma(g \otimes g)|n\rangle_{\widehat{V}}|n'\rangle_{\widehat{V}'}, \qquad (5.5)$$

and rewrite this identity in terms of generating functions (5.1).

5.2   The Case of KP/Toda τ-Functions

We do not present here the standard derivation of Hirota equations in the free-fermion formalism, because it is both well known and easily recognizable in the general picture from the previous subsection. Instead we describe here its slight variation—starting from the fundamental representations of $SL(\infty)$. The reason why this case is the closest one to the standard integrable hierarchies is that, in variance with generic Verma modules for group $G \neq SL(2)$, the fundamental representations are generated by subset of the *mutually commuting* operators, not by entire set of generators from maximal nilpotent subalgebra. We describe the basic construction for $G = SL(n)$, since in this case the (finite) Grassmannian construction is the most similar to the conventional infinite-dimensional $(G = \widehat{U(1)})$ situation.

The Lie algebra $SL(n)$ is generated by operators $T_{\pm\vec{\alpha}}$ and Cartan operators $H_{\vec{\beta}}$, such that $[H_{\vec{\beta}}, T_{\pm\vec{\alpha}}] = \pm\frac{1}{2}(\vec{\alpha}\vec{\beta})T_{\pm\vec{\alpha}}$. All elements of all representations are eigenfunctions of $H_{\vec{\beta}}$, $H_{\vec{\beta}}|\vec{\lambda}\rangle = \frac{1}{2}(\vec{\beta}\vec{\lambda})|\vec{\lambda}\rangle$. The highest weight of representation $F^{(k)}$ is $\vec{\mu}_k$. Vectors $\vec{\mu}_k$'s are "dual" to the *simple* roots $\vec{\alpha}_i$, $i = 1, \ldots, r$: $(\vec{\mu}_i\vec{\alpha}_j) = \delta_{ij}$, and $\vec{\rho} = \frac{1}{2}\sum_{\vec{\alpha}>0}\vec{\alpha} = \sum_i\vec{\mu}_i$.

There are as many as $r \equiv \text{rank}\,G = n - 1$ fundamental representations of $SL(n)$. Let us begin with the simplest fundamental representation $F \equiv F^{(1)}$—the n-tuple, which consists of the states

$$\psi_i = T_{-(i-1)}\cdots T_{-2}T_{-1}\psi_1, \quad i = 1,\ldots,n. \tag{5.6}$$

Moreover

$$T_{-i}\psi_j = \delta_{ij}\psi_{i+1}, \tag{5.7}$$

and the weights are given by

$$\vec{\lambda}(\psi_i) = \vec{\mu}_1 - \vec{\alpha}_1 - \cdots - \vec{\alpha}_{i-1},$$

where $\vec{\mu}_1$ is the highest weight of $F^{(1)}$. Here $T_{\pm i} \equiv T_{\pm\vec{\alpha}_i}$ are generators, associated with the simple roots. Let us denote the corresponding basis in Cartan algebra $H_i = H_{\vec{\alpha}_i}$, and $H_i|\vec{\lambda}\rangle = \frac{1}{2}(\vec{\alpha}_i\vec{\lambda})|\vec{\lambda}\rangle = \lambda_i|\vec{\lambda}\rangle$. Then

$$\lambda_i^{(j)} \equiv \lambda_i(\psi_j) = \frac{1}{2}(\delta_{ij} - \delta_{i,j-1}).$$

This formula, together with (5.6) and defining commutation relations between the positive and negative simple-root generators implies that $\|\psi_i\|^2 = 1$, and, with the help of the classical comultiplication formula, $\Delta(T) = T \otimes I + I \otimes T$, one immediately observes that the antisymmetric combinations $\psi_{[1}\cdots\psi_{k]}$ are all the highest-weight vectors (i.e., are annihilated by all $\Delta_k(T_{+i})$ and, thus by all the $\Delta_k(T_{+\vec{\alpha}})$). These combinations are the highest vectors of all the other fundamental representations $F^{(k)}$, which are thus skew powers of $F = F^{(1)}$:

$$F^{(k)} = \{\Psi^{(k)}_{i_1 \ldots i_k} \sim \psi_{[i_1} \cdots \psi_{i_k]}\}. \tag{5.8}$$

From this description it is clear that $0 \leq k \leq n$; moreover $F^{(0)}$ and $F^{(n)}$ are respectively the singlet and dual singlet representations.

According to (5.7) one can also describe all the states of $F^{(1)}$ in terms of a single generator T_-, which is a sum of those for all the r *simple* roots of G, $T_- = \sum_{i=1}^{r} T_{-\vec{\alpha}_i}$:

$$\psi_i = T_-^{i-1} |0\rangle_F, \quad i = 1, \ldots, n.$$

Looking at the explicit form (5.8) of the states in $F^{(k)}$ it is easy to see that they can be all generated from the highest-weight one, $\Psi_{12\cdots k}$, by the operators

$$R_k(T_-^i) \equiv T_-^i \otimes I \otimes \cdots \otimes I + I \otimes T_-^i \otimes \cdots \otimes I + I \otimes I \otimes \cdots \otimes T_-^i. \tag{5.9}$$

These operators obviously commute with each other. For given k exactly k of them (with $i = 1, \ldots, k$) are independent. However, they are neither (linear combinations of) the generators of Lie algebra acting in $F^{(k)}$, nor even their algebraic functions (note that $R_k(T_-^i) \neq \left(R_k(T_-)\right)^i$). If one wants to make clear that $F^{(k)}$ is indeed a representation of G, it is better to say that it is generated by another set of operators,

$$\Delta^{k-1}(\mathcal{T}_i), \quad \mathcal{T}_i \equiv \sum_{\vec{\alpha}:h(\vec{\alpha})=i} T_{-\vec{\alpha}}. \tag{5.10}$$

The "height" $h(\vec{\alpha})$ is the number of items in the linear decomposition of the root $\vec{\alpha}$ in simple roots $\vec{\alpha}_i$ (in $F^{(1)}$ where all the generators $T_{\vec{\alpha}}$ are represented by $n \times n$ matrices, $\mathcal{T}_i$ are matrices with all zero entries except for units at the ith subdiagonal). In particular, $T_- = \mathcal{T}_1$. Operators (5.10) are obviously generators of G. Instead their mutual commutativity is somewhat less transparent. Since both sets (5.10) and (5.9) generate $F^{(k)}$ it is a matter of convenience which of them is used in particular considerations. In dealing with KP/Toda hierarchies the explicitly commuting set (5.9) is more convenient. It is exactly the lack of such equivalence of two sets that makes consideration of KP/Toda hierarchies more subtle in the quantum ($q \neq 1$) case; see Section 5.4 below.

The intertwining operators of interest for us are

$$I_{(k)} : F^{(k+1)} \to F^{(k)} \otimes F,$$
$$I^*_{(k)} : F^{(k-1)} \to F^* \otimes F^{(k)},$$

and

$$\Gamma_{k|k'} : F^{(k+1)} \otimes F^{(k'-1)} \to F^{(k)} \otimes F^{(k')}.$$

Here

$$F^* = F^{(r)} = \{\psi^i \sim \epsilon^{ii_1\cdots i_r}\psi_{[i_1}\cdots\psi_{i_r]}\},$$

$$I_{(k)}: \Psi^{(k+1)}_{i_1\cdots i_{k+1}} = \Psi^{(k)}_{[i_1\cdots i_k}\psi_{i_{k+1}]},$$

$$I^*_{(k)}: \Psi^{(k-1)}_{i_1\cdots i_{k-1}} = \Psi^{(k)}_{i_1\cdots i_{k-1}i}\psi^i,$$

and $\Gamma_{k|k'}$ is constructed with the help of embedding $I \to F \otimes F^*$, induced by the pairing $\psi_i\psi^i$: the basis in linear space $F^{(k+1)} \otimes F^{(k'-1)}$, induced by $\Gamma_{k|k'}$ from that in $F^{(k)} \otimes F^{(k')}$ is

$$\Psi^{(k)}_{[i_1\cdots i_k}\Psi^{(k')}_{i_{k+1}]i'_1\cdots i'_{k'-1}}.$$

Operation Γ can be now rewritten in terms of matrix elements

$$g^{(k)}\begin{pmatrix} i_1\cdots i_k \\ j_1\cdots j_k \end{pmatrix} \equiv \langle\Psi_{i_1\cdots i_k}|g|\Psi_{j_1\cdots j_k}\rangle = \det_{1\leq a,b\leq k} g^{i_a}_{j_b} \qquad (5.11)$$

as follows:

$$g^{(k)}\begin{pmatrix} i_1\cdots i_k \\ j_1\cdots j_k \end{pmatrix} g^{(k')}\begin{pmatrix} i'_1\cdots i'_k \\ j_{k+1}]j'_1\cdots j'_{k'-1} \end{pmatrix}$$
$$= g^{(k+1)}\begin{pmatrix} i_1\cdots i_k[i'_{k'} \\ j_1\cdots j_{k+1} \end{pmatrix} g^{(k'-1)}\begin{pmatrix} i'_1\cdots i'_{k'-1}] \\ j'_1\cdots j'_{k'-1} \end{pmatrix}. \qquad (5.12)$$

This is the explicit expression for eq. (5.4) in the case of fundamental representations, and it is certainly identically true for any $g^{(k)}$ of the form (5.11).[22]

Let us note that one can use the minors (5.11) to construct local coordinates in the Grassmannian. Bilinear Plucker relations satisfied by these coordinates are nothing but defining equations of the Grassmannian consisting of all the k-dimensional vector subspaces of n-dimensional vector space. Parametrizing determinants (5.11) by time variables (see (5.18)), one gets a set of bilinear differential equations on the generating function of these Plucker coordinates, which is just a τ-function [57].

Now let us introduce time variables and rewrite (5.12) in terms of τ-functions. We shall denote time variables through s_i, $\bar{s}_i$, $i = 1, \ldots, r$

---

[22]To see this directly it is enough to rewrite the l.h.s. of (5.12) as

$$g^{(k)}\begin{pmatrix} i_1\cdots i_k \\ [j_1\cdots j_k \end{pmatrix} g^{[i'_k}_{j_{k+1}]}g^{(k'-1)}\begin{pmatrix} i'_1\cdots i'_{k-1}] \\ j'_1\cdots j'_{k-1} \end{pmatrix}$$

(expansion of the determinant $g^{(k')}$ in the first column) and now the first two factors can be composed into $g^{(k+1)}$ (expansion of the determinant $g^{(k+1)}$ in the first row), thus giving the r.h.s. of (5.12).

in order to emphasize their difference from generic $t_{\tilde{\alpha}}$, $\bar{t}_{\tilde{\alpha}}$ labeled by all the positive roots $\tilde{\alpha}$ of G. Note that in order to have a closed system of equations we need to introduce all the r times s_i for all $F^{(k)}$ (though $\tau^{(k)}$ actually depends only on k-independent combinations of these).

Since the highest weight of representation $F^{(k)}$ is identified as

$$|0\rangle_{F^{(k)}} = |\Psi^{(k)}_{1\cdots k}\rangle,$$

we have

$$\tau^{(k)}(t, \bar{t} \mid g) = \langle \Psi^{(k)}_{1\cdots k} | \exp\left(\sum_i t_i R_k(T^i_+)\right) g$$
$$\times \exp\left(\sum_i \bar{t}_i R_k(T^i_-)\right) |\Psi^{(k)}_{1\cdots k}\rangle. \quad (5.13)$$

Now

$$\exp\left(\sum_i t_i R_k(T^i)\right) = \exp\left(R_k\left(\sum_i t_i T^i\right)\right)$$
$$= \left(\exp\left(\sum_i t_i T^i\right)\right)^{\otimes k} = \left(\sum_j P_j(t) T^j\right)^{\otimes k}, \quad (5.14)$$

where we used the definition of Schur polynomials

$$\exp\left(\sum_i t_i z^i\right) = \sum_j P_j(t) z^j. \quad (5.15)$$

An essential property of Schur polynomials is that

$$\frac{\partial}{\partial t_i} P_j(t) = \left(\frac{\partial}{\partial t_1}\right)^i P_j(t) = P_{j-i}(t). \quad (5.16)$$

Because of (5.14), we can rewrite the r.h.s. of (5.13) as

$$\tau^{(k)}(t, \bar{t} \mid g) = \sum_{\substack{i_1,\ldots,i_k \\ j_1,\ldots,j_k}} P_{i_1}(t) \cdots P_{i_k}(t) \langle \Psi^{(k)}_{1+i_1, 2+i_2, \ldots, k+i_k} | g$$
$$\times |\Psi^{(k)}_{1+j_1, 2+j_2, \ldots, k+j_k}\rangle P_{j_1}(\bar{t}) \cdots P_{j_k}(\bar{t})$$
$$= \det_{1 \leq \alpha, \beta \leq k} H^{\alpha}_{\beta}(t, \bar{t}), \quad (5.17)$$

where

$$H^{\alpha}_{\beta}(t, \bar{t}) = \sum_{i,j} P_{i-\alpha}(t) g^i_j P_{j-\beta}(\bar{s}). \quad (5.18)$$

This formula can be considered as including infinitely many times s_i and $\bar{s}_i$, and it is only due to the finiteness of matrix $g^i_j \in \mathrm{SL}(n)$ that the H-matrix is additionally constrained

$$\left(\frac{\partial}{\partial t_1}\right)^n H_\beta^\alpha = 0,$$

$$\vdots \tag{5.19}$$

$$\frac{\partial}{\partial t_i} H_\beta^\alpha = 0, \quad \text{for } i \geq n.$$

The characteristic property of H_β^α is that it satisfies the following "shift" relations (see (5.16)):

$$\frac{\partial}{\partial t_i} H_\beta^\alpha = H_\beta^{\alpha+i}, \quad \frac{\partial}{\partial \bar{t}_i} H_\beta^\alpha = H_{\beta+i}^\alpha. \tag{5.20}$$

Coming back to bilinear relation (5.12), it can be easily rewritten in terms of H-matrix: it is enough to convolute them with Schur polynomials. For the sake of convenience let us denote $H\left(\begin{smallmatrix} \alpha_1 \cdots \alpha_k \\ \beta_1 \cdots \beta_k \end{smallmatrix}\right) = \det_{1 \leq a,b \leq k} H_{\beta_b}^{\alpha_a}$. In accordance with this notation $\tau^{(k)} = H\left(\begin{smallmatrix} 1 \cdots k \\ 1 \cdots k \end{smallmatrix}\right)$, while bilinear equation turns into

$$H\left(\begin{matrix} \alpha_1 \cdots \alpha_k \\ \beta_1 \cdots \beta_k \end{matrix}\right) H\left(\begin{matrix} \alpha_k' \alpha_1' \cdots \alpha_{k-1}' \\ \beta_{k+1}] \beta_1' \cdots \beta_{k-1}' \end{matrix}\right)$$
$$= H\left(\begin{matrix} \alpha_1 \cdots \alpha_k [\alpha_k' \\ \beta_1 \cdots \beta_k \beta_{k+1} \end{matrix}\right) H\left(\begin{matrix} \alpha_1' \cdots \alpha_{k-1}]' \\ \beta_1' \cdots \beta_{k-1}' \end{matrix}\right).$$

Just like the original (5.12) these are just matrix identities, valid for any H_β^α. However, after the switch from g to H we, first, essentially represented the equations in the n-independent form and, second, opened the possibility to rewrite them in terms of time derivatives.

For example, in the simplest case of

$$\alpha_i = i, \quad i = 1, \ldots, k',$$
$$\beta_i = i, \quad i = 1, \ldots, k+1,$$
$$\alpha_i' = i, \quad i = 1, \ldots, k-1, \quad \alpha_k' = k+1,$$
$$\beta_i' = i, \quad i = 1, \ldots, k-1$$

we get

$$H\left(\begin{matrix} 1 \cdots k \\ 1 \cdots k \end{matrix}\right) H\left(\begin{matrix} k+1, 1 \cdots k-1 \\ k+1, 1 \cdots k-1 \end{matrix}\right)$$
$$- H\left(\begin{matrix} 1 \cdots k-1, k \\ 1 \cdots k-1, k+1 \end{matrix}\right) H\left(\begin{matrix} k+1, 1 \cdots k-1 \\ k, 1 \cdots, k-1 \end{matrix}\right)$$
$$= H\left(\begin{matrix} 1 \cdots k+1 \\ 1 \cdots k+1 \end{matrix}\right) H\left(\begin{matrix} 1 \cdots k-1 \\ 1 \cdots k-1 \end{matrix}\right)$$

(all other terms arising in the process of symmetrization vanish). This in turn can be represented through τ-functions:

$$\partial_1 \bar{\partial}_1 \tau^{(k)} \cdot \tau^{(k)} - \bar{\partial}_1 \tau^{(k)} \partial \tau^{(k)} = \tau^{(k+1)} \tau^{(k-1)}. \tag{5.21}$$

This is the usual lowest Toda-lattice equation. For finite n the set of solutions is labeled by $g \in \mathrm{SL}(n)$ (rather than $\mathrm{SL}(\infty)$) as a result of additional constraints (5.19).

We can now illustrate the ambiguity of the definition of the τ-function, or, to put it differently, the ambiguity in the choice of time variables. Equation (5.21) is actually a corollary of *two* statements: the basic identity (5.12) and the particular choice of evolution operators in eq. (5.2), which in the case of (5.13) implies (5.18) with P's being ordinary Schur polynomials (5.15). At least, in this simple situation (of fundamental representations of $\mathrm{SL}(n)$) one could define τ-function not by eq. (5.2), but just by eq. (5.17), with

$$H_\beta^\alpha(t, \bar{t}) \to \mathcal{H}_\beta^\alpha(t, \bar{t}) = \sum_{i,j} \mathcal{P}_{i-\alpha}(t) g_j^i \mathcal{P}_{j-\beta}(\bar{t})$$

with *any* set of independent functions (not even polynomials) $\mathcal{P}_\alpha$. Such

$$\tau_{\mathcal{P}}^{(k)} = \det_{1 \le \alpha, \beta \le k} \mathcal{H}_\beta^\alpha \tag{5.22}$$

still remains a generating function for all matrix elements of $G = \mathrm{SL}(n)$ in the representation $F^{(k)}$. This freedom should be kept in mind when dealing with "generalized τ-functions." As a simple example, one can take $\mathcal{P}_\alpha(t)$ to be q-Schur polynomials,

$$\prod_i e_q(t_i z^i) = \sum_j P_j^{(q)}(t) z^j, \quad \text{or} \quad \prod_i e_{q^i}(t_i z^i) = \sum_j \widehat{P}_j^{(q)}(t) z^j,$$

which satisfy

$$D_{t_i} P_j^{(q)}(t) = (D_{t_1})^i P_j^{(q)}(t) = P_{j-i}^{(q)}(t).$$

where D are finite-difference operators. Then instead of (5.20) we would have:

$$D_{t_i} \mathcal{H}_\beta^\alpha = \mathcal{H}_\beta^{\alpha+i}, \quad D_{\bar{t}_i} \mathcal{H}_\beta^\alpha = \mathcal{H}_{\beta+i}^\alpha$$

and

$$\tau_{P(q)}^{(k)}(t, \bar{t} \mid g) = \det_{1 \le \alpha, \beta \le k} D_{t_1}^{\alpha-1} D_{\bar{t}_1}^{\beta-1} \mathcal{H}_1^1(t, \bar{t}).$$

So the defined τ-function satisfies a *difference* rather than the differential equations [58, 59]:

$$\tau^{(k)} \cdot D_{t_1} D_{\bar{t}_1} \tau^{(k)} - D_{t_1} \tau^{(k)} \cdot D_{\bar{t}_1} \tau^{(k)} = \tau^{(k-1)} \cdot M_{t_1}^+ M_{\bar{t}_1}^+ \tau^{(k+1)}.$$

$$\vdots$$

We emphasize, however, that this is just another description of the SL(n), not the SL$_q(n)$ τ-function, if it is interpreted as a generating function of matrix elements. In particular, this τ-function takes c- rather than q-number values. Still, as concerns its *times-*, not g-dependence; it has something to do with the SL$_q(n)$ group, in the spirit of relation between q-hypergeometric functions and quantum groups (see, e.g., Ref. 60).

5.3 Example of SL(2)$_q$

Construction from Section 5.1 is immediately applicable to the case of quantum groups. The only thing one should keep in mind is that our definition (5.1) gives τ as an element of a "coordinate ring" $\mathcal{A}(G_q)$, not just a c-number. If one wants to obtain a c-number τ-function for $q \neq 1$ it is necessary to restrict the construction further to particular representation of a coordinate ring (this last step will not be discussed in this paper). We present here in full detail the simplest possible example of SL(2)$_q$ [55].

5.3.1 Bilinear Identities

To begin with, fix the notations. We consider generators T_+, T_-, and T_0 of $U_q(\mathrm{SL}(2))$ with commutation relations

$$q^{T_0} T_{\pm} q^{-T_0} = q^{\pm 1} T_{\pm},$$

$$[T_+, T_-] = \frac{q^{2T_0} - q^{-2T_0}}{q - q^{-1}},$$

and comultiplication

$$\Delta(T_{\pm}) = q^{T_0} \otimes T_{\pm} + T_{\pm} \otimes q^{-T_0},$$

$$\Delta(q^{T_0}) = q^{T_0} \otimes q^{T_0}.$$

Verma module V_λ with highest-weight λ (not obligatory half-integer), consists of the elements

$$|n\rangle_\lambda \equiv T_-^n |0\rangle_\lambda, \quad n \geq 0,$$

such that

$$T_- |n\rangle_\lambda = |n + 1\rangle_\lambda,$$

$$T_0 |n\rangle_\lambda = (\lambda - n)|n\rangle_\lambda,$$

$$T_+ |n\rangle_\lambda \equiv b_n(\lambda)|n - 1\rangle_\lambda,$$

$$b_n(\lambda) = [n][2\lambda + 1 - n], \quad [x] \equiv \frac{q^x - q^{-x}}{q - q^{-1}},$$

$$\|n\|_\lambda^2 \equiv {}_\lambda\langle n \mid n\rangle_\lambda = \frac{[n]!\,\Gamma_q(2\lambda + 1)}{\Gamma_q(2\lambda + 1 - n)} \overset{\lambda \in \mathbb{Z}/2}{=} \frac{[2\lambda]!\,[n]!}{[2\lambda - n]!}.$$

Now,

$$\left(\Delta(T_-)\right)^n = q^{nT_0} \otimes T_-^n + [n]T_- q^{(n-1)T_0} \otimes T_-^{n-1} q^{-T_0}$$
$$+ \cdots + [n]T_-^{n-1} q^{T_0} \otimes T_- q^{-(n-1)T_0} + T_-^n \otimes q^{-nT_0}. \quad (5.23)$$

Let us manifestly derive equations (5.5) taking for W an irreducible spin-$\frac{1}{2}$ representation of $U_q(SL(2))$. Then $\widehat{V} = V_{\lambda \pm 1/2}$, $V = V_\lambda$ and the highest weights of $\widehat{V}$ in $W \otimes V$ or $V \otimes W$ are[23]:

$$|0\rangle_{\lambda+1/2} = |+\rangle|0\rangle_\lambda, |+\rangle \equiv |0\rangle_{1/2}, \text{ or } |0\rangle_\lambda||+\rangle,$$
$$|0\rangle_{\lambda-1/2} = |+\rangle|1\rangle_\lambda - q^{(\lambda+1/2)}[2\lambda]|-\rangle|0\rangle_\lambda, |-\rangle \equiv |1\rangle_{1/2},$$
$$\text{or } (q \to q^{-1})|1\rangle_\lambda|+\rangle - q^{-(\lambda+1/2)}[2\lambda]|0\rangle_\lambda|-\rangle.$$

The entire Verma module is generated by the action of $\Delta(T_-)$:

$$|n\rangle_{\lambda+1/2} = \left(\Delta(T_-)\right)^n |0\rangle_{\lambda+1/2}$$
$$\to q^{n/2}(|+\rangle|n\rangle_\lambda + q^{-(\lambda+1/2)}[n]|-\rangle|n-1\rangle_\lambda) \text{ or}$$
$$q^{-n/2}(|n\rangle_\lambda|+\rangle + q^{(\lambda+1/2)}[n]|n-1\rangle_\lambda|-\rangle),$$
$$|n\rangle_{\lambda-1/2} = \left(\Delta(T_-)\right)^n |0\rangle_{\lambda-1/2}$$
$$\to q^{n/2}(|+\rangle|n+1\rangle_\lambda + q^{(\lambda+1/2)}[n-2\lambda]|-\rangle|n\rangle_\lambda) \text{ or}$$
$$q^{-n/2}(|n+1\rangle_\lambda|+\rangle + q^{-(\lambda+1/2)}[n-2\lambda]|n\rangle_\lambda|-\rangle).$$

Step 2 to be made in accordance with our general procedure is to project the tensor product of two different W's onto singlet state $S = |+\rangle|-\rangle - q|-\rangle|+\rangle$[24]:

$$(A|+\rangle + B|-\rangle) \otimes (|+\rangle C + |-\rangle D) \to AD - qBC.$$

With our choice of W we can now consider two different cases:

(A) both $\widehat{V} = V_{\lambda-1/2}$ and $\widehat{V}' = V_{\lambda'-1/2}$, or

(B) $\widehat{V} = V_{\lambda-1/2}$ and $\widehat{V}' = V_{\lambda'+1/2}$:

Case A.

$$|n\rangle_{\lambda-1/2}|n'\rangle_{\lambda'-1/2} \to q^{(n'-n-1)/2}([n'-2\lambda']q^{\lambda'}|n+1\rangle_\lambda|n'\rangle_{\lambda'}$$
$$- [n-2\lambda]q^{-\lambda}|n\rangle_\lambda|n'+1\rangle_{\lambda'}). \quad (5.24)$$

---

[23]Hereafter we omit the symbol of tensor product from the notations of the states $|+\rangle \otimes |0\rangle_\lambda$, etc.

[24]This state is a singlet of $U_q(SL(2))$. In the case of $U_q(GL(2))$ one should account for the $U(1)$ noninvariance of S. This is the origin of the factor $\det_q g$ at the r.h.s. of the final equation (5.29).

Case B.

$$|n\rangle_{\lambda+1/2}|n'\rangle_{\lambda'-1/2} \rightarrow q^{(n'-n-1)/2}([n'-2\lambda']q^{\lambda'}|n\rangle_\lambda|n'\rangle_{\lambda'}$$
$$- [n]q^{+\lambda+1}|n-1\rangle_\lambda|n'+1\rangle_{\lambda'}). \quad (5.25)$$

Now we proceed to the step 3. Consider any "group element," i.e., an element g from some extension of $U_q(G)$, which possesses the property

$$\Delta(g) = g \otimes g,$$

and take matrix elements of the formula (5.4),

$$_{\lambda'}\langle k'|_\lambda\langle k|(g \otimes g\Gamma = \Gamma g \otimes g)|n\rangle_{\tilde\lambda}|n'\rangle_{\tilde\lambda'}. \quad (5.26)$$

The action of operator Γ can be represented as

$$\Gamma|n\rangle_{\tilde\lambda}|n'\rangle_{\tilde\lambda'} = \sum_{l,l'}|l\rangle_\lambda|l'\rangle_{\lambda'}\Gamma(l,l' \mid n,n'), \quad (5.27)$$

and in these terms (5.26) turns into:

$$\sum_{m,m'}\Gamma(k,k' \mid m,m')\frac{\|k\|_\lambda^2\|k'\|_{\lambda'}^2}{\|m\|_{\tilde\lambda}^2\|m'\|_{\tilde\lambda'}^2}\langle m|g|n\rangle_{\tilde\lambda}\langle m'|g|n'\rangle_{\tilde\lambda'}$$
$$= \sum_{l,l'}\langle k|g|l\rangle_\lambda\langle k'|g|l'\rangle_{\lambda'}\Gamma(l,l' \mid n,n'). \quad (5.28)$$

In order to rewrite this as a difference equation, we use our definition of τ-function:

$$\tau_\lambda(t,\bar t \mid g) \equiv \langle\lambda|e_q(tT^+)ge_q(\bar tT^-)|\lambda\rangle = \sum_{m,n}\langle m|g|n\rangle_\lambda\frac{t^m}{[m]!}\frac{\bar t^n}{[n]!}.$$

Then, one can write down the generating formula for the equation (5.28), using the manifest form (5.24)–(5.25) of matrix elements $\Gamma(l,l' \mid n,n')$:

Case A.

$$\sqrt{M_{\bar t}^- M_{\bar t'}^+}(q^{\lambda'}D_{\bar t}^{(0)}\bar t'D_{\bar t'}^{(2\lambda')} - q^{-\lambda}\bar tD_{\bar t}^{(2\lambda)}D_{\bar t'}^{(0)})\tau_\lambda(t,\bar t \mid g)\tau_{\lambda'}(t',\bar t' \mid g)$$
$$= [2\lambda][2\lambda'](\det_q g)(q^{-(\lambda+1/2)}t' - q^{(\lambda'+1/2)}t)$$
$$\times \tau_{\lambda-1/2}(t,\bar t \mid g)\tau_{\lambda'-1/2}(t',\bar t' \mid g). \quad (5.29)$$

Here $D_{\bar t}^{(\alpha)} \equiv (q^{-\alpha}M_t^+ - q^\alpha M_t^-)/((q-q^{-1})t)$ and $M^\pm$ are multiplicative shift operators, $M_t^\pm f(t) = f(q^{\pm 1}t)$.

Case B.

$$\sqrt{M_{\bar{t}}^- M_{\bar{t}'}^+} \left(q^{\lambda'} \bar{t}' D_{\bar{t}'}^{(2\lambda')} - q^{(\lambda+1)} \bar{t} D_{\bar{t}'}^{(0)} \right) \tau_\lambda(t, \bar{t}|g) \tau_{\lambda'}(t', \bar{t}'|g)$$

$$= \frac{[2\lambda']}{[2\lambda+1]} \sqrt{M_t^- M_{t'}^+} (q^{\lambda'} t D_t^{(2\lambda+1)} - q^\lambda t' D_t^{(0)})$$

$$\times \tau_{\lambda+1/2}(t, \bar{t} \mid g) \tau_{\lambda'-1/2}(t', \bar{t}' \mid g). \quad (5.30)$$

Let us note that the derivation of these equations can be presented in the form that looks even closer to conventional free-fermion formalism. It is possible to represent operator Γ in component form as $E_1^R \otimes E_2^L - q E_2^R \otimes E_1^L$, where E_i's are components of the vertex operator (given by fixing different vectors from W). Then the equation (5.4) can be rewritten

$$_{\hat{V}}\langle 0|e_q(tT_+) E_1^R g e_q(\bar{t}T_-)|0\rangle_V \cdot {}_{\hat{V}'}\langle 0|e_q(tT_+) E_2^L g e_q(\bar{t}T_-)|0\rangle_{V'}$$

$$- q_{\hat{V}}\langle 0|e_q(tT_+) E_2^R g e_q(\bar{t}T_-)|0\rangle_V \cdot {}_{\hat{V}'}\langle 0|e_q(tT_+) E_1^L g e_q(\bar{t}T_-)|0\rangle_{V'}$$

$$= {}_{\hat{V}}\langle 0|e_q(tT_+) g E_1^R e_q(\bar{t}T_-)|0\rangle_V \cdot {}_{\hat{V}'}\langle 0|e_q(tT_+) g E_2^L e_q(\bar{t}T_-)|0\rangle_{V'}$$

$$- q_{\hat{V}}\langle 0|e_q(tT_+) g E_2^R e_q(\bar{t}T_-)|0\rangle_V \cdot {}_{\hat{V}'}\langle 0|e_q(tT_+) g E_1^L e_q(\bar{t}T_-)|0\rangle_{V'}. \quad (5.31)$$

We can easily obtain commutation relations of E_i's with generators of algebra as well as their action on vacuum states. Then it is straightforward to commute E_i's with q-exponentials in the expression (5.31) and represent the result of the commutation by the action of difference operators. Of course, the results (5.29) and (5.30) are reproduced in this way.

5.3.2   Solution to Bilinear Identities

In this particular case (of $\mathrm{SL}(2)_q$) one can easily evaluate the τ-function explicitly, and let us use this possibility to show how bilinear equations are satisfied. The fact that our τ-function is operator-valued will be of course of principal importance.

Let us begin from the case of $\lambda = \frac{1}{2}$. Then

$$\tau_{1/2}(t, \bar{t} \mid g) = \langle +|g|+\rangle + \bar{t}\langle +|g|-\rangle + t\langle -|g|+\rangle + t\bar{t}\langle -|g|-\rangle$$
$$= a + b\bar{t} + ct + dt\bar{t},$$

where a, b, c, and d are elements of the matrix

$$T = \begin{pmatrix} a & b \\ c & d \end{pmatrix}$$

with the commutation relations dictated by $TTR = RTT$ [61]

$$ab = qba, \quad ac = qca, \quad bd = qdb,$$
$$cd = qdc, \quad bc = cb, \quad ad - da = (q - q^{-1})bc. \quad (5.32)$$

If b or c or both are nonvanishing, $\tau_{1/2}(t, \bar{t} \mid g)$ with different values of time variables t, $\bar{t}$ do not commute. Still such $\tau_{1/2}(t, \bar{t} \mid g)$ does satisfy the same bilinear identity (5.29); moreover, for this to be true it is essential that commutation relations (5.32) be exactly what they are. Indeed, the l.h.s. of the equation (5.29) is equal to

$$
\begin{aligned}
- q^{1/2}\sqrt{M_{\bar{t}}^-}(b + dt)\sqrt{M_{\bar{t}'}^+}(a + ct') &+ q^{-1/2}\sqrt{M_{\bar{t}}^-}(a + ct)\sqrt{M_{\bar{t}'}^+}(b + dt') \\
= (q^{-1/2}ab - q^{1/2}ba) &+ (q^{-1/2}cd - q^{1/2}dc)tt' \\
&+ (q^{-1/2}cb - q^{1/2}da)t + (q^{-1/2}ad - q^{-1/2}bc)t' \\
= (q^{-1/2}t' - q^{1/2}t)\det{}_q g, &\qquad\qquad (5.33)
\end{aligned}
$$

which coincides with the r.h.s. of the equation (5.29).

To perform the similar check for any half-integer-spin representation, let us note that the corresponding τ-function can be easily written in terms of $\tau_{1/2}$. Indeed,

$$
\begin{aligned}
|n\rangle_\lambda &= (q^{T_0} \otimes T_- + T_- \otimes q^{-T_0})^n |0\rangle_{\lambda - 1/2} \otimes |0\rangle_{1/2} \\
&= q^{-n/2}(|n\rangle_{\lambda - 1/2} \otimes |0\rangle_{1/2} + [n]q^\lambda |n - 1\rangle_{\lambda - 1/2} \otimes |1\rangle_{1/2}), \\
{}_\lambda\langle n| &= {}_{\lambda - 1/2}\langle 0| \otimes {}_{1/2}\langle 0| = (q^{T_0} \otimes T_+ + T_+ \otimes q^{-T_0})^n \\
&= q^{-n/2}({}_{\lambda - 1/2}\langle n| \otimes {}_{1/2}\langle 0| + [n]q^\lambda {}_{\lambda - 1/2}\langle n - 1| \otimes {}_{1/2}\langle 1|).
\end{aligned}
$$

Thus

$$
\begin{aligned}
{}_\lambda\langle k|g|n\rangle_\lambda = q^{-(k+n)/2}[&{}_{\lambda - 1/2}\langle k|g|n\rangle_{\lambda - 1/2}\langle +|g|+\rangle \\
&+ q^\lambda[n]{}_{\lambda - 1/2}\langle k|g|n - 1\rangle_{\lambda - 1/2}\langle +|g|-\rangle \\
&+ q^\lambda[k]{}_{\lambda - 1/2}\langle k - 1|g|n\rangle_{\lambda - 1/2}\langle -|g|+\rangle \\
&+ q^{2\lambda}[k][n]{}_{\lambda - 1/2}\langle k - 1|g|n - 1\rangle_{\lambda - 1/2}\langle -|g|-\rangle]
\end{aligned}
$$

or, in terms of generating (τ-)functions,

$$
\tau_\lambda(t, \bar{t} \mid g) = \sqrt{M_t^- M_{\bar{t}}^-}\left(\tau_{\lambda - 1/2}(t, \bar{t} \mid g)(a + q^\lambda \bar{t}b + q^\lambda tc + q^{2\lambda}t\bar{t}d)\right).
$$

Applying this procedure recursively we get

$$
\begin{aligned}
\tau_\lambda(t, \bar{t} \mid g) &= \tau_{\lambda - 1/2}(q^{-1/2}t, q^{-1/2}\bar{t} \mid g)\tau_{1/2}(q^{\lambda - 1/2}t, q^{\lambda - 1/2}\bar{t} \mid g) \\
&\stackrel{\text{if } \lambda \in \mathbb{Z}/2}{=} \tau_{1/2}(q^{1/2 - \lambda}t, q^{1/2 - \lambda}\bar{t} \mid g) \\
&\qquad\times \tau_{1/2}(q^{3/2 - \lambda}t, q^{3/2 - \lambda}\bar{t} \mid g) \cdots \tau_{1/2}(q^{\lambda - 1/2}t, q^{\lambda - 1/2}\bar{t} \mid g),
\end{aligned}
$$

i.e., for half-integer λ τ_λ is a polynomial of degree 2λ in a, b, c, d.

For example,

$$
\begin{aligned}
\tau_1(t, \bar{t} \mid g) &= \tau_{1/2}(q^{-1/2}t, q^{-1/2}\bar{t} \mid g)\tau_{1/2}(q^{1/2}t, q^{1/2}\bar{t} \mid g) \\
&= (a + q^{-1/2}\bar{t}b + q^{-1/2}tc + q^{-1}t\bar{t}d)(a + q^{1/2}\bar{t}b + q^{1/2}tc + qt\bar{t}d) \\
&= a^2 + (q^{1/2}ab + q^{-1/2}ba)\bar{t} + (q^{1/2}ac + q^{-1/2}ca)t + b^2\bar{t}^2 \\
&\qquad + (qad + bc + cb + q^{-1}da)t\bar{t} + c^2t^2 + (q^{1/2}bd + q^{-1/2}db)t\bar{t}^2 \\
&\qquad + (q^{1/2}cd + q^{-1/2}dc)t^2\bar{t} + d^2t^2\bar{t}^2
\end{aligned}
$$

Using the relations like

$$q^{1/2}ab + q^{-1/2}ba = [2]q^{1/2}ba = [2]q^{-1/2}ab, \text{ etc.,}$$

one gets for this case

$$\tau_1(t, \bar{t} \mid g) = a^2 + [2]q^{-1/2}ab\bar{t} + [2]q^{-1/2}act + b^2\bar{t}^2 + ([2]qbc + [2]da)t\bar{t}$$
$$+ c^2t^2 + [2]q^{1/2}dbt\bar{t}^2 + [2]q^{1/2}dct^2\bar{t} + d^2t^2\bar{t}^2.$$

With this explicit expression, one can trivially make the calculations similar to (5.33) in order to check manifestly equation (5.29) for $\lambda = 1$, $\lambda' = \frac{1}{2}$, 1 and equation (5.30) for $\lambda = \lambda' = \frac{1}{2}$.

Thus, we showed explicitly (for the case of $\mathrm{SL}_q(2)$) that the quantum bilinear identities have as many solutions as the classical ones, provided the τ-function is allowed to take values in the noncommutative ring $A(G)$.

5.4   Comments on the Quantum Deformation of KP/Toda τ-Functions

As we saw in the previous subsection, the generic construction is easily applicable to quantum groups. Still the problem of quantum KP/Toda hierarchies deserves separate consideration and is not yet fully resolved. The problem is, that the evolution operator $U(t)$ in (5.2) is usually constructed from all the operators of the algebra, not just from a commuting set—as it happens in particular case of fundamental representations of GL(N). As a result, generic evolution of τ with variation of t's is not described as a set of *commuting* flows; rather they form a closed, but nontrivial, algebra. This manifests itself also in the fact that naturally the number of independent time variables is rather close to dimension than to the rank of the group. The problem of quantum deformation of KP/Toda hierarchy is to find a deformation which, while dealing with τ-function for quantum group, is still describable in terms of few time variables. If at all resolvable this is the problem of a clever choice of the weight functions $s^R_{m, \overline{m}}(t, \bar{t})$ in (5.1). Following Ref. 56 we shall now demonstrate that the problem *is* resolvable in principle, though at the moment it is a quantum deformation of somewhat nonconventional description of KP/Toda system (with evolution, introduced differently from that in Section 5.2, and it is not just a change of time variables: the transformation is representation R-dependent).

According to Ref. 62 parametrization of group elements which allows the most straightforward quantum deformation involves only *simple* roots $\pm\vec{\alpha}_i$, $i = 1, \ldots, r_G$:

$$g = g_U g_D g_L,$$

$$g_U = \prod_s^{<} e^{\theta_s T_{i(s)}}, \quad g_L = \prod_s^{>} e^{\chi_s T_{-i(s)}}, \quad g_D = \prod_{i=1}^{r_G} e^{\vec{\phi}\vec{H}} \tag{5.34}$$

Every particular simple root $\vec{\alpha}_i$ can appear several times in the product, and there are different parametrizations of the group elements of such type, depending on the choice of the set $\{s\}$ and the mapping $i(s)$. Quantum deformation of *such* formula is especially simple because comultiplication rule is especially simple for generators, associated with *simple* roots:

$$\Delta(T_i) = T_i \otimes q^{-2H_i} + I \otimes T_i,$$
$$\Delta(T_{-i}) = T_{-i} \otimes I + q^{2H_i} \otimes T_{-i}$$

For $q \neq 1$ any expression of the form (5.34) remains just the same, provided exponents in g_U and g_L are understood as q-exponents (in the simply laced case, $q^{\|\vec{\alpha}_i\|^2/2}$-exponents in general), and parameters ψ, χ, $\vec{\phi}$ become non-commuting generators of the "coordinate ring" $\mathcal{A}(G_q)$. Actually they form a kind of a very simple Heisenberg-like algebra:

$$\theta_s \theta_{s'} = q^{-\vec{\alpha}_{i(s)}\vec{\alpha}_{i(s')}} \theta_{s'} \theta_s, \quad s < s',$$
$$\chi_s \chi_{s'} = q^{-\vec{\alpha}_{i(s)}\vec{\alpha}_{i(s')}} \chi_{s'} \chi_s, \quad s < s',$$
$$q^{\vec{\beta}\vec{\phi}}\theta_s = \theta_s q^{\vec{\beta}\vec{\phi}} q^{\vec{\beta}\vec{\alpha}_{i(s)}},$$
$$q^{\vec{\beta}\vec{\phi}}\chi_s = \chi_s q^{\vec{\beta}\vec{\phi}} q^{\vec{\beta}\vec{\alpha}_{i(s)}}$$

These relations imply that $\Delta(g) = g \otimes g$.

The simplest possible assumption about evolution operators would be to say that, just as it was in the case of the standard KP/Toda theory (see Section 5.2), $U(t)$ is always an object of the type g_U, while $\overline{U}(\bar{t})$ is of the type g_L. However, these are no longer group elements:

$$\Delta(g_U) \neq g_U \otimes g_U, \quad \Delta(g_L) \neq g_L \otimes g_L,$$

because of the lack of factors g_D. Still the simplest possibility is to insist on identification of U and $\bar{U}$ as objects of the type g_U and g_L, respectively, and explicitly investigate implications of the failure of (5.3). As result one obtains instead of (5.3)

$$\Delta(U(\xi)) = U_L^{(2)}(\xi) \cdot U_R^{(2)}(\xi),$$

where

$$U(\xi) = \prod_s{}^< \mathcal{E}_q(\xi_s T_{i(s)}),$$
$$U_L^{(2)} = \prod_s{}^< \mathcal{E}_q(\xi_s T_{i(s)} \otimes q^{-2H_{i(s)}}) \neq I \otimes U(\xi),$$
$$U_R^{(2)} = \prod_s{}^< \mathcal{E}_q(\xi_s I \otimes T_{i(s)}) = I \otimes U(\xi)$$

and this has some simply accountable implications for determinant formulas for quantum τ-functions.

In Section 5.2 we essentially used an evolution operator of the type

$$U(\xi) = \prod_{1 \le i \le N}^{r} \prod_{1 \le j < i} \exp(\xi_{ij} T_{i-j}) \tag{5.35}$$

where ξ_{ij} are certain functions of only N-independent variables t. While (5.35) is trivial to deform in the direction of $q \ne 1$, it is a separate (yet unresolved) problem to find such reduction to only N-variables, consistent with the commutation relations between ξ_{ij},

$$\xi_{ij}\xi_{i'j'} = q^{-\vec{\alpha}_{i-j}\vec{\alpha}_{i'-j'}} \xi_{i'j'}\xi_{ij}, \quad \{i,j\} < \{i',j'\}.$$

One can instead use a much simpler evolution,

$$\widehat{U}(\xi) = \prod_{i=1}^{r_G}{}^{<} \exp(\xi_i T_i). \tag{5.36}$$

This is enough to generate all the states of any fundamental representation from the corresponding vacuum (highest vector) state, but $\langle \mathrm{vac}_{F_n}|\widehat{U}(\xi)$ has nothing to do with the usual $\langle \mathrm{vac}_{F_n}|U(t)$, where $U(t)$ is given by (5.35). It can be better to say that identification $\langle \mathrm{vac}_{F_n}|U^{(A)}(\xi) = \langle \mathrm{vac}_{F_n}|U(t)$ defines a relation $\xi_i(t)$, which explicitly depends on n.

One can of course build the theory of KP/Toda hierarchies in terms of ξ-variables instead of conventional t-variables, but it *cannot* be obtained by just change of time variables; the whole construction will look different. Instead this new construction is immediately deformed to the case of $q \ne 1$; instead of (5.36) we just write

$$\widehat{U}(\xi) = \prod_{i=1}^{r_G}{}^{<} \mathcal{E}_q(\xi_i T_i) \tag{5.37}$$

where ξ's are noncommuting variables,

$$\xi_i \xi_j = q^{-\vec{\alpha}_i \vec{\alpha}_j} \xi_j \xi_i, \quad i < j,$$

and it is easy to derive the quantum counterpart of any statement of the classical ($q = 1$) theory once it is formulated for ξ-parametrization.

In what follows we first briefly describe the conventional KP/Toda hierarchy in this nonstandard parametrization, then consider the corresponding quantum deformation and derive the substitute of determinant formulas for $\tau_n \equiv \tau_{F_n}$ in the case of $q \ne 1$.

5.4.1 On the Modified KP/Toda Hierarchy

Our first purpose is to demonstrate that all the main ingredients of description of the classical KP/Toda hierarchy, as described in Section 5.2, are preserved if evolution (5.37) is used instead of (5.35), in particular, there are determinant formulas and a hierarchy of differential equations.

From now on we denote the τ-function associated with the evolution (5.37) through $\hat{\tau}(\xi, \bar{\xi} \mid g)$. This τ-function is linear in each time variable ξ_i; hence, it satisfies simpler determinant formulas and simpler hierarchy of equations. Indeed, now we have

$$\hat{\tau}_1(\xi, \bar{\xi} \mid g) \equiv \langle 0_{F_1} | \widehat{U}(\xi) g \widehat{\bar{U}}(\bar{\xi}) | 0_{F_1} \rangle = \sum_{k, \bar{k} \geq 0} s_k \bar{s}_{\bar{k}} \langle k | g | \bar{k} \rangle$$

where $s_k = \xi_1 \xi_2 \cdots \xi_k$, $s_0 = 1$, and

$$
\begin{aligned}
\hat{\tau}_1^{m\bar{m}}(\xi, \bar{\xi} \mid g) &\equiv \langle m_{F_1} | \widehat{U}(\xi) g \widehat{\bar{U}}(\bar{\xi}) | \bar{m}_{F_1} \rangle \\
&= \frac{1}{s_m \bar{s}_{\bar{m}}} \sum_{\substack{k \geq m \\ \bar{k} \geq \bar{m}}} s_k \bar{s}_{\bar{k}} \langle k | g | \bar{k} \rangle \\
&= \frac{1}{s_m \bar{s}_{\bar{m}}} \sum_{\substack{k \geq m \\ \bar{k} \geq \bar{m}}} \frac{\partial}{\partial \log s_k} \frac{\partial}{\partial \log \bar{s}_{\bar{k}}} \tau_1(\xi, \bar{\xi} \mid g) \\
&= \frac{1}{s_{m-1} \bar{s}_{\bar{m}-1}} \frac{\partial}{\partial \xi_m} \frac{\partial}{\partial \bar{\xi}_{\bar{m}}} \tau_1(\xi, \bar{\xi} \mid g).
\end{aligned}
$$

Thus,[25]

$$
\begin{aligned}
\hat{\tau}_{n+1} = \det_{0 \leq m, \bar{m} \leq n} \hat{\tau}_1^{m\bar{m}} &= \left(\prod_{m=1}^n s_m \bar{s}_{\bar{m}} \right)^{-1} \det_{m\bar{m}} \left(\sum_{\substack{k \geq m \\ \bar{k} \geq \bar{m}}} s_k \bar{s}_{\bar{k}} \langle k | g | \bar{k} \rangle \right) \\
&= \frac{1}{s_n \bar{s}_n} \sum_{k, \bar{k} \geq n} s_k \bar{s}_{\bar{k}} \det_{0 \leq m, \bar{m} \leq n-1} \begin{pmatrix} g_{m\bar{m}} & g_{m\bar{k}} \\ g_{k\bar{m}} & g_{m\bar{m}} \end{pmatrix} \\
&\equiv \frac{1}{s_n \bar{s}_n} \sum_{k, \bar{k} \geq n} s_k \bar{s}_{\bar{k}} \mathcal{D}_{k\bar{k}}^{(n)}.
\end{aligned}
\qquad (5.38)
$$

---

[25] One can compare determinant representations (5.22) and (5.38) to find the connection between different coordinates t and ξ. For every given n the variables s_k are some functions of $P_j(t)$. For example, in the simplest case of the first fundamental representation $F^{(1)}$ we have $\tau_1(t \mid g) = \hat{\tau}_1(\xi \mid g)$ and $s_k = P_k(t)$, $\partial/\partial t_k = \sum_i s_{i-k} \partial/\partial s_i$. However, identification of $\tau_n(t)$ and $\hat{\tau}_n(\xi)$ with $n \neq 1$ will lead to different relations between ξ and t. Thus the two different evolutions are *not* related just by a change of time variables, relation is representation-dependent, and cannot be lifted to the actual KP/Toda case (when $n = \infty$). Two evolutions provide two equally nice, but not just equivalent descriptions of the same hierarchy.

5.4.2 q-Determinant-Like Representation

In this section we demonstrate how the technique developed in the previous sections is deformed to the quantum case and, in particular, obtain q-determinant-like deformation of (5.38). Our evolution operator (5.37) satisfies the following comultiplication rule

$$\Delta^{n-1}(U\{T_i\}) = \prod_{m=1}^{n} U^{(m)},$$

where

$$U^{(m)} = U\{\; I \otimes \ldots I \otimes \xi_i T_i \otimes q^{-2H_i} \otimes \cdots \otimes q^{-2H_i}\}$$

and T_i appears at the mth place in the tensor product. Similarly

$$\overline{U}^{(m)} = \overline{U}\{q^{2H_i} \otimes \cdots \otimes q^{2H_i} \otimes T_{-i} \otimes I \otimes \cdots \otimes I\}.$$

Now let us transform the operator-valued q-factors into c-number ones. Let

$$H_i|\bar{j}_{F_1}\rangle = h_{i,\bar{j}}|\bar{j}_{F_1}\rangle, \quad \langle j_{F_1}|H_i = h_{i,j}\langle j_{F_1}|$$

(in fact, for SL(N) $2h_{i,i-1} = +1$, $2h_{i,i} = -1$, all the rest are vanishing). Then

$$\hat{\tau}_n^{j_1\cdots j_n \bar{j}_1 \cdots \bar{j}_n}(\xi_i, \bar{\xi}_i \mid g)$$

$$\equiv \left(\bigotimes_{m=1}^{n} \langle j_m|\right) \Delta^{n-1}(U) g^{\otimes n} \Delta^{n-1}(\overline{U}) \left(\bigotimes_{m=1}^{n} |\bar{j}_m\rangle\right)$$

$$= \prod_{m=1}^{n} \langle j_m|U\{\xi_i T_i q^{-2\sum_{l=m+1}^{n} h_{i,j_l}}\} g \overline{U}\{\bar{\xi}_i T_{-i} q^{2\sum_{l=1}^{m-1} h_{i,\bar{j}_l}}\} |\bar{j}_m\rangle$$

$$= \prod_{m=1}^{n} \hat{\tau}_1^{j_m \bar{j}_m}(\xi_i q^{-2\sum_{l=m+1}^{n} h_{i(s),j_l}}, \bar{\xi}_i q^{2\sum_{l=1}^{m-1} h_{i(s),\bar{j}_l}}). \tag{5.39}$$

In order to get the analogue of (5.22), one should replace antisymmetrization by q-antisymmetrization, since, in the quantum case, fundamental representations are described by q-antisymmetrized vectors. We define q-antisymmetrization as a sum over all permutations,

$$([1,\ldots,k]_q) = \sum_{P}(-q)^{\deg P}(P(1),\ldots,P(k)),$$

where

$$\deg P = \#\ \text{of inversions in}\ P.$$

Then, q-antisymmetrizing (5.39) with $j_k = k - 1$, $\bar{j}_{\bar{k}} = \bar{k} - 1$, one finally gets

$$\tau_n(\xi, \bar{\xi} \mid g) = \sum_{P,P'} (-q)^{\deg P + \deg P'}$$
$$\times \prod_{m=0}^{n-1} \tau_1^{P(m)P'(\bar{m})} (\xi_s q^{-2\sum_{l=m+1}^{n-1} h_{i(s),P(l)}}, \bar{\xi}_s q^{2\sum_{\bar{l}=0}^{m-1} h_{i(s),P'(\bar{l})}}).$$

This would be just a q-determinant if there were no q-factors that twist the time variables.[26]

To make this expression more transparent let us consider the simplest example of the second fundamental representation:

$$\tau_2 = \tau_1^{00}(q\xi_1, q^{-1}\xi_2, \xi_i; \bar{\xi}_1, \bar{\xi}_2, \bar{\xi}_i)\tau_1^{11}(\xi_1, \xi_2, \xi_i; q\bar{\xi}_1, \bar{\xi}_2, \bar{\xi}_i)$$
$$- q\tau_1^{01}(q\xi_1, q^{-1}\xi_2, \xi_i; \bar{\xi}_1, \bar{\xi}_2, \bar{\xi}_i)\tau_1^{10}(\xi_1, \xi_2, \xi_i; q^{-1}\bar{\xi}_1, q\bar{\xi}_2, \bar{\xi}_i)$$
$$- q\tau_1^{10}(q^{-1}\xi_1, \xi_2, \xi_i; \bar{\xi}_1, \bar{\xi}_2, \bar{\xi}_i)\tau_1^{01}(\xi_1, \xi_2, \xi_i; q\bar{\xi}_1, \bar{\xi}_2, \bar{\xi}_i)$$
$$+ q^2\tau_1^{11}(q^{-1}\xi_1, \xi_2, \xi_i; \bar{\xi}_1, \bar{\xi}_2, \bar{\xi}_i)\tau_1^{00}(\xi_1, \xi_2, \xi_i; q^{-1}\bar{\xi}_1, q\bar{\xi}_2, \bar{\xi}_i).$$

This can be written in a more compact form with the help of operators

$$\mathbb{D}_i^L \equiv D_i \otimes I, \qquad\qquad \mathbb{D}_i^R \equiv \prod_j M_j^{-\vec{\alpha}_i \vec{\alpha}_j} \otimes D_i,$$

$$\bar{\mathbb{D}}_i^L \equiv \bar{D}_i \otimes \prod_j \bar{M}_j^{-\vec{\alpha}_i \vec{\alpha}_j}, \qquad \bar{\mathbb{D}}_i^R \equiv I \otimes \bar{D}_i.$$

These operators have simple commutation relations:

$$\mathbb{D}_i^L \mathbb{D}_j^R = q^{\vec{\alpha}_i \vec{\alpha}_j} \mathbb{D}_j^R \mathbb{D}_i^L,$$
$$\bar{\mathbb{D}}_i^L \bar{\mathbb{D}}_j^R = q^{\vec{\alpha}_i \vec{\alpha}_j} \bar{\mathbb{D}}_j^R \bar{\mathbb{D}}_i^L.$$

Then,

$$\tau_2 = (M_1^- \otimes \bar{M}_1^+)(\mathbb{D}_1^R - q\mathbb{D}_1^L) \cdot (\bar{\mathbb{D}}_1^R - q\bar{\mathbb{D}}_1^L)\tau_1 \otimes \tau_1.$$

6   Conclusion

These notes combine presentation of some well-established facts with that of more recent and sometime disputable speculations. There are all reasons

---

[26]Remember that the q-determinant is defined as

$$\det_q A \sim A_{[1}^{[1} \cdots A_{n]_q}^{n]_q} = \sum_{P,P'} (-q)^{\deg P + \deg P'} \prod_a A_{P'(a)}^{P(a)}$$

Note that this is not obligatory the same as $A_{[1}^1 \cdots A_{n]_q}^n$. It is the same only for peculiar commutation relations of the matrix elements A_i^j.

to believe that further developments will prove that the theory of generalized τ-functions and nonperturbative partition functions can become a flourishing branch of mathematical physics with applications well beyond the present modest scope of topological theories and $c < 1$ string models and with profound relations to other fields of the string theory. It is also important that there are plenty of "small problems" at all the levels of this theory, which are enjoyable to think about.

Acknowledgments: I am indebted to my coauthors and friends for cooperation during our work on the subject of matrix models and integrable systems.

It is a pleasure to thank Gordon Semenoff and Luc Vinet for the invitation to present this material at the Banff school, as well as for their hospitality and support.

The text was partly prepared during my stay at CRM, Montréal.

7   REFERENCES

1. A. Morozov. Integrability and matrix models. *UFN*, 164 (1): 3–62, 1994.

2. A. Morozov. The string theory: what is this? *UFN*, 162 (8): 84–175, 1992.

3. M. L. Kontsevich. Intersection theory on the moduli space of curves. *Funk. Anal. & Prilozh.*, 25: 50–57, 1991. Russian.

4. S. Kharchev, A. Marshakov, A. Mironov, A. Morozov, and A. Zabrodin. Unification of all string models with $c < 1$. *Phys. Lett.*, 275B (3-4): 311–314, 1992; S. Kharchev, A. Marshakov, A. Mironov, A. Morozov, and A. Zabrodin. Towards unified theory of $2d$ gravity. *Nucl. Phys.*, B380 (1-2): 181–240, 1992.

5. R. Dijkgraaf. Intersection theory, integrable hierarchies and topological field theory. In *New Symmetry Principles in Quantum Field Theory*, (Cargèse, 1991), volume B295 of *NATO ASI*, 1992. Plenum, New York, pages 95–158.

6. M. Kontsevich. Intersection theory on the moduli space of curves. *Commun. Math. Phys.*, 147 (1): 1–23, 1992. Russian.

7. C. Itzykson and J.-B. Zuber. Combinatorics of the modular group. II. the Kontsevich integrals. *Int. J. Mod. Phys.*, A7 (1): 5661–5705, 1992.

8. M. Adler and P. van Moerbeke. A matrix integral solution to two-dimensional w_p-gravity. *Commun. Math. Phys.*, 147 (1): 25–56, 1992.

9. A. Gerasimov, Yu. Makeenko, A. Marshakov, A.Mironov, A. Morozov, and A. Orlov. Matrix models as integrable systems: from universality to geometrodynamical principle of string theory. *Mod. Phys. Lett.*, A6 (33): 3079–3090, 1991.

10. A. Marshakov, A. Mironov, and A. Morozov. Generalized matrix models as conformal field theories. Discrete case. *Phys. Lett.*, 265B (1-2): 99–107, 1991; A. Mironov and S. Pakuliak. Double-scaling limit in the matrix models of a new type. Technical Report FIAN/TD/05-92, P. N. Lebedev Physical Institute, 1992; A. Mironov and S. Pakuliak. On the continuum limit of conformal matrix models. *Int. J. Mod. Phys.*, A8: 3107–3137, 1993; S. Kharchev, A. Marshakov, A. Mironov, A. Morozov, and S. Pakuliak. Conformal matrix models as an alternative to conventional multi-matrix models. *Nucl. Phys.*, B404 (3): 717–750, 1993.

11. E. Brézin, C. Itzykson, G. Parisi, and J.-B.Zuber. Planar diagrams. *Commun. Math. Phys.*, 59 (1): 35–51, 1978.

12. I. Frenkel and V. Kac. Basic representations of affine Lie algebras and dual resonance models. *Invent. Math.*, 62 (1): 23–66, 1980/81; G. Segal. Unitary representations of some infinite-dimensional groups. *Commun. Math. Phys.*, 80 (3): 301–342, 1981.

13. A. Gerasimov, A. Morozov, M. Olshanetsky, A. Marshakov, and S. Shatashvili. Wess-Zumino-Witten model as a theory of free fields. *Int. J. Mod. Phys.*, A5 (13): 2495–2589, 1990.

14. M. Bershadsky and H. Ooguri. Hidden SL(n) symmetry in conformal field theories. *Commun. Math. Phys.*, 126 (1): 49–83, 1989.

15. M. Bershadsky. Conformal field theories via Hamiltonian reduction. *Commun. Math. Phys.*, 139 (1): 71–82, 1991.

16. F. A. Bais, T. Tjin, and P. van Driel. Covariantly coupled chiral algebras. *Nucl. Phys.*, B357 (2-3): 632–654, 1991.

17. L. Feher, L. O'Raifeartaigh, P. Ruelle, I. Tsutsui, and A. Wipf. On the general structure of Hamiltonian reductions of WZW theory. Technical Report UdeM-LPN-TH-71/91, Lab. de physique nucléaire, Univ. de Montréal, 1991.

18. A. Zamolodchikov. Infinite extra symmetries in two-dimensional conformal quantum field theory. *Teoret. Mat. Fiz.*, 63 (3): 347–359, 1985. Russian; V. Fateev and S. Lykyanov. The models of two-dimensional conformal quantum theory with Z_n symmetry. *Int. J. Mod. Phys.*, A3 (2): 507–520, 1988.

19. M. Adler, A. Morozov, T. Shiota, and P. van Moerbeke. New matrix model solutions to the Kac-Schwarz problem. In *Theory of Elementary Particles*, (Buckow, 1995), volume 49 of *Nucl. Phys. B Proc. Suppl.*, 1996. pages 201–212.

20. V. Kazakov and A. Migdal. Induced gauge theory at large N. *Nucl. Phys.*, B397 (1-2): 214–238, 1993; S. Kharchev, A. Marshakov, A. Mironov, and A. Morozov. Generalized Kazakov-Migdal-Kontsevich model: Group theory aspects. *Int. J. Mod. Phys.*, A10 (14): 2015–2051, 1995.

21. A. Migdal. Exact solution of induced lattice gauge theory at large-N. *Mod. Phys. Lett.*, A8 (4): 359–371, 1993.

22. C. Itzykson and J.-B. Zuber. The planar approximation II. *J. Math. Phys.*, 21 (3): 411–421, 1980.

23. Harish-Chandra. Differential operators on a semisimple Lie algebra. *Amer. J. Math.*, 79: 87–120, 1957.

24. M. Semenov-Tyan-Shanskii. Harmonic analysis on Riemann symmetric spaces of negative curvature and scattering theory. *Izv. Akad. Nauk USSR*, 40: 562–591, 1976. Russian.

25. A. Alekseev, L. Faddeev, and S. Shatashvili. Quantization of symplectic orbits of compact lie groups by means of the functional integral. *J. Geom. Phys.*, 5 (3): 391–406, 1988.

26. A. Alekseev and S. Shatshvili. Path integral quantization of the coadjoint orbits of the Virasoro group and 2-d gravity. *Nucl. Phys.*, B323 (3): 719–733, 1989.

27. J. Duistermaat and G. Heckman. On the variation in the cohomology of the symplectic form of the reduced phase space. *Invent. Math.*, 69 (2): 259–268, 1982.

28. M. Blau, E. Keski-Vakkuri, and A. Niemi. Path integrals and geometry of trajectories. *Phys. Lett.*, 246B (1-2): 92–98, 1990; E. Keski-Vakkuri, A. Niemi, G. Semenoff, and O. Tirkkonen. Topological quantum theories and integrable models. *Phys. Rev.*, D44 (12): 3899–3905, 1991; A. Hietamäki, A. Morozov, A. Niemi, and K. Palo. Geometry of $n = 1/2$ supersymmerty and the Atiyah-Singer index theorem. *Phys. Lett.*, 263B (3-4): 417–424, 1991; A. Morozov, Niemi, and K. Palo. Supersymmetry and loop space geometry. *Phys. Lett.*, 271B (3-4): 365–371, 1991; A. Morozov, A. Niemi, and K. Palo. Supersymplectic geometry of supersymmetric quantum field theories. *Nucl. Phys.*, B377 (1-2): 295–338, 1992.

29. E. Witten. Two-dimensional gauge theories revisited. *J. Geom. Phys.*, 9 (4): 303–368, 1992.

30. A. Niemi and O. Tirkkonen. On exact evaluation of path integrals. *Ann. Phys.*, 235 (2): 318–349, 1994.

31. I. Kogan, A. Morozov, G. Semenoff, and N. Weiss. Area law and continuum limit in "induced QCD." *Nucl. Phys.*, B395 (3): 547–580, 1993.

32. A. Morozov. Pair correlation in the Itzykson-Zuber integral. *Mod. Phys. Lett.*, A7 (37): 3503–3507, 1992.

33. S. Shatashvili. Correlation functions in the Itzykson-Zuber model. *Commun. Math. Phys.*, 154 (2): 421–432, 1993.

34. A. Matytsin. On the large-N limit of the Itzykson-Zuber integral. *Nucl. Phys.*, B411 (2-3): 805–820, 1994.

35. A. Mironov, A. Morozov, and G. Semenoff. Unitary matrix integral in the framework of generalized Kontsevich model. I. Brézin-Gross-Witten model. Technical Report ITEP-M6/93, ITEP, Moscow, 1993.

36. S. Kharchev, A. Marshakov, A. Mironov, and A. Morozov. Landau-Ginzburg topological theories in the framework of GKM and equivalent hierarchies. *Mod. Phys. Lett.*, A8 (11): 1047–1061, 1993.

37. V. Kac and A. Schwarz. Geometric interpretation of the partition function of 2D gravity. *Phys. Lett.*, B257 (3-4): 329–334, 1991.

38. A. Marshakov and S. Kharchev. Topological versus nontopological theories and $p-q$ duality in $c \leq 1$ 2d gravity models. In *Proceedings of International Workshop on String Theory, Quantum Gravity and the Unification of Fundamental Interactions*, (Rome, 1992), 1992. pages 331–346.

39. D. Gross and M. Newman. Unitary and Hermitian matrices in an external field. *Phys. Lett.*, 266B (3-4): 291–297, 1991.

40. A. Marshakov, A. Mironov, and A. Morozov. From Virasoro constraints in Kontsevich's model to w-constraints in two-matrix models. *Mod. Phys. Lett.*, A7 (15): 1345–1359, 1992.

41. L. Chekhov and Yu. Makeenko. A hint on the external field problem for matrix models. *Phys. Lett.*, 278B (3): 271–278, 1992.

42. S. Kharchev, A.Marshakov, A.Mironov, and A. Morozov. Generalized Kontsevich model versus Toda hierarchy and discrete matrix models. *Nucl. Phys.*, B397 (1-2): 339–378, 1993.

43. E. Witten. On the Kontsevich model and other models of two-dimensional gravity. In *Proc. of the XXth International Conference on Differential Geometric Methods in Theoretical Physics*, (New York, 1991), 1992. World Scientific, River Edge, NJ, pages 176–216; A. Marshakov, A. Mironov, and A. Morozov. On the equivalence of topological and quantum 2D gravity. *Phys. Lett.*, 274B (3-4): 280–288, 1992.

44. M. Fukuma, H. Kawai, and R. Nakayama. Continuum Schwinger-Dyson equations and universal structures in two quantum gravity. *Int. J. Mod. Phys.*, A6 (8): 1385–1406, 1991; R. Dijkgraaf, E. Verlinde, and H. Verlinde. Loop equations and Virasoro constraints in nonperturbative two-dimensional quantum gravity. *Nucl. Phys.*, B348 (3): 435–456, 1991.

45. A. Mikhailov. Ward identities and w constraints in the generalized Kontsevich model. *Int. J. Mod. Phys.*, A9 (6): 873–889, 1994.

46. M. Sato, T. Miwa, and M. Jimbo. Holonomic qantum fields. I. *Publ. RIMS, Kyoto Univ.*, 14 (1): 223–267, 1978; M. Sato, T. Miwa, and M. Jimbo. Holonomic quantum fields. II. the Remann-Hilbert problem. *Publ. RIMS, Kyoto Univ.*, 15 (1): 201–278, 1979; M. Sato, T. Miwa, and M. Jimbo. Holonomic quantum fields. III. *Publ. RIMS, Kyoto Univ.*, 15 (2): 577–629, 1979; M. Sato, T. Miwa, and M. Jimbo. Holonomic quantum fields. IV. *Publ. RIMS, Kyoto Univ.*, 15 (3): 871–972, 1979; M. Sato, T. Miwa, and M. Jimbo. Holonomic quantum fields. V. *Publ. RIMS, Kyoto Univ.*, 16 (2): 531–584, 1980.

47. A. Morozov, M. Shifman, and A. Turbiner. Continuous Sugawara-like realization of $c = 1$ conformal models. *Int. J. Mod. Phys.*, A5 (15): 2953–2991, 1990.

48. M. Sato. Soliton equations as dynamical systems on infinite-dimensional Grassmann manifolds. *RIMS Kokyuroku*, 439: 30–40, 1981; M. Sato and Y. Sato. Soliton equations as dynamical systems on infinite-dimensional Grassmann manifold. In *Nonlinear Partial Differential Equations in Applied Science*, (Tokyo, 1982), volume 81 of *North-Holland Math. Stud.*, 1983. North-Holland, Amsterdam, pages 259–271; G. Segal and G. Wilson. Loop groups and equation of KdV type. *Publ. I.H.E.S.*, 61: 5–65, 1985.

49. V. Kac. *Infinite-Dimensional Lie Algebras*, 2nd edition. Cambridge Univ. Press, Cambridge, 1985.

50. A. Orlov and E. Shulman. Additional symmetries for integrable equations and conformal representation. *Lett. Math. Phys.*, 12 (3): 171–179, 1986; P. Grinevich and A. Orlov. Virasoro action on Riemann surfaces, Grassmannians, $\det \bar{\partial}_J$ and Segal-Wilson τ-function.

In *Problems of Modern Quantum Field Theory*, (Alushta, 1989), Res. Rep. Phys., 1989. Springer-Verlag, Berlin.

51. S. Kharchev, A. Marshakov, A. Mironov, A. Orlov, and A. Zabrodin. Matrix models among integrable theories: forced hierarchies and operator formalism. *Nucl. Phys.*, B366 (3): 569–601, 1991.

52. K. Ueno and K. Takasaki. Toda lattice theory. In *Group Representations and Systems of Differential Equations*, (Tokyo, 1982), volume 4 of *Adv. Studies in Pure Math.*, 1984. North-Holland, Amsterdam, pages 1–95.

53. M. Bowick, A. Morozov, and D. Shevitz. Reduced unitary matrix models and the hierarchy of τ-functions. *Nucl. Phys.*, B354 (2-3): 496–530, 1991.

54. A. Gerasimov, A. Marshakov, A. Mironov, A. Morozov, and A. Orlov. Matrix models of two-dimensional gravity and Toda theory. *Nucl. Phys.*, B357 (2-3): 565–618, 1991.

55. A. Gerasimov, S. Khoroshkin, D. Lebedev, A. Mironov, and A. Morozov. Generalized Hirota equations and representation theory. I. The case of SL(2) and $SL_q(2)$. *Int. J. Mod. Phys.*, A10 (18): 2589–2614, 1995.

56. S. Kharchev, A. Mironov, and A. Morozov. Nonstandard KP evolution and quantum τ-function. *Teoret. Mat. Fiz.*, 104: 129–143, 1995.

57. Y. Ohta, J. Satsuma, D. Takahashi, and T. Tokihiro. An elementary introduction to Sato theory. Recent developments in soliton theory. *Prog. Theor. Phys. Suppl*, 94: 210–241, 1988.

58. K. Kajiwara and J. Satsuma. q-difference version of the two-dimensional Toda lattice equation. *J. Phys. Soc. Japan*, 60 (12): 3986–3989, 1991; K. Kajiwara, Y. Ohta, and J. Satsuma. q-discrete Toda molecule equation. *Phys. Lett. A*, 180 (3): 249–256, 1993.

59. A. Mironov, A. Morozov, and L. Vinet. On a c-number quantum τ-function. *Theor. Math. Phys.*, 100 (1): 890–889, 1994.

60. R. Floreanini and L. Vinet. On the quantum group and quantum algebra approach to q-special functions. *Lett. Math. Phys.*, 27 (3): 179–190, 1993.

61. N. Yu. Reshetikhin, L. A. Takhtajan, and L. D. Faddeev. Quantization of Lie groups and Lie algebras. *Algebra i Analiz*, 1 (1): 178–206, 1989. Russian.

62. A. Morozov and L. Vinet. Free-field representation of group element for simple quantum groups. Technical Report CRM-2202, CRM, Montréal, 1994.

6

Localization, Equivariant Cohomology, and Integration Formulas

Antti J. Niemi

ABSTRACT In this chapter we review the derivation of the Duistermaat-
Heckman integration formula and its path integral generalizations, and
explain the underlying formalism of equivariant cohomology. We evaluate
the quantum mechanical partition function for a general integrable model
by localizing onto an ordinary integral of an equivariant characteristic class.
We also describe the Mathai-Quillen formalism and its equivariant loop
space extensions. We show how certain standard relations in classical Morse
theory can be derived from this formalism, and generalize these relations
to the infinite-dimensional and equivariant context. We also explain how
Poincaré supersymmetric quantum field theories can be formulated using
equivariant cohomology in the loop space.

In this chapter we shall describe why and how certain Hamiltonian phase
space path integrals can be evaluated exactly using geometrical methods
based on the localization principle. By an exact evaluation of a path integral
we here mean, that the quantum mechanical partition function localizes
into an ordinary integral either over the original classical phase space or
over the moduli space of the classical solutions to Hamilton's equations of
motion.

The first to observe that certain path integrals can be localized was
Semenov-Tjan-Šhanskiĭ [1], and independently Duistermaat and Heckman
[2]. These authors found, that for certain Hamiltonians H finite-dimensional
phase space integrals of the form

$$\int \omega^n e^{-\beta H}, \tag{0.1}$$

where ω^n denotes the Liouville measure, can be *localized* to the critical
points of H whenever the canonical flow of H determines the action of
$U(1) \sim S^1$ on the phase space, i.e., whenever H can be viewed as a Cartan

generator of some Lie algebra on the phase space. The integration formula presented in Refs. 1 and 2 coincides with that obtained when (0.1) is evaluated by WKB approximation, except that now the summation is over *all* critical points of the Hamiltonian H, not just over its local minima as in the WKB approximation.

Subsequently, it has been observed [3–5] that the localization of the integral (0.1) can be understood in terms of *equivariant cohomology*, hence it is also intimately related to the concept of *equivariant characteristic classes* [6]. The integration formula by Duistermaat and Heckman has also been generalized to certain infinite-dimensional cases, and applied in particular to the evaluation of the Atiyah-Singer index theorem [5, 7].

A formal generalization of the Duistermaat-Heckman integration formula for generic bosonic phase space path integrals has been presented in Ref. 8. The derivation is based on loop space equivariant cohomology, and the ensuing integration formula assumes that the classical Hamiltonian generates the action of S^1 on the phase space with isolated critical points. This integration formula again coincides with that obtained from the WKB approximation, except that again the summation extends over all critical trajectories of the classical action, not just over its local minima. A further generalization has been presented in Ref 9. This integration formula relates the path integral to equivariant characteristic classes, and the final result can be viewed as an equivariant version of the Atiyah-Singer index theorem. In particular, in this generalization the final result is *not* a discrete sum over the critical trajectories of the action, but an integral over the original phase space. Consequently it is applicable also in cases, where the standard WKB approximation does not work, for example [10] if the critical trajectories of the classical action coalesce at points in the phase space.

The previous integration formulas are all based on the assumption, that the classical Hamiltonian determines the global action of $S^1 \sim \mathrm{U}(1)$ on the phase space. A generalization to Hamiltonians that are either quadratic functions of such U(1) generators, or even quadratic functions of arbitrary generators of some non-Abelian Lie algebra, has been presented in Ref. 11, and an extension to *a priori* arbitrary functions of such generators has been presented in Ref. 12. The original integrals are now localized to integrals over (some submanifolds of) the original phase space, and the final result can be quite different from the WKB approximation. As a consequence, these more general integration formulas provide new nonperturbative methods for the exact evaluation of a large class of phase space integrals. They can also be applied to certain infinite dimensional functional integrals such as two-dimensional Yang-Mills theory [11].

In the present chapter we shall review the derivation of these integration formulas, and explain the underlying framework of equivariant cohomology and Mathai-Quillen formalism. We shall first introduce some basic concepts from symplectic geometry, followed by a derivation of the Duistermaat-

Heckman integration formula. We then show, how their integration formula can be generalized to infinite dimensions, first by evaluating the Atiyah-Singer index theorem for a Dirac operator on a compact Riemannian manifold, and then by evaluating the quantum mechanical path integral that represents the partition function of a generic integrable Hamiltonian system. After this, we then extend our techniques to consider more general phase space integrals that can be related to Morse theory. In particular, we show how localization methods can be applied to derive equivariant generalizations of the standard relation between Poincaré-Hopf and Gauss-Bonnet-Chern theorems. Finally, we argue that Poincaré supersymmetric quantum field theories can also be represented in our framework of equivariant cohomology, suggesting that the methods we have presented here could provide exact tools to study supersymmetric quantum field theories.

1 Symplectic Geometry

We shall first review some basic concepts in symplectic geometry [13], which is the geometrical framework for Hamiltonian dynamics. In this formalism, one views the classical phase space $\mathcal{M}$ as a symplectic manifold as follows: We shall assume that instead of the standard momentum and position variables p_a, q^a we have some arbitrary local coordinates z^a ($a = 1, \ldots, 2n$) on $\mathcal{M}$. Darboux theorem ensures that in a local neighborhood on $\mathcal{M}$ we can always find a coordinate transformation such that $z^a \to p_a, q^a$ with the standard Poisson bracket, but in general the z^a are arbitrary functions of p_a's and q^a's and their Poisson brackets are nontrivial,

$$\{z^a, z^b\} = \omega^{ab}(z).$$

Here ω^{ab} is nondegenerate so that we can introduce the inverse matrix

$$\omega^{ac}\omega_{cb} = \delta^a_b.$$

The matrix elements of this inverse matrix then define components of a two-form which is called the symplectic two-form

$$\omega = \frac{1}{2}\omega_{ab}dz^a \wedge dz^b.$$

The Jacobi identity for Poisson brackets implies that this two-form is closed, so that we can locally represent it as an exterior derivative of a one-form ϑ,

$$d\omega = 0 \implies \omega = d\vartheta = \partial_a \vartheta_b dz^a \wedge dz^b.$$

The one-form ϑ is called the symplectic one-form.

Canonical transformations are exactly coordinate transformations that leave ω intact:

$$\vartheta \xrightarrow{\psi} \vartheta + d\psi \implies \omega \xrightarrow{\psi} \tilde{\omega} \equiv \omega.$$

In terms of Darboux coordinates i.e., standard position and momentum variables this gives the familiar

$$p_a dq^a = \vartheta \xrightarrow{\psi} \vartheta + d\psi = \tilde{\vartheta} = P_a dQ^a \implies p_a dq^a - P_a Q^a = d\psi.$$

The symplectic two-form defines a natural volume element on $\mathcal{M}$, the Liouville measure

$$\omega^n = \omega \wedge \cdots \wedge \omega.$$

The classical partition function is

$$Z = \int \omega^n e^{-\beta H},$$

where H is some Hamiltonian, i.e., an arbitrary function on $\mathcal{M}$. It determines a Hamiltonian flow on the phase space, which is identified as the motion generated by the corresponding Hamiltonian vector field $\mathcal{X}_H$ defined by

$$\omega(\mathcal{X}_H, \cdot) + dH = 0 \iff \mathcal{X}_H^a = \omega^{ab}\partial_b H.$$

The Poisson brackets of two Hamiltonian functions can then be represented as

$$\{H, G\} = \omega^{ab}\partial_a H \partial_b G = \mathcal{X}_H^a \partial_a G = \omega_{ab}\mathcal{X}_H^a \mathcal{X}_G^b = \omega(\mathcal{X}_H, \mathcal{X}_G).$$

Using this vector field, we can introduce internal multiplication (contraction)

$$\{H, G\} = \mathcal{X}_H^a \partial_a G = i_H dG.$$

This is a nilpotent operator on the exterior algebra $\Lambda(\mathcal{M})$

$$i_H \colon \Lambda^k \to \Lambda^{k-1}, \quad i_H^2 = 0.$$

Combining this with the exterior derivative d we obtain the equivariant exterior derivative

$$d_H = d + i_H,$$

which maps $\Lambda^k \to \Lambda^{k+1} \oplus \Lambda^{k-1}$, i.e., even forms to odd forms and vice versa. By identifying even forms with bosons and odd with fermions, we recognize this as a supersymmetry operator. The supersymmetry algebra defines the Lie derivative

$$d i_H + i_H d = d_H^2 = \mathcal{L}_H$$

and for Poisson brackets,

$$\{H, G\} = \omega^{ab}\partial_a H \partial_b G = \mathcal{X}_H^a \partial_a G = \mathcal{L}_H G.$$

Finally, we introduce the invariant subspace $\Lambda_{\text{inv}}(\mathcal{M}) \in \Lambda(\mathcal{M})$ as the subspace of forms that satisfy

$$\mathcal{L}_H \xi = 0.$$

2   Equivariant Cohomology

We shall now shortly describe the concept of equivariant cohomology, to the extend we need it in these lectures. For a more thorough discussion from the present point of view, see Refs. 14 and 15.

We assume there is an action of a compact group $\mathcal{G}$ on our symplectic manifold, $\mathcal{G} \times \mathcal{M} \to \mathcal{M}$. The corresponding equivariant cohomology is essentially the cohomology of $\mathcal{M}$ mod $(\mathcal{G})$.

If the action of $\mathcal{G}$ is free so that $\mathcal{M}/\mathcal{G}$ is a manifold, the $\mathcal{G}$-equivariant cohomology is is just ordinary cohomology,

$$H_{\mathcal{G}}^*(\mathcal{M}) \sim H^*(\mathcal{M}/\mathcal{G}).$$

However, if the action of $\mathcal{G}$ is not free, we need to develop methods for computing it. Three different models have been introduced,

- Cartan model

- Weil model

- BRST model

To describe the relevant aspects, we assume the Lie algebra of $\mathcal{G}$ is realized by vector fields

$$[\mathcal{X}_\alpha, \mathcal{X}_\beta] = f_{\alpha\beta\gamma}\mathcal{X}_\gamma.$$

The corresponding Lie derivatives are

$$\mathcal{L}_\alpha = di_\alpha + i_\alpha d$$

and they also provide a representation of the Lie algebra,

$$[\mathcal{L}_\alpha, \mathcal{L}_\beta] = f_{\alpha\beta\gamma}\mathcal{L}_\gamma.$$

This generates $\mathcal{G}$-action on $\Lambda(\mathcal{M})$.

If the action of the group elements is symplectic, it preserves the symplectic structure. This we can rephrase by

$$\mathcal{L}_\alpha\omega = 0. \tag{2.1}$$

In the following we shall assume that

$$H^1(\mathcal{M}, R) = 0.$$

Using this, we conclude from (2.1) that $i_\alpha\omega$ is exact. Hence we can define a *momentum map* $H\colon \mathcal{M} \to \mathbf{g}^*$: If we select $\{\phi^\alpha\}$ to be a symmetric basis for the dual Lie algebra $\mathbf{g}^*$, we can write

$$H = \phi^\alpha H_\alpha$$

where now

$$i_\alpha \omega = -dH_\alpha$$

and the Hamiltonian functions H_α satisfy the Lie algebra

$$\{H_\alpha, H_\beta\} = f_{\alpha\beta\gamma} H_\gamma + \kappa_{\alpha\beta}$$

where $\kappa_{\alpha\beta}$ is a two-cocycle. In the following we shall only consider the simplest case of $\mathcal{G} \sim \mathrm{U}(1) \sim S^1$ so that we have only one Hamiltonian function H and the basis of the dual Lie algebra consist of a single element ϕ, which we can view as a real parameter. The corresponding Hamiltonian vector field

$$\mathcal{X}_H = \omega^{ab} \partial_b H \partial_a$$

is then a generator of the symplectic $\mathrm{U}(1)$ action, and we can construct the pertinent equivariant exterior derivative operator

$$d_H = d + \phi i_H.$$

The Lie derivative is

$$d_H^2 = \phi(d i_H + i_H d) = \phi \mathcal{L}_H.$$

The cohomology of d_H is the $H^*_{\mathrm{U}(1)}(\mathcal{M})$ equivariant cohomology that we are interested in. Of particular interest in the following will be the observation, that the linear combination of the Hamiltonian H and the symplectic two-form ω defines a closed element in this cohomology, as a consequence of the Hamiltonian equations of motion,

$$(d + \phi i_H)(\omega + \phi H) = 0.$$

3 Duistermaat-Heckman Integration Formula

We shall now proceed to apply the previous formalism to derive integration formulas, first for ordinary phase space integrals and then for quantum mechanical path integrals. In finite dimensions these integration formulas were first considered by Duistermaat and Heckman, and we shall first explain the theorem they derived:

We are interested in evaluating the classical partition function

$$Z = \int \omega^n \exp\{i\phi H\}, \tag{3.1}$$

where H is a Hamiltonian that determines the symplectic action of $\mathrm{U}(1)$ on the phase space. We shall first consider the special case where the critical points of H, i.e., points where $dH = 0$ are isolated and nondegenerate. In this case, the Duistermaat-Heckman integration formula states, that the integral (3.1) localizes to the critical points of H,

$$Z = Z_{DH} = \sum_{dH=0} \exp\left\{i\frac{\pi}{4}\eta_H\right\} \frac{\sqrt{\det\|\omega_{ab}\|}}{\sqrt{\det\|\partial_{ab}H\|}} \exp\{i\phi H\}. \tag{3.2}$$

Here η_H is the η-invariant of the matrix $\partial_{ab}H$, i.e., if we denote the dimensions of the positive and negative eigenspaces of the matrix $\partial_{ab}H$ at a critical point p by $\dim T_p^+$ and $\dim T_p^-$,

$$\eta_H = \dim T_p^+ - \dim T_p^-.$$

In the following we shall always include this phase factor in the definition of the determinants that emerge from our computations.

Notice that this integration formula coincides with the standard WKB formula, except that now we sum over *all* critical points of H, not just over its local minima. In order to prove it we consider first the following more general integral

$$Z_\lambda = \int \omega^n \exp\{i\phi H + \lambda d_H \psi\}. \tag{3.3}$$

Here λ is a real parameter and ψ is an arbitrary element on the exterior algebra of the phase space. Notice that for $\lambda = 0$ this reduces to the original integral (3.1). We shall now argue, that (3.3) is in fact independent of λ, provided

$$\mathcal{L}_H \psi = 0. \tag{3.4}$$

More specifically, we shall establish that if (3.4) is satisfies, then

$$Z_\lambda = Z_{\lambda+\delta\lambda}$$

where $\delta\lambda$ is a variation of the parameter. After having established this, we shall then show that when $\lambda \to \infty$ the integral localizes to the critical points according to (3.2). Since for $\lambda = 0$ (3.3) reduces to our original integral (3.1), the theorem then follows from the λ independence.

In order to show the λ independence of (3.3), we represent the integration measure using anticommuting $\mathbf{c}^a$,

$$\int \omega^n = \int d^{2n}z\sqrt{\det\|\omega_{ab}\|} = \int dz\, d\mathbf{c}\, \exp\left\{\frac{1}{2}\mathbf{c}^a\omega_{ab}\mathbf{c}^b\right\}.$$

In particular, we identify these anticommuting variables with the basis of one-forms on the exterior algebra,

$$\mathbf{c}^a \sim dz^a.$$

We also introduce the corresponding basis of internal multiplication

$$\mathbf{i}_a \mathbf{c}^b = \delta_a^b$$

so that the equivariant exterior derivative can be written as

$$d_H = \mathbf{c}^a \partial_a + \omega^{ab} \mathbf{i}_a \equiv \mathbf{c}^a \partial_a + \mathcal{X}_H^a \mathbf{i}_a$$

and the Lie derivative is

$$\mathcal{L}_H = \mathcal{X}_H^a \partial_a + \mathbf{c}^a \partial_a \mathcal{X}_H^b \mathbf{i}_b.$$

Consider now our integral

$$Z_\lambda = \int dz \, d\mathbf{c} \, \exp\{i\phi(H + \omega) + \lambda d_H \psi\}.$$

In order to show that this is independent of λ, we select an infinitesimal functional $\delta\psi$ such that $\mathcal{L}_H \delta\psi = 0$ and introduce the following change of variables,

$$z^a \to z^a + \delta z^a = z^a + \delta\psi d_H z^a = z^a + \delta\psi \mathbf{c}^a,$$
$$\mathbf{c}^a \to \mathbf{c}^a + \delta\mathbf{c}^a = \mathbf{c}^a + \delta\psi d_H \mathbf{c}^a = \mathbf{c}^a + \delta\psi \mathcal{X}_H^a.$$

Under this change of variables the integrand remains intact,

$$i\phi(H + \omega) + \lambda d_H \psi \to i\phi(H + \omega) + \lambda d_H \psi.$$

However, since $\delta\psi$ is not a constant we obtain a nontrivial Jacobian,

$$dzd\mathbf{c} \to (1 + d_H \delta\psi)dzd\mathbf{c} \sim \exp\{d_H(\delta\psi)\}dzd\mathbf{c}.$$

Consequently we find for the partition function under this change of variables

$$Z_\lambda \to \int dz \, d\mathbf{c} \, \exp\{i\phi(H + \omega) + \lambda d_H \psi + d_H(\delta\psi)\}$$

and in particular, if we select

$$\delta\psi = \delta\lambda \psi$$

we find that the only effect is to shift $\lambda \to \lambda + \delta\lambda$,

$$Z_\lambda = \int dz \, d\mathbf{c} \, \exp\{i\phi(H + \omega) + (\lambda + \delta\lambda)d_H \psi\} = Z_{\lambda + \delta\lambda}.$$

In order to construct a ψ with the required properties, we shall assume that on $\mathcal{M}$ there exists a metric tensor g_{ab} which is H-invariant,

$$\mathcal{L}_H g = 0. \tag{3.5}$$

Later on, we shall comment on the existence of such a metric tensor. If we select

$$\psi = i_H g = g_{ab} \mathcal{X}_H^a c^b$$

the condition (3.5) ensures that we indeed have

$$\mathcal{L}_H \psi = 0.$$

Explicitly, we have for the additional term in (3.3),

$$d_H \psi = g_{ab} \mathcal{X}_H^a \mathcal{X}_H^b + \frac{1}{2} c^a [\partial_a (g_{bc} \mathcal{X}_H^c) - \partial_b (g_{ac} \mathcal{X}_H^c)] c^b \equiv K + \Omega$$

and in particular, our general arguments imply that the original partition function coincides with

$$Z = \int dz \, dc \, \exp\{i\phi(H + \omega) - \frac{\lambda}{2}(K + \Omega)\} \tag{3.6}$$

independently of λ. We shall now evaluate this as $\lambda \to \infty$. In this limit, we use

$$\delta(\alpha x) = \frac{1}{|\alpha|} \delta(x) = \lim_{\lambda \to \infty} \sqrt{\frac{\lambda}{2\pi}} \exp\left\{-\frac{\lambda}{2}(\alpha x)^2\right\}$$

which localizes the partition function

$$Z \to \int dz \, dc \, \frac{\sqrt{\det \|\Omega_{ab}\|}}{\sqrt{\det \|g_{ab}\|}} \delta(\mathcal{X}_H) e^{i\phi(H+\omega)} = \sum_{dH=0} \frac{\sqrt{\det \|\omega_{ab}\|}}{\sqrt{\det \|\partial_{ab} H\|}} \exp\{i\phi H\}.$$

This is the finite-dimensional Duistermaat-Heckman integration formula, in the nondegenerate case.

We shall now shortly comment the previous derivation: In a central role is the existence of a $\mathcal{L}_H$-invariant metric on $\mathcal{M}$,

$$\mathcal{L}_H g = 0. \tag{3.7}$$

One can easily show, that *locally* such a metric *always* exists: Simply select Darboux coordinates $\phi^a \sim p_a$, q^a, which always exists in (small) neighborhoods of $\mathcal{M}$, so that the Hamiltonian H in this neighborhood coincides with the coordinate p_1, for example. The invariant metric g_{ab} in this neighborhood is then simply the Cartesian metric. However, the *global* existence of the metric g_{ab} is possible only for special Hamiltonians: If the Hamiltonian can be viewed as a generator of a Lie group that acts canonically, then we can construct an invariant metric simply by averaging *any* metric over the group. On the other hand, since the isometry group of a metric on a compact manifold is compact, we conclude that if an invariant metric exists, the Hamiltonian H must be a generator of a compact Lie group acting on the phase space. Without loss of generality, we may then assume that H is an element of U(1), i.e., it generates the action of a circle (more generally torus) on the phase space. This means in particular, that we can identify H with an action variable in the sense of integrable models.

Consider now the integral (3.6): The effective action that we have there is

$$i\phi(H + w) - \frac{\lambda}{2}(K + \Omega)$$

where

$$K = g_{ab}\mathcal{X}_H^a \mathcal{X}_H^b$$

can be viewed as another Hamiltonian function, and

$$\Omega = \frac{1}{2}\mathbf{c}^a[\partial_a(g_{bc}\mathcal{X}_H^c) - \partial_b(g_{ac}\mathcal{X}_H^c)]\mathbf{c}^b \tag{3.8}$$

which is called the Riemannian momentum map, defines a closed two-form on the phase space. Indeed, we can identify (H, ω) and (K, Ω) as a bi-Hamiltonian structure in the sense that the corresponding Hamiltonian equations of motion coincide,

$$\dot{z}^a = \{z^a, H\}_\omega = \{z^a, K\}_\Omega.$$

This existence of a bi-Hamiltonian structure is consistent with the classical integrability of (H, ω)

Finally, we comment shortly on the asserted metric dependence of our final result, as described in Ref. 10. These authors consider the harmonic oscillator, and conclude that there is an arbitrary parameter in the partition function when evaluated by localization methods. This parameter arises in their computation essentially as follows: The phase space of a harmonic oscillator is the plane $\mathbb{R}^2$. In polar coordinates, the authors in Ref. 10 argue that a solution to the condition (3.7) is (essentially) given by the metric

$$ds^2 = dr^2 + c \cdot r^2 d\phi^2$$

with c their parameter. The authors in Ref. 10 then find that the partition function of a harmonic oscillator depends on c. The resolution of this parameter dependence is however very simple: *Only for $c = 1$ does the metric correspond to a metric on the plane*, hence *only for $c = 1$ do we have a solution of (3.7)*. For other values of c it defines a metric on a cone which is not even continuous at $r = 0$ as can be seen easily, if Cartesian coordinates are introduced: The cone has a tip. Consequently *only for $c = 1$ do we actually have a smooth Riemannian metric on the phase space of a harmonic oscillator*. In particular, there is no evidence or arguments supporting the residual metric dependence asserted in Ref 10.

4 Degeneracies

We shall now proceed to investigate the degenerate case, which is important when we generalize the Duistermaat-Heckman integration formula to path integrals. Now the set of points where $dH = 0$ does not consist of isolated points, but defines a submanifold $\mathcal{M}_0$ which we call the critical submanifold. It is the moduli space of critical points of H. We also introduce the normal bundle $\mathcal{N}_\perp$ in $\mathcal{M}$, so that we have

$$\mathcal{M} = \mathcal{M}_0 \cup \mathcal{N}_\perp.$$

In local neighborhoods we can then divide local coordinates accordingly,

$$z^a = \hat{z}^a + \delta z^a \quad \text{and} \quad \mathbf{c}^a = \hat{\mathbf{c}}^a + \delta \mathbf{c}^a$$

where now

$$\hat{z}^a, \ \hat{\mathbf{c}}^a \in \mathcal{M}_0 \quad \text{and} \quad \delta z^a, \ \delta \mathbf{c}^a \in \mathcal{N}_\perp$$

and in particular, we have by definition

$$\mathcal{X}_H^a(\hat{z}) = 0 \quad \text{and} \quad \Omega_{ab}(\hat{z})\hat{\mathbf{c}}^b = 0.$$

We now consider the partition function (3.6) for a degenerate H,

$$Z = \int dz\, d\mathbf{c}\, \exp\{i\phi(H + \omega) - \lambda(K + \Omega)\}.$$

If λ is very large, the integrand vanishes rapidly outside of $\mathcal{M}_0$, and can be consequently extended over entire T^*M. We now change variables as follows,

$$z^a = \hat{z}^a + \delta z^a \rightarrow \hat{z}^a + \lambda^{-1/2}\delta z^a,$$

$$\mathbf{c}^a = \hat{\mathbf{c}}^a + \delta \mathbf{c}^a \rightarrow \hat{\mathbf{c}}^a + \lambda^{-1/2}\delta \mathbf{c}^a.$$

The supersymmetry, determined by our equivariant cohomology ensures that the corresponding Jacobian is trivial. As $\lambda \to \infty$, we find

$$\frac{\lambda}{2}\Omega \xrightarrow{\lambda \to \infty} \frac{1}{2}\delta \mathbf{c}^a \Omega_{ab}(\hat{z})\delta \mathbf{c}^b + \frac{1}{2}\delta z^a \Omega_a^e R_{ebcd}\hat{\mathbf{c}}^c\hat{\mathbf{c}}^d\delta z^b + \mathcal{O}\left(\frac{1}{\sqrt{\lambda}}\right)$$

and

$$\frac{\lambda}{2}K \xrightarrow{\lambda \to \infty} \frac{1}{2}\delta z^a \Omega_a{}^c \Omega_{cb}\delta z^b + \mathcal{O}\left(\frac{1}{\sqrt{\lambda}}\right).$$

By taking $\lambda \to \infty$ and repeating the steps that we used earlier in derivation of the (nondegenerate) Duistermaat-Heckman integration formula, we then find that the partition function localizes to

$$Z = \int\limits_{\mathcal{M}_0} dz\, d\mathbf{c}\, \frac{\exp\{i\phi(H + \omega)\}}{\mathrm{Pf}(\Omega_b^a + R_{bcd}^a \mathbf{c}^c \mathbf{c}^d)} \tag{4.1}$$

where we have defined the metric connection

$$\Gamma^a_{bc} = \frac{1}{2}g^{ad}(\partial_c g_{dc} + \partial_c g_{db} - \partial_d g_{bc})$$

and the Riemann tensor

$$R^a_{bcd} = \partial_c \Gamma^a_{db} - \partial_d \Gamma^a_{cb} + \Gamma^a_{ce}\Gamma^e_{db} - \Gamma^a_{de}\Gamma^e_{ba}$$

in the usual manner. The result (4.1) is the degenerate version of the Duistermaat-Heckman integration formula.

5   Equivariant Characteristic Classes

The two terms that appear in the integrand in (4.1) are examples of equivariant characteristic classes [6]: Equivariant characteristic classes are simply generalizations of ordinary characteristic classes to equivariant cohomology. To understand them, we observe that since

$$d_H(H + \omega) = (d + i_H)(H + \omega) = 0$$

we also have

$$d_H \exp\{H + \omega\} = 0$$

which defines an equivariant version of the Chern class. Indeed, we have the following parallelism between the formalism of symplectic geometry and that of $U(1)$ gauge theories,

$$dF = 0 \iff d\omega = 0,$$
$$F = dA \iff \omega = d\vartheta.$$

If we now also generalize ordinary de Rham cohomology to equivariant cohomology by

$$d \to d_H = d + i_H$$

we find that the conventional Chern class in $U(1)$ gauge theories can be generalized to its equivariant version

$$\exp\left\{\frac{1}{4\pi}F\right\} \to \exp\{H + \omega\}.$$

In order to define additional equivariant characteristic classes that appear in localization formulas, we introduce a covariant extension of the equivariant exterior derivative

$$D_H = d + \Gamma + i_H = d_H + \Gamma$$

where Γ is the (metric) connection one-form. Following the discussion of ordinary characteristic classes, see, e.g., Ref. 16, we then conclude that since

$$D_H(\Omega^a_b + R^a_b) = 0$$

where Ω_{ab} is the Riemannian momentum map we introduced in (3.8), we can generalize the $\hat{A}$-genus into its equivariant version which is equivariantly closed,

$$d_H\sqrt{\det\left[\frac{(\Omega^a{}_b + R^a_b)/2}{\sinh[(\Omega^a{}_b + R^a_b)/2]}\right]} = 0.$$

Similarly, we can introduce the equivariantly closed equivariant Todd class

$$d_H\sqrt{\det\left[\frac{\Omega^a_b + R^a_b}{1 + \exp\{(\Omega^a{}_b + R^a_b)/2\}}\right]} = 0$$

and the equivariantly closed equivariant Euler class

$$d_H\sqrt{\det[\Omega^a_b + R^a_b]} = 0.$$

All these equivariant characteristic classes appear in localization formulas.

6   Loop Space

We shall now proceed to generalize the previous integration formulas to path integrals. For this, we introduce the loop space $L\mathcal{M}$, which is the space of T-periodic trajectories (loops) on the phase space parametrized by "time" t

$$z^a \to z^a(t); \quad z^a(0) = z^a(T).$$

On this loop space we introduce loop space symplectic geometry as follows. We first extend the exterior derivative to loop space by

$$d = \int_0^T dt\, dz^a(t)\, \frac{\delta}{\delta z^a(t)} \equiv dz^a\, \frac{\delta}{\delta z^a}$$

where in the last step we have dropped the explicit integration symbol. In the following we shall use this "loop space summation convention" extensively, and it will always be clear from the context when an integration is understood.

The symplectic structure on $L\mathcal{M}$ is defined by lifting the symplectic two-form on $\mathcal{M}$ by

$$\widehat{\Omega} = \int dt\, dt'\, \frac{1}{2}\omega_{ab}(t,t')dz^a(t) \wedge dz^b(t').$$

It defines a closed symplectic two-form on the loop space,

$$d\widehat{\Omega} = 0 \implies \widehat{\Omega} = d\Theta = d\left\{ \int_0^T dt\, \vartheta_a(t)dz^a(t) \right\}.$$

If S_B is a functional on the loop space, we can then define the corresponding loop space Hamiltonian vector field by

$$\frac{\delta S_B}{\delta z^a} = \widehat{\Omega}_{ab}\mathcal{X}_S^b$$

and introduce corresponding loop space equivariant cohomology by

$$d_S = d - \mathbf{i}_S = d - \mathcal{X}_S^a \mathbf{i}_a. \tag{6.1}$$

Similarly, we can extend to loop space all other quantities and concepts we have introduced in the finite-dimensional case.

In the following we shall be interested in evaluating the path integral

$$Z = \int [dz^a]\, \sqrt{\det \|\widehat{\Omega}_{ab}\|}\, \exp\{iS_B\} \tag{6.2}$$

where for S_B we select the standard action for Hamiltonian H,

$$S_B = \int_0^T \{\vartheta_a \dot{z}^a - H(z)\}.$$

The integral (6.2) is the quantum mechanical path integral representation of the partition function for H, now presented in a general coordinate system. In particular, it is (formally!) invariant under canonical transformations on $\mathcal{M}$. In order to interpret it geometrically in terms of loop space equivariant cohomology, we again introduce anticommuting variables $\mathbf{c}^a(t)$ and get

$$Z = \int [dz^a]\,[d\mathbf{c}^a]\, \exp\{iS_B + i\mathbf{c}^a\widehat{\Omega}_{ab}\mathbf{c}^b\}$$

$$= \int [dz^a]\,[d\mathbf{c}^a]\, \exp\{iS_B + iS_F\} \tag{6.3}$$

generalizing (3.1). We then observe that this action admits a supersymmetry determined by our loop space equivariant exterior derivative (6.1):

$$d_S = \mathbf{c}^a \partial_a - \mathcal{X}_S^a \mathbf{i}_a = \mathbf{c}^a \partial_a - (\omega_{ab}\dot{z}^b - \partial_a H)\mathbf{i}_a.$$

The action of this operator determines the following supersymmetry transformation

$$d_S z^a = \mathbf{c}^a \quad \text{and} \quad d_S \mathbf{c}^a = -\mathcal{X}_S^a$$

which leaves the action in (6.3) invariant,

$$d_S(S_B + S_F) = 0$$

generalizing our earlier finite-dimensional supersymmetry

$$d_H(H + \omega) = 0$$

to loop space.

7 Example: Atiyah-Singer Index Theorem

As a first example we shall derive the Atiyah-Singer index theorem [16]. For this, we consider the eigenvalue problem for a Dirac equation on a compact even dimensional Riemannian manifold $\mathcal{M}$,

$$D\Psi = \gamma^\mu D_\mu \Psi = \begin{pmatrix} 0 & D_+ \\ D_- & 0 \end{pmatrix} \begin{pmatrix} \Psi_- \\ \Psi_+ \end{pmatrix}$$

$$= \gamma^\mu \left(\partial_\mu + \frac{1}{8} \omega_{\mu jk} [\gamma^j, \gamma^k] \right) \Psi = \mathcal{E}\Psi. \tag{7.1}$$

Here we have introduced vielbeins e_μ^i that transform the coordinate basis to an orthonormal basis, e.g., the metric tensor $g_{\mu\nu}$ on $\mathcal{M}$ is related to a local flat metric by $g_{\mu\nu} = \eta_{ij} e_\mu^i e_\nu^j$ and local (flat) γ-matrices to coordinate basis γ-matrices by $\gamma^i = e_\mu^i \gamma^\mu$. The inverse vielbein is defined by $E_i^\mu e_\mu^j = \delta_i^j$ and the relation between the spin connection and the metric connection is

$$\omega_{\mu j}^i = e_\nu^i (\partial_\mu E_j^\nu + \Gamma_{\mu\lambda}^\nu E_j^\lambda).$$

The Atiyah-Singer index theorem counts the difference in the number of positive and negative chirality zero modes of (7.1),

$$\text{Index}\, D = \text{Dim}\,\text{Ker}\, D_+ - \text{Dim}\,\text{Ker}\, D_- = \lim_{\beta\to\infty} \text{Tr}\{\gamma^c e^{-\beta D^2}\}.$$

In order to evaluate it using path integrals, we introduce the cotangent bundle $T^*\mathcal{M}$ with the Poisson bracket

$$\{p_\mu, x^\nu\} = \delta_\mu^\nu.$$

Similarly, we realize the γ-matrices canonically with (anticommuting) Poisson brackets

$$\{\psi^i, \psi^j\} = \eta^{ij} \implies \{\psi^\mu, \psi^\nu\} = g^{\mu\nu}.$$

The zero-mode equation for D can then be identified canonically as a graded constraint

$$S = \psi^\mu \left(p_\mu + \frac{1}{4}\omega_{\mu jk}, \psi^j \psi^k \right) = 0.$$

The corresponding constraint algebra coincides with the standard $N = \frac{1}{2}$ supersymmetry algebra

$$\{S, S\} = \mathcal{H} = g^{\mu\nu} \left(p_\mu + \frac{1}{4}\omega_{\mu ij} \psi^i \psi^j \right), \left(p_\nu + \frac{1}{4}\omega_{\nu kl} \psi^k \psi^l \right),$$

$$\{S, \mathcal{H}\} = \{\mathcal{H}, \mathcal{H}\} = 0,$$

and the problem of evaluating the Atiyah-Singer index becomes a problem of evaluating the following (proper-time gauge) BRST gauge fixed path integral for this constrained system (see Ref. 17 for details)

$$Z = \int [d(\text{Liouville})] \exp(iS + i\{Q_{\text{BRST}}, \vartheta\})$$

$$= \int [dx^\mu] [d\psi^\mu] \exp\left\{ i \int_0^\beta \frac{1}{2} g_{\mu\nu} \dot{x}^\mu \dot{x}^\nu \right.$$

$$\left. + \frac{1}{2}\psi^\mu (g_{\mu\nu}\partial_t + \dot{x}^\rho g_{\mu\sigma}\Gamma^\sigma_{\rho\nu})\psi^\nu \right\}. \quad (7.2)$$

Here we are interested in explaining, how this path integral is evaluated. For this we introduce a loop space by defining the following loop space exterior derivative

$$d = \psi^\mu \frac{\delta}{\delta x^\mu}. \quad (7.3)$$

We then observe, that the fermionic part of our action determines a loop space (pre)symplectic two-form

$$\Omega = \frac{1}{2}\Omega_{\mu\nu}\psi^\mu\psi^\nu = \frac{1}{2}\psi^\mu(g_{\mu\nu}\partial_t + \dot{x}^\rho g_{\mu\sigma}\Gamma^\sigma_{\rho\nu})\psi^\nu, \quad d\Omega = 0$$

while the bosonic part determines a Hamiltonian functional with Hamiltonian vector field

$$\partial_\mu \left(\frac{1}{2}g_{\nu\sigma}\dot{x}^\nu\dot{x}^\sigma \right) \equiv \mathcal{X}_\mu = g_{\mu\nu}\ddot{x}^\nu + g_{\mu\sigma}\Gamma^\sigma_{\rho\nu}\dot{x}^\rho\dot{x}^\nu,$$

$$\mathcal{X}_\mu = \Omega_{\mu\nu}\mathcal{X}^\nu \implies \mathcal{X}^\mu = \dot{x}^\mu(t). \quad (7.4)$$

We next define an extended phase space with loop space Poisson brackets

$$\{\pi_\mu(t), x^\nu(t')\} = \delta^\nu_\mu(t - t'),$$

$$\{\bar\psi_\mu(t), \psi^\nu(t')\} = \delta^\nu_\mu(t - t'),$$

and equivariantize (7.3) using the Hamiltonian vector field in (7.4)

$$Q = d + i_S = \psi^\mu \pi_\mu + \mathcal{X}^\mu \bar\psi_\mu = \psi^\mu \pi_\mu + \dot x^\mu \bar\psi_\mu.$$

The action of this operator on our fields then determine the following $N = \frac{1}{2}$ supersymmetry transformation

$$\delta x^\mu = \psi^\mu, \quad \delta \psi^\mu = \dot x^\mu$$

and the supersymmetry algebra closes to the Lie derivative

$$\frac{1}{2}\{Q,Q\} = \mathcal{L}_S = \dot x^\mu \pi_\mu + \dot\psi_\mu \bar\psi_\mu = \partial_t.$$

From our general arguments we conclude that we can extend the path integral (7.2) into

$$Z = \int [dx^\mu][d\psi^\mu] \exp(iS) = \int [dx^\mu][d\psi^\mu] \exp(iS + i\{Q,\vartheta\})$$

provided

$$\mathcal{L}_S \vartheta = \partial_t \int_0^T \vartheta = \vartheta(T) - \vartheta(0) = 0$$

that is, the functional ϑ is *single*-valued in the loop space. We select

$$\vartheta = \frac{1}{2} g_{\mu\nu} \dot x^\mu \psi^\nu.$$

This coincides with the loop space symplectic one-form, since

$$d\vartheta = d\left(\frac{1}{2} g_{\mu\nu} \dot x^\mu \psi^\nu\right) = \Omega.$$

Furthermore, we observe that the *entire* action can be represented as a Q-differential

$$\{Q,\vartheta\} = (d + i_S)\vartheta = (\psi^\mu \partial_\mu + \dot x^\mu \bar\psi_\mu)\left(\frac{1}{2} g_{\mu\nu} \dot x^\mu \psi^\nu\right)$$

$$= \frac{1}{2} g_{\mu\nu} \dot x^\mu \dot x^\nu + \frac{1}{2}\psi^\mu(\partial_t + \dot x^\sigma g_{\mu\rho}\Gamma^\rho_{\sigma\nu})\psi^\nu = i_S\vartheta + \Omega$$

and the path integral becomes simply

$$Z \equiv Z_\lambda = \int [dx^\mu][d\psi^\mu] \exp\left\{ i\int_0^\beta \frac{\lambda}{2} g_{\mu\nu}\dot x^\mu \dot x^\nu + \frac{\lambda}{2}\psi^\mu(\partial_t + \dot x^\sigma g_{\mu\rho}\Gamma^\rho_{\sigma\nu})\psi^\nu \right\}$$

$$= \int [dx^\mu][d\psi^\mu] \exp\left(i\int_0^\beta \lambda\{Q,\vartheta\} \right)$$

independently of the localization parameter λ ($\neq 0$). To evaluate this we introduce the decomposition

$$x^\mu(t) = x_0^\mu + x_t^\mu,$$
$$\psi^\mu(t) = \psi_0^\mu + \psi_t^\mu,$$

where x_0, ψ_0 is the constant mode of a loop $x(t)$, $\psi(t)$, e.g., in a Fourier decomposition, and x_t, ψ_t are the fluctuation modes. The corresponding decomposition of the path integral measure is

$$[dx^\mu]\,[d\psi^\mu] = dx_0^\mu\,d\psi_0^\mu \prod_t dx_t^\mu\,d\psi_t^\mu.$$

We change variables (Jacobian is trivial by supersymmetry)

$$x_t^\mu \to \frac{1}{\sqrt{\lambda}} x_t^\mu,$$
$$\psi_t^\mu \to \frac{1}{\sqrt{\lambda}} \psi_t^\mu,$$

and take the $\lambda \to \infty$ limit. In this limit the action becomes

$$S \to \int_0^\beta \frac{1}{2} g_{\mu\nu}(x_0)\dot{x}_t^\mu \dot{x}_t^\nu + \frac{1}{2}\psi_t^i \eta_{ij}\partial_t\psi_t^j + \frac{1}{2}R_{ij\mu\nu}(x_0)\psi_0^i\psi_0^j x_t^\mu \dot{x}_t^\nu + \mathcal{O}\left(\frac{1}{\sqrt{\lambda}}\right)$$

and evaluating the integrals over x_t^μ and ψ_t^i we obtain

$$\Longrightarrow Z = \int_\mathcal{M} dx\,d\psi\,\sqrt{\det\left|\frac{iR_{\mu\nu\rho\sigma}\psi^\rho\psi^\sigma/(4\pi)}{\sinh(iR_{\mu\nu\rho\sigma}\psi^\rho\psi^\sigma/(4\pi))}\right|} = \int_\mathcal{M} \hat{A}(R).$$

This is the Atiyah-Singer index for a spin complex [16].

8   Duistermaat-Heckman in Loop Space

We shall now proceed to generalize the previous computation to include a nontrivial Hamiltonian flow on the phase space $\mathcal{M}$. This will give us the Duistermaat-Heckman integration formula in the loop space [8]. We are interested in evaluating the partition function

$$\text{Tr}\{e^{-iTH}\} = Z$$

$$= \int_{PBC} [dz^a]\,[dc^a] \exp\left\{i\int_0^T \vartheta_a \dot{z}^a - H + \frac{1}{2}\mathbf{c}^a\omega_{ab}\mathbf{c}^b\right\} \quad (8.1)$$

for a quantum mechanical system determined by the Hamiltonian H. For this we introduce the loop space Hamiltonian vector field

$$\mathcal{X}_S^a = \dot{z}^a - \omega^{ab}\partial_b H$$

and the loop space equivariant exterior derivative

$$d_S = \mathbf{c}^a \partial_a + \mathcal{X}_S^a \mathbf{i}_a \quad (\mathbf{i}_a \mathbf{c}^b = \delta_a^b)$$

and the corresponding Lie derivative

$$\mathcal{L}_S = d_S^2 = d\mathbf{i}_S + \mathbf{i}_S d.$$

From our earlier discussion we know that if we select

$$\mathcal{L}_S \psi = (d\mathbf{i}_S + \mathbf{i}_S d)\psi = (\mathcal{X}_S^a \partial_a + \mathbf{c}^a \partial_a \mathcal{X}_S^b \bar{\mathbf{c}}_b)\psi = 0 \qquad (8.2)$$

then

$$Z_\psi = \int [dz]\,[d\mathbf{c}]\, \exp\Big\{ i \int_0^T \vartheta_a \dot{z}^a - H + \frac{1}{2}\omega + d_S\psi \Big\}$$

is (formally) independent of ψ and coincides with the original partition function (8.1),

$$Z = Z_\psi.$$

In order to construct a ψ that satisfies (8.2), we again assume that H generates an isometry, that is there exists an H-invariant metric

$$\mathcal{L}_H g = 0. \qquad (8.3)$$

This implies that

$$\psi_\lambda = \frac{\lambda}{2} g_{ab} \dot{z}^a \mathbf{c}^b$$

where λ is a parameter, is a $\mathcal{L}_S$-invariant functional in $L\mathcal{M}$. Consequently our original partition function (8.1) coincides with

$$Z = Z_\lambda$$
$$= \int [dz]\,[d\mathbf{c}]\, \exp\Big\{ i \int_0^T \vartheta_a \dot{z}^a - H + \frac{1}{2}\omega + \frac{\lambda}{2} d_S(g_{ab}\dot{z}^a \mathbf{c}^b) \Big\} \qquad (8.4)$$

independently of λ. Explicitly,

$$S = \frac{\lambda}{2} g_{ab} \dot{z}^a \dot{z}^b + \Big(\vartheta_a - \frac{\lambda}{2} g_{ab} \mathcal{X}_H^b \Big) \dot{z}^a - H$$
$$+ \frac{\lambda}{2} \mathbf{c}^a (g_{ab}\partial_t + \dot{z}^c g_{bd}\Gamma_{ac}^d)\mathbf{c}^b + \frac{1}{2}\mathbf{c}^a \omega_{ab}\mathbf{c}^b.$$

In order to evaluate (8.4), we again introduce the decomposition

$$z^a(t) = z_0^a + z_t^a,$$
$$\mathbf{c}^a(t) = \mathbf{c}_0^a + \mathbf{c}_t^a,$$

and define the path integral measure accordingly,

$$[dz^a][d\mathbf{c}^a] = dz_0^a d\mathbf{c}_0^a \prod_t dz_t^a d\mathbf{c}_t^a.$$

We then introduce the following change of variables

$$z_t^a \to \frac{1}{\sqrt{\lambda}} z_t^a,$$

$$\mathbf{c}_t^a \to \frac{1}{\sqrt{\lambda}} \mathbf{c}_t^a.$$

Our equivariant cohomology (i.e., supersymmetry) ensures that the corresponding Jacobian is trivial.

In the $\lambda \to \infty$ limit we can evaluate the integrals over the fluctuation modes z_t^a and $\mathbf{c}_t^a$, and we find

$$Z = \int dz_0^a \, d\mathbf{c}_0^a \, \frac{\exp\{-iTH + iT\mathbf{c}_0^a \omega_{ab} \mathbf{c}_0^b / 2\}}{\sqrt{\det[g_{ab}\partial_t - (\Omega_{ab} + R_{abcd}\mathbf{c}_0^c \mathbf{c}_0^d)/2]}}$$

where we have again introduced the Riemannian momentum map

$$\Omega_{ab} = \partial_b(g_{ac}\mathcal{X}_H^c) - \partial_a(g_{bc}\mathcal{X}_H^c).$$

The determinant can be evaluated, e.g., by ζ-function method, and the result is

$$Z = \int dz \, d\mathbf{c} \, \exp\left\{-iTH + i\frac{T}{2}\mathbf{c}^a \omega_{ab}\mathbf{c}^b\right\} \sqrt{\det\left[\frac{T(\Omega_b^a + R_b^a)/2}{\sinh[T(\Omega_b^a + R_b^a)]}\right]}$$

$$= \int \mathrm{Ch}(H + \omega)\hat{A}(\Omega + R) \tag{8.5}$$

which is the Duistermaat-Heckman integration formula in the loop space.

Notice that if we take the limit where the Hamiltonian vanishes (or $T \to 0$), our partition function reduces to

$$\mathrm{Tr}\{e^{-iTH}\} \to \dim[\mathcal{H}_\mathcal{M}]$$

i.e., an integer that counts the number of states in our quantum mechanical Hilbert space. Indeed, if we set $H = 0$ the previous computation reduces to our computation of the Atiyah-Singer index: The pertinent action becomes

$$S \xrightarrow{H \to 0} \frac{\lambda}{2} g_{ab}\dot{z}^a \dot{z}^b + \vartheta_a \dot{z}^a + \frac{\lambda}{2} \mathbf{c}^a (g_{ab}\partial_t + \dot{z}^c \Gamma_{acb})\mathbf{c}^b + \frac{1}{2}\mathbf{c}^a \omega_{ab}\mathbf{c}^b$$

and if we identify

$$\vartheta_a \sim A_\mu$$

$$\omega_{ab} \sim F_{\mu\nu}$$

by comparing with (7.2) we recognize that this describes Dirac operator in gauge and gravitational background, and for $H = 0$ we recognize in (8.5) the correct result for its index.

In the case of the Atiyah-Singer index, we found that the action could be represented in an equivariantly closed form. This is also true in the present case, using (8.3) we find that the entire action is equivariantly closed

$$S = (d + \mathbf{i}_{\dot{z}} + \mathbf{i}_H)\left(\frac{\lambda}{2}g_{ab}\dot{z}^a\mathbf{c}^b + \vartheta_a\mathbf{c}^a\right).$$

Consequently we are dealing with (equivariant) cohomological topological theory [18], and the partition function can be identified as a path integral representation of the Lefschetz number of a Dirac operator [9].

9   General Integrable Models

In the previous evaluation of the path integral we have assumed that the Hamiltonian flow is an isometry on the phase space so that we can find a metric tensor that obeys (8.3). We have pointed out, that this means we have an integrable model where the Hamiltonian is simply an action variable, $H(\phi) \sim I$. In the case of general integrable models, the Hamiltonian is a nontrivial *function* of the action variables, that label invariant torii on the phase-space and are conjugate to the angle variables,

$$\{I_a, \theta^b\}_\omega = \delta_a^b, \quad H = H(I_a).$$

Action variables can be always viewed as generators in a Cartan subalgebra of some non-Abelian Lie algebra $\mathcal{G}$ with canonical Poisson bracket realization,

$$\{G_i, G_j\}_\omega = \omega^{ab}\partial_a G_i \partial_b G_j = f_{ij}{}^k G_k,$$

and integrable models are special cases of path integrals, where the Hamiltonian is some (generic) function of Lie algebra generators,

$$Z = \int [dz^a] \sqrt{\|\omega_{ab}\|} \exp\left\{ i \int_0^T dt \left[\vartheta_a \dot{z}^a - H(G_i)\right] \right\}.$$

By introducing a c-number source $j_i(t)$ and using standard identities we may always represent these integrals as

$$Z = \exp\left\{-i \int_0^T dt\, H\left[\frac{\delta}{i\delta j_i(t)}\right]\right\}$$

$$\times \int [dz^a] \sqrt{\|\omega_{ab}\|} \exp\left\{ i \int_0^T dt\, (\vartheta_a \dot{z}^a - j_i G_i) \right\}\Bigg|_{j=0}.$$

Consequently the path integral we need to evaluate is

$$Z_0(j) = \int [dz^a] \sqrt{\|\omega_{ab}\|} \exp\left\{ i \int_0^T dt \, (\vartheta_a \dot{z}^a - j_i G_i) \right\}$$
$$= \int [dz^a][d\mathbf{c}^a] \exp\left\{ i \int_0^T dt \, (\vartheta_a \dot{z}^a + \frac{1}{2}\mathbf{c}^a \omega_{ab} \mathbf{c}^b - j_i G_i) \right\}.$$

Consider now the following infinitesimal variation of the bosonic terms

$$\delta(\vartheta_a \dot{z}^a - j_i G_i) = \delta z^a (\omega_{ab} \dot{z}^b - j_i \partial_a G_i).$$

In particular, if we select

$$\delta z^a = \epsilon_i \{ G_i, z^a \}_\omega$$

where ϵ_i are infinitesimal z^a-independent parameters, this corresponds to the canonical conjugation

$$z^a \to e^{-\epsilon_i G_i} z^a e^{\epsilon_i G_i} = z^a + \epsilon_i \{z^a, G_i\} + \frac{1}{2}\epsilon_i \epsilon_j \{\{z^a, G_i\}, G_j\} + \cdots$$

which leaves $\mathbf{c}^a \omega_{ab} \mathbf{c}^b$ invariant and yields

$$\delta(\vartheta_a \dot{z}^a - j_i G_i) = -\dot{\epsilon}_i G_i - f_{ijk}\epsilon_j j_k G_i.$$

Consequently its only effect is a shift

$$j_i \to j_i + \dot{\epsilon}_i + f_{ijk}\epsilon_j j_k.$$

Here we recognize the structure of a time-dependent non-Abelian gauge transformation, with identifications

$$G_i \leftrightarrow D_i^{ab} E_i^b,$$
$$j_i \leftrightarrow A_0^a.$$

In particular, for integrable Hamiltonians that only depend on the action variables, i.e., generators that are in the Abelian subalgebra

$$\{I_i, I_j\} = 0$$

the relevant gauge transformation is a time-dependent Abelian gauge transformation

$$j_i \to j_i + \dot{\epsilon}_i.$$

As a consequence for integrable models the relevant path integral is

$$Z_0(j) = \int [dz^a][d\mathbf{c}^a] \exp\left\{ i \int_0^T dt \left(\vartheta_a \dot{z}^a + \frac{1}{2}\mathbf{c}^a \omega_{ab} \mathbf{c}^b - j_i I_i \right) \right\}$$

and if we decompose the source into its constant and t-dependent modes,

$$j_i(t) = j_{i0} + j_{it}$$

using Abelian gauge invariance we then conclude that it is sufficient to evaluate

$$Z_0(j) = Z_0(j_0) = \int [dz^a]\,[dc^a]\,\exp\left\{i\int_0^T dt\left(\vartheta_a\dot{z}^a + \frac{1}{2}\mathbf{c}^a\omega_{ab}v^b - j_{io}I_i\right)\right\}$$

that is, we can "gauge transform" away the t-dependent modes of j_i, so that the path integral depends only on the constant mode of $j_i(t)$. But the effective Hamiltonian that appears here,

$$H = j_{i0}I_i$$

is just a generator of U(1) on the phase space, and consequently our previous results apply. In particular, we can evaluate the path integral using our previous results, and we find

$$Z_0(j_0) = \int_{\mathcal{M}} dz\,d\mathbf{c}\,\exp\left\{-iTj_{i0}I_{i0} + i\frac{T}{2}\mathbf{c}^a\omega_{ab}\mathbf{c}^b\right\}$$

$$\times \sqrt{\det\left[\frac{T[j_{i0}(\Omega_{i0})_{ab} + R_{ab}]/2}{\sinh[T[j_{i0}(\Omega_{i0})_{ab} + R_{ab}]/2]}\right]}$$

$$= \int_{\mathcal{M}} \mathrm{Ch}(j_{i0}I_{i0} + \omega)\hat{A}(j_{i0}\Omega_{i0} + R).$$

Consequently for a general integrable model we have obtained the following integration formula for the quantum mechanical partition function [12]

$$Z = \exp\left\{-iTH\left[\frac{1}{i}\frac{\partial}{\partial j_{i0}}\right]\right\}\int_{\mathcal{M}} \mathrm{Ch}(j_{i0}I_{i0} + \omega)\hat{A}(j_{i0}\Omega_{i0} + R)\Bigg|_{j_{i0}=0}.$$

Notice that this is an *ordinary* integral, and in particular the derivatives that appear here are now *ordinary* derivatives w.r.t. the parameter j_{i0}.

10   Mathai-Quillen Formalism

We shall now proceed to explain the Mathai-Quillen formalism [19], which is an important formal framework to understand integration formulas, and their relations to Morse theory. For this, we first consider the Lie derivative along a Hamiltonian vector field

$$\mathcal{L}_H = d\mathbf{i}_H + \mathbf{i}_H d = \mathcal{X}_H^a\partial_a + c^a\partial_a\mathcal{X}_H^b\mathbf{i}_b.$$

We introduce its canonical realization, by defining Poisson brackets

$$\{p_a, z^b\} = \{\bar{\mathbf{c}}_a, \mathbf{c}^b\} = \delta_a^b.$$

The canonical realization is then

$$\mathcal{L}_H = \mathcal{X}_H^a p_a + \mathbf{c}^a \partial_a \mathcal{X}_H^b \bar{\mathbf{c}}_b.$$

To relate this to the Mathai-Quillen formalism [20], we introduce a super-manifold $S^*\mathcal{M}$ and the corresponding exterior algebra $\Lambda(S^*\mathcal{M})$ as follows: We interpret the commuting z^a and the anticommuting $\bar{\mathbf{c}}_a$ as coordinates on $S^*\mathcal{M}$, and $\mathbf{c}^b$ and p_a as the corresponding basis one-forms. We then define the nilpotent exterior derivative

$$d = \mathbf{c}^a \frac{\partial}{\partial z^a} + p_a \frac{\partial}{\partial \bar{\mathbf{c}}_a}.$$

We also introduce interior multiplication (contraction) on $\Lambda(S^*\mathcal{M})$,

$$\mathbf{i}_a \mathbf{c}^b = \pi^b p_a = \delta_a^b$$

and consider the following conjugation of d,

$$d \xrightarrow{\Phi} e^{-\Phi} d e^{\Phi} = d + [d, \Phi] + \frac{1}{2!}[[d, \Phi], \Phi] + \cdots. \tag{10.1}$$

Since this conjugation is invertible, it preserves the cohomological structure. We select

$$\Phi = -\Gamma_{ab}^c \pi^a \mathbf{c}^b \bar{\mathbf{c}}_c$$

where Γ is a metric connection. This yields

$$d = \mathbf{c}^a \frac{\partial}{\partial z^a} + (p_a + \Gamma_{ab}^c \mathbf{c}^b \bar{\mathbf{c}}_c)\frac{\partial}{\partial \bar{\mathbf{c}}_a} + \left(\Gamma_{ab}^c p_c \mathbf{c}^b - \frac{1}{2} R_{adb}^c \mathbf{c}^b \mathbf{c}^d \bar{\mathbf{c}}_c\right)\pi^a.$$

On the variables, the action of this d is

$$dz^a = \mathbf{c}^a,$$
$$dp_a = \Gamma_{ab}^c p_c \mathbf{c}^b - \frac{1}{2} R_{adb}^c \mathbf{c}^b \mathbf{c}^d \bar{\mathbf{c}}_c,$$
$$d\mathbf{c}^a = 0,$$
$$d\bar{\mathbf{c}}_a = p_a + \Gamma_{ab}^c \mathbf{c}^b \bar{\mathbf{c}}_c.$$

Here we recognize the supersymmetric transformation laws of standard $N = 1$ de Rham supersymmetric quantum mechanics, with p_a the auxiliary field. The formalism that we have developed here is equivalent to that introduced by Mathai and Quillen.

11 Short Review of Morse Theory

We shall now shortly review certain aspects of classical Morse theory [21], that will be important for us in the sequel. We are particularly interested in the relation between the Poincaré-Hopf theorem and the Gauss-Bonnet-Chern theorem, which gives a relation between the critical points of a function defined on a compact manifold and the Euler class of the manifold. The Mathai-Quillen formalism was originally introduced to explain this relation.

In Morse theory, we are interested in the critical points p of a smooth function F,

$$dF\big|_p = 0$$

defined on a compact oriented $2n$-dimensional manifold $\mathcal{M}$. We assume, that $p \in \mathcal{M}$ are isolated and nondegenerate.

The Poincaré-Hopf theorem states, that the Euler class $\mathcal{X}$ of $\mathcal{M}$ is related to these critical points by

$$\mathcal{X}(\mathcal{M}) \equiv \sum_{i=0}^{2n}(-1)^i \dim[H^i(\mathcal{M};R)] = \sum_p \text{sign}\left(\det\left\|\frac{\partial^2 F}{\partial x^\mu \partial x^\nu}\right\|\right)$$

where $H_i(\mathcal{M};R)$ are the de Rham cohomology classes of $\mathcal{M}$ with real coefficients.

The Gauss-Bonnet-Chern theorem states, that the Euler class can also be represented as an integral of the Pfaffian of the Riemann curvature two-form over $\mathcal{M}$, i.e., the Euler character,

$$\mathcal{X}(\mathcal{M}) = \int_{\mathcal{M}} \text{Pf}(R_b^a).$$

To verify this, we consider following integral on $S^*\mathcal{M}$

$$Z = \int_{\mathcal{M}} dz\, dp\, d\mathbf{c}\, d\bar{\mathbf{c}}\ \exp\{d\psi\}.$$

Here the exterior derivative is now the one that we have constructed when we explained the Mathai-Quillen formalism,

$$d = \mathbf{c}^a \frac{\partial}{\partial z^a} + (p_a + \Gamma^c_{ab}\mathbf{c}^b\bar{\mathbf{c}}_c)\frac{\partial}{\partial \bar{\mathbf{c}}_a} + \left(\Gamma^c_{ab}p_c\mathbf{c}^b - \frac{1}{2}R^c_{adb}\mathbf{c}^b\mathbf{c}^d\bar{\mathbf{c}}_c\right)\pi^a.$$

Since $d^2 = 0$, our standard arguments imply that Z is invariant under arbitrary local variations of an arbitrary functional ψ. We first select

$$\psi = \frac{1}{2}\mathcal{X}_H^a\bar{\mathbf{c}}_a$$

so that

$$d\psi = p_a \mathcal{X}_H^a + \mathbf{c}^a \nabla_a \mathcal{X}_H^b \bar{\mathbf{c}}_b.$$

Substituting in Z and evaluating the integrals, this gives

$$Z = \sum_{dH=0} \text{sign}(\det \|\partial_{ab} H\|)$$

which is the quantity that appears in the Poincaré-Hopf theorem.

On the other hand, if we select instead

$$\psi = g^{ab} p_a \bar{\mathbf{c}}_b$$

so that

$$d\psi = g^{ab} p_a p_b - \frac{1}{2} R_{adb}^c \mathbf{c}^b \mathbf{c}^c \bar{\mathbf{c}}_d g^{ae} \bar{\mathbf{c}}_e$$

we find the Euler class:

$$Z = \int_{\mathcal{M}} dz \, d\mathbf{c} \, \text{Pf} \left[\frac{1}{2} R_{bcd}^a \mathbf{c}^d \mathbf{c}^c \right].$$

Consequently we have established the relation between Poincaré-Hopf and Gauss-Bonnet-Chern theorems using the Mathai-Quillen formalism.

12   Equivariant Mathai-Quillen Formalism

In order to include the effect of a nontrivial Hamiltonian flow in the previous discussion, we shall now generalize the Mathai-Quillen formalism to the equivariant context [20]. For this we again consider the Lie derivative

$$\mathcal{L}_H = \mathcal{X}_H^a p_a + \mathbf{c}^a \partial_a \mathcal{X}_H^b \bar{\mathbf{c}}_b.$$

In particular, its canonical action on the variables is

$$\mathcal{L}_H z^a = \mathcal{X}_H^a,$$
$$\mathcal{L}_H \mathbf{c}^a = \mathbf{c}^b \partial_b \mathcal{X}_H^a,$$
$$\mathcal{L}_H \bar{\mathbf{c}}_a = -\partial_a \mathcal{X}_H^b \bar{\mathbf{c}}_b,$$
$$\mathcal{L}_H p_a = -\partial_a \mathcal{X}_H^b p_b - \mathbf{c}^b \partial_{ab} \mathcal{X}_H^c \bar{\mathbf{c}}_c.$$

In order to reproduce these transformation laws on $\Lambda(S^* \mathcal{M})$ we introduce the following vector field

$$\mathbf{i}_\mathcal{X} = \mathcal{X}_H^a \mathbf{i}_a - \bar{\mathbf{c}}_b \partial_a \mathcal{X}_H^b \pi^a$$

where $\mathbf{i}_a$ and π^a are contractions w.r.t. $\mathbf{c}^a$ and p_a. We then define the corresponding equivariant exterior derivative

$$Q_H = d + i_\mathcal{X}$$

and the corresponding Lie-derivative

$$L_\mathcal{X} = d i_\mathcal{X} + i_\mathcal{X} d$$
$$= \mathcal{X}_H^a \partial_a - \partial_a \mathcal{X}_H^b \bar{\mathbf{c}}_c \frac{\partial}{\partial \bar{\mathbf{c}}_c} + \mathbf{c}^b \partial_b \mathcal{X}_H^a \mathbf{i}_a - (p_b \partial_a \mathcal{X}_H^b + \partial_{ab} \mathcal{X}_H^c \mathbf{c}^b \bar{\mathbf{c}}_c) \pi^a.$$

The action of this Lie-derivative on the fields then reproduces the canonical action of $\mathcal{L}_H$ on $\Lambda(S^* \mathcal{M})$

We introduce again the canonical conjugation (10.1) with

$$\Phi = -\Gamma_{ab}^c \pi^a \mathbf{c}^b \bar{\mathbf{c}}_c.$$

This gives

$$Q_\mathcal{X} \to e^{-\Phi} Q_\mathcal{X} e^{\Phi} = \mathbf{c}^a \frac{\partial}{\partial z^a} + (p_a + \Gamma_{ab}^c \mathbf{c}^b \bar{\mathbf{c}}_c) \frac{\partial}{\partial \bar{\mathbf{c}}_a} + \mathcal{X}_H^a \mathbf{i}_a$$
$$+ \left(\Gamma_{ba}^c p_c \mathbf{c}^b - \frac{1}{2} R_{adb}^c \mathbf{c}^b \mathbf{c}^d \bar{\mathbf{c}}_c - \mathcal{X}_H^b \Gamma_{ab}^c \bar{\mathbf{c}}_c - \partial_a \mathcal{X}_H^b \bar{\mathbf{c}}_b \right) \pi^a.$$

For the Lie-derivative this conjugation yields

$$L_\mathcal{X} \to \mathcal{X}_H^a \partial_a - \partial_a \mathcal{X}_H^b \bar{\mathbf{c}}_b \frac{\partial}{\partial \bar{\mathbf{c}}_a} + \mathbf{c}^b \partial_b \mathcal{X}_H^a \mathbf{i}_a - p_b \partial_a \mathcal{X}_H^b \pi^a$$
$$- \mathbf{c}^b \bar{\mathbf{c}}_c (\mathcal{X}_H^d \partial_d \Gamma_{ab}^c + \partial_b \mathcal{X}_H^d \Gamma_{ad}^c + \partial_a \mathcal{X}_H^d \Gamma_{db}^c - \Gamma_{ab}^d \partial_d \mathcal{X}_H^c + \partial_{ab} \mathcal{X}_H^c) \pi^a.$$

Here we identify in the last term

$$\mathcal{X}_H^d \partial_d \Gamma_{ab}^c + \partial_b \mathcal{X}_H^d \Gamma_{ad}^c + \partial_a \mathcal{X}_H^d \Gamma_{db}^c - \Gamma_{ab}^d \partial_d \mathcal{X}_H^c + \partial_{ab} \mathcal{X}_H^c$$

the Lie-derivative of Γ along $\mathcal{X}_H$ on $\mathcal{M}$ and it vanishes only if the Hamiltonian determines an isometry: In this way, we again meet our earlier condition that the Hamiltonian flow leaves the metric tensor invariant, now formulated in terms of the metric connection.

As before, we assume that the Hamiltonian flow determines an isometry so that we can set

$$\mathcal{L}_H \Gamma_a = 0.$$

This implies that $L_\mathcal{X}$ simplifies to

$$L_\mathcal{X} = \mathcal{X}_H^a \partial_a - \partial_a \mathcal{X}_H^b \bar{\mathbf{c}}_b \frac{\partial}{\partial \bar{\mathbf{c}}_a} + \mathbf{c}^b \partial_b \mathcal{X}_H^a \mathbf{i}_a - p_b \partial_a \mathcal{X}_H^b \pi^a.$$

In particular, the corresponding Lie-transformation laws of all our variables are now *manifestly covariant*:

$$L_\chi z^a = \mathcal{X}_H^a,$$
$$L_\chi \bar{c}_a = -\partial_a \mathcal{X}_H^b \bar{c}_b,$$
$$L_\chi c^a = c^b \partial_b \mathcal{X}_H^a,$$
$$L_\chi p_a = -p_b \partial_a \mathcal{X}_H^b.$$

The formalism that we have here developed can be viewed as an equivariant generalization of the Mathai-Quillen formalism.

13   Equivariant Morse Theory

We shall now apply the previous equivariant version of the Mathai-Quillen formalism to derive an equivariant version of Morse theory [20]: We are interested in evaluating the integral

$$Z_\psi = \int dz\, dp\, d\mathbf{c}\, d\bar{\mathbf{c}}\, \exp\{i\phi(H + \omega) + Q_\chi \psi\}$$

where H is again a Hamiltonian that generates the symplectic action of $U(1)$. Notice that the measure here is invariant: It is the Liouville measure on $\Lambda(S^*\mathcal{M})$. Furthermore, since both H and ω only depend on z^a, c^a, all dependence on p_a and $math\bar{b}fc$ comes from the last term. Our standard arguments imply, that if

$$L_\chi \psi = 0 \tag{13.1}$$

then the integral is invariant under local variations of ψ, provided these variations are also in the subspace (13.1).

$$Z_\psi = Z_{\psi + \delta\psi}.$$

However, since all p_a and $\bar{c}$ dependence resides in the Q_χ contribution, we can not set $\psi = 0$—the integral would not be defined. Hence the present integral does *not* coincide with the partition function we evaluated in Section 2.

In order to evaluate (13), we need to construct functionals ψ that satisfy (13.1). For this we observe that since H generates an isometry, L_χ-transformations are generally covariant and *any* generally covariant quantity constructed from our variables satisfies (13.1). In particular, (13) is invariant under *any* local variation of ψ which is generally covariant.

We first select

$$\psi = \tfrac{1}{2} \mathcal{X}_H^a \bar{c}_a$$

which is generally covariant. Explicitly, we obtain

$$Q_\chi \psi = p_a \mathcal{X}_H^a + \mathbf{c}^a \nabla_a \mathcal{X}_H^b \bar{\mathbf{c}}_b$$

and our integral becomes

$$Z = \int dq \, dp \, d\mathbf{c} \, d\bar{\mathbf{c}} \exp\{i\phi(H + \omega) + p_a \mathcal{X}_H^a + \mathbf{c}^a \nabla_a \mathcal{X}_H^b \bar{\mathbf{c}}_b\}.$$

We again assume H only admits isolated nondegenerate critical points. We can then evaluate the integrals, and we find

$$Z = \int dq \, \exp\{i\phi H\} \cdot \delta(\mathcal{X}_H^a) \det \|\partial_a \mathcal{X}_H^b\|$$

$$= \sum_{dH=0} \exp\{i\phi H\} \cdot \text{sign}(\det \|\partial_{ab} H\|). \qquad (13.2)$$

We recognize this result as an equivariant version of the Poincaré-Hopf theorem.

Next we select

$$\psi = g^{ab} p_a \bar{\mathbf{c}}_b.$$

As a generally covariant quantity it automatically obeys the Lie-derivative condition (13.1). We now find for the last term in (13)

$$Q_\chi \psi = g^{ab} p_a p_b - \frac{1}{2} R^c_{adb} \mathbf{c}^b \mathbf{c}^d \bar{\mathbf{c}}_c g^{ae} \bar{\mathbf{c}}_e - \nabla_c (\partial_b \mathcal{X}_H^a) g^{bc} \bar{\mathbf{c}}_a \bar{\mathbf{c}}_c$$

and our integral becomes

$$Z_\psi = \int dz \, dp \, d\mathbf{c} \, d\bar{\mathbf{c}} \exp\{i\phi(H + \omega) + Q_\chi \psi\}$$

$$= \int dz \, d\mathbf{c} \, \exp\{i\phi(H + \omega)\} \cdot \text{Pf}\left[\nabla_b \mathcal{X}_H^a + \frac{1}{2} R^a_{bcd} \mathbf{c}^c \mathbf{c}^d\right]. \qquad (13.3)$$

This we identify as an equivariant version of the Gauss-Bonnet-Chern theorem.

Combining our results (13.2) and (13.3) we then have

$$\sum_{dH=0} \exp\{i\phi H\} \cdot \text{sign}(\det \|\partial_{ab} H\|)$$

$$= \int dz \, d\mathbf{c} \, \exp\{i\phi(H + \omega)\} \cdot \text{Pf}\left[\nabla_b \mathcal{X}_H^a + \frac{1}{2} R^a_{bcd} \mathbf{c}^c \mathbf{c}^d\right]$$

which generalizes our results derived in Section 11. to the equivariant context.

A generalization to the degenerate case can be derived by generalizing our earlier discussion in Section 4.

14 Loop Space and Morse Theory

We shall now proceed to generalize the previous discussion of equivariant Morse theory to path integrals [20]. As before, we introduce the loop space by lifting our canonical phase space variables to be loop space variables. We shall first consider the case, where the is no Hamiltonian; as in the case of the Atiyah-Singer index theorem, we consider the natural action of a circle, defined by

$$x^a(t) \to x^a(t + \tau)$$

for all variables in the superloop space $L(S^*\mathcal{M})$. The relevant superloop space equivariant exterior derivative is

$$Q_t = \mathbf{c}^a \frac{\partial}{\partial z^a} + p_a \frac{\partial}{\partial \bar{\mathbf{c}}_a} + \dot{z}^a \mathbf{i}_a + \dot{\bar{\mathbf{c}}}_a \pi^a$$

and the corresponding Lie derivative in $L(S^*\mathcal{M})$ is

$$\mathbf{L}_t = Q_t^2 = \dot{z}^a \frac{\partial}{\partial z^a} + \dot{\bar{\mathbf{c}}}_a \frac{\partial}{\partial \bar{\mathbf{c}}_a} + \dot{\mathbf{c}}^a \mathbf{i}_a + \dot{p}_a \pi^a \equiv \frac{\partial}{\partial t}.$$

We again introduce the conjugation (10.1), with

$$\Phi = -\Gamma^c_{ab} \pi^a \mathbf{c}^b \bar{\mathbf{c}}_c.$$

For Q_t this gives

$$Q_t \to e^\Phi Q_t e^{-\Phi} = \mathbf{c}^a \frac{\partial}{\partial z^a} + (p_a + \Gamma^c_{ab} \mathbf{c}^b \bar{\mathbf{c}}_c) \frac{\partial}{\partial \bar{\mathbf{c}}_a}$$
$$+ \dot{z}^a \mathbf{i}_a + \left(\dot{\bar{\mathbf{c}}}_a - \dot{z}^b \Gamma^c_{ab} \bar{\mathbf{c}}_c + \Gamma^c_{ab} p_c \mathbf{c}^b - \frac{1}{2} R^c_{abd} \mathbf{c}^d \mathbf{c}^b \bar{\mathbf{c}}_c \right) \pi^a$$

while the Lie derivative remains intact,

$$\mathbf{L}_t \to e^\Phi \mathbf{L}_t e^{-\Phi} \equiv \mathbf{L}_t.$$

We wish to evaluate the path integral

$$Z = \int [dz] [d\mathbf{c}] [dp] [d\bar{\mathbf{c}}] \exp\left\{ i \int Q_t \psi \right\}$$

which is of the standard form of a (cohomological) topological path integral [18]. The Lie-derivative condition

$$\mathbf{L}_t \psi = \int_0^T dt \partial_t \psi = \psi(T) - \psi(0)$$

is again satisfied by *any* ψ which is single valued in $L(S^*\mathcal{M})$. According to our standard arguments the path integral is invariant under arbitrary local variations of ψ, provided these variations are also single valued in $L(S^*\mathcal{M})$. We select

$$\psi = \frac{\lambda}{2} g_{ab}\dot{z}^a \mathbf{c}^b + g^{ab} p_a \bar{\mathbf{c}}_b$$

which clearly satisfies our Lie-derivative condition, with g_{ab} an arbitrary Riemannian metric. For the action this gives

$$S = \int Q_t \psi = \int \frac{\lambda}{2} g_{ab}\dot{z}^a \dot{z}^b + \frac{\lambda}{2} \mathbf{c}^a (g_{ab}\partial_t + g_{bd}\dot{z}^c\Gamma^d_{ca})\mathbf{c}^b + \frac{1}{2} g^{ab} p_a p_b$$
$$- \frac{1}{2} R^{ca}_{db}\mathbf{c}^b \mathbf{c}^d \bar{\mathbf{c}}_c \bar{\mathbf{c}}_a + \frac{1}{2}(\bar{\mathbf{c}}_e g^{ea})(g_{ab}\partial_t - g_{bd}\dot{z}^c\Gamma^d_{ca})(g^{bf}\bar{\mathbf{c}}_f).$$

We again define loop space coordinates by separating the constant mode,

$$z^a(t) = z^a_0 + z^a_t,$$
$$\mathbf{c}^a(t) = \mathbf{c}^a_0 + \mathbf{c}^a_t,$$
$$p_a(t) = p_{a0} + p_{at},$$
$$\bar{\mathbf{c}}_a(t) = \bar{\mathbf{c}}_{a0} + \bar{\mathbf{c}}_{at},$$

and define the path integral measure by

$$[dz]\,[d\mathbf{c}]\,[dp]\,[d\bar{\mathbf{c}}] = dz^a_0\, d\mathbf{c}^a_0\, dp_{a0}\, d\bar{\mathbf{c}}_{a0} \prod_t dz^a_t\, d\mathbf{c}^a_t\, dp_{at}\, d\bar{\mathbf{c}}_{at}.$$

We change variables to

$$z^a_t \rightarrow \frac{1}{\sqrt{\lambda}} z^a_t,$$
$$\mathbf{c}^a_t \rightarrow \frac{1}{\sqrt{\lambda}} \mathbf{c}^a_t,$$

and set $\lambda \rightarrow \infty$ and evaluate integrals. This yields

$$Z = \int dz\, d\mathbf{c}\; \mathrm{Pf}(R_{abcd}\mathbf{c}^c\mathbf{c}^d) \equiv \chi(\mathcal{M}) \tag{14.1}$$

which is the Euler character of $\mathcal{M}$.

We now introduce an arbitrary smooth vector field on $\mathcal{M}$ with local components V^a. We assume the zeroes of V are isolated and nondegenerate. We then select

$$\psi = \frac{\lambda}{2} g_{ab}\dot{z}^a \mathbf{c}^b + (\dot{z}^a + V^a)\bar{\mathbf{c}}_a$$

which satisfies our Lie derivative condition, as a generally covariant quantity. For the action this gives

$$S = \int Q_t \psi = \int \frac{\lambda}{2} g_{ab} \dot{z}^a \dot{z}^b + \frac{\lambda}{2} \mathbf{c}^a (g_{ab} \partial_t + g_{bd} \dot{z}^c \Gamma_{ca}^d) \mathbf{c}^b + p_a (\dot{z}^a + V^a)$$
$$+ p_a (\dot{z}^a + V^a) + \bar{\mathbf{c}}_a (\delta_b^a \nabla_t + \nabla_b V^a) \mathbf{c}^b.$$

We define variables as before, and evaluate integrals in the $\lambda \to \infty$ limit. The result is

$$Z = \sum_{dV=0} \text{sign}(\det \|\nabla_a V^b\|). \tag{14.2}$$

Combining (14.1) and (14.2) we then conclude, that we have derived the standard relation between Poincaré-Hopf and Gauss-Bonnet-Chern theorems, now for an arbitrary - not necessarily gradient or Hamiltonian - vector field V,

$$\sum_{dV=0} \text{sign}(\det \|\nabla_a V^b\|) = \int dz\, d\mathbf{c}\, \text{Pf}(R_{abcd} \mathbf{c}^c \mathbf{c}^d).$$

15   Loop Space and Equivariant Morse Theory

We shall now proceed to generalize the previous discussion to the equivariant context [20]. For this, we introduce the classical action

$$S = \int \vartheta_a \dot{z}^a - H$$

of a Hamiltonian system, and interpret it as a loop space Hamiltonian functional, with the corresponding loop space Hamiltonian vector field

$$\mathcal{X}_S^a = \dot{z}^a - \mathcal{X}_H^a.$$

We equivariantize Q_t to account for the nontrivial flow induced by H,

$$Q_S = d + \mathbf{i}_S = \mathbf{c}^a \frac{\partial}{\partial z^a} + p_a \frac{\partial}{\partial \bar{\mathbf{c}}_a} + (\dot{z}^a - \mathcal{X}_H^a)\mathbf{i}_a + (\dot{\bar{\mathbf{c}}}_a - \partial_a \mathcal{X}_H^b \bar{\mathbf{c}}_b)\pi^a$$

and for simplicity we again assume that the Hamiltonian generates an isometry, that is we can select the connection on $\mathcal{M}$ so that it is H-invariant

$$\mathcal{L}_H \Gamma_a = 0.$$

We introduce conjugation by

$$\Phi = -\Gamma_{bc}^a \pi^b \mathbf{c}^c \bar{\mathbf{c}}_a$$

which yields

$$Q_S \to e^{\Phi} Q_S e^{-\Phi} = \mathbf{c}^a \frac{\partial}{\partial z^a} + (p_a + \Gamma^c_{ab} \mathbf{c}^b \bar{\mathbf{c}}_c) \frac{\partial}{\partial \bar{\mathbf{c}}_a} + (\dot{z}^a - \mathcal{X}^a_H) i_a$$

$$+ \left\{ \Gamma^c_{ab} p_c \mathbf{c}^b - \frac{1}{2} R^c_{adb} \mathbf{c}^b \mathbf{c}^d \bar{\mathbf{c}}_c - (\dot{z}^b - \mathcal{X}^b_H) \Gamma^c_{ab} \bar{\mathbf{c}}_c \right. $$
$$\left. + (\delta^c_a \partial_t - \partial_a \mathcal{X}^c_H) \bar{\mathbf{c}}_c \right\} \pi^a$$

and for the Lie derivative

$$L_S = \partial_t + L_{\mathcal{X}}$$

$$\to e^{\Phi} L_S e^{-\Phi} = \partial_t + \mathcal{X}^a_H \frac{\partial}{\partial z^a} + \mathbf{c}^a \partial_a \mathcal{X}^b_H i_b - \partial_a \mathcal{X}^b_H \bar{\mathbf{c}}_b \frac{\partial}{\partial \bar{\mathbf{c}}_a} - p_b \partial_a \mathcal{X}^b_H \pi^a.$$

We are interested in using this formalism to evaluate the partition function

$$Z = \int [dz] [dp] [d\mathbf{c}] [d\bar{\mathbf{c}}] \exp \left\{ i \int \vartheta_a \dot{z}^a - H + \frac{1}{2} \mathbf{c}^a \omega_{ab} \mathbf{c}^b + Q_S \psi \right\}.$$

By construction, this remains invariant under a local variation of ψ provided such a variation satisfies the condition

$$L_S \psi = 0. \tag{15.1}$$

An example of a ψ that obeys this condition is given by

$$\psi_1 = g^{ab} p_a \bar{\mathbf{c}}_b$$

since the metric is assumed to satisfy

$$\mathcal{L}_H g = 0.$$

In our action this gives

$$Q_S \psi_1 = g^{ab} p_a p_b - \frac{1}{2} R^d_{acb} \mathbf{c}^b \mathbf{c}^c \bar{\mathbf{c}}_d g^{ae} \bar{\mathbf{c}}_e - \bar{\mathbf{c}}_a (g^{ab} \partial_t + \partial_c \mathcal{X}^a_H g^{cb}) \bar{\mathbf{c}}_b$$
$$+ \bar{\mathbf{c}}_a g^{ad} (\dot{z}^c - \mathcal{X}^c_H) \Gamma^b_{cd} \bar{\mathbf{c}}_b.$$

We also introduce the following functional

$$\psi_2 = \frac{\lambda}{2} g_{ab} (\dot{x}^a - \mathcal{X}^a_H) \mathbf{c}^b$$

which also satisfies the condition (15.1). This now gives

$$Q_S \psi_2 = \frac{\lambda}{2} g_{ab} (\dot{z}^a - \mathcal{X}^a_H)(\dot{z}^b - \mathcal{X}^b_H) + \frac{\lambda}{2} \mathbf{c}^a \partial_a (g_{bc} \dot{z}^b - g_{bc} \mathcal{X}^b_H) \mathbf{c}^c.$$

We substitute both ψ_1 and ψ_2 in Z and assuming classical solutions of our Hamiltonian equations of motion are nondegenerate, in the $\lambda \to \infty$ limit we get equivariant loop space version of the quantity that appears in the Poincaré-Hopf theorem,

$$Z = \sum_{\delta S = 0} \text{sign}(\det \|\delta_{ab} S\|) \exp\{iS\}.$$

Next we introduce

$$\psi_3 = \frac{\lambda}{2} g_{ab} \dot{z}^a \mathbf{c}^b.$$

This gives

$$Q_S \psi_3 = \frac{\lambda}{2} (\dot{z}^a - \mathcal{X}_H^a) g_{ab} \dot{z}^b + \frac{\lambda}{2} \mathbf{c}^a (g_{ab} \partial_t + \dot{z}^c g_{ad} \Gamma_{cb}^d) \mathbf{c}^b.$$

We now substitute ψ_1 and ψ_3 in Z. In the $\lambda \to \infty$ limit this yields the equivariant version of the quantity that appears in the Gauss-Bonnet-Chern theorem,

$$Z = \int_{\mathcal{M}} dz\, d\mathbf{c}\, \exp\left\{-iT\left(H + \frac{1}{2} \mathbf{c}^a \omega_{ab} \mathbf{c}^b\right)\right\} \text{Pf}\left[\frac{1}{2}(\Omega_b^a + R_{bcd}^a \mathbf{c}^c \mathbf{c}^d)\right].$$

Combining we then have the following loop space version of Poincaré-Hopf and Gauss-Bonnet-Chern relation in equivariant Morse theory

$$\sum_{\delta S = 0} \text{sign}(\det \|\delta_{ab} S\|) \exp\{iS\}$$

$$= \int_{\mathcal{M}} dz\, d\mathbf{c}\, \exp\left\{-iT\left(H + \frac{1}{2} \mathbf{c}^a \omega_{ab} \mathbf{c}^b\right)\right\} \text{Pf}\left[\frac{1}{2}(\Omega^a{}_b + R_{bcd}^a \mathbf{c}^c \mathbf{c}^d)\right].$$

16   Poincaré Supersymmetry and Equivariant Cohomology

We shall conclude these lectures by explaining, how the previous formalism can also be applied to formulate general Poincaré supersymmetric quantum field theories in terms of equivariant cohomology [17]. We have already seen, how this is correct in the case of both $N = \frac{1}{2}$ and $N = 1$ supersymmetric quantum mechanics. It turns out, that these observation generalize to *arbitrary* Poincaré supersymmetric quantum field theories in any number of dimensions.

As an example, we first consider the following supersymmetric ($N = 1$) quantum mechanics

$$S_B + S_F = \int \frac{1}{2} \dot{q}^2 - \frac{1}{2} W_q^2 + \frac{1}{2}(\theta_1 \dot{\theta}_1 + \theta_2 \dot{\theta}_2) - \theta_2 W_{qq} \theta_1$$

which is invariant under the supersymmetry transformation

$$\delta q = \alpha\theta_1 + \beta\theta_2,$$
$$\delta\theta_1 = -\alpha\dot{q} + \beta W_q,$$
$$\delta\theta_2 = -\beta\dot{q} - \alpha W_q.$$

This is a realization of the $N = 1$ supersymmetry algebra

$$\{Q, Q^\dagger\} = H,$$
$$\{Q, Q\} = \{Q^\dagger, Q^\dagger\} = \{Q, H\} = \{Q^\dagger, H\} = 0$$

which is seen directly, if we introduce the following canonical realization

$$\eta = \frac{1}{\sqrt{2}}(\theta_1 + i\theta_2),$$

$$\bar{\eta} = \frac{1}{\sqrt{2}}(\theta_1 - i\theta_2),$$

$$\{p, q\} = \{\eta, \bar{\eta}\} = 1,$$

$$Q = \frac{1}{\sqrt{2}}\eta(p + iW_q),$$

$$Q^\dagger = \frac{1}{\sqrt{2}}\bar{\eta}(p - iW_q),$$

$$\{Q, Q^\dagger\} = H = \frac{1}{2}p^2 + \frac{1}{2}W_q^2 + i\bar{\eta}W_{qq}\eta,$$

and in particular,

$$S_B + S_F = \int p\dot{q} + \bar{\eta}\dot{\eta} - H.$$

In order to relate this to loop space symplectic geometry, we introduce the canonical conjugation

$$p \to e^{-\Phi}pe^\Phi = p + \{p, \Phi\} + \frac{1}{2}\{\{p, \Phi\}, \Phi\} + \cdots$$

and similarly for other variables. We select

$$\Phi = -iW(q).$$

This gives for the supercharges

$$Q \xrightarrow{\Phi} e^{-\Phi}Qe^\Phi = \frac{1}{\sqrt{2}}\eta p,$$

$$Q^\dagger \xrightarrow{\Phi} e^{-\Phi}Q^\dagger e^\Phi = \frac{1}{\sqrt{2}}\bar{\eta}(p - 2iW_q).$$

We redefine $p \to p + \dot{q}$. For the action we then obtain the following auxiliary field representation

$$S_B + S_F = \int \frac{1}{2}\dot{q}^2 + \frac{1}{2}p^2 + pW_q + \frac{1}{2}(\theta_1\dot{\theta}_1 + \theta_2\dot{\theta}_2) - \theta_2 W_{qq}\theta_1.$$

This we can interpret in terms of loop space equivariant cohomology. For this, we introduce superloop space with equivariant exterior derivative

$$d_S = d + i_S = \theta_1 \frac{\delta}{\delta q} + p\frac{\delta}{\delta\theta_2} - \dot{q} \cdot i_{\theta_1} - \dot{\theta}_2 \cdot i_p.$$

We define the superloop space one-form

$$\vartheta = -\frac{1}{2}\dot{q}\theta_1 + \frac{1}{2}\theta_2 p$$

and the superloop space zero-form

$$\Lambda = \theta_2 W_q.$$

We then find, that the action can be represented as

$$S_B + S_F = d_S(\vartheta + \Lambda)$$
$$= \int \frac{1}{2}\dot{q}^2 + \frac{1}{2}p^2 + pW_q + \frac{1}{2}(\theta_1\dot{\theta}_1 + \theta_2\dot{\theta}_2) - \theta_2 W_{qq}\theta_1.$$

As another example we consider the Wess-Zumino model in two dimensions,

$$S = \int \frac{1}{2}(\partial_\mu A)^2 - \frac{1}{2}W_A^2 + \frac{i}{2}\bar{\psi}\gamma^\mu\partial_\mu\psi - \frac{1}{2}W_{AA}\bar{\psi}\psi.$$

We use the representation $\gamma^0 = \sigma^2$ and $\gamma^1 = i\sigma^1$ of the γ-matrices, and introduce the chiral components

$$\psi = \begin{pmatrix} \theta_+ \\ \theta_- \end{pmatrix} \qquad \bar{\psi} = \psi^T\gamma^0 = i\begin{pmatrix} \theta_- & -\theta_+ \end{pmatrix}.$$

Using the canonical realization

$$\{P_x, A_y\} = \delta(x - y),$$
$$\{\theta_\pm(x), \theta_\pm(y)\} = -i\delta(x - y)$$

the supercharges then become

$$Q = (P + \partial_x A + iW_A)\theta_+ + i(P - \partial_x A + iW_A)\theta_-,$$
$$Q^\dagger = (P + \partial_x A - iW_A)\theta_+ - i(P - \partial_x A - iW_A)\theta_-,$$

with the $(N = 1)$ supersymmetry algebra

$$\{Q, Q^\dagger\} = -4iH$$

with Hamiltonian

$$H = \frac{1}{2}P^2 + \frac{1}{2}(\partial_x A)^2 + \frac{1}{2}W_A^2 - \frac{i}{2}\theta_+\partial_x\theta_+ + \frac{i}{2}\theta_-\partial_x\theta_- - i\theta_+\theta_-W_{AA}$$

and the action is

$$S_B + S_F = \int P\dot{A} + \frac{i}{2}(\theta_+\dot{\theta}_+ + \theta_-\dot{\theta}_-) - H.$$

We again introduce conjugations

$$P \xrightarrow{\Phi} e^{i\Phi}Pe^{-i\Phi} = P - i\{P, \Phi\} - \frac{1}{2}\{\{P, \Phi\}, \Phi\} + \cdots$$

and similarly for other variables, with $\Phi = W(A)$. For the supercharges we get

$$Q \xrightarrow{\Phi} e^{i\Phi}Qe^{-i\Phi} = (P + \partial_x A)\theta_+ + i(P - \partial_x A)\theta_-,$$

$$Q^\dagger \xrightarrow{\Phi} e^{i\Phi}Q^\dagger e^{-i\Phi} = (P + \partial_x A - 2iW_A)\theta_+ - i(P - \partial_x A - 2iW_A)\theta_-.$$

Shifting $P \to P + \dot{A}$ we then obtain the auxiliary field representation

$$S_B + S_F = \int \frac{1}{2}\dot{A}^2 + \frac{1}{2}P^2 - P(\partial_x A - W_A)$$
$$+ \frac{i}{2}(\theta_+\partial_+\theta_+ + \theta_-\partial_-\theta_-) + i\theta_+\theta_-W_{AA}.$$

If we now define a superloop space with equivariant exterior derivative

$$d_S = d + i_S = \psi_1\frac{\delta}{\delta A} + iP\frac{\delta}{\delta\psi_2} + i\dot{A}i_{\psi_1} + \dot{\psi}_2 \cdot i_P$$

and introduce the superloop space one-form

$$\vartheta = -\frac{i}{2}\psi_2\dot{A} - \frac{i}{2}P\psi_1$$

and superloop space zero-form

$$\Lambda = i(\partial_x A - W_A)\psi_2$$

we find that the action can be represented as

$$d_S(\vartheta + \Lambda) = \int \frac{1}{2}\dot{A}^2 + \frac{1}{2}P^2 - P(\partial_x A - W_A)$$
$$+ \frac{i}{2}(\psi_1\dot{\psi}_1 + \psi_2\dot{\psi}_2) - i\psi_2\partial_x\psi_1 + i\psi_1W_{AA}\psi_2$$

when we identify

$$\begin{pmatrix} \psi_1 \\ \psi_2 \end{pmatrix} = \frac{1}{\sqrt{2}} \begin{pmatrix} 1 & 1 \\ 1 & -1 \end{pmatrix} \begin{pmatrix} \theta_+ \\ \theta_- \end{pmatrix}.$$

These examples can be generalized to higher dimensions and to include supersymmetric gauge theories [17]. It has also been established, that a loop space interpretation can be given at the level of an arbitrary $N = 1$ supermultiplet [22], which implies that this interpretation of Poincaré-supersymmetry in terms of loop space equivariant cohomology is model independent. These results then suggest, that loop space equivariant cohomology could provide an effective tool for exact, nonperturbative analysis of supersymmetric quantum field theories.

Acknowledgments: I wish to thank the organizers of the Banff meeting, and in particular G. Semenoff and L. Vinet for giving me an opportunity to present these discussions.

Work supported by NFR Grant F-AA/FU 06821-308 and by Göran Gustafsson's Foundation for Science and Medicine

17   References

1. M. A. Semenov-Tjan-Shanskiĭ. A certain property of the Kirillov integral. In *Differential Geometry, Lie Group, and Mechanics*, volume 37 of *Mat. Ind. Steklov (LOMI)*, pages 53–65, 1973. Russian; M. A. Semenov-Tjan-Shanskiĭ. Harmonic analysis on Riemannian symmetric spaces of negative curvature, and scattering theory. *Izv. Akad. Nauk.*, 40 (3): 562–592, 1976.

2. J. J. Duistermaat and G. Heckman. On the variation in the cohomology of the symplectic form of the reduced phase space. *Invent. Math.*, 69 (2): 259–268, 1982; J. J. Duistermaat and G. Heckman. Addendum to "On the variation in the cohomology of the symplectic form of the reduced phase space." *Invent. Math.*, 72 (1): 153–158, 1983.

3. N. Berline and M. Vergne. Zéros d'un champ de vecteurs et classes caractéristiques équivariantes. *Duke Math. J.*, 50 (2): 539–549, 1983. French; N. Berline and M. Vergne. The equivariant index and Kirillov's character formula. *Amer. J. Math.*, 107 (5): 1159–1190, 1985.

4. M. F. Atiyah and R. Bott. The moment map and equivariant cohomology. *Topology*, 23 (1): 1–28, 1984; M. F. Atiyah and R. Bott. A Lefschetz fixed point formula for elliptic complexes. I. *Ann. Math.*, 86: 374–407, 1967; M. F. Atiyah and R. Bott. A Lefschetz fixed

point formula for elliptic complexes. II. *Ann. Math.*, 88: 451–491, 1968; M. Atiyah and I. Singer. The index of elliptic operators III. *Ann. Math.*, 87: 546–604, 1968; M. Atiyah and I. Singer. The index of elliptic operators IV. *Ann. Math.*, 93: 119–138, 1971.

5. J.-M. Bismut. Index theorem and equivariant cohomology on the loop space. *Commun. Math. Phys.*, 98 (2): 213–237, 1985; J.-M. Bismut. Localization formulas, superconnections, and the index theorem for families. *Commun. Math. Phys.*, 103 (1): 127–166, 1986; J.-M. Bismut. The infinitesimal Lefschetz formulas: A heat equation proof. *J. Funct. Anal.*, 62 (3): 435–457, 1985.

6. N. Berline, E. Getzler, and M. Vergne. *Heat Kernels and Dirac Operators*, volume 298 of *Grundlehren der Mathematischen Wissenschaften*. Springer-Verlag, Berlin, 1992.

7. M. F. Atiyah. Circular symmetry and stationary-phase approximation. In *Colloquim in Honor of Laurent Schwartz*. I, volume 131 of *Astérisque*, pages 43–59, 1985.

8. M. Blau, E. Keski-Vakkuri, and A. J. Niemi. Path integrals and geometry of trajectories. *Phys. Lett.*, B246 (1-2): 92–98, 1990; A. J. Niemi and P. Pasanen. Orbit geometry, group representations and topological quantum field theories. *Phys. Lett.*, B253 (3-4): 349–356, 1991; E. Keski-Vakkuri, A. J. Niemi, G. Semenoff, and O. Tirkkonen. Topological quantum theories and integrable models. *Phys. Rev.*, D44 (12): 3899–3905, 1991.

9. A. J. Niemi and O. Tirkkonen. Cohomological partition functions for a class of bosonic theories. *Phys. Lett.*, B293 (3-4): 339–343, 1992.

10. H. M. Dykstra, J. D. Lykken, and E. J. Raiten. Exact path integrals by equivariant localization. *Phys. Lett.*, B302 (2-3): 223–229, 1993.

11. E. Witten. Two-dimensional gauge theories revisited. *J. Geom. Phys.*, 9 (4): 303–368, 1992.

12. A. J. Niemi and O. Tirkkonen. On exact evaluation of path integrals. *Ann. Phys.*, 235 (2): 318–349, 1994; A. J. Niemi and K. Palo. On quantum integrability and the Lefschetz number. *Mod. Phys. Lett.*, A8 (24): 2311–2321, 1993.

13. V. I. Arnol'd. *Mathematical Methods of Classical Mechanics*, volume 60 of *Graduate Texts in Mathematics*. Springer-Verlag, New York-Heidelberg, 1978.

14. J. Kalkman. BRST model for equivariant cohomology and representatives for the equivariant Thorn class. *Commun. Math. Phys.*, 153

(3): 447–463, 1993; J. Kalkman. BRST model applied to symplectic geomerty. hep-th/9308132.

15. A. J. Niemi and O. Tirkkonen. Equivariance, BRST symmetry, and superspace. *J. Math. Phys.*, 35 (12): 6418–6433, 1994.

16. T. Eguchi, P. Gilkey, and A. Hanson. Gravitation, gauge theories and differential geometry. *Phys. Rep.*, 66 (6): 213–393, 1980.

17. A. Hietamäki, A. Yu. Morozov, A. J. Niemi, and K. Palo. Geometry of $n = \frac{1}{2}$ supersymmetry and the Atiyah-Singer index theorem. *Phys. Lett.*, B263 (3-4): 417–424, 1991; A. Yu. Morozov, A. J. Niemi, and K. Palo. Supersymmetry and loop space geometry. *Phys. Lett.*, B271 (3-4): 365–371, 1991.

18. D. Birmingham, M. Blau, M. Rakowski, and G. Thompson. Topological field theory. *Phys. Rep.*, 209 (4-5): 129–340, 1991.

19. V. Mathai and D. Quillen. Superconnections, Thorn classes, and equivariant differential forms. *Topology*, 25 (1): 85–110, 1986; M. Blau. The Mathai-Quillen formalism and topological field theory. *J. Geom. Phys.*, 11 (4-5): 95–127, 1993; S. Cordes, G. Moore, and S. Ramgoolan. Lectures on $2D$ Yang-Mills theory, equivariant cohomology and topological field theory. hep-th/9411210.

20. A. J. Niemi and K. Palo. Equivariant Morse theory and quantum integrability. hep-th/9406068; A. J. Niemi and K. Palo. On the characterization of classical dynamical systems using supersymmetric nonlinear σ-models. hep-th/9412023.

21. J. Milnor. *Morse Theory*, volume 51 of *Ann. Math. Studies*. Princeton University Press, Princeton, NJ, 1963.

22. K. Palo. Symplectic geometry of supersymmetry and nonlinear sigma model. *Phys. Lett.*, B321 (1-2): 61–65, 1994.

7

Systems of Calogero-Moser Type

S. N. M. Ruijsenaars

ABSTRACT We survey results on Galilei- and Poincaré-invariant Calogero-Moser and Toda N-particle systems, both in the context of classical mechanics and of quantum mechanics. Special attention is given to integrability issues and interconnections between the various models. Action-angle and joint eigenfunction transforms are also considered, and some novel results on $N = 2$ eigenfunctions of hyperbolic Askey-Wilson type and of relativistic elliptic type are sketched.

1 Introduction

This chapter is concerned with a class of finite-dimensional integrable dynamical systems, both at the classical and at the quantum level. The systems model N interacting point particles on a line or a ring. Thus the classical state space is a $2N$-dimensional symplectic manifold, and the quantum state space a Hilbert space $L^2(G, dx_1 \cdots dx_N)$, with G equal to the classical configuration space.

The systems admit both a nonrelativistic (Galilei-invariant) and a relativistic (Poincaré-invariant) version. The nonrelativistic systems were introduced in the seventies and are known as Calogero-Moser (or Calogero-Sutherland) and Toda systems. They can be tied in with the root system A_{N-1} and also admit integrable versions for the remaining root systems. At the classical level all of these systems yield N Poisson commuting independent Hamiltonians with a polynomial dependence on the particle momenta $p_1, \ldots, p_N$, so that the quantum versions are partial differential operators (PDOs).

The systems just delineated were surveyed in the early eighties by Olshanetsky and Perelomov, both in the classical [1] and in the quantum context [2]. These surveys contain extensive lists of references, and are to a large extent concerned with the relations of the systems to group theory, Lie algebra theory, and harmonic analysis on symmetric spaces.

Integrable relativistic generalizations (corresponding once more to the root system A_{N-1}) were first introduced in Ref. 3 at the classical and in Ref. 4 at the quantum level. The Poisson commuting classical Hamiltonians have an exponential dependence on the particle momenta, so that the quantum versions are analytic difference operators (AΔOs). The relativistic systems and their relations to the nonrelativistic A_{N-1} systems and various well-known solitonic field theories and spin systems were surveyed in Ref. 5.

Quite recently, the relativistic systems with pair interactions of the trigonometric type were also shown to admit a generalization to the root systems B_N, C_N, D_N, and BC_N. More precisely, in Ref. 6 van Diejen introduces a quite general integrable quantum system that can be specialized to all of the root systems mentioned above. In further work [7, 8] he proposes even more general systems with elliptic interactions, but for these systems integrability has not yet been completely proved. At any rate, the latter systems encompass by specialization virtually all systems of Calogero-Moser and Toda type that are known to be integrable, including external field couplings that go beyond the root system machinery.

We shall deal here exclusively with Galilei- and Poincaré-invariant models. In particular, no external fields and root systems other than A_{N-1} will be treated, and we also omit from consideration thermodynamical aspects, discretizations, R-matrix formulations, internal degrees of freedom, and Haldane-Shastry chains. The discussion is addressed principally to theoretical physicists at the graduate student/postdoc level, but we believe it should also be accessible to mathematicians interested in the systems from various viewpoints different from physics. Our emphasis is on expounding the integrability of the systems and their interconnections, and on providing a conceptual understanding of the transforms that diagonalize the commuting dynamics—the action-angle and joint eigenfunction transforms. In doing so, we have tried to use as few ingredients as possible without losing mathematical precision.

We should mention, though, that the models involved can be viewed from a great many angles, and a lot of subfields of mathematics and theoretical physics can be brought to bear on them. Correspondingly, our bare-handed approach will possibly be viewed as a liability rather than an asset by some experts—but these notes are not primarily written for experts. We shall return to various related matters toward the end of this section.

Section 2 is concerned with the nonrelativistic Calogero-Moser and Toda systems in the context of classical mechanics. The mathematics involved in getting a solid grasp of what classical integrability is all about has been particularly well expounded in Refs. 9–11 (in order of increasing sophistication). In Section 2.1 we have summarized some material that can be found (in far greater detail) in these sources. In the process, we introduce various concepts and notation that will reappear later on.

In Section 2.2 we introduce the Calogero-Moser systems. The pair potential characterizing the systems is a rational, hyperbolic, trigonometric, or elliptic function; cf. (2.18)–(2.21) respectively. The former three choices may be viewed as degenerate cases of the latter; cf. (2.22).

Having in mind readers who are not familiar with elliptic functions, we would like to mention that we use very little of the extensive lore on this subject. Indeed, from a pragmatic viewpoint one may regard (2.63) as a definition of the key function $s(x; \omega, \omega')$ in terms of elementary functions. If one now takes (2.64) for granted, and uses (2.50) and (2.52) to introduce the Weierstrass $\mathcal{P}$-function in terms of $s(x)$, then almost all further properties we need readily follow. But if need be, we can highly recommend Ref. 12 to read up on elliptic functions (and, more generally, the classic special functions entering in Section 6.3).

The rational and hyperbolic systems (denoted type I and II, respectively) live on an unambiguous phase space, and each initial state is a scattering state. By contrast, the trigonometric and elliptic systems (denoted type III and IV, respectively) give rise to three distinct phase spaces. The choice of state space depends on whether one views the particles as moving on a line or on a ring, and—in the latter case—on whether one wants to regard the particles as distinguishable or indistinguishable. In all three cases, the internal motion is oscillatory.

In the nonrelativistic case the existence of integrals (conserved quantities) for the defining dynamics is most easily seen via a so-called Lax pair. We sketch the Lax pair formalism in some detail in Section 2.2 and present Lax matrices for each of the four types of pair potentials; cf. (2.32) (type I–III) and (2.51) (type IV). We show how the Lax matrix can be used to deduce that the rational and hyperbolic interactions lead to a scattering of soliton type. That is, the asymptotic momenta are conserved and the position shifts factorize as if independent pair collisions were taking place; cf. (2.47)–(2.48).

In Section 2.3 we introduce the periodic and nonperiodic Toda systems (denoted type V and VI, respectively). In the former, oscillatory motion takes place, whereas the latter lead to soliton scattering. Once more, the integrals can be taken to be power traces or symmetric functions of Lax matrices, given by (2.57) (type V) and (2.59) (type VI). The Toda systems may be viewed as limits of Calogero-Moser systems, as encoded in the connection diagram (2.62). We detail the arrows in this diagram for the various Lax matrices and, therefore, for all of the Poisson commuting integrals at once.

In Section 3 we present and discuss Poincaré-invariant generalizations of the Galilei-invariant systems from Section 2. Section 3.1 begins by recalling the pertinent space-time symmetry groups and their representation in the classical mechanics description of N free equal mass particles. It is then shown how a natural Ansatz to introduce interactions in the relativistic context can only be made to work provided the "pair potentials" that enter are natural generalizations of the potentials characterizing the Calogero-Moser and Toda systems from Section 2. The structure of the time and space translation generators suggests candidates for Poisson commuting integrals, and these functions (given by (3.19), combined with (3.10), (3.11), and (3.15)–(3.17)) are indeed in involution. Thus, integrability follows as a bonus from Poincaré invariance.

In the relativistic case, soliton scattering for the type I, II, and VI systems can be shown without using a Lax matrix; moreover, no Lax matrix is needed to handle the connections in the diagram (2.62)—all of which are detailed in Section 3.1. However, in Section 3.2 we show that the commuting Hamiltonians can be tied in with Lax matrices. These matrices are given by (3.33)–(3.36) (type I–III), (3.41)–(3.44) (type VI), (3.45)–(3.47) (type IV), and (3.53)–(3.58) (type V). Cauchy's identity (3.32) and its elliptic generalization (3.49) are the key to the connection between the Poisson commuting functions (3.19) and the symmetric functions of these matrices.

In the nonrelativistic limit $\beta \to 0$ (with $\beta = 1/c$, c denoting the speed of light) the Lax matrices reduce to their counterparts from Section 2, as detailed in (3.38). (Up to diagonal similarity transformations in some cases.) The resulting relation (3.39) between the nonrelativistic and relativistic Hamiltonians yields nonrelativistic integrability as a corollary of the functional equations (3.20) encoding relativistic integrability.

Section 4 deals with the quantum versions of the nonrelativistic and relativistic systems. Section 4.1 has an introductory character. First, we present an algebraic notion of quantum integrability, which is tied to the systems at hand. Specifically, at the nonrelativistic level it amounts to requiring the existence of N commuting independent PDOs, including the defining dynamics. As a PDO the latter is unambiguously determined—in contrast to the AΔO quantization of the defining relativistic dynamics, which exhibits ordering ambiguities. In the relativistic case, therefore, we speak of an integrable quantization whenever the ordering in the classical Hamiltonians is such that the corresponding quantizations commute as AΔOs.

The remainder of Section 4.1 prepares the ground for a reinterpretation (by means of unitary joint eigenfunction transforms) of the commuting PDOs and AΔOs as commuting Hilbert space operators. This problem is particularly difficult for the AΔOs, and only partial solutions have been obtained to date. It is not widely appreciated what is involved here; in fact, even in the commuting PDO case there are no general results ensuring that a unitary joint eigenfunction transform exists. For AΔOs this existence problem is greatly aggravated by multiplier ambiguities, as explained in Section 4.1. We have also summarized some of the Hilbert space notions that are essential for a mathematically sound analysis—notions that are, unfortunately, still regarded as outlandish in theoretical physics.

No Hilbert space lore is needed to understand Sections 4.2 and 4.3, however. Here, we address the issue of quantum integrability at the nonrelativistic and relativistic levels, respectively; this issue has an algebraic rather than an analytic character. As it happens, it is actually more straightforward to establish quantum integrability in the relativistic than in the nonrelativistic case. Indeed, with the ordering choice exhibited by (4.35) (type I–IV) and (4.38) (type V and VI) quantum commutativity comes down to the functional equations (4.36) and (4.39), respectively. We supplement the direct proofs of nonrelativistic quantum integrability discussed in Section 4.2 with a novel proof at the end of Section 4.3. Here we derive nonrelativistic integrability indirectly, as a corollary of relativistic integrability.

Section 5 is concerned with action-angle transforms—canonical maps $(x, p) \mapsto (\hat{x}, \hat{p})$ that conjugate the Poisson commuting Hamiltonians $H_k(x, p)$ to Hamiltonians $H_k(\hat{p})$, $k = 1, \ldots, N$. Thus the new Hamiltonians depend only on the actions $\hat{p}_1, \ldots, \hat{p}_N$. Therefore, they give rise to evolutions of the angles $\hat{x}_1, \ldots, \hat{x}_N$ that are linear in the respective evolution parameters ("times"). As a result, the commuting dynamics are simultaneously "diagonalized": The action-angle map is the classical analog of the quantum joint eigenfunction transform.

Whenever the commuting Hamiltonians generate complete flows, the Liouville-Arnol'd theorem ensures the existence of action-angle maps on suitable invariant submanifolds of phase space. Unfortunately, this existence theorem is of quite limited practical use, but it does guarantee that one is not wasting time in searching for explicit diagonalizing canonical transformations. The theorem is already sketched at the end of Section 2.1, but for a good understanding of its subject matter it is important to study concrete examples.

We present various elementary examples in Section 5.1. In particular, this enables us to link up action-angle maps and the wave maps from scattering theory, in a very simple setting where phase diagrams can be used. In Section 5.2 we elaborate on this link, showing in particular how the wave map formalism for a large class of repulsive pair potentials entails that all of these potentials give rise to integrable systems. Therefore, it is crucial to single out the systems of type I, II, and VI (for which the wave maps exist on all of phase space) by the extra feature of soliton scattering. (The wave maps are rather well known in quantum mechanics (usually as Møller operators), but not so in classical mechanics. The reader might consult Ref. 10 for the wave map formalism in the latter context. More generally, in Ref. 13 wave maps form the starting point for a great variety of contexts in which scattering takes place, including classical mechanics.)

In Section 5.3 we detail the construction of an action-angle map for the relativistic type II system. This map is, roughly speaking, an interpolation of the incoming and outgoing wave maps. From the construction and its specialization to the nonrelativistic type II system and the type I systems, one readily deduces some highly remarkable duality properties. Specifically, the inverse of the action-angle map serves as an action-angle map for dual systems living on the action-angle phase space, and these dual systems are once more Calogero-Moser systems—as encoded in (5.24). Further spinoffs include a rather explicit picture of an extensive class of evolutions and, as a consequence, a complete elucidation of their long-time asymptotics; cf. (5.42)–(5.46).

Toward the end of Section 5.3 we also take a brief look at action-angle maps for type III systems. There is much more geometry involved here, since oscillatory motion and partial equilibria are present. On the other hand, the dual systems are once more characterized by a solitonic long-time asymptotics. In this case, however, each of the dual dynamics (of which (5.51) is the simplest representative) gives rise to a codimension-one subvariety containing states that do not have a free asymptotics (due to coinciding velocities, roughly speaking).

In Section 6 we consider eigenfunctions of the PDOs and AΔOs at hand, with a bias toward their suitability as kernels of diagonalizing unitaries. This is a subject where many of the key questions are still open. For instance, no eigenfunctions at all are known for the Toda AΔOs. Open questions abound for the type II and IV systems, too. Even in the type III case, where transforms in terms of multivariable orthogonal polynomials are known both for the commuting PDOs [14, 15] and for the commuting AΔOs [16, 17], the polynomials are not known in a sufficiently explicit way to establish the duality properties expected from the classical level (save for $N = 2$ [5]).

In our survey [5] we have already written on eigenfunctions vs. Hilbert space aspects, and our choice of topics for Section 6 supplements the discussion that can be found there. In particular, we do not reconsider the connections with harmonic analysis on homogeneous spaces associated with classical and quantum groups and algebras—a subject that has mushroomed considerably over the past few years. Instead, we mention a few recent references (among many) that are concerned with this theme from various partly complementary viewpoints, namely Refs. 18–22; further literature can be found there.

Section 6.1 has an introductory character. We specify the PDOs and AΔOs to be discussed and reappraise the problem of their Hilbert integrability—the question whether and when they can be defined as commuting Hilbert space operators via a unitary joint eigenfunction transform. As simple examples, we present two such transforms for the type I and II PDOs and AΔOs.

In Section 6.2 we sketch how multivariable orthogonal polynomials emerge as joint eigenfunctions for the type III PDOs and AΔOs. As we see it, the key idea dates back to Sutherland's paper [14]: All of the operators at issue take a triangular form w.r.t. a suitable partial order on a well-known orthonormal base for the symmetric subspace of $L^2(\mathbb{T}^N)$ ("free boson eigenstates"), with $\mathbb{T}^N$ the N-torus. In this case, the multiplier ambiguity for the AΔO eigenfunctions can be ignored, since any nontrivial multiplier would spoil the polynomial character of the latter. (But when one tries to solve the "band" problem, one needs nonpolynomial eigenfunctions interpolating the polynomials, and so the ambiguity reappears.)

We also use the type III eigenfunctions to illustrate how the nonrelativistic "anyon" particles turn into fermions at the relativistic level. Moreover, at the end of Section 6.2 we compare the quantum and classical type III transforms, reading off exactness of semiclassical quantization.

In Section 6.3 we sketch some of our (hitherto unpublished) results on eigenfunctions for the $N = 2$ relativistic type II and IV systems. As it turns out, the integral representation for the former which we have found admits a natural generalization to four coupling constants instead of one, and then yields eigenfunctions for the hyperbolic version of the trigonometric Askey-Wilson AΔO [23]. These novel eigenfunctions have a great many remarkable properties. In particular, they are not only self-dual (in a sense generalizing the self-duality of the relativistic type II system), but they are also simultaneous eigenfunctions for two commuting Askey-Wilson type AΔOs acting on the *same* side of the duality picture.

The Askey-Wilson polynomials can be obtained from these functions by analytic continuation and discretization of one of the two pertinent variables. The self-duality mentioned above already left its footprints for these polynomials, but the second commuting AΔO for the continuous variable has a trivial action on the polynomials. This is because the latter are periodic, with the period equal to the step size of the relevant AΔO.

The Askey-Wilson polynomials can be tied in with compact quantum groups; more generally, various q-special functions correspond to quantum (Hopf) algebras. (See for example Refs. 24–27 and references given there.) It is therefore natural to expect that functions of the type occurring in Section 6.3 are again related to algebraic objects. Candidates include noncompact quantum groups, whose representation theory is still in its infancy.

Section 6.3 is concluded with a description of type IV $N = 2$ eigenfunctions corresponding to special coupling constants. The relativistic eigenfunctions generalize the Lamé functions, represented in a form that can be found in Ref. 12. Among other things, we have not yet been able to show that these functions give rise to an orthonormal base for the pertinent Hilbert space, as we do expect. (This is one reason why Ref. 28, already promised in Ref. 5, has not been published yet.)

We were originally planning an additional section on relations with various well-known infinite-dimensional integrable systems. However, this had to be omitted, so as to keep these lecture notes (and the time spent in writing them) within bounds. Let us, therefore, finish this introduction by just mentioning some of these relations and a few references.

First, as regards the classical versions of the systems, these have already been compared to various soliton field theories and soliton lattices in Ref. 5; cf. also Refs. 3, 29, 30. These infinite-dimensional integrable systems include the Korteweg-de Vries, modified KdV, sine-Gordon, nonlinear Schrödinger, Boussinesq, Hirota-Satsuma and Landau-Lifshitz (XYZ) equations, and the infinite Toda lattice. What emerges from these results is an intimate relation between the N-soliton solutions and the N-particle relativistic Calogero-Moser systems for special parameter values. More specifically, provided the N-soliton scattering maps arising in the infinite systems are suitably parametrized, they coincide with the N-particle scattering maps of the pertinent Calogero-Moser system. Moreover, for some of the infinite systems (including the KdV, modified KdV, and sine-Gordon equations), the N-soliton solutions themselves can be obtained via suitable N-particle dynamics, which gives rise to a natural notion of soliton space-time trajectories.

There is meanwhile considerable evidence that the soliton-particle correspondence turns into physical equivalence at the quantum level (i.e., the same scattering and bound-state structure occurs for the quantum-mechanical particles as for the field- and lattice-theoretic solitons). In particular, it can be shown that the $N = 2$ transforms of type II and IV from Section 6.3 have the expected properties on the sine-Gordon and XYZ lines, respectively (cf. also Refs. 5, 31). However, in the absence of explicit N-particle relativistic type II and IV eigenfunction transforms with all of the required properties (such as unitarity and factorized scattering), the equivalence remains a conjecture whose plausibility can be disputed.

Finally, we would like to mention a novel theme of more recent vintage than those surveyed in Chapter 4 of Ref. 5. This concerns eigenfunctions of quantum Calogero-Moser models vs. solutions to equations of Knizhnik-Zamolodchikov type. This relation is currently under active study; most of the relevant literature can be traced from the recent references [19, 20, 22, 32].

2   Classical Nonrelativistic Calogero-Moser and Toda Systems

2.1   Background: Classical Mechanics/Symplectic Geometry

As mentioned previously, the N-particle systems at issue describe one-dimensional particles. The simplest mathematical description of this physical situation is to let each particle position vary freely over $\mathbb{R}$. An initial state of the system is then encoded in a position vector $x = (x_1, \ldots, x_N) \in \mathbb{R}^N$ and a momentum vector $p = (p_1, \ldots, p_N) \in \mathbb{R}^N$, whereas the time evolution is given by Hamilton's equations

$$\dot{x}_j = \frac{\partial H}{\partial p_j}, \quad \dot{p}_j = -\frac{\partial H}{\partial x_j}, \quad j = 1, \ldots, N, \qquad (2.1)$$

with $H(x, p)$ the Hamiltonian (energy function) of the system.

As a first example of this setting, consider a Hamiltonian of the form

$$H = \sum_{j=1}^{N} \frac{p_j^2}{2m_j} + U(x)$$

where $U(x)$ (the potential energy) is a smooth real-valued function on $\mathbb{R}^N$. When $U(x)$ vanishes, one is left with the sum of the kinetic energies of N particles with mass m_j, and the solution to (2.1) is obviously given by $x_j(t) = x_{0,j} + tp_{0,j}/m_j$, $p_j(t) = p_{0,j}$, $j = 1, \ldots, N$. Thus, particle j moves with uniform velocity $p_{0,j}/m_j$ along the line, without "seeing" the remaining particles. More generally, when $U(x)$ is of the form

$$U(x) = \sum_{j=1}^{N} V_j(x_j)$$

the particles move independently of each other in external fields. Then the ODE system (2.1) decouples and one is left with solving Newton's ODE $F(y) = m\ddot{y}$, where $F(y) = -V'(y)$ is the force field. Since $m\dot{y}^2/2 + V(y)$ is time independent, qualitative features of the motion can be read off from a plot of the contour lines $p^2/2m + V(x) = E$ in the (x,p)-plane (phase diagram).

Of most interest for the present lectures is the special case where $U(x)$ is a sum of pair potentials,

$$U(x) = \sum_{\substack{j,k=1 \\ j<k}}^{N} V_{jk}(x_j - x_k).$$

This may be viewed as a one-dimensional analog of the three-dimensional gravitational N-body problem. In the latter situation the pair potentials $V_{jk}(y)$, $y \in \mathbb{R}^3$, are proportional to $1/|y|$, and so the potential energy diverges when collisions occur. For the Calogero-Moser systems, too, the pair potential is singular at the origin. To avoid such singularities at least for initial states, one should restrict the range of variation of the system position vector x. Thus, one chooses initial positions in an open subset G of $\mathbb{R}^N$—the system's *configuration space*. Then the space of initial states—the system's *phase space*—is given by the set

$$\Omega = \{(x,p) \in \mathbb{R}^{2N} \mid x \in G\}. \tag{2.2}$$

As just sketched, the choice of phase space Ω (the kinematics) depends on the system Hamiltonian H (the dynamics). Assuming from now on that H is a smooth function on Ω, it follows from standard ODE lore that the system (2.1) with initial value $u_0 \in \Omega$ has a unique solution $u(t) \in \Omega$ for some t-interval $(-T_-, T_+)$ around 0. But even in the simplest case where $U(x)$ is smooth on $\mathbb{R}^N$ and correspondingly one can take $\Omega = \mathbb{R}^{2N}$, this local solution need not be extendible to a global solution on $(-\infty, \infty)$: one or more of the particles may escape to infinity in finite time. In the "next simplest" case where a collision set is discarded, it can also happen that collisions do occur after a finite time.

The first question to answer, therefore, is whether for a given $u_0 \in \Omega$ a global solution $u(t) \in \Omega$, $t \in \mathbb{R}$, exists. If this is the case, one can ask questions about the long-time characteristics of the trajectory, such as whether it stays in bounded subsets of Ω (corresponding to an N-body bound state), or whether it exhibits a free motion

$$u(t) \sim (x^{\pm} + tv^{\pm}, p^{\pm}), \quad t \to \pm\infty,$$

for asymptotic times (corresponding to a scattering state).

Of course, such questions can be more easily dealt with when one is able to solve Hamilton's equations (2.1) in a sufficiently explicit way. But this appears impossibly difficult for most Hamiltonians of physical interest, and certainly so for the N-body gravitational Hamiltonian. Accordingly, the latter context gives rise to simple qualitative questions that are wide open even three centuries after Newton.

The Calogero-Moser and Toda Hamiltonians are notable exceptions to this rule. An important property of these Hamiltonians is that they are integrable—a notion we shall discuss shortly. First, however, we would like to recall some geometric formalism that makes it possible to handle (finite-dimensional) classical mechanics in a mathematically precise and concise way.

As a quite general-state space on which Hamiltonian mechanics can be defined, one can take a $2N$-dimensional differentiable manifold Ω equipped with a nondegenerate closed 2-form ω—a *symplectic manifold* $\langle \Omega, \omega \rangle$. The cotangent bundle of an N-dimensional differentiable manifold G can be equipped with such a form in a natural way, and we shall mostly specialize to this setting. In particular, viewing the phase space (2.2) as the cotangent bundle to the open set $G \subset \mathbb{R}^N$, this *symplectic form* reads

$$\omega = \sum_{j=1}^{N} dx_j \wedge dp_j. \tag{2.3}$$

More generally, fixing a point u_0 in a symplectic manifold $\langle \Omega, \omega \rangle$, there exist coordinates $(x(u), p(u)) \in \mathbb{R}^{2N}$ for u in a neighborhood $\mathcal{U}$ of u_0 such that ω takes the form (2.3) on $\mathcal{U}$ (Darboux's theorem); such coordinates are referred to as *canonical* (or *symplectic*) coordinates.

Since ω is nondegenerate, a 1-form α on Ω gives rise to a vector field $X^{(\alpha)}$ on Ω, uniquely determined by requiring

$$\omega(X^{(\alpha)}, X) = \alpha(X)$$

for arbitrary vector fields X. To obtain a dynamics on the symplectic manifold Ω one can now start from any real-valued smooth function H on Ω and introduce the associated *Hamiltonian vector field* $X_H \equiv X^{(dH)}$; then the time evolution is governed by

$$\dot{u} = X_H(u), \quad u \in \Omega \tag{2.4}$$

and the corresponding flow $\Omega \to \Omega$, $u_0 \mapsto u(t)$ is written $\exp(tX_H)$, or briefly e^{tH}. As before, this flow is a priori only locally defined, with the time interval depending on u_0. When all trajectories stay in Ω for all times, the flow is called *complete*. Showing completeness for Hamiltonians of physical interest may be quite difficult. In purely mathematical work, however, one often assumes that Ω is compact. Then the completeness problem evaporates, since in that case any Hamiltonian flow is complete.

In the special (noncompact) setting (2.2), (2.3), we have $H = H(x, p)$ and

$$dH = \sum_{j=1}^{N}\left(\frac{\partial H}{\partial x_j}dx_j + \frac{\partial H}{\partial p_j}dp_j\right), \tag{2.5}$$

$$X_H = \sum_{j=1}^{N}\left(\frac{\partial H}{\partial p_j}\frac{\partial}{\partial x_j} - \frac{\partial H}{\partial x_j}\frac{\partial}{\partial p_j}\right). \tag{2.6}$$

Thus, (2.4) can be rewritten as

$$(\dot{x}, \dot{p})^t = S\nabla H, \quad S \equiv \begin{pmatrix} 0 & 1_N \\ -1_N & 0 \end{pmatrix},$$

which amounts to (2.1).

Returning to the general case, we introduce the *Poisson bracket*

$$\{\cdot, \cdot\}: C^\infty(\Omega) \times C^\infty(\Omega) \to C^\infty(\Omega), \quad (F, G) \mapsto \{F, G\} \equiv \omega(X_F, X_G).$$

(In Sections 2, 3, and 5, we use the symbol $C^\infty(M)$ to denote the space of *real-valued* smooth functions on M.) Recalling the above definition of Hamiltonian vector fields, this can also be written

$$\{F, G\} = dF(X_G) = -dG(X_F) = -X_F(G) = X_G(F).$$

It is easily checked that the Poisson bracket equips $C^\infty(\Omega)$ with a Lie algebra structure: $\{\cdot, \cdot\}$ is bilinear, antisymmetric, and satisfies the Jacobi identity. The relation to the Lie bracket $[\cdot, \cdot]$ (anticommutator) on the Hamiltonian vector fields is given by

$$[X_F, X_G] = X_{-\{F, G\}}.$$

In the special case (2.2), (2.3) one can write, more concretely,

$$\{F, G\} = \sum_{j=1}^{N}(\partial_{x_j}F\partial_{p_j}G - \partial_{p_j}F\partial_{x_j}G) = \nabla F \cdot S\nabla G$$

whence one obtains the *canonical commutation relations*

$$\{x_i, x_j\} = \{p_i, p_j\} = 0, \quad \{x_i, p_j\} = \delta_{ij}, \quad i, j = 1, \ldots, N.$$

Fixing a Hamiltonian $H \in C^\infty(\Omega)$, one can now characterize the functions $\mathcal{I}_H \subset C^\infty(\Omega)$ that are conserved under the H flow e^{tH}—the so-called *integrals*: One has

$$I \in \mathcal{I}_H \iff \{I, H\} = 0. \qquad (2.7)$$

Indeed, this follows from

$$\frac{dI}{dt} = X_H(I) = \{I, H\}$$

where the argument $\exp(tX_H)(u_0)$ is suppressed. From antisymmetry it is immediate that $H \in \mathcal{I}_H$. More generally, assuming $I_1, \ldots, I_k \in \mathcal{I}_H$, any $f \in C^\infty(\mathbb{R}^k)$ gives rise to a function $f(I_1, \ldots, I_k) \in \mathcal{I}_H$, as is easily verified. Thus, $\mathcal{I}_H$ is an (infinite-dimensional) algebra. Using the Jacobi identity one infers that $\mathcal{I}_H$ is also a Lie algebra w.r.t. the Poisson bracket.

We proceed by discussing symplectic maps. First, let us assume that Ω is an open subset of $\mathbb{R}^{2N}$ with coordinates (x, p) and symplectic form (2.3). A diffeomorphism

$$\Phi \colon \Omega \to \widehat{\Omega}, \quad (x, p) \mapsto (\hat{x}, \hat{p}), \qquad (2.8)$$

onto an open subset $\widehat{\Omega}$ of $\mathbb{R}^{2N}$ is then called a *canonical transformation* when the functions $\hat{x}_1(x, p), \ldots, \hat{p}_N(x, p) \in C^\infty(\Omega)$ satisfy the canonical commutation relations

$$\{\hat{x}_i, \hat{x}_j\} = \{\hat{p}_i, \hat{p}_j\} = 0, \ \{\hat{x}_i, \hat{p}_j\} = \delta_{ij}, \quad i, j = 1, \ldots, N.$$

Equivalently, for any $(x, p) \in \Omega$ the Jacobian matrix $(D\Phi)(x, p)$ belongs to the symplectic group $\mathrm{Sp}(2N, \mathbb{R})$, i.e.,

$$(D\Phi)\mathcal{S}(D\Phi)^t = \mathcal{S}. \qquad (2.9)$$

A third equivalent definition can be used to introduce a coordinate-free generalization. To state this definition, we equip $\widehat{\Omega}$ with the symplectic form

$$\widehat{\omega} \equiv \sum_{j=1}^{N} d\hat{x}_j \wedge d\hat{p}_j.$$

Then the above map (2.8) is a canonical transformation iff

$$\Phi^* \widehat{\omega} = \omega. \qquad (2.10)$$

A *symplectic map* (or *symplectomorphism*) between two symplectic manifolds $\langle \Omega, \omega \rangle$ and $\langle \widehat{\Omega}, \widehat{\omega} \rangle$ is now a diffeomorphism Φ from Ω onto $\widehat{\Omega}$ such that

(2.10) holds true. Equivalently, one can require that Φ preserve Poisson brackets. That is, letting

$$F \equiv \Phi^* \widehat{F}, \quad \widehat{F} \in C^\infty(\widehat{\Omega}),$$

one should have

$$\{F, G\} = \{\widehat{F}, \widehat{G}\}, \quad \forall \widehat{F}, \widehat{G} \in C^\infty(\widehat{\Omega}).$$

A complete Hamiltonian flow is readily shown to yield a 1-parameter group of symplectic maps $\exp(tH) \colon \langle \Omega, \omega \rangle \to \langle \Omega, \omega \rangle$.

For canonical transformations one often interprets the functions $\hat{x}(x, p)$ and $\hat{p}(x, p)$ as new coordinates on Ω, which have the special property that the symplectic form ω given by (2.3) can be written as $\sum_j d\hat{x}_j \wedge d\hat{p}_j$; Thus, these coordinates are *canonical*, just as the coordinates x, p; cf. the definition below (2.3). The key property of canonical transformations is that they leave Hamilton's equations invariant. That is, setting

$$\widehat{H}(\hat{u}) \equiv H\big(\mathcal{E}(\hat{u})\big), \quad \mathcal{E} \equiv \Phi^{-1}, \hat{u} \equiv (\hat{x}, \hat{p}),$$

and assuming $\hat{u}(t)$ solves the Hamilton equations

$$\frac{d\hat{u}}{dt} = \mathcal{S}\nabla_{\hat{u}}\widehat{H}(\hat{u})$$

in the new coordinates, one gets a solution $u(t) \equiv \mathcal{E}\big(\hat{u}(t)\big)$ to the Hamilton equations

$$\frac{du}{dt} = \mathcal{S}\nabla_u H(u), \quad u = (x, p),$$

in the original coordinates. (Indeed, this follows from the chain rule and the canonicity property (2.9).) Reformulated in the coordinate-free setting of symplectic maps, this state of affairs amounts to a commutative diagram

$$
\begin{array}{ccc}
u \in \Omega & \xleftarrow{\;\mathcal{E}\;} & \hat{u} \in \widehat{\Omega} \\
{\scriptstyle \exp(tH)}\big\uparrow & & \big\uparrow{\scriptstyle \exp(t\widehat{H})} \\
u_0 \in \Omega & \xrightarrow[\;\Phi\;]{} & \hat{u}_0 \in \widehat{\Omega}
\end{array}
\tag{2.11}
$$

i.e., to the equality

$$e^{tH} = \mathcal{E} \circ e^{t\widehat{H}} \circ \Phi, \quad \mathcal{E} \equiv \Phi^{-1}, \widehat{H} \equiv \mathcal{E}^*(H). \tag{2.12}$$

We are now prepared to discuss the notion of integrability. Fixing $H \in C^\infty(\Omega)$, one calls H an *integrable* Hamiltonian when there exist integrals $I_1 = H, I_2, \ldots, I_N$ that are independent and in involution. That is, their gradients $dI_1, \ldots, dI_N$ are linearly independent on an open dense subset of Ω and they mutually (Poisson) commute.

Let us first illustrate this definition with a trivial, yet instructive example. Consider the Hamiltonian $H(x, p) \equiv p_1$ on the phase space (2.2). Then any $I \in C^\infty(\Omega)$ that does not depend on x_1 belongs to $\mathcal{I}_H$. In particular, H is integrable in the above sense, since one can choose, for instance, I_k equal to x_k or p_k for $k = 2, \ldots, N$. Each of these 2^{N-1} choices leads to distinct maximal Abelian subalgebras of $\mathcal{I}_H$ (which is itself non-Abelian), consisting of functions $f(I_1, \ldots, I_N)$, $f \in C^\infty(\mathbb{R}^N)$. (As is readily seen, a symplectic form cannot vanish on a k-dimensional subspace for $k > N$; this is why the subalgebras cannot be enlarged without violating commutativity.)

The definition of integrability just exemplified can be widely found in the physics literature. It is however not strong enough to guarantee the applicability of the Liouville-Arnol'd theorem. Moreover, it does not single out Calogero-Moser Hamiltonians among N-particle Hamiltonians of the form

$$H = \frac{1}{2} \sum_{j=1}^N p_j^2 + g^2 \sum_{\substack{j,k=1 \\ 1j<k}}^N V(x_j - x_k), \quad g > 0, \qquad (2.13)$$

with a repulsive pair potential $V(x)$, since any such Hamiltonian is integrable. In this subsection we do not elaborate on the latter assertion (this is deferred to Section 5.2), but we do want to discuss the Liouville-Arnol'd theorem. Accordingly, we should first sharpen the above definition of integrability.

We shall henceforth call a Hamiltonian *Liouville integrable* iff (i) it is integrable in the above sense; (ii) the flows generated by $I_1, \ldots, I_N$ are complete. To appreciate the additional restriction, it is important to be aware of the fact that any Hamiltonian H on a symplectic manifold $\langle \Omega, \omega \rangle$ is integrable in a neighborhood $\mathcal{U}$ of any point $u_0 \in \Omega$ for which $dH(u_0) \neq 0$. Indeed, it can be proved that there exist canonical coordinates $x(u)$, $p(u)$ on $\mathcal{U}$ such that $H(u) = p_1(u)$ on $\mathcal{U}$; therefore, the example just discussed applies. Of course, the crux is that typically $\mathcal{U}$ will not be complete under the flows $\exp(t_j I_j)$.

Let us now assume that a given $H \in C^\infty(\Omega)$ is Liouville integrable. Fixing a point $u_0 \in \Omega$ at which the gradients $dI_1, \ldots, dI_N$ are linearly independent, consider

$$M(u_0) \equiv \{u(t) \mid t \in \mathbb{R}^N\}, \quad u(t) \equiv \exp(t_1 I_1) \cdots \exp(t_N I_N)(u_0). \qquad (2.14)$$

This is a well-defined subset of Ω, since the flows are complete on Ω. Moreover, linear independence of the gradients entails that $M(u_0)$ is an N-dimensional submanifold. Since the flows commute, the set $\{t \in \mathbb{R}^N \mid u(t) = u_0\}$ is a discrete subgroup of $\mathbb{R}^N$. From this one infers that $M(u_0)$

is diffeomorphic to $\mathbb{T}^k \times \mathbb{R}^{N-k}$ for some $k \in \{0, 1, \ldots, N\}$. (In particular, when $M(u_0)$ is compact, one must have $k = N$.)

These somewhat sketchy remarks are the starting point for the Liouville-Arnol'd theorem. This theorem asserts that on suitable disjoint, open and connected submanifolds Ω_i, $i = 1, 2, \ldots$ there exist canonical coordinates $x(u)$, $p(u)$ such that the commuting Hamiltonians $I_1 = H$, I_2, $\ldots$, I_N depend solely on p_1, $\ldots$, p_N (the so-called *action variables*), whereas the *angle variables* x_1, $\ldots$, x_N vary over $\mathbb{R}$ or the torus $\mathbb{T}^1$. Thus the flow $\exp(t_j I_j)$, $j \in \{1, \ldots, N\}$, amounts to a translation of the angle variables that is linear in the evolution parameter t_j.

Of course, our description of this theorem is incomplete as long as we do not define the qualifier "suitable." We shall have more to say about this in Section 5. For the time being we mention that "suitable" includes first of all the assumption that the gradients dI_1, $\ldots$, dI_N are linearly independent on Ω_i and that the commuting flows are complete on Ω_i. This assumption already entails that Ω_i is a union of N-dimensional submanifolds of the form (2.14). As a second assumption, these submanifolds of Ω_i should all be diffeomorphic.

For our purposes it is convenient to reinterpret the existence of action-angle coordinates on the submanifold Ω_i as the existence of a symplectic map

$$\Phi_i \colon \langle \Omega_i, \omega \rangle \to \langle \widehat{\Omega}_i, \widehat{\omega}_i \rangle, \quad u \mapsto (\hat{x}, \hat{p})$$

onto a new manifold of the form

$$\widehat{\Omega}_i = M_i \times A_i \tag{2.15}$$

with

$$M_i \equiv \mathbb{T}^{k_i} \times \mathbb{R}^{N-k_i}, \quad k_i \in \{0, 1, \ldots, N\}, \tag{2.16}$$

and A_i an open connected subset of $\mathbb{R}^N$; the symplectic form on $\widehat{\Omega}_i$ reads

$$\widehat{\omega}_i = \sum_{j=1}^{N} d\hat{x}_j \wedge d\hat{p}_j. \tag{2.17}$$

Here, $\hat{p}$ varies over A_i, $\hat{x}_1$, $\ldots$, $\hat{x}_{k_i}$ over $(-\pi, \pi]$ (so $\mathbb{T}^1$ is viewed as $\mathbb{R}/2\pi\mathbb{Z}$) and $\hat{x}_{k_i+1}$, $\ldots$, x_N over $\mathbb{R}$. This reinterpretation is notationally and conceptually useful, since our starting point differs considerably from the abstract, coordinate-free setting of the theorem. Indeed, our Hamiltonian H is typically given as a concrete function of canonical coordinates x, p, whose range of variation serves to *define* Ω. Whenever one can show that $H(x, p)$ is Liouville integrable, one should try and concretize the submanifolds Ω_i and action-angle maps Φ_i, and obtain in particular explicit functions of $\hat{p}$ for the transformed integrals. In Section 5 we shall illustrate this general program with several concrete examples.

2.2   Calogero-Moser Systems

The Calogero-Moser systems are dynamical systems defined by N-particle Hamiltonians of the form (2.13) with a special choice of pair potential $V(x)$. One can distinguish four different types, denoted by I—IV:

I.
$$\frac{1}{x^2}$$
(rational)   (2.18)

II.
$$\frac{\mu^2}{4}\,\mathrm{sh}^{-2}\left(\frac{\mu x}{2}\right),\quad \mu > 0$$
(hyperbolic)   (2.19)

III.
$$\frac{\mu^2}{4}\,\sin^{-2}\left(\frac{\mu x}{2}\right),\quad \mu > 0$$
(trigonometric)   (2.20)

IV.
$$\mathcal{P}(x;\omega,\omega'),\quad \omega,-i\omega' > 0$$
(elliptic)   (2.21)

Here, $\mathcal{P}$ is the Weierstrass function, a doubly periodic meromorphic function with primitive periods 2ω, $2\omega'$ and double poles at the period lattice. The type I–III systems can be viewed as limiting cases of the type IV system, since one has

$$\mathcal{P}(x;\omega,\omega') = \begin{cases} 1/x^2, & \omega = \infty, \omega' = i\infty, \\ \nu^2/3 + \nu^2/\,\mathrm{sh}^2\,\nu x, & \omega = \infty, \omega' = i\pi/2\nu, \\ -\nu^2/3 + \nu^2/\sin^2\,\nu x, & \omega = \pi/2\nu, \omega' = i\infty. \end{cases} \quad (2.22)$$

Consider first the type I and II cases. Discarding the collision sets $x_i = x_j$ from $\mathbb{R}^N$, one obtains an open set $\mathbb{R}^N_{\neq}$ with $N!$ connected components, corresponding to the various particle orderings. On $\mathbb{R}^N_{\neq}$ the potential energy is a smooth function. Fixing an initial point $(x_0, p_0) \in \mathbb{R}^N_{\neq} \times \mathbb{R}^N$, the energy $E_0 = H(x_0, p_0)$ is conserved along the corresponding orbit $\big(x(t), p(t)\big)$. Since the potential is positive, this leads to an upper bound on $|p(t)|$ and a nonzero lower bound on the particle distances. Therefore, no collisions can occur, and escape to infinity in finite time is excluded, since $|\dot{x}_j(t)| = |p_j(t)| \leq C$ along the orbit. Hence the flow is complete, and we may as well restrict attention to the phase space

$$\Omega \equiv G \times \mathbb{R}^N, \quad G \equiv \{x \in \mathbb{R}^N \mid x_N < \cdots < x_1\}, \quad \text{(I, II)} \quad (2.23)$$

with its canonical form (2.3).

Next, consider the system of type III, often called the Sutherland system. In this case one can distinguish three different versions of the Hamiltonian. First, one can again avoid the singularities of the potential by deleting the sets $x_i \equiv x_j \pmod{2\pi/\mu}$ from $\mathbb{R}^N$. This yields an infinite number of connected components, each giving rise to phase spaces on which the Hamiltonian flow is complete. (This follows in the same way as before from energy conservation.) We restrict attention to one of these,

$$\widetilde{\Omega} \equiv G \times \mathbb{R}^N, \quad G \equiv \left\{ x \in \mathbb{R}^N \;\middle|\; \begin{array}{l} x_N < \cdots < x_1, \\ x_1 - x_N < 2\pi/\mu \end{array} \right\}. \quad \text{(III)} \quad (2.24)$$

This choice amounts to a fixed ordering and minimal distances between the particles. We equip $\widetilde{\Omega}$ with the form (2.3), and so obtain once more a symplectic manifold of the cotangent bundle type.

The choice of phase space just delineated is mathematically acceptable, but appears unnatural from a physical standpoint. One would rather like to view the x_j's as encoding angular positions, so that the type III Hamiltonian describes N particles on a ring that interact via an $1/d^2$ potential, where d is the distance between particles as measured in the plane of the ring.

This interpretation can be mathematically implemented as follows. Consider the map

$$\Gamma \colon (x_1, \ldots, x_N, p_1, \ldots, p_N)$$
$$\mapsto \left(x_N + \frac{2\pi}{\mu}, x_1, \ldots, x_{N-1}, p_N, p_1, \ldots, p_{N-1} \right). \quad (2.25)$$

Restricted to $\widetilde{\Omega}$, this map is bijective, has no fixed points, and is symplectic. As such, it generates a symplectic $\mathbb{Z}$-action on $\widetilde{\Omega}$, and we may divide out this action to obtain a new symplectic manifold

$$\Omega \equiv \widetilde{\Omega}/\mathbb{Z} \quad (2.26)$$

equipped with the quotient form, again denoted by ω.

A simple choice of coordinates for Ω reads

$$\Omega \simeq \{ (x,p) \in \mathbb{R}^{2N} \mid x \in F_N \} \quad (2.27)$$

where F_N is defined by

$$F_N \equiv \left\{ x \in \mathbb{R}^N \;\middle|\; -\frac{\pi}{\mu} < x_N < \cdots < x_1 \le \frac{\pi}{\mu} \right\}. \quad (2.28)$$

This choice is in accordance with the above-mentioned physical picture of particles occupying distinct positions on a ring, with the particles viewed as indistinguishable. Indeed, an initial state of this physical system can be uniquely encoded in N phase factors $\exp(ix_j/\mu)$, $j = 1, \ldots, N$, $x \in F_N$, regarded as positions on the ring $S^1 \subset \mathbb{C} \simeq \mathbb{R}^2$, and in associated momenta $p_1, \ldots, p_N \in \mathbb{R}$.

From a mathematical point of view the choice (2.26) and coordinatization (2.27) are also natural. Indeed, the type III Hamiltonian on $\widetilde{\Omega}$ is invariant under the $\mathbb{Z}$-action generated by Γ (i.e., it takes the same values on orbits of Γ). Thus, its (complete) flow on $\widetilde{\Omega}$ descends to a well-defined and complete flow on Ω. Moreover, $F_N \times \mathbb{R}^N$ is a fundamental set for the $\mathbb{Z}$-action on $\widetilde{\Omega}$. That is, for a given $(\tilde{x}, \tilde{p}) \in \widetilde{\Omega}$, there exist uniquely determined $x \in F_N$, $l \in \{1, \ldots, N\}$ and $m \in \mathbb{Z}$ such that

$$x_1 = \tilde{x}_l + \frac{2\pi m}{\mu}$$

$$\vdots$$

$$x_{N-l+1} = \tilde{x}_N + \frac{2\pi m}{\mu}$$

$$x_{N-l+2} = \tilde{x}_1 + \frac{2\pi(m-1)}{\mu}$$

$$\vdots$$

$$x_N = \tilde{x}_{l-1} + \frac{2\pi(m-1)}{\mu}$$

and then p is given by

$$p_1 = \tilde{p}_l$$

$$\vdots$$

$$p_{N-l+1} = \tilde{p}_N$$
$$p_{N-l+2} = \tilde{p}_1$$

$$\vdots$$

$$p_N = \tilde{p}_{l-1}.$$

At this point it should be emphasized that the coordinatization (2.27), (2.28) is a set-theoretic one: each Γ-orbit in $\widetilde{\Omega}$ is labeled by a unique $(x,p) \in F_N \times \mathbb{R}^N$. There are no globally defined smooth coordinates on Ω, just as no global smooth coordinate function exists on $\mathbb{T}^1$. Of course, on the open dense coordinate patch $\{x_1 < \pi/\mu\}$ the coordinates *are* smooth, and on this patch the quotient form is given by (2.3). However, one cannot simply discard the subset $\{x_1 = \pi/\mu\}$, since the H flow is not complete on the submanifold $\{x_1 < \pi/\mu\}$.

Alternatively, one can view the particles on the ring as distinguishable. This can be encoded mathematically by employing the phase space

$$\Omega' \equiv \widetilde{\Omega}/\mathbb{Z}' \tag{2.29}$$

where $\mathbb{Z}'$ denotes the $\mathbb{Z}$-action generated by

$$\Gamma^N : (x,p) \mapsto \left(x_1 + \frac{2\pi}{\mu}, \ldots, x_N + \frac{2\pi}{\mu}, p\right).$$

One way to coordinatize Ω' is to take

$$\Omega' \simeq \{(x,p) \in \mathbb{R}^{2N} \mid x \in F'_N\}$$

where

$$F'_N \equiv \left\{ x \in G \ \middle| \ \sum_{j=1}^N x_j \in \left(-\frac{\pi N}{\mu}, \frac{\pi N}{\mu}\right] \right\}$$

Indeed, for a given $(\tilde{x}, \tilde{p}) \in \tilde{\Omega}$ one can take

$$x_j = \tilde{x}_j + \frac{2\pi m}{\mu}, \quad p_j = \tilde{p}_j, \quad j = 1, \ldots, N, \tag{2.30}$$

where $m \in \mathbb{Z}$ is uniquely determined.

These three interpretations of the type III Hamiltonian illustrate that the canonical coordinates x and p in Hamilton's equations (2.1) may have several meanings, depending on what one intends to model physically. The same ambiguity occurs for the type IV systems, since the graph of $\mathcal{P}(x)$ on $(0, 2\omega)$ has the same features as that of $1/\sin^2(\pi x/2\omega)$. Thus, replacing μ by π/ω in the above, one obtains three mathematically and physically distinct versions of the elliptic system.

For all of the above Hamiltonians $H(x, p)$, the function

$$P \equiv \sum_{j=1}^{N} p_j$$

belongs to $\mathcal{I}_H$; cf. (2.7). This expresses conservation of total momentum under the H flow. (Reciprocally, the conservation of H under the P flow

$$e^{yP}(x, p) = (x + (y, \ldots, y), p)$$

amounts to translational invariance of the potential energy.) Thus the Hamiltonian H is integrable for $N = 2$, but this clearly holds true for *any* Hamiltonian of the form

$$H(x_1, x_2, p_1, p_2) = \frac{p_1^2}{2} + \frac{p_2^2}{2} + U(x_1 - x_2).$$

The Calogero-Moser Hamiltonians, however, *remain* integrable for $N > 2$. The existence of N independent integrals can be most easily seen via a so-called *Lax pair* formulation of the Hamilton equations. This concept is also crucial in the context of infinite-dimensional integrable systems, such as the sine-Gordon field theory. We continue by describing it in a rather general form that can be applied to the latter context, too.

Specifically, suppose that $L(t)$, $M(t)$, $t \in \mathbb{R}$, are two families of (linear) operators on a (separable, complex) Hilbert space $\mathcal{H}$ such that

$$\dot{L} = [M, L], \quad \forall t \in \mathbb{R}. \tag{2.31}$$

(When $\dim \mathcal{H} = \infty$ and $L(t)$, $M(t)$ are unbounded operators, one needs additional assumptions. Such technicalities will be ignored here.) Then the spectrum of $L(t)$ is constant in t. This assertion can be quickly proved by using linear ODE methods, as follows.

We begin by noting that the ODE $\dot{U} = MU$ has a unique solution $U(t)$ such that $U(0) = 1$. Specifically, iteration of the corresponding integral equation

$$U(t) = 1 + \int_0^t ds\, M(s)U(s)$$

yields the solution

$$U(t) = 1 + \sum_{n=1}^{\infty} \int_0^t ds_1 \cdots \int_0^{s_{n-1}} ds_n\, M(s_1) \cdots M(s_n).$$

(Depending on subcultures, this formula is referred to variously as the Volterra expansion/variation of constants formula/Dyson expansion/time-ordered exponential/product integral.) Defining now the operator family

$$K(t) \equiv U(t)L(0)U(t)^{-1} \tag{2.32}$$

we see that $K(t)$ is similar to $L(0)$, and so its spectrum equals that of $L(0)$. To complete the proof, it therefore remains to show that $K(t)$ equals $L(t)$. Since this is true for $t = 0$, we need only prove $\dot{K} = [M, K]$. (Indeed, this linear ODE again has a solution that is uniquely determined by its initial value.) But this readily follows from (2.32) by using $0 = (UU^{-1})\dot{} = U(U^{-1})\dot{} + M$.

For a finite-dimensional Hilbert space $\mathcal{H} = \mathbb{C}^l$ a second proof applies; this proof will also lead to the conserved Hamiltonians we are looking for. Let us define the *power traces*

$$H_k = \frac{1}{k}\operatorname{Tr} L^k, \quad k = 1, 2, \ldots. \tag{2.33}$$

Using cyclicity of the trace, we then get

$$\dot{H}_k = \operatorname{Tr}(\dot{L}L^{k-1}) = \operatorname{Tr}(ML^k - LML^{k-1}) = 0$$

so that the traces are t-independent. Now write

$$\det(1_l + \lambda L) = \sum_{k=0}^{\infty} \lambda^k S_k.$$

Thus, S_k equals the sum of all kth order principal minors of L for $k \le l$ and vanishes for $k > l$. These *symmetric functions* of L are polynomials in the power traces H_k (and vice versa), so we may conclude $\dot{S}_k = 0$. Consequently, $L(t)$ has t-independent spectrum, as claimed.

We mention in passing that the polynomial relations between the S_k and the H_k boil down to the well-known Newton identities that connect the coefficients of a polynomial with the sums of its root powers. The relations can be explicitly determined from the formal power series identity

$$1 + \lambda S_1 + \lambda^2 S_2 + \lambda^3 S_3 + \cdots = \exp(\lambda H_1 - \lambda^2 H_2 + \lambda^3 H_3 - \cdots). \tag{2.34}$$

(To see that this holds for an arbitrary $l \times l$ matrix L, first take $L = \mathrm{diag}(\lambda_1, \ldots, \lambda_l)$; then (2.34) is clear from the Taylor series $\ln(1 + x) = -\sum(-x)^j$, $|x| < 1$. But then (2.34) follows for diagonalizable L, since traces and determinants are invariant under similarity. Now diagonalizable L are dense in $M_l(\mathbb{C})$, so the polynomial identities obtained by equating powers of λ indeed hold true for arbitrary $l \times l$ matrices.)

To apply this isospectrality result to a Hamiltonian H on a symplectic manifold $\langle \Omega, \omega \rangle$, it suffices to construct $M_l(\mathbb{C})$-valued functions L and M on Ω such that the pair $K(t) \equiv K(u(t))$, $K = L, M$, satisfies (2.31). (Of course, $u(t)$ denotes the H flow; the abuse of notation occurring here is standard.) Indeed, whenever this can be done, the power traces of L (or, equivalently, its symmetric functions) will generate a subalgebra of $\mathcal{I}_H$ (cf. the paragraph containing (2.7)). At this point it should be emphasized that there are no general methods for obtaining Lax pairs: They have to be pulled out of a hat. Moreover, when they exist, they are highly nonunique.

For the above Calogero-Moser Hamiltonians Lax pairs were found some twenty years ago. As it has turned out, a suitable choice for the *Lax matrix* L not only yields N integrals in involution, but also yields a key tool for constructing action-angle maps. We continue by detailing such a choice of L and an associated M for the hyperbolic case, and then exploit L to derive various results of physical interest. Subsequently, the three remaining cases will be discussed briefly.

Our type II Lax pair is defined by taking $l = N$ and

$$L_{jk} = \delta_{jk}p_j + ig(1 - \delta_{jk})\frac{\mu}{2\,\mathrm{sh}\,\mu(x_j - x_k)/2}, \quad j, k = 1, \ldots, N, \quad (2.35)$$

$$M_{jk} = \frac{ig\mu^2}{4}\left(-\delta_{jk}\sum_{l \neq j}\frac{1}{\mathrm{sh}^2\,\mu(x_j - x_l)/2}\right.$$

$$\left. + (1 - \delta_{jk})\frac{\mathrm{ch}\,\mu(x_j - x_k)/2}{\mathrm{sh}^2\,\mu(x_j - x_k)/2}\right). \quad (2.36)$$

To verify that (2.31) is obeyed for this choice, we first calculate

$$\dot{L}_{jj} = \dot{p}_j = -\partial_{x_j}H, \quad (2.37)$$

$$[M, L]_{jj} = \sum_{k \neq j}(M_{jk}L_{kj} - L_{jk}M_{kj}) = \frac{g^2\mu^2}{4}\sum_{k \neq j}\frac{\mathrm{ch}\,\mu(x_j - x_k)/2}{\mathrm{sh}^3\,\mu(x_j - x_k)/2}. \quad (2.38)$$

Recalling (2.13) and (2.19), we see that the two right-hand sides are indeed equal. The off-diagonal elements yield

$$\dot{L}_{jk} = \frac{-ig\mu^2}{4}\frac{\mathrm{ch}\,\mu(x_j - x_k)/2}{\mathrm{sh}^2\,\mu(x_j - x_k)/2}(\dot{x}_j - \dot{x}_k)$$

which should be compared to

$$[M, L]_{jk} \equiv M_{jk}L_{kk} - L_{jj}M_{jk} + R_{jk}$$

$$= \frac{ig\mu^2}{4} \frac{\operatorname{ch}\mu(x_j - x_k)/2}{\operatorname{sh}^2\mu(x_j - x_k)/2}(p_k - p_j) + R_{jk}. \tag{2.39}$$

From this one infers that it remains to show $R_{jk} = 0$. To this end, one can combine the terms with summation index $l \neq j$, k and use the elementary functional equation

$$(\operatorname{ch} a \operatorname{sh} b - \operatorname{sh} a \operatorname{ch} b)\operatorname{sh}(a + b) + \operatorname{sh}^2 a - \operatorname{sh}^2 b = 0 \tag{2.40}$$

with $a = \mu(x_j - x_l)/2$ and $b = \mu(x_l - x_k)/2$.

The upshot is, that the power traces H_k and symmetric functions S_k of the Lax matrix (2.35) are conserved under the H flow. Obviously, we have

$$H_1 = \operatorname{Tr} L = P, \quad H_2 = \tfrac{1}{2}\operatorname{Tr} L^2 = H,$$

and

$$H_k = \frac{1}{k}\sum_{j=1}^{N} p_j^k + U_k(x, p), \quad k > 2, \tag{2.41}$$

where U_k has degree $< k$ in $p_1, \dots, p_N$.

From this one readily infers that the gradients of $H_1, \dots, H_N$ are linearly independent on an open dense subset of Ω. (Fixing $\lambda > 0$, the determinant of the $N \times N$ matrix with rows $\lambda^{-k+1}(\nabla_p H_k)(x, \lambda p)$ is a real-analytic function on Ω. Hence it either vanishes identically or on a closed nowhere dense subvariety of Ω. Assuming the first case applies, one obtains a contradiction: taking $\lambda \to \infty$, the determinant converges to a Vandermonde determinant, which is nonzero for $p_1, \dots, p_N$ distinct.) In fact, the gradients are independent on all of Ω, as we will see later on.

Next, we turn to showing involutivity of the integrals. One way to prove this property proceeds via the long-time asymptotics of the flow. The first step of the argument (which is in essence due to Moser [33]) consists in showing that for any initial point $(x_0, p_0) \in \Omega$ one gets

$$x_j(t) \sim x_j^{\pm} + tp_j^{\pm}, \ p_j(t) \sim p_j^{\pm}, \quad t \to \pm\infty, \tag{2.42}$$

with

$$p_N^- > \cdots > p_1^-, \quad p_N^+ < \cdots < p_1^+. \tag{2.43}$$

In view of the repulsive character of the interparticle forces, this result is very plausible. The complete proof (which is quite subtle) exploits this physical intuition, and for brevity we skip it.

The second step is easy. Recalling first

$$C_{kl} \equiv \{H_k, H_l\} \in \mathcal{I}_H \tag{2.44}$$

(cf. the paragraph containing (2.7)), we obtain

$$C_{kl}(x_0, p_0) = C_{kl}\big(x(t), p(t)\big), \quad \forall t \in \mathbb{R}. \tag{2.45}$$

Now each term in C_{kl} contains at least one factor $1/\operatorname{sh}[\mu(x_i - x_j)/2]$, so taking $t \to \infty$ in (2.45) and using (2.42), we deduce $C_{kl} = 0$, as desired.

Combining the asymptotics (2.42) with the properties of the Lax matrix, we can easily arrive at further conclusions of physical interest. First, using (2.35) we deduce

$$L\big(x(t), p(t)\big) \sim \operatorname{diag}(p_1^{\pm}, \dots, p_N^{\pm}), \quad t \to \pm\infty. \tag{2.46}$$

Second, combining (2.46) and isospectrality, we obtain

$$p_j^+ = p_{N-j+1}^-, \quad j = 1, \dots, N \text{ (conservation of momenta)}. \tag{2.47}$$

Third, let us assume that the scattering map

$$S \colon (x^-, p^-) \mapsto (x^+, p^+)$$

is a canonical transformation. (This is plausible, since it amounts to a limit of t-dependent canonical transformations; cf. Section 5.2.) If we then define Δ_j by setting

$$x_{N-j+1}^+ = x_j^- + \Delta_j$$

and use (2.47), it follows that $\partial \Delta_j / \partial x_k^- = 0$. Thus, Δ_j depends only on p^- and can be determined by choosing $x_1^-, \dots, x_N^-$ such that the collisions take place approximately pairwise. But then one clearly gets

$$\Delta_j(p^-) = \sum_{k \neq j} \delta(p_j^-, p_k^-) \text{ (factorization)} \tag{2.48}$$

where $\delta(p, p')$ denotes the position shift incurred in a 2-particle collision with asymptotic momenta p, p'.

The two properties (2.47) and (2.48) are the hallmark of *soliton scattering*: the Calogero-Moser particles scatter just as the soliton solutions to various two-dimensional integrable PDEs (such as the Korteweg-de Vries and sine-Gordon equations). In Section 5 we shall elaborate on the above findings, and show in particular that the maps $(x, p) \mapsto (x^{\pm}, p^{\pm})$ may be viewed as action-angle maps. Here, we only add that completeness of the flow $\exp(t_k H_k)$ (and hence Liouville integrability of H) is easily established. Indeed, conservation of H yields an upper bound on $|p(t_k)|$ and a nonzero lower bound on particle distances. Using these bounds one infers that $|dx_j/dt_k|$ is bounded above, so escape to infinity in finite time cannot occur either.

Next, we take a brief look at the remaining cases. First, the above discussion applies verbatim to the type I system: One need only send μ to 0 in the equations (2.35), (2.36), and (2.38)–(2.39). Taking now $\mu \to i\mu$ in these equations, they clearly apply when H is taken to be the Sutherland Hamiltonian (2.13), (2.20), and $\widetilde{\Omega}$ (2.24) is chosen as phase space. Changing (2.40) accordingly, it follows once more that one is dealing with an isospectral flow. As is easily checked, the trigonometric Lax matrix satisfies

$$L\big(\Gamma(x,p)\big) = S_-^t \, L(x,p) S_-, \quad (x,p) \in \widetilde{\Omega},$$

where Γ is the generator (2.25) and S_- the antiperiodic shift,

$$S_- \equiv \begin{pmatrix} 0 & 1 & \dots & 0 \\ \vdots & \vdots & \ddots & \vdots \\ 0 & 0 & \dots & 1 \\ -1 & 0 & \dots & 0 \end{pmatrix}.$$

From this similarity relation we deduce that the power traces on $\widetilde{\Omega}$ descend to smooth functions on the quotient manifolds Ω (2.26) and Ω' (2.29). Independence and involutivity of $H_1, \dots, H_N$ on all three phase spaces now follow by analytic continuation in μ from the hyperbolic case.

Of course, no scattering takes place in the Sutherland case. Choosing $\widetilde{\Omega}$ as phase space, the center of mass

$$\frac{1}{N} \sum_{j=1}^{N} x_j \equiv X \in C^\infty(\widetilde{\Omega})$$

moves uniformly along the line under the H flow. The distances between the particles are bounded above by $2\pi/\mu$, and one is therefore dealing with an oscillatory motion (classical N-particle "molecule").

The function X is *not* invariant under Γ, so it does not descend to a smooth function on Ω and Ω'. (The physical picture of particles on a ring does not allow for an unambiguous center of mass position *on* the ring either.) The X flow on $\widetilde{\Omega}$, given by

$$e^{aX}(x,p) = \left(x, p - \left(\frac{a}{N}, \dots, \frac{a}{N} \right) \right) \tag{2.49}$$

is invariant under Γ, however. Therefore, it descends to a smooth symplectic flow on Ω and Ω'. Accordingly, the associated vector fields are locally, but not globally Hamiltonian. This reflects the fact that the quotient manifolds are not simply connected. More specifically, (2.26) and (2.29) entail $\pi_1(\Omega) = \pi_1(\Omega') = \mathbb{Z}$, since $\widetilde{\Omega}$ is convex and hence simply connected.

Replacing $2\pi/\mu$ by 2ω, the last two paragraphs also apply to the type IV system. In this case, too, a Lax pair formulation exists. Here, however, it turns out to be crucial to let L and M depend on an additional *spectral parameter* $\lambda \in \mathbb{C}$. Such a Lax pair was introduced by Krichever [34], who used the Lax matrix for constructing an action-angle map. His Lax matrix is expressed in terms of the Weierstrass σ-function $\sigma(x; \omega, \omega')$, but both here and later on it is convenient to trade $\sigma(x)$ for the 2ω-antiperiodic function $s(x)$ defined by

$$s(x; \omega, \omega') \equiv \sigma(x; \omega, \omega') \exp\left(-\eta(\omega, \omega')\frac{x^2}{2\omega}\right). \tag{2.50}$$

Moreover, we only detail the Lax matrix: We choose

$$L_{jk} = \delta_{jk} p_j + ig(1 - \delta_{jk})\frac{s(x_j - x_k + \lambda)}{s(\lambda)s(x_j - x_k)}, \quad j, k = 1, \ldots, N. \tag{2.51}$$

This matrix is a similarity transform of Krichever's Lax matrix, so it yields the same symmetric functions and power traces. In particular, using the well-known identity

$$\frac{\sigma(x + \lambda)\sigma(x - \lambda)}{\sigma^2(x)\sigma^2(\lambda)} = \mathcal{P}(\lambda) - \mathcal{P}(x) \tag{2.52}$$

we obtain

$$H_2 = \tfrac{1}{2}\operatorname{Tr} L^2 = H - \tfrac{1}{2}g^2 N(N - 1)\mathcal{P}(\lambda).$$

Moreover, in this case one readily verifies

$$L\big(\Gamma(x, p)\big) = S_+^t L(x, p)S_+, \quad (x, p) \in \widetilde{\Omega},$$

where Γ is given by (2.25) with $\mu \to \pi/\omega$ and S_+ is the periodic shift,

$$S_+ \equiv \begin{pmatrix} 0 & 1 & \cdots & 0 \\ \vdots & \vdots & \ddots & \vdots \\ 0 & 0 & \cdots & 1 \\ 1 & 0 & \cdots & 0 \end{pmatrix}. \tag{2.53}$$

Therefore, the functions $H_k \in C^\infty(\widetilde{\Omega})$ descend to smooth functions on Ω and Ω'.

From our later account of the relativistic version of the type IV system it will transpire that the symmetric functions $S_k(\lambda)$ of L (2.51) (and hence its power traces, too) are polynomials in g, p_j and $\mathcal{P}(x_j - x_k)$, $j, k = 1, \ldots, N$, with λ-dependent coefficients that are real for λ purely imaginary and not equal to $2k\omega'$, $k \in \mathbb{Z}$; furthermore, $S_k(\lambda_1)$ and $S_l(\lambda_2)$ commute also when $\lambda_1 \neq \lambda_2$. At this point, this is very far from obvious, of course. But the linear independence of $dH_1, \ldots, dH_N$ on open dense subsets of the phase spaces $\widetilde{\Omega}, \Omega$ and Ω' follows in the same way as before, and once we know H is integrable, we can deduce its Liouville integrability (viz., completeness of the H_k flows, $k = 1, \ldots, N$) just as for the type III case.

As we have seen above, the type I–III Hamiltonians H may be viewed as specializations of the type IV Hamiltonian H; cf. (2.22). The relation of $L(\mathrm{IV})$ (2.51) to $L(\mathrm{II})$ (2.35) and its associated type III ($\mu \to i\mu$) and type I ($\mu \to 0$) versions remains to be explained, however. It is convenient to do so at the end of the next subsection.

2.3   Toda Systems

The Toda systems are defined by Hamiltonians of the form

$$H = \frac{1}{2} \sum_{j=1}^{N} p_j^2 + U(x) \tag{2.54}$$

with $U(x)$ given by

V. $a^2 \left(\sum_{j=2}^{N} e^{\mu(x_j - x_{j-1})} + e^{\mu(x_1 - x_N)} \right)$, $a, \mu > 0$ (periodic Toda) (2.55)

VI. $\sum_{j=2}^{N} e^{\mu(x_j - x_{j-1})}$, $\mu > 0$ (nonperiodic Toda) (2.56)

In both cases $U(x)$ is smooth on $\mathbb{R}^N$, so as phase space we should take

$$\Omega \equiv G \times \mathbb{R}^N, \quad G \equiv \mathbb{R}^N. \tag{V, VI}$$

Since $U(x)$ is positive, energy conservation yields an upper bound on $|p(t)|$, and so escape to infinity in finite time cannot occur. Hence, the H flows of type V and VI are complete on Ω. In the former case energy conservation also yields an upper bound on particle distances, so one gets an oscillatory motion for arbitrary initial states ("Toda molecule"). In the latter case the interparticle forces are repulsive, and so each initial state is a scattering state.

The existence of N independent integrals can be established once more via a Lax pair formulation of the H flow. Here, we only detail the Lax matrix L. For the periodic Toda system it is important to let L depend on a spectral parameter $w \in \mathbb{C}^*$. Specifically, one can take

$$L(\mathrm{V})_{jk} = \delta_{jk} p_j + \delta_{j,k-1} + a^2 \delta_{j,k+1} \exp\big(\mu(x_j - x_{j-1})\big)$$
$$- (ia)^N w \delta_{jN} \delta_{k1} - a^2 (ia)^{-N} w^{-1} \delta_{j1} \delta_{kN} \exp\big(\mu(x_1 - x_N)\big) \tag{2.57}$$

so that

$$H_1 = \operatorname{Tr} L = P, \quad H_2 = \tfrac{1}{2} \operatorname{Tr} L^2 = H, \tag{2.58}$$

(When $N = 2$, one should add the constant $a^2(w + 1/w)$ to H.) For the nonperiodic case we choose

$$L(\mathrm{VI})_{jk} = \delta_{jk} p_j + \delta_{j,k-1} + \delta_{j,k+1} \exp\big(\mu(x_j - x_{j-1})\big) \tag{2.59}$$

and then (2.58) holds true again.

Using the argument below (2.41), one deduces once more that the power traces $H_1, \ldots, H_N$ are independent. Taking for granted that $H_3, \ldots, H_N \in \mathcal{I}_H$, one also infers completeness of the flows $\exp(t_k H_k)$ from H being conserved. For the nonperiodic Toda case one can again exploit the repulsive character of the forces to prove (2.42) and (2.43), and then involutivity and the soliton scattering properties (2.47) and (2.48) follow in the same way as for the hyperbolic case.

Next, we substitute

$$x_j \to x_j - 2j\mu^{-1}\ln a, \quad j = 1, \ldots, N, \tag{2.60}$$

in the Lax matrix $L(V)(x,p)$, yielding a new matrix-valued function $L^{(a)}(x,p)$ on Ω. Since (2.60) may be viewed as a canonical transformation on Ω, the power traces of $L^{(a)}$ still Poisson commute. Now we clearly have

$$\lim_{a \to 0} L^{(a)} = L(VI) \tag{2.61}$$

so the latter Hamiltonians converge to the power traces of $L(VI)$. Therefore, the nonperiodic Toda system may be obtained as a limit of the periodic one.

To conclude this section, we detail similar limit relations between the Lax matrices for the six types of systems introduced above. These can be encoded in the following hierarchy:

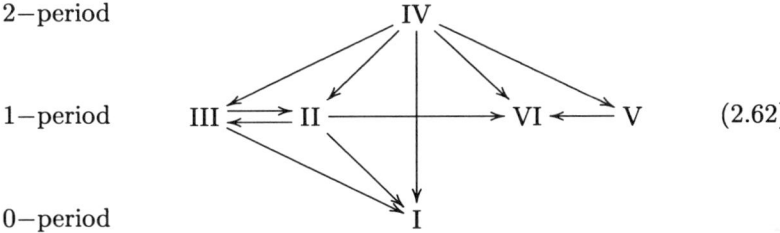

$$\tag{2.62}$$

The transition V → VI was specified in the previous paragraph. Moreover, the relations between the type I–III Lax matrices were already described in the previous subsection: $L(II)$ and $L(III)$ (and hence H and its commuting integrals) are related by analytic continuation in μ, and taking $\mu \to 0$ yields $L(I)$. The relation between $H(IV)$ and $H(I)$–$H(III)$ can be read off from (2.22), but the transition $L(IV) \to L(I)$–$L(III)$ is not obvious from (2.51) and (2.35)—if only because (2.51) depends on a spectral parameter and (2.35) does not. To elucidate this transition (and also for later purposes) we need more information on $s(x)$.

First, using the product representations for the Weierstrass σ-function, Legendre's relation, and (2.50), we obtain

$$s(x; \omega, \omega') = \frac{2\omega}{\pi} \sin\left(\frac{\pi x}{2\omega}\right) \prod_{l=1}^{\infty} \frac{(1 - q^{2l}\exp(i\pi x/\omega))(x \to -x)}{(1 - q^{2l})^2}, \tag{2.63}$$

$$s(x; \omega, \omega') = \frac{2\omega'}{i\pi} \exp\left(-\frac{i\pi x^2}{4\omega\omega'}\right) \text{sh}\left(\frac{i\pi x}{2\omega'}\right)$$

$$\times \prod_{l=1}^{\infty} \frac{(1 - \tilde{q}^{2l} \exp(i\pi x/\omega'))(x \to -x)}{(1 - \tilde{q}^{2l})^2} \tag{2.64}$$

where

$$q \equiv \exp\left(\frac{i\pi\omega'}{\omega}\right), \quad \tilde{q} \equiv \exp\left(-\frac{i\pi\omega}{\omega'}\right).$$

Hence we have

$$s(x; \omega, \omega') = \begin{cases} x, & \omega = \infty, \omega' = i\infty, \\ \text{sh}(\nu x)/\nu, & \omega = \infty, \omega' = i\pi/2\nu, \\ \sin(\nu x)/\nu, & \omega = \pi/2\nu, \omega' = i\infty. \end{cases} \tag{2.65}$$

Thus, we obtain a hyperbolic Lax matrix depending on a spectral parameter λ by substituting $\omega' = i\pi/\mu$ in (2.51) and sending ω to ∞. If we now take $\text{Re}\,\lambda \to \infty$ in this matrix, we obtain a similarity transform of (2.35). Substituting $\omega = \pi/\mu$ and letting $\omega' \to i\infty$ yields a trigonometric Lax matrix depending on λ; taking $\text{Im}\,\lambda \to \infty$ in the latter yields a similarity transform of the previous $L(\text{III})$. Finally, putting $\omega' = i\omega$ and letting $\omega \to \infty$ leads to a rational Lax matrix depending on λ; taking $\lambda \to \infty$, one gets the previous $L(\text{I})$.

The transition IV $\to$ V is quite nonobvious already at the level of the defining Hamiltonian H. To our knowledge, for H this transition was first pointed out by Inozemtsev [35]. Here, we detail a more general limit $L(\text{IV}) \to L(\text{V})$, which dates back to Ref. 5. To begin with, ω' should be replaced by $i\pi/\mu$, and a position shift

$$x_j \to x_j - \frac{2j\omega}{N}, \quad j = 1, \dots, N, \tag{2.66}$$

should be made in (2.51). Then the power traces of the new Lax matrix (denoted by $L^{(s)}$) are smooth Poisson commuting functions on the shifted phase space

$$\widetilde{\Omega}^{(s)} \equiv G^{(s)} \times \mathbb{R}^N$$

where

$$G^{(s)} \equiv \left\{ x \in \mathbb{R}^N \,\middle|\, \begin{array}{l} x_N - 2(N-1)\dfrac{\omega}{N} < \cdots < x_2 - 2\dfrac{\omega}{N} < x_1 \\ x_1 - x_N < 2\dfrac{\omega}{N} \end{array} \right\} \tag{2.67}$$

Now we substitute

$$g \to a\mu^{-1} \exp\left(\frac{\mu\omega}{N}\right), \tag{2.68}$$

$$\lambda \to \omega - \frac{\delta}{\mu}, \quad \delta \equiv -i\pi + \ln w,$$

in $L^{(s)}$. Then it follows from a long, but straightforward calculation using (2.64) that the resulting matrix (still denoted $L^{(s)}$) satisfies

$$\lim_{\omega \to \infty} L^{(s)}_{jk} = L(\mathrm{V})_{jk} \left(ia \exp\left(\frac{\delta}{N} \right) \right)^{k-j} \exp\left(-\frac{\mu(x_j - x_k)}{2} \right).$$

(The infinite product in (2.64) may be omitted for $k - j < N/2$, and contributes at most one factor otherwise.)

Note that the shifted configuration space $G^{(s)}$ (2.67) converges to the Toda configuration space $\mathbb{R}^N$ for $\omega \to \infty$. From a physical point of view, the new positions can be regarded as deviations from equilibrium positions. The distance $2\omega/N$ between successive equilibrium positions is taken to ∞, and the coupling strength is simultaneously taken to ∞ in such a way that a finite interaction persists in the limit.

Next, we specify the transition II → VI. To this end we substitute

$$x_j \to x_j + 2j\mu^{-1} \ln \epsilon, \quad j = 1, \ldots, N, \tag{2.69}$$

$$g \to \frac{1}{\mu\epsilon}, \tag{2.70}$$

in $L(\mathrm{II})(x, p)$ (2.35), yielding a new matrix $L^{(\epsilon)}(x, p)$ on a shifted phase space whose definition will be clear from the previous transition. Then one easily checks

$$\lim_{\epsilon \to 0} L^{(\epsilon)}_{jk} = L(\mathrm{VI})_{jk}(-i)^{j-k} \exp\left(-\frac{\mu(x_j - x_k)}{2} \right). \tag{2.71}$$

(For the defining Hamiltonians this transition was already pointed out by Sutherland [36], as we recently learned. Independently, we obtained the more general transition (2.71) and its relativistic generalization in 1985.)

Finally, we describe how (a similarity transform of) $L(\mathrm{VI})$ can be reached directly from $L(\mathrm{IV})$: Substituting

$$x_j \to x_j - 2j\left(\frac{\omega}{N} - \frac{\ln \omega}{\mu} \right), \quad j = 1, \ldots, N, \tag{2.72}$$

$$g \to \frac{\exp(\mu\omega/N)}{\mu\omega}, \tag{2.73}$$

$$\lambda \to \omega,$$

in $L(\mathrm{IV})$, the resulting matrix $\tilde{L}^{(s)}$ obeys

$$\lim_{\omega \to \infty} \tilde{L}^{(s)}_{jk} = L(\mathrm{VI})_{jk}(-i)^{j-k} \exp\left(-\frac{\mu(x_j - x_k)}{2} \right).$$

(Once more, this can be verified by using the product representation (2.64) for $s(x)$.)

3   Relativistic Versions at the Classical Level

3.1   The Defining Dynamics and its Commuting Integrals

In two space-time dimensions the nonrelativistic (Galilei) and relativistic (Poincaré) symmetry groups are semidirect products of boosts

$$(t, x) \mapsto \begin{cases} (t, x + vt) & \text{(Galilei)} \\ \left((t + \frac{vx}{c^2})(1 - \frac{v^2}{c^2})^{-1/2}, (x + vt)(1 - \frac{v^2}{c^2})^{-1/2}\right) & \text{(Lorentz)} \end{cases} \tag{3.1}$$

(where c denotes the speed of light) and space-time translations

$$(t, x) \mapsto (t + a_0, x + a_1).$$

Clearly, the translations are 1-parameter diffeomorphism groups generated by vector fields

$$X_t \equiv \partial_t, \quad X_s \equiv \partial_x,$$

whereas the Galilei boosts can be written

$$\exp(v X_b)(t, x), \quad X_b \equiv t \partial_x. \tag{Galilei}$$

To obtain a 1-parameter group of Lorentz boost diffeomorphisms, one should introduce the *rapidity* $\theta \in \mathbb{R}$ by setting

$$\frac{v}{c} = \text{th}\left(\frac{\theta}{c}\right).$$

Then the Lorentz boost (3.1) reads

$$(t, x) \mapsto \left(t \, \text{ch}\left(\frac{\theta}{c}\right) + \frac{x}{c} \, \text{sh}\left(\frac{\theta}{c}\right), x \, \text{ch}\left(\frac{\theta}{c}\right) + ct \, \text{sh}\left(\frac{\theta}{c}\right)\right)$$

and so can be rewritten

$$\exp(\theta X_b)(t, x), \quad X_b \equiv \frac{x}{c^2} \partial_t + t \partial_x. \tag{Lorentz}$$

The vector fields X_t, X_s, and X_b give rise to the space-time Lie algebras

$$[X_t, X_s] = 0, \quad [X_t, X_b] = X_s, \quad [X_s, X_b] = \begin{cases} 0, & \text{(Galilei)} \\ X_t/c^2. & \text{(Poincaré)} \end{cases}$$

Thus, the Galilei group and its Lie algebra may be viewed as deformations of the Poincaré group and its Lie algebra: The former result by taking $c \to \infty$ in the latter.

Next, we observe that for *any* potential $V(x)$ the Hamiltonians

$$H \equiv \frac{1}{2m} \sum_{j=1}^{N} p_j^2 + \sum_{1 \leq j < k \leq N} V(x_j - x_k), \tag{3.2}$$

$$P \equiv \sum_{j=1}^{N} p_j, \tag{3.3}$$

$$B \equiv -m \sum_{j=1}^{N} x_j, \tag{3.4}$$

represent the Galilei Lie algebra. More precisely, one has

$$\{H, P\} = 0, \quad \{H, B\} = P, \quad \{P, B\} = Nm,$$

so that one obtains a central extension. But constant Hamiltonians generate trivial flows (the corresponding Hamiltonian vector field vanishes; cf. (2.6)), so we do obtain a faithful representation at the group level (assuming H, P, and B generate complete flows).

Consider now the functions

$$H \equiv mc^2 \sum_{j=1}^{N} \text{ch}\left(\frac{p_j}{mc}\right),$$

$$P \equiv mc \sum_{j=1}^{N} \text{sh}\left(\frac{p_j}{mc}\right),$$

$$B \equiv -m \sum_{j=1}^{N} x_j.$$

Clearly, these satisfy the Poincaré Lie algebra

$$\{H, P\} = 0, \quad \{H, B\} = P, \quad \{P, B\} = \frac{H}{c^2}.$$

Physically, they describe a system of N relativistic free (equal rest mass) particles in a slightly unorthodox way. Namely, instead of using the customary 1-particle momentum k (in terms of which the kinetic energy reads $(k^2 c^2 + m^2 c^4)^{1/2}$) and its canonically conjugate position y, we are using the rapidity variable p/m and the variable x canonically conjugate to p. The variables p and x then have the dimension of momentum and position, respectively.

Now recall this century's most widely known formula, $E = Mc^2$. In words, this formula says that M is determined not only by rest mass, but also by kinetic energy and any other form of energy due to interactions.

Therefore, a quite simple way to take particle interactions into account for the above energy function is to replace it by

$$H = \sum_{j=1}^{N} M_j c^2, \quad M_j \equiv m \operatorname{ch}\left(\frac{p_j}{mc}\right) V_j(x), \tag{3.5}$$

with

$$V_j(x) \equiv \prod_{k \neq j} f(x_j - x_k). \tag{3.6}$$

Indeed, the function $V_j(x)$ (potential) then encodes the change in mass of particle j due to its interaction with the remaining particles, just as the function $\operatorname{ch}(p_j/mc)$ encodes the change in mass due to its motion.

We should now ensure that the altered H is still the time translation generator of (a phase space representation of) the Poincaré group. In contrast to the Galilei case (3.2)–(3.4), this is no longer true when both P and B are left unchanged. In fact, already translation invariance (viz., $\{H, P\} = 0$) is violated when P is left unchanged and $f(x)$ is not constant. The simplest choice is, therefore, to keep B unchanged and change P accordingly: It must read

$$P = mc \sum_{j=1}^{N} \operatorname{sh}\left(\frac{p_j}{mc}\right) V_j(x) \tag{3.7}$$

to yield the desired Poisson brackets $\{H, B\} = P$ and $\{P, B\} = H/c^2$.

But now we still have to satisfy the translation invariance constraint $\{H, P\} = 0$. Assuming $f(x)$ is an even function, and calculating the Poisson bracket for (3.5) and (3.7), we see that it vanishes iff $f(x)$ satisfies the functional equation

$$\sum_{j=1}^{N} \partial_j \prod_{k \neq j} f^2(x_j - x_k) = 0. \tag{3.8}$$

Since $f(x)$ is assumed to be even, this yields no constraint for $N = 2$. But for $N = 3$ (3.8) can be rewritten

$$\begin{vmatrix} f^2(u) & f(u)f'(u) & 1 \\ f^2(v) & f(v)f'(v) & 1 \\ f^2(u+v) & -f(u+v)f'(u+v) & 1 \end{vmatrix} = 0$$

and this functional equation is known to be valid iff

$$f^2(x) = a + b\mathcal{P}(x), \quad a, b \in \mathbb{C}, \tag{3.9}$$

where $\mathcal{P}$ is the Weierstrass function encountered in the previous sections.

More generally, (3.8) turns out to be valid for $N > 3$ when (3.9) holds [3]. (For $N > 3$ it is not known, however, whether other solutions to (3.8) exist.) Therefore, we obtain relativistic analogs of the Hamiltonians of type I–IV when we take

$$f(x) = \left(s^2 \left(\frac{ig}{mc} \right) \left[\mathcal{P} \left(\frac{ig}{mc} \right) - \mathcal{P}(x) \right] \right)^{1/2}, \quad g > 0. \qquad \text{(IV)} \quad (3.10)$$

(The radicand is positive, and we take positive square roots whenever this is the case.) In view of (2.22) and (2.65) this specializes to

$$f(x) = \begin{cases} (1 + g^2/m^2c^2x^2)^{1/2}, & \text{(I)}, \\ \left(1 + \sin^2(\nu g/mc)/\operatorname{sh}^2(\nu x)\right)^{1/2}, & \text{(II)}, \\ \left(1 + \operatorname{sh}^2(\nu g/mc)/\sin^2(\nu x)\right)^{1/2}, & \text{(III)}. \end{cases} \qquad (3.11)$$

We shall henceforth take $m = 1$, just as we did in the nonrelativistic setting (2.13). For type I–III it is then routine to verify that one has

$$\lim_{c \to \infty} (H_{\text{rel}} - Nc^2) = H_{\text{nr}}, \qquad (3.12)$$

$$\lim_{c \to \infty} P_{\text{rel}} = P_{\text{nr}}. \qquad (3.13)$$

Here, H_{nr} denotes the nonrelativistic Hamiltonian (2.13) with $V(x)$ given by (2.18)–(2.20), and H_{rel} denotes the relativistic Hamiltonian (3.5), with $V_j(x_1, \ldots, x_N)$ and $f(x)$ given by (3.6) and (3.11), respectively; similarly, P_{nr} and P_{rel} denote (3.3) and (3.7), respectively. To obtain the corresponding limits for type IV, one needs

$$\mathcal{P}(\epsilon) = \epsilon^{-2} + O(\epsilon^2), \quad s(\epsilon) = \epsilon - \frac{\epsilon^3 \eta}{2\omega} + O(\epsilon^5), \quad \epsilon \to 0.$$

Then it is clear that (3.13) still holds, whereas (3.12) should be replaced by

$$\lim_{c \to \infty} (H_{\text{rel}} - Nc^2) = \frac{1}{2} \sum_{j=1}^{N} p_j^2 + g^2 \sum_{\substack{j,k=1 \\ j<k}}^{N} \left(\mathcal{P}(x_j - x_k) + \frac{\eta}{\omega} \right)$$

$$= H_{\text{nr}} + \frac{g^2 N(N-1)\eta}{2\omega}.$$

(To check that this is consistent with the type I–III limits (3.12), one should use

$$\lim_{\omega \to \infty} \frac{\eta(\omega, i\pi/2\nu)}{\omega} = -\frac{\nu^2}{3},$$

$$\lim_{\omega' \to i\infty} \frac{\eta(\pi/2\nu, \omega')}{\omega} = \frac{\nu^2}{3};$$

cf. also (2.22).)

Next, let us assume that the potential $V_j(x)$ in (3.5) and (3.7) has a "nearest neighbor" structure,

$$V_j(x) = f_T(x_{j+1} - x_j)f_T(x_j - x_{j-1})$$

so as to mimic the interactions in the Toda systems. Then one easily verifies that one has $\{H, P\} = 0$ iff

$$\sum_{j=1}^{N} \partial_j \left(f_T^2(x_{j+1} - x_j) f_T^2(x_j - x_{j-1}) \right) = 0. \tag{3.14}$$

Now it is not an easy matter to verify the functional equations (3.8), even for the rational case. In contrast, it is very simple to check that (3.14) is satisfied when one takes

$$f_T^2(x) = a + be^{\mu x}, \quad a, b \in \mathbb{C},$$

with the convention

$$x_0 \equiv x_N, \qquad x_{N+1} \equiv x_1 \qquad \text{(V)} \quad (3.15)$$

$$x_0 \equiv \infty, \qquad x_{N+1} \equiv -\infty \qquad \text{(VI)} \quad (3.16)$$

Indeed, the terms in (3.14) simply cancel in pairs. Choosing

$$f_T(x) \equiv (1 + a^2 c^{-2} e^{\mu x})^{1/2}, \quad \begin{cases} a > 0 & \text{(V)} \\ a = 1 & \text{(VI)} \end{cases} \tag{3.17}$$

it is also easy to verify (3.12) and (3.13).

The mathematical upshot of the above physical reasoning is that insistence on the structure (3.5)–(3.7) for the Hamiltonians H and P, together with the requirement $\{H, P\} = 0$, has led us to a 1-parameter generalization of the commuting Hamiltonians H and P from Section 2. The former Hamiltonians can be rewritten

$$H = \frac{S_1 + S_{-1}}{2\beta^2}, \quad P = \frac{S_1 - S_{-1}}{2\beta}, \tag{3.18}$$

where we have introduced a (more convenient) new parameter

$$\beta \equiv \frac{1}{c}$$

and the functions

$$S_{\pm k} \equiv \sum_{\substack{I \subset \{1,\ldots,N\} \\ |I|=k}} \exp\left(\pm \beta \sum_{i \in I} p_i \right)$$

$$\cdot \begin{cases} \prod_{\substack{i \in I \\ j \notin I}} f(x_i - x_j) & \text{(I–IV)} \\ \prod_{\substack{i \in I \\ i+1 \notin I}} f_T(x_{i+1} - x_i) \prod_{\substack{i \in I \\ i-1 \notin I}} f_T(x_i - x_{i-1}) & \text{(V, VI)} \end{cases} \tag{3.19}$$

with $k = 1, \ldots, N$. Since H and P commute, the "light cone Hamiltonians" S_1 and S_{-1} commute as well.

Next, choosing $|I| = N$ in (3.19), we clearly get Hamiltonians

$$S_{\pm N} = \exp\big(\pm\beta(p_1 + \cdots + p_N)\big)$$

that commute with $S_{\pm k}, k < N$, since only position differences occur in (3.19). It is therefore natural to conjecture that all of the functions S_k, $\pm k \in \{1, \ldots, N\}$, Poisson commute. Now with some perseverance, it can be verified that the commutators $\{S_k, S_l\}$ vanish for any $k, l \in \{\pm 1, \ldots, \pm N\}$ and $N > 1$ iff the functions f and f_T satisfy the functional equations

$$0 = \sum_{\substack{I \subset \{1, \ldots, N\} \\ |I| = k}} \left(\sum_{i \in I} \partial_i\right)$$

$$\cdot \begin{cases} \prod_{\substack{i \in I \\ j \notin I}} f^2(x_i - x_j) & \text{(I–IV)} \\ \prod_{\substack{i \in I \\ i+1 \notin I}} f_T^2(x_{i+1} - x_i) \prod_{\substack{i \in I \\ i-1 \notin I}} f_T^2(x_i - x_{i-1}) & \text{(V, VI)} \end{cases} \quad (3.20)$$

for any $k \in \{1, \ldots, N\}$ and $N > 1$. For $k = 1$ these equations clearly reduce to the functional equations (3.8) and (3.14) encoding relativistic invariance. But even for the Toda case, it is not easy to prove directly that these functional equations are satisfied for $k > 1$, too. A direct proof that the equations (3.20) *are* valid in the Toda case can be found in Appendix A of [37]. As we shall see in Section 4.3, they are valid for the $\mathcal{P}$-function, too, but in that case no direct proof is known.

Let us now restrict attention to $S_1, \ldots, S_N$. (One easily verifies that

$$S_{-k} = \frac{S_{N-k}}{S_N}, \quad k = 1, \ldots, N. \quad (3.21)$$

In particular, H and P may be viewed as functions of $S_1, \ldots, S_N$.) These functions yield smooth Poisson commuting Hamiltonians on the type-dependent phase spaces described in Section 2.2. (Notice that for type III and IV the functions (3.19) are manifestly invariant under the map (2.25) and its type IV analog.) In all cases the functions $V_j(x)$ are bounded away from 0, just as the functions $\mathrm{ch}(\beta p_j)$. Therefore, conservation of H yields an upper bound on $|p|$ and a nonzero lower bound on particle distances, and so completeness of the flows $\exp(t_j S_j)$, $j = 1, \ldots, N$, follows in the same way as in the nonrelativistic setting. However, the scaling argument below (2.41) no longer applies, so independence of $S_1, \ldots, S_N$ must be shown by other means.

A simple argument yielding independence for all cases at once now follows. View the determinant of the $N \times N$ matrix with rows $\nabla_p S_k$ as a polynomial in $e_j = \exp(\beta p_j)$, $j = 1, \ldots, N$, with x-dependent coefficients. Expanding the determinant, the product of the diagonal elements yields in particular a monomial $e_1^N e_2^{N-1} \cdots e_N$ with a positive coefficient. A moment's thought now shows that none of the remaining $N! - 1$ products can yield such a monomial. Therefore, the determinant cannot vanish identically, and independence results.

From the preceding three paragraphs we deduce that the light cone Hamiltonian S_1 is Liouville integrable. (Of course, this is true for H, too, but it is more convenient to focus on S_1.) Since we know its commuting integrals explicitly, it might seem irrelevant to try and find a Lax pair formulation for the flow it generates. It turns out, however, that a special choice of Lax matrix L makes it possible to derive some crucial results, as will become clear later on. (We do not need a Lax *pair*, however. Note in this connection that M depends on the dynamics S_k one selects, whereas L encodes all dynamics of interest simultaneously.)

Before we detail Lax matrices, we would like to settle two issues for which a Lax matrix is not needed—in contrast to the nonrelativistic situation. First, specializing to the systems of type I, II, and VI, the repulsive character of the interparticle potentials $f(x_j - x_k)$ and $f_T(x_j - x_k)$, respectively, can be exploited to prove that the S_1 dynamics leads to soliton scattering. Specifically, Moser's argument can be adapted to the S_1 flow [3], yielding here

$$x_j(t) \sim x_j^{\pm} + \beta t \exp(\beta p_j^{\pm}), \quad p_j(t) \sim p_j^{\pm}, \quad t \to \pm\infty, \qquad (3.22)$$

and, once more, (2.43). Therefore, the functions $s_k : t \mapsto S_k\big(x(t), p(t)\big)$ reduce to symmetric functions of matrices $L^{\pm} \equiv \mathrm{diag}\big(\exp(\beta p_1^{\pm}), \ldots, \exp(\beta p_N^{\pm})\big)$ for $t \to \pm\infty$. Since $s_k(t)$ is t-independent, the roots of the polynomial $\lambda \mapsto |L^- - \lambda 1_N|$ equal those of $\lambda \mapsto |L^+ - \lambda 1_N|$. Hence, conservation of momenta results; more in detail, (2.47) follows upon using (2.43). Then the factorization (2.48) of the asymptotic position shifts follows in the same way as before.

We point out in passing that (3.22) also entails that involutivity of S_1, $\ldots$, S_N is a consequence of $S_1, \ldots, S_N$ being integrals of the S_1 flow. (Equivalently, for the systems of type I, II, and VI—and, by analytic continuation, for type III, too—the functional equations (3.20) follow once the special cases (3.8) and (3.14), respectively, are proved.) Indeed, the argument in the paragraph containing (2.44) can easily be adapted to the systems involved.

The second issue we wish to address is the connection diagram (2.62), with regard to the functions $S_1, \ldots, S_N$. First of all, the connections between the type I–IV functions are obvious from (2.22) and (2.65); cf. (3.10) and (3.11). Second, we detail the transition V $\to$ VI: once again, one need only substitute (2.60) and take $a \to 0$ to obtain the nonperiodic Toda S_k from the periodic ones.

Third, we consider the transition II $\to$ VI. To this end, we note first that the choice $\beta \mu g \in (0, 2\pi)$ is not the only one yielding a positive potential

$$f(x) = \left(1 + \frac{\sin^2(\beta\mu g/2)}{\operatorname{sh}^2(\mu x/2)}\right)^{1/2}. \qquad \text{(II) (3.23)}$$

Indeed, we can allow $\beta\mu g = \pi - id$, $d \in \mathbb{R}$, so as to let $\sin^2(\beta\mu g/2)$ vary over $(1, \infty)$, too. Taking now

$$\beta\mu g \to \pi - 2i\ln\left(\frac{\beta}{\epsilon}\right) \qquad (3.24)$$

and substituting (2.69) in $f(x_j - x_k)$, we obtain a function f_{jk} with limits

$$f_{jk} \to \begin{cases} 1, & |j - k| > 1, \\ (1 + \beta^2\exp[\mu(x_j - x_{j-1})])^{1/2}, & k = j - 1, \\ (1 + \beta^2\exp[\mu(x_{j+1} - x_j)])^{1/2}, & k = j + 1, \end{cases} \qquad (3.25)$$

for $\epsilon \to 0$. Hence, these substitutions ensure that the hyperbolic S_k converge to the nonperiodic Toda S_k for $\epsilon \to 0$.

Fourth, we study the IV$\to$V limit. To this end we take $\omega' \equiv i\pi/\mu$ and consider the elliptic potential, rewritten as

$$f(x) = \left(\frac{s(x + i\beta g)s(x - i\beta g)}{s^2(x)}\right)^{1/2}; \qquad \text{(IV) (3.26)}$$

cf. (3.10), (2.50), and (2.52). Using (2.64) this becomes

$$f(x) = \rho(g)f_{\mathrm{II}}(x)\left(\prod_{l=1}^{\infty} \frac{(1 - 2\tilde{q}^{2l}\operatorname{ch}\mu(x + i\beta g) + \tilde{q}^{4l})(g \to -g)}{(1 - 2\tilde{q}^{2l}\operatorname{ch}\mu x + \tilde{q}^{4l})^2}\right)^{1/2} \qquad (3.27)$$

where f_{II} denotes the hyperbolic potential (3.23), and where

$$\rho(g) \equiv \exp\left(\frac{\mu\beta^2 g^2}{4\omega}\right), \quad \tilde{q} \equiv \exp(-\mu\omega). \qquad (3.28)$$

Now let us substitute (2.66) and

$$i\beta g \to \frac{2\omega}{N} + \frac{1}{\mu}(i\pi + 2\ln(a\beta)) \qquad (3.29)$$

in $f(x_j - x_k)$, and study the three factors that arise from (3.27) for $\omega \to \infty$. First, the calculation in the previous paragraph (taking $\epsilon = \exp(-\mu\omega/N)$) shows that the f_{II} factor yields (3.25) with β replaced by $a\beta$. Second, the infinite product converges to 1 unless $\{j, k\} = \{1, N\}$; in the latter case only the $l = 1$ factor in the numerator does not converge to 1, but rather to $1 + a^2\beta^2\exp\mu(x_1 - x_N)$.

Third, consider the constant prefactor $\rho(g)$ (3.28) with the substitution (3.29) in force. Clearly, the resulting factor is not real, and it goes to 0 for $\omega \to \infty$. However, this is easily remedied: We need only replace $f(x)$ by

$$f_r(x) \equiv \frac{f(x)}{\rho(g)} \qquad (3.30)$$

to obtain renormalized functions $S_{r,k}$ with the desired type V limits when (2.66) and (3.29) are substituted. (These functions are, moreover, positive for $\beta\mu g = \pi - id$, $d \in \mathbb{R}$.) This entails, in particular, that the involutivity of the periodic Toda S_k may be viewed as a corollary of the involutivity of the elliptic S_k.

Finally, we specify the IV $\to$ VI transition. To this purpose we substitute (2.72) and

$$i\beta g \to \frac{2\omega}{N} + \frac{\kappa}{\mu}, \quad \kappa \equiv i\pi + 2\ln\left(\frac{\beta}{\omega}\right), \tag{3.31}$$

in the above $S_{r,k}$ and take $\omega \to \infty$. This is readily verified to yield the type VI S_k; in this case the infinite product in (3.27) converges to 1, and taking $\epsilon = \omega\exp(-\mu\omega/N)$ one can invoke (3.25) once again.

3.2   Lax Matrices and Their Interrelationships

Next, we turn to explicit Lax matrices, as promised above. Among other things, these will enable us to clarify the connection between the relativistic S_k and their nonrelativistic counterparts.

We begin by specializing to the hyperbolic case Ii. Here, the key ingredient to arrive at the desired Lax matrix is an identity due to Cauchy, which is useful in various other contexts. It reads

$$\det\left(\frac{1}{w_j - z_k}\right)_{j,k=1}^{N} = \prod_{j=1}^{N}\frac{1}{w_j - z_j}\prod_{\substack{j,k=1\\j<k}}^{N}\frac{(w_j - w_k)(z_j - z_k)}{(w_j - z_k)(z_j - w_k)} \tag{3.32}$$

and can be proved, e.g., via induction on N. Substituting

$$w_j \to \exp\mu\left(x_j + \frac{i\beta g}{2}\right), \quad z_k \to \exp\mu\left(x_k - \frac{i\beta g}{2}\right), \quad \beta,\mu,g > 0$$

in the matrix occurring at the left-hand side of (3.32) yields a matrix C with elements

$$C_{jk} = \exp\left(-\mu\frac{x_j + x_k}{2}\right)\frac{\operatorname{sh}(i\beta\mu g/2)}{\operatorname{sh}\mu(x_j - x_k + i\beta g)/2} \tag{3.33}$$

and determinant

$$|C| = \exp\left(-\mu\sum_j x_j\right)\prod_{j<k}f^{-2}(x_j - x_k) \tag{3.34}$$

where f is given by (3.23).

If we now set

$$L_{jk} \equiv e_j C_{jk} e_k \tag{3.35}$$

where

$$e_j \equiv \exp\left(\frac{\mu x_j + \beta p_j}{2}\right) \prod_{l \neq j} f(x_j - x_l)^{1/2}, \tag{3.36}$$

then it follows from (3.34) that the principal minor with indices $\{i_1, \ldots, i_k\}$ $\equiv I$ is given by

$$
\begin{aligned}
L(I) &= \left(\prod_{i \in I} e_i^2\right) C(I) \\
&= \left(\prod_{i \in I} \exp(\beta p_i)\right) \left(\prod_{i \in I} \prod_{l \neq i} f(x_i - x_l)\right) \prod_{\substack{j < k \\ j, k \in I}} f^{-2}(x_j - x_k) \\
&= \left(\prod_{i \in I} \exp(\beta p_i)\right) \prod_{\substack{i \in I \\ l \notin I}} f(x_i - x_l). \tag{3.37}
\end{aligned}
$$

Therefore, the kth symmetric function of L reads

$$S_k = \sum_{|I| = k} \exp\left(\beta \sum_{i \in I} p_i\right) \prod_{\substack{i \in I \\ j \notin I}} f(x_i - x_j).$$

Thus, it equals the above type II Hamiltonian S_k (3.19), as anticipated by our notation.

Recalling now the nonrelativistic hyperbolic Lax matrix (2.35), we obtain, using obvious notation,

$$L_{\mathrm{rel}} = 1_N + \beta L_{\mathrm{nr}} + O(\beta^2), \quad \beta \to 0. \tag{3.38}$$

Taking β to 0 in the determinant of the matrix $\beta^{-1}(L_{\mathrm{rel}} - 1_N) - \lambda 1_N$, we deduce

$$S_{k,\mathrm{nr}} = \lim_{\beta \to 0} \beta^{-k} \sum_{l=0}^{k} (-)^{k+l} \binom{N-l}{N-k} S_{l,\mathrm{rel}} \tag{3.39}$$

Thus, the involutivity of the nonrelativistic type II functions follows from that of their relativistic counterparts. (Note that L_{rel} is holomorphic at $\beta = 0$, so that the $\beta \to 0$ limit may be interchanged with the partials in the Poisson brackets involved.)

Of course, we can now obtain type III and I Lax matrices from (3.35) by taking $\mu \to i\mu$ and $\mu \to 0$, respectively; then (3.38) and (3.39) hold true again. We can also derive a type VI Lax matrix from (3.35) by substituting (3.24) and (2.69) in the similarity transform

$$\tilde{L}_{jk} \equiv \beta^{k-j} e_j^2 C_{jk} \tag{3.40}$$

and taking $\epsilon \to 0$. Indeed, using (3.25), this limit is easily calculated, yielding a matrix

$$L(\text{VI})_{jk} = \beta^{k-j} b_j E_{jk} \tag{3.41}$$

where

$$b_j \equiv \exp(\beta p_j)(1 + \beta^2 \exp[\mu(x_{j+1} - x_j)])^{1/2}$$
$$\times (1 + \beta^2 \exp[\mu(x_j - x_{j-1})])^{1/2}, \tag{3.42}$$

$$E_{jk} \equiv \begin{cases} 1, & k - j = N - 1, \ldots, 1, 0, \\ a_j, & k - j = -1, \\ 0, & k - j = -2, \ldots, -N + 1, \end{cases} \tag{3.43}$$

$$a_j \equiv \beta^2 \exp[\mu(x_j - x_{j-1})](1 + \beta^2 \exp[\mu(x_j - x_{j-1})])^{-1}, \tag{3.44}$$

(with the convention (3.16) in effect). It follows from the above that the symmetric functions of $L(\text{VI})$ are indeed the type VI functions already defined. (It is not hard to verify this directly.) Comparing (3.41)–(3.44) and (2.59), it is also clear that (3.38) holds true once more, and so (3.39) follows for the nonperiodic Toda case, too.

We proceed by introducing a type IV Lax matrix: We choose

$$L_{jk} \equiv d_j D_{jk} \tag{3.45}$$

where

$$d_j \equiv \exp(\beta p_j) \prod_{l \neq j} f(x_j - x_l), \quad f(x) \equiv \left(\frac{s(x + i\beta g)s(x - i\beta g)}{s^2(x)} \right)^{1/2}, \tag{3.46}$$

$$D_{jk} \equiv \frac{s(x_j - x_k + \lambda)}{s(\lambda)} \frac{s(i\beta g)}{s(x_j - x_k + i\beta g)}. \tag{3.47}$$

It is immediate that this matrix is again related to its nonrelativistic analog (2.51) via (3.38), and that its trace equals the above elliptic Hamiltonian S_1. It is not at all immediate, but true, that all of its symmetric functions Σ_k are proportional to the above elliptic S_k (3.19). Specifically, we have

$$\Sigma_k = s(\lambda)^{-k} s(\lambda - i\beta g)^{k-1} s(\lambda + (k-1)i\beta g) S_k, \quad k = 1, \ldots, N. \tag{3.48}$$

This assertion follows from a calculation similar to (3.37) and an identity generalizing Cauchy's identity (3.32), viz.,

$$\det \left(\frac{s(q_j - r_k + \lambda)}{s(q_j - r_k + \gamma)} \right)_{j,k=1}^{N}$$
$$= s(\lambda - \gamma)^{N-1} s\left(\lambda + (N-1)\gamma + \sum_j (q_j - r_j) \right)$$
$$\cdot \prod_{i<j} s(q_i - q_j) s(r_j - r_i) \prod_{i,j} s(q_i - r_j + \gamma)^{-1}. \tag{3.49}$$

This identity follows from (2.50) and Theorem B2 in Ref. 4. (Substantially the same identity was already proved (in a different way) by Frobenius [38]. We learned this fact from a paper by Raina [39], who supplies yet a third proof.)

Since (3.38) holds true with L_{rel} given by (3.45) and L_{nr} given by (2.51), we may conclude that (3.39) holds true, with $S_{l,\mathrm{rel}}$ equal to Σ_l (3.48). Now the x-dependence of the Σ_l occurs solely in factors $\mathcal{P}(x_j - x_k)$ (recall (3.10)), and the λ-dependent prefactor in (3.48) is real for λ purely imaginary and not equal to $2l\omega'$, $l \in \mathbb{Z}$. (This is because $s(x)$ is purely imaginary for x purely imaginary; cf. (2.63).) Therefore, the assertions concerning the nonrelativistic S_k made below (2.53) readily follow from (3.39).

Specializing the Lax matrix (3.45) to the rational, hyperbolic and trigonometric contexts, we obtain Lax matrices of type I, II, and III depending on a spectral parameter $\lambda \in \mathbb{C}$. Clearly, the previous Lax matrices (cf. (3.35)) result by taking $\lambda \to \infty$, $\mathrm{Re}\,\lambda \to \infty$ and $\mathrm{Im}\,\lambda \to \infty$, respectively (up to diagonal similarity factors).

Next, we use the above elliptic Lax matrix to arrive at a periodic Toda Lax matrix. Recall we already analyzed the IV $\to$ V transition for the functions S_k; cf. (3.26)–(3.30). From these findings we infer that we should start from a renormalized Lax matrix

$$L_r \equiv \rho(g)^{-N+1} L \tag{3.50}$$

with the substitutions $\omega' = i\pi/\mu$, (2.66) and (3.29). The symmetric functions of L_r are then given by

$$\Sigma_{r,k} = \frac{s(\lambda - i\beta g)^{k-1} s(\lambda + (k-1)i\beta g)}{\rho(g)^{k(k-1)} s(\lambda)^k} S_{r,k}, \quad k = 1, \ldots, N, \tag{3.51}$$

with $S_{r,k}$ defined below (3.30). Now the latter substitutions ensure that $S_{r,k}$ has the desired type V limit S_k (3.19) for $\omega \to \infty$. To get a finite limit for the prefactor in (3.51), we also substitute

$$\lambda \to \frac{2\omega}{N} + \frac{2}{\mu} \ln(a\beta) - \frac{N}{\mu} \ln(ia\beta) - \frac{1}{\mu} \ln w.$$

Using (2.64) with $\omega' = i\pi/\mu$, we see that then the limit $\omega \to \infty$ of (3.51) is given by

$$\Sigma_k = \begin{cases} (1 + (ia\beta)^N w)^{k-1} S_k, & k = 1, \ldots, N-1, \\ (1 + (ia\beta)^N w)^{N-1} (1 + (ia\beta)^N w^{-1}) S_N, & k = N. \end{cases} \tag{3.52}$$

(The infinite products do not contribute, save for the $l = 1$ term coming from the factor $s(\lambda + (k-1)i\beta g)$ when $k = N$.)

The above substitutions also ensure that L_r (3.50) has a finite limit for $\omega \to \infty$. Specifically, the limit can be written

$$\lim_{\omega \to \infty} L_{r,jk} = L(\mathrm{V})_{jk} \left(ia \exp\left[\frac{i\pi + \ln w}{N} \right] \right)^{k-j}.$$

The relativistic periodic Toda Lax matrix thus obtained reads explicitly

$$L(\mathrm{V})_{jk} = \beta^{k-j} b_j E_{jk}. \tag{3.53}$$

Here, one has

$$b_j = \exp(\beta p_j)(1 + a^2\beta^2 \exp[\mu(x_{j+1} - x_j)])^{1/2}$$
$$\times (1 + a^2\beta^2 \exp[\mu(x_j - x_{j-1})])^{1/2}, \tag{3.54}$$

$$E_{1N} = \frac{1 - (ia\beta)^{-N} w^{-1} a^2\beta^2 \exp[\mu(x_1 - x_N)]}{1 + a^2\beta^2 \exp[\mu(x_1 - x_N)]}, \tag{3.55}$$

$$E_{jk} = 1, \quad k - j = N - 2, \ldots, 1, 0, \tag{3.56}$$

$$E_{j,j-1} = \frac{-(ia\beta)^N w + a^2\beta^2 \exp[\mu(x_j - x_{j-1})]}{1 + a^2\beta^2 \exp[\mu(x_j - x_{j-1})]}, \tag{3.57}$$

$$E_{jk} = -(ia\beta)^N w, \quad k - j = -2, \ldots, -N + 1, \tag{3.58}$$

with the convention (3.15) in force. Its symmetric functions are given by (3.52)—a fact that is quite hard to see directly from (3.53)–(3.58).

It *is* immediate from the latter formulas that (3.38) holds true; cf. (2.57). From (3.39) it then follows that the nonrelativistic periodic Toda functions S_k are w-independent for $k < N$, whereas for $k = N$ the w-dependence is given by an additive term $(ia)^N(w^{-1} + (-)^N w)$. It is also plain that when we substitute (2.60) in (3.53)–(3.58), yielding a new matrix $L^{(a)}$, then (2.61) holds true once more; cf. (3.41)–(3.44).

To complete our account of the connection diagram (2.62) at the level of relativistic Lax matrices, it remains to specify the direct transition IV → VI. To this end we take ω' equal to $i\pi/\mu$ and substitute (2.72), (3.31), and

$$\lambda \to \frac{2\omega}{N}$$

in L_r. Then the resulting matrix $\tilde{L}_r$ fulfils

$$\lim_{\omega \to \infty} \tilde{L}_{r,jk} \left(\exp\left(\frac{\kappa}{N}\right) \right)^{j-k} = L(\mathrm{VI})_{jk} \beta^{j-k}$$

where κ is defined by (3.31).

We conclude this subsection with some remarks. First, we would like to point out that the function s may be replaced by the function $\tilde{s}$ obtained from the right-hand side of (2.64) by omitting the exponential factor. Of course, this makes no difference for the type I–III systems, since $\tilde{s}(x; \omega, \omega')$ obeys (2.65), too. Moreover, for the nonrelativistic elliptic Lax matrix (2.51) the substitution $s \to \tilde{s}$ amounts to a similarity transformation, so that the symmetric functions do not change. But in (3.45) this substitution leads to a similarity transform of L_r, and not of L; cf. (3.46),

(3.47) and (3.50). Therefore, no renormalization is needed to obtain Toda limits; moreover, several similarity factors become simpler. Nevertheless, we have opted for using s, since this is the simplest choice at the quantum level. (Note that $\tilde{s}(x; \omega, \omega')$ is not 2ω-antiperiodic, in contrast to $s(x; \omega, \omega')$.)

Second, we recall that our account of limit transitions has dealt with the connections encoded in the diagram (2.62) both for the nonrelativistic and for the relativistic systems. Viewing the two diagrams as stacked on top of each other, we have also detailed the vertical limits $\mathcal{S}_{\mathrm{rel}} \to \mathcal{S}_{\mathrm{nr}}$, $\mathcal{S} = \mathrm{I}$–VI. We have, however, not analyzed skew limits, thus far. One of these is of particular interest (especially at the quantum level), viz., the limit $\mathrm{II}_{\mathrm{rel}} \to \mathrm{VI}_{\mathrm{nr}}$. In contrast to the detours via the $\mathrm{II}_{\mathrm{nr}}$ or $\mathrm{VI}_{\mathrm{rel}}$ systems, one should keep βg *fixed*; specifically, setting

$$\frac{\beta \mu g}{2} \equiv \tau \in (0, \pi), \tag{3.59}$$

$$x_j \to x_j + 2j\mu^{-1} \ln\left(\frac{\beta}{2 \sin \tau}\right), \tag{3.60}$$

in $\tilde{L}$ (3.40), one obtains a matrix $L^{(\tau)}$ for which one easily verifies

$$L^{(\tau)} = 1_N + \beta L^{(\tau)}(\mathrm{VI}_{\mathrm{nr}}) + O(\beta^2), \quad \beta \to 0,$$

where

$$L^{(\tau)}(\mathrm{VI}_{\mathrm{nr}})_{jk} \equiv L(\mathrm{VI}_{\mathrm{nr}})_{jk}(1 - e^{-2i\tau})^{k-j}.$$

Consequently, the type II relativistic S_k (with (3.59) and (3.60) in force) give rise to the type VI nonrelativistic S_k as detailed by (3.39).

Third, we would like to point out that there exist additional systems (denoted $\mathrm{III}_{\mathrm{b}}$ and IV_{b}) that are related to the above $\mathrm{III}_{\mathrm{rel}}$ and $\mathrm{IV}_{\mathrm{rel}}$ systems via the substitution $\beta \to i\beta$. To obtain real-valued Hamiltonians, one should consider, e.g.,

$$I_k \equiv \frac{S_k + S_{-k}}{2}, \quad k = 1, \ldots, N.$$

Moreover, taking $\omega \equiv \pi/\mu$ so as to handle both cases at once, one should work with a configuration space

$$G \equiv \left\{ x \in \mathbb{R}^N \ \middle| \ \begin{matrix} x_j - x_{j+1} > \beta g, j = 1, \ldots, N-1, \\ x_1 - x_N < 2\pi/\mu - \beta g \end{matrix} \right\}. \tag{$\mathrm{III}_{\mathrm{b}}, \mathrm{IV}_{\mathrm{b}}$}$$

This is necessary to keep the potentials $f(x_j - x_k)$ positive, since one now has

$$f(x) = (s^2(\beta g)[\mathcal{P}(\beta g) - \mathcal{P}(x)])^{1/2}, \quad \beta, g > 0.$$

Consequently, one needs to require

$$N < \frac{2\pi}{\beta\mu g}$$

for G to be nonempty.

The Hamiltonians $I_1, \ldots, I_N$ on the phase space $\widetilde{\Omega} \equiv G \times \mathbb{R}^N$ are not only invariant under the $\mathbb{Z}$-action generated by Γ (2.25), but also under the $\mathbb{Z}^N$-action $p \mapsto p + 2\pi k/\beta$, $k \in \mathbb{Z}^N$. Quotienting out the resulting action of $\mathbb{Z} \ltimes \mathbb{Z}^N$ on $\widetilde{\Omega}$ yields a second phase space Ω; alternatively, one can consider quotients by proper subgroups, yielding phase spaces interpolating between the maximal and minimal phase spaces $\widetilde{\Omega}$ and Ω, respectively. The flows generated by the commuting Hamiltonians $I_1, \ldots, I_{N-1}$ and their quotients are not complete, but this can be remedied. Specifically, a suitable interpolating phase space Ω^c can be densely embedded in a phase space of the form $\mathbb{R}^2 \times \mathbb{P}^{N-1}$, on which all flows *are* complete. Here, $\mathbb{P}^k$ denotes complex projective space, viewed as a compact symplectic manifold by equipping it with the symplectic form derived from the Fubini-Study metric. Using (2.63), it is readily verified that this completion for the III$_b$ case (which is detailed in Ref. 40) is also appropriate for the IV$_b$ system. (Of course, inclusion of these systems in the diagram (2.62) would lead to a further proliferation of arrows.)

Last but not least, one can obtain new systems of considerable physical interest via analytic continuation of the positions. A detailed study of these soliton-antisoliton systems can be found in Ref. 30.

4   Quantum Calogero-Moser and Toda Systems

4.1   Background: Quantum Mechanics/Hilbert Space Theory

We choose as our starting point a classical phase space Ω of the cotangent bundle type (2.2), and a smooth real-valued Hamiltonian $H(x, p)$ on $\Omega \simeq G \times \mathbb{R}^N$. Then the state space of the associated quantum system consists of the unit vectors in the Hilbert space $L^2(G, dx)$ of square-integrable complex-valued functions; when ψ is a unit vector, the integral of $|\psi(x)|^2$ over a subset B of G is interpreted as the probability to find the system position vector in the set B.

The quantization of the dynamics $H(x, p)$ is obtained by means of the canonical quantization substitution

$$p_j \to -i\hbar \frac{\partial}{\partial x_j} \equiv \hat{p}_j, \quad j = 1, \ldots, N, \tag{4.1}$$

where $\hbar > 0$ denotes Planck's constant. Since the quantum operators $\hat{p}_j$ and x_j do not commute, this substitution may lead to ordering ambiguities.

For the Calogero-Moser and Toda Hamiltonians (2.13), (2.18)–(2.21) and (2.54)–(2.56) this is not the case, however: (4.1) yields an unambiguous (linear) partial differential operator (PDO) of the form

$$H = \frac{1}{2} \sum_{j=1}^{N} \hat{p}_j^2 + U(x) \tag{4.2}$$

with $U(x)$ real-analytic on the classical configuration space G.

We shall say that a PDO of the form (4.2) is *integrable* when there exist independent PDOs $I_1 = H$, $I_2(x,\hat{p})$, ... , $I_N(x,\hat{p})$ that commute pairwise. Here, independent means by definition that there are no polynomial relations between I_1, ... , I_N. Moreover, the notation $I_k(x,\hat{p})$ is meant to indicate that a specific ordering in the so-called *symbol* $I_k(x,p)$ has been chosen whenever ordering ambiguities are present. It should be observed that this definition does not involve any Hilbert space notion: Commutativity of PDOs is an algebraic issue. (Of course, one does need some smoothness conditions on the coefficients.) Presently, we shall introduce a stronger notion of Hilbert integrability, which is tied to Hilbert space theory.

Next, consider the Hamiltonians (3.18)–(3.19), taking $f(x) = f_T(x) = 1$ at first (free particles). Then the substitution (4.1) yields exponential dependence on $\hat{p}_j$. The obvious way to interpret this is in the sense of *analytic difference operators* (AΔOs). Thus, for instance,

$$(\exp(a\hat{p}_j)\psi)(x_1,\dots,x_N) \equiv \psi(x_1,\dots,x_j - i\hbar a,\dots,x_N), \quad a \in \mathbb{R}, \tag{4.3}$$

where one insists that ψ admit analytic continuation off the real axis—just as PDOs should act on differentiable functions.

Consider now the interacting case f, $f_T \neq 1$. Then one still has $S_{\pm N} = \exp(\pm\beta(p_1 + \cdots + p_N))$, so the quantum versions of $S_{\pm N}$ are the same AΔOs as before. But for $|k| < N$ an interpretation of $S_k = \sum_{|I|=|k|} T_I$ as an AΔO is ambiguous: the AΔO $T_I(x,\hat{p})$ depends on the ordering of x- and p-dependent factors specified for its symbol $T_I(x,p)$. In particular, the AΔO $H(x,\hat{p})$ associated with the Hamiltonian H that defines the relativistic version of the model at hand is not uniquely determined—by contrast to the PDO (4.2).

In view of this state of affairs we shall say that the Hamiltonians S_k, $k = \pm 1$, ... , $\pm N$, defined by (3.19) admit an *integrable quantization* when there exist pairwise commuting AΔOs $S_k(x,\hat{p})$ with symbols $S_k(x,p)$ equal to S_k (as functions on the classical phase space). Whenever this is the case, we shall use (3.18) to include quantum versions of H and P in the commutative AΔO algebra generated by $S_{\pm 1}(x,\hat{p})$, ... , $S_{\pm N}(\hat{p})$. As we will see in Section 4.3, all of the type I–VI systems do admit integrable quantizations. (Anticipating the outcome, we mention that one should first choose an ordering different from the one in (3.19) before substituting (4.1).)

Once more, these definitions do not involve Hilbert space: With appropriate analyticity assumptions on the x-dependent coefficients understood, commutativity of AΔOs is an algebraic issue. In Sections 4.2 and 4.3 we deal with (nonrelativistic) PDOs and (relativistic) AΔOs in a purely algebraic way, and the remainder of this subsection has no relevance for the issues addressed there.

The following is, however, crucial for Section 6, where we aim to understand the PDOs and AΔOs as well-defined Hilbert space operators. Let us begin by recalling some standard fare. For a given quantum Hamiltonian H on the Hilbert space $L^2(\mathbb{R}^N, dx)$ (say) the time evolution is encoded in the time-dependent Schrödinger equation

$$i\hbar\dot{\Psi} = H\Psi. \tag{4.4}$$

To solve this, one first reduces it to the time-independent Schrödinger equation

$$H\psi = E\psi \tag{4.5}$$

by setting

$$\Psi(t, x) = \exp\left(-\frac{itE}{\hbar}\right)\psi(x).$$

Solutions of (4.5) for which ψ is square-integrable correspond to bound states; when no such solutions exist, one should try and find bounded solutions and construct square-integrable solutions to (4.4) by building wave packets.

For a Hamiltonian of the form (4.2), the simplest example of the latter situation is the free case $U = 0$. Taking first $N = 1$ and $E > 0$, there exist two linearly independent solutions to (4.5), viz., the plane waves $\exp(\pm ipx/\hbar)$, with $p^2/2 = E$. These are bounded, in contrast to the two solutions of this form for $E < 0$: then p is purely imaginary and the solution increases exponentially for $x \to \infty$ or $x \to -\infty$.

The Hilbert space $L^2(\mathbb{R}, dx)$ can be spanned by the former solutions, in the sense that wave packet superpositions

$$\psi(x) \equiv \int_{-\infty}^{\infty} dp \, \exp\left(\frac{ipx}{\hbar}\right)\phi(p), \quad \phi \in C_0^\infty(\mathbb{R}),$$

are dense. (At the quantum level we are dealing with complex-valued functions. Accordingly, C^∞ stands for smooth and *complex-valued* in Sections 4 and 6—in contrast with our convention at the classical level. Recall the subscript 0 means compact support.) Then one has

$$(H\psi)(x) = \int_{-\infty}^{\infty} dp \, \exp\left(\frac{ipx}{\hbar}\right)\frac{p^2}{2}\phi(p)$$

and the time-dependent equation (4.4) is solved by

$$\left(\exp\left(-\frac{itH}{\hbar}\right)\psi\right)(x) = \int_{-\infty}^{\infty} dp \,\exp\left(\frac{ipx}{\hbar} - \frac{iEt}{\hbar}\right)\phi(p), \quad E \equiv \frac{p^2}{2}.$$

Moreover, a suitable normalization ensures that one obtains a unitary operator from the spectral representation space $L^2(\mathbb{R}, dp)$ onto $L^2(\mathbb{R}, dx)$. Specifically, one should choose

$$(\mathcal{E}\phi)(x) \equiv (2\pi\hbar)^{-1/2} \int_{-\infty}^{\infty} dp \,\exp\left(\frac{ipx}{\hbar}\right)\phi(p) \tag{4.6}$$

so as to obtain a unitary $\mathcal{E}$. This operator is a simple example of an *eigenfunction transform*—a unitary operator that diagonalizes H via suitably normalized (not necessarily square-integrable) eigenfunctions.

For $N > 1$ and $U = 0$ one need only work with tensor products to obtain the corresponding quantities. The resulting transform amounts to Fourier transformation in N variables,

$$(\mathcal{E}\phi)(x) = (2\pi\hbar)^{-N/2} \int dp \,\exp\left(ip \cdot \frac{x}{\hbar}\right)\phi(p). \tag{4.7}$$

(From now on, we omit the integration region when it equals $\mathbb{R}^k$ for some $k \in \mathbb{N}$.) It diagonalizes not only PDOs with constant coefficients, but also constant coefficient AΔOs. Indeed, from (4.3) one gets

$$\exp(a\hat{p}_j)\exp\left(ip \cdot \frac{x}{\hbar}\right) = \exp(ap_j)\exp\left(ip \cdot \frac{x}{\hbar}\right).$$

Now functions of the form (4.7) with $\phi \in C_0^\infty(\mathbb{R}^N)$ extend to entire functions in $x_1, \dots, x_N$, and so the quantum versions

$$S_k \equiv \sum_{|I|=\pm k} \exp(\pm\beta\hat{p}_I), \quad \pm k = 1, \dots, N, \tag{4.8}$$

of the functions (3.19) with $f = f_T = 1$ satisfy

$$(S_k\mathcal{E}\phi)(x) = (2\pi\hbar)^{-N/2} \int dp \,\exp\left(ip \cdot \frac{x}{\hbar}\right)e_k(p)\phi(p), \quad \phi \in C_0^\infty(\mathbb{R}^N) \tag{4.9}$$

with

$$e_k(p) \equiv \sum_{|I|=\pm k} \exp(\pm\beta p_I) \tag{4.10}$$

The above is quite elementary and well known, but it differs in spirit and notation from the Dirac approach employed in most quantum mechanics textbooks. We shall have more to say about these differences shortly, but first we would like to introduce and discuss the notion of Hilbert integrability. Roughly speaking, we shall say that commuting PDOs and AΔOs (of the above-defined integrable kind) are Hilbert integrable when they can be simultaneously diagonalized as real-valued multiplication operators.

To be specific, let us first assume that $I_1 = H(x, \hat{p})$, $x \in G$, is an integrable PDO with commuting PDOs $I_2(x, \hat{p}), \ldots, I_N(x, \hat{p})$. Then we call H *Hilbert integrable* when there exist joint eigenfunctions

$$I_k E(x, p) = M_k(p) E(x, p), \quad p \in \Lambda, k = 1, \ldots, N, \qquad (4.11)$$

(with Λ a subset of $\mathbb{R}^N$ and M_k a real-valued continuous function) and a unitary operator

$$\mathcal{E} \colon \hat{\mathcal{H}} \equiv L^2(\Lambda, \mu) \to \mathcal{H} \equiv L^2(G, dx),$$
$$\phi(p) \mapsto \int_\Lambda d\mu(p) \, E(x, p) \phi(p), \qquad (4.12)$$

where μ is a measure on Λ. Similarly, starting from an integrable quantization $S_{\pm 1}, \ldots, S_{\pm N}$ of the type I–VI Hamiltonians (3.19) (as defined above), we shall say that the commuting AΔOs are *Hilbert integrable* when there exist joint eigenfunctions

$$S_k E(x, p) = M_k(p) E(x, p), \quad p \in \Lambda, \pm k = 1, \ldots, N, \qquad (4.13)$$

(with $\Lambda \subset \mathbb{R}^N$, and M_k real-valued and continuous) and a unitary of the form (4.12). In either case, we define the commuting PDOs/AΔOs as commuting Hilbert space operators by pulling back the multiplication operators M_k to $\mathcal{H}$ via $\mathcal{E}$.

Starting from integrable PDOs $I_1, \ldots, I_N$ or AΔOs $S_{\pm 1}, \ldots, S_{\pm N}$, there is no simple method to establish whether they are Hilbert integrable—and if they are, the unitary $\mathcal{E}$ and the resulting commuting operators on $\mathcal{H}$ need not be unique, even in the free case. In this connection, a key question is: which solutions to the above time-independent Schrödinger equations are relevant in the Hilbert space context?

For PDOs much is known about this question, whereas for AΔOs of the type occurring in our models very little is known. The different character of the two classes of operators can already be gleaned from simple operators H on $L^2(\mathbb{R})$. Taking $H = \hat{p}$, the solution of (4.5) (unique up to a multiplicative constant) contributes to the eigenfunction transform (4.6) for all $E = p \in \mathbb{R}$. For $H = \hat{p}^2$, however, we should restrict attention to $E = p^2 > 0$, and then we need the two linearly independent solutions $\exp(\pm i E^{1/2} x/\hbar)$ for the eigenfunction transform, which is again given by (4.6). (The constant solution for $E = 0$ is bounded, too, but its contribution to (4.6) may be ignored.) Now take $H = \hat{p}^{2k}$, $k > 1$. Then we need again $E = p^{2k} > 0$ to get the two bounded solutions making up the diagonalizing transform (4.6), but now we have to discard $2k - 2$ linearly independent unbounded solutions for all $E > 0$.

For the AΔO $H = \exp(a\hat{p})$ the situation is essentially different, though. Of course, the plane wave $\exp(ipx/\hbar)$ is the only solution to (4.5) for $E = e^{ap} > 0$ that is needed to construct a unitary eigenfunction transform (4.6). Here, however, one is throwing away an *infinite-dimensional* vector space of *bounded* solutions to (4.5)! Indeed, one can multiply the plane wave by any function that is entire (say), has period $ia\hbar$, and is bounded on $\mathbb{R}$. The functions $\exp\big(iq\,\mathrm{ch}(2\pi x/a\hbar + r)\big)$, q, $r \in \mathbb{R}$, have all of these properties, for example. One can also construct bound states by taking instead $q \in i(0, \infty)$, or allow meromorphic multipliers of the form $\mathrm{ch}(2\pi x/a\hbar + ir)/\mathrm{c.c.}$, $r \in \mathbb{R}$.

A closely related phenomenon can be illustrated by the type II AΔOs. As we will see in Section 4.3, their coefficients have period $2i\pi/\mu$ in x_1, ..., x_N, just as the classical functions (3.19). Therefore, they commute with the free AΔOs

$$\tilde{S}_k \equiv \sum_{|I|=\pm k} \exp(\pm 2\pi(\mu\hbar)^{-1}\hat{p}_I), \quad \pm k = 1, \ldots, N. \tag{4.14}$$

(Worse yet, they commute with any AΔO obtained from $\tilde{S}_k$ via multiplication of the shift monomials by arbitrary x-dependent coefficients, as long as the latter have period $i\beta\hbar$ in x_1, ..., x_N!) But the plane waves that diagonalize the $\tilde{S}_k$ are not in any sense eigenfunctions of the type II AΔOs (unless the latter have constant coefficients, too).

From these simple observations it can already be seen that it is quite difficult to interpret the relevant AΔOs as well-defined commuting Hilbert space operators. We shall return to this problem in Section 6. This subsection will be concluded by sketching some material from Hilbert space theory, which puts the above in its proper mathematical context. (The notions to be introduced will reappear in Section 6.) Before doing so, we insert a sociological aside.

All of what follows can already be found in von Neumann's classic [41], which was written more than sixty years ago. The subject matter summarized below should have become bread and butter for theoretical physicists, but is actually still widely ignored—if not taboo. Indeed, almost all of the standard textbooks on quantum mechanics still contain a brand of Hilbert space theory that is considered antediluvian by functional analysts—to put it kindly. Up-to-date accounts oriented toward theoretical physics can be found for instance in Refs. 13, 42–45, and with such lucid and elegant sources available, one need not spend undue effort in learning the basics. From a purely pragmatic standpoint, too, the analyst's mathematical framework for quantum mechanics is most useful, as it pins down the essential difficulties and prevents tilting at windmills.

With the preaching and advertising out of the way, let us turn to less contentious matters. First of all, it should be recalled that quantum mechanical Hamiltonians are typically unbounded operators, whose definition must include the specification of a dense subspace of the Hilbert space as a domain to act on. (As a consequence of unboundedness, they do not admit a continuous linear extension to all of Hilbert space.) Now for any (linear) operator H defined on a dense domain $\mathcal{D}$ in the (separable, complex) Hilbert space $\mathcal{H}$ one can define the *adjoint* H^* on a definition domain $\mathcal{D}^*$, as follows: A vector $\phi \in \mathcal{H}$ belongs to $\mathcal{D}^*$ iff there exists a vector $\phi^* \in \mathcal{H}$ such that

$$(\phi^*, \psi) = (\phi, H\psi), \quad \forall \psi \in \mathcal{D}.$$

Then the action of H^* on ϕ is defined by

$$H^*\phi \equiv \phi^*.$$

The definition domain $\mathcal{D}^*$ of the adjoint H^* need not be dense in $\mathcal{H}$; in fact, it may even consist solely of the zero vector. From now on we assume that the operator H is *symmetric*, i.e., that we have

$$(\phi, H\psi) = (H\phi, \psi), \quad \forall \phi, \psi \in \mathcal{D}.$$

Then $\mathcal{D}^*$ obviously contains $\mathcal{D}$, and so is dense in $\mathcal{H}$; moreover, on $\mathcal{D}$ the adjoint H^* coincides with H. For instance, the type I–VI defining Hamiltonians (4.2) are symmetric on the dense subspace $C_0^\infty(G) \subset L^2(G)$—as is readily checked.

Next, we turn to an example that we will use to motivate the notions of self-adjointness and self-adjoint extensions. Specifically, we start from

$$\mathcal{H} = L^2([0,\pi]), \quad \mathcal{D} = C_0^\infty((0,\pi)), \quad H = -\frac{d^2}{dx^2}. \qquad (4.15)$$

(This setting actually arises for the center-of-mass, $g = 0$, $N = 2$, type III and IV Hamiltonians; we are putting $\hbar = 1$ for the remainder of this subsection.) Obviously, H is symmetric. Here, $\mathcal{D}^*$ contains the space $C^2([0,\pi])$, and on this space H^* acts again as $-d^2/dx^2$. (Use the above definition and integration by parts to check this.) In particular, the functions $\psi_p \equiv \exp(ixp)$, $p \in \mathbb{C}$, belong to $C^2([0,\pi])$, and one has $H^*\psi_p = p^2\psi_p$. Therefore, the adjoint has spectrum $\mathbb{C}$.

Now this is bad news for physics, since the spectrum should correspond to the physically measurable energies of the system—which are real. This is why one should work with Hamiltonians that are not just symmetric, but *self-adjoint* (s.a.) or at least *essentially self-adjoint* (e.s.a.). By definition, a symmetric H is s.a. iff $\mathcal{D}^*$ equals $\mathcal{D}$, and is e.s.a. iff H^* is s.a. If H is symmetric, but not e.s.a., one should try and extend H to an operator H_e on a definition domain $\mathcal{D}_e$ satisfying $\mathcal{D} \subset \mathcal{D}_e \subset \mathcal{D}^*$, such that H_e is s.a.

A symmetric operator need not have any s.a. extensions, but when it commutes with a conjugation (i.e., an anti-unitary with square the identity), then it does admit s.a. extensions. When H is e.s.a., it also admits s.a. extension, viz., to the operator H^*; in this case the s.a. extension is unique.

Returning to our example (4.15), we see that H commutes with complex conjugation, so it admits s.a. extensions. These are not unique, however, so that H is not e.s.a. Indeed, three distinct s.a. extensions of H are well known: We can enlarge the definition domain by allowing smooth functions with $\psi(0) = \psi(\pi)$, $\psi'(0) = \psi'(\pi)$ (periodic boundary conditions), or with $\psi(0) = \psi(\pi) = 0$ (Dirichlet b.c.), or with $\psi'(0) = \psi'(\pi) = 0$ (Neumann b.c.). These extensions yield e.s.a. operators with a complete set of bound state eigenfunctions, namely, linear combinations of $\exp(\pm ipx)$ for appropriate $p \in [0, \infty)$. In this way one obtains three distinct real point spectra. Thus, the physics depends on what s.a. extension is chosen.

To summarize, one should insist on quantum-mechanical Hamiltonians being s.a. or at least e.s.a., so as to ensure reality of the spectrum. Self-adjointness is also sufficient for applicability of the spectral theorem, to which we now turn. This theorem can be presented in several guises (cf. Ref. 42, Chapters VII, VIII), but in our context of quantum-integrable systems one of these is particularly useful. Crudely speaking, it says that any s.a. operator H with dense definition domain $\mathcal{D}$ in an abstract Hilbert space $\mathcal{H}$ can be unitarily transformed to a real-valued multiplication operator on a concrete space of square-integrable functions. More precisely, the multiplication operators arising in this way can be defined as follows. Let $\widehat{\mathcal{H}}$ be the Hilbert space

$$\widehat{\mathcal{H}} \equiv \bigoplus_{j=1}^{m} L^2(\mathbb{R}, \mu_j), \quad m \le \infty, \tag{4.16}$$

where $\mu_1, \mu_2, \ldots$ are measures on the real line. Then the operator $M : \widehat{\mathcal{D}} \to \widehat{\mathcal{H}}$ with domain

$$\widehat{\mathcal{D}} \equiv \{\psi \in \widehat{\mathcal{H}} \mid (E\psi_1(E), E\psi_2(E), \ldots) \in \widehat{\mathcal{H}}\}$$

and action

$$(M\psi)_j(E) \equiv E\psi_j(E), \quad \psi \in \widehat{\mathcal{D}}, \ j = 1, 2, \ldots,$$

is s.a. Now the spectral theorem says that for any s.a. operator $H : \mathcal{D} \to \mathcal{H}$ there exists a space $\widehat{\mathcal{H}}$ of the above form and a unitary $\mathcal{E}$ from $\widehat{\mathcal{H}}$ onto $\mathcal{H}$ satisfying

$$\mathcal{E}(\widehat{\mathcal{D}}) = \mathcal{D}, \quad H\mathcal{E} = \mathcal{E}M.$$

Thus, H is diagonalized by the unitary $\mathcal{E}$.

Of course, the choice of spectral representation space $\widehat{\mathcal{H}}$ is highly non-unique: parts of the measures in (4.16) can be reshuffled among the summands, and the normalization of the measures is not fixed either. But the closure of the union of the supports of the measures is unique: It is the *spectrum* of H. More generally, the measures encode all spectral properties of H. For example, whenever one of the measures μ_j assigns nonzero weight to a point $E_0 \in \mathbb{R}$, H has a bound state with eigenvalue E_0; continuous parts of the spectrum correspond to continuous parts of the measures; the spectral multiplicity of $E \in \mathbb{R}$ equals the limit of the number of measures assigning nonzero weight to the interval $(E - \epsilon, E + \epsilon)$ as $\epsilon \downarrow 0$, etc.

As a matter of fact, it is often more convenient to diagonalize H as a nonlinear function of a spectral variable; we have already seen several concretizations of this more general form of the spectral theorem. If need be, one can easily convert this to the previous form. For instance, the Fourier transformation (4.6) diagonalizes all operators $\hat{p}^l$, $l \in \mathbb{N}^*$, and $\exp(a\hat{p})$, $a \in \mathbb{R}^*$, simultaneously; then one should transform to $L^2([0,\infty), dE) \oplus L^2([0,\infty), dE)$ for even l and to $L^2(\mathbb{R}, dE)$ otherwise.

The spectral theorem makes it possible to define and work with functions of H. In particular, one way to obtain the quantum time evolution is by pulling back the unitary multiplication operator $\exp(-itE)$ on $\hat{\mathcal{H}}$:

$$\exp(-itH) = \mathcal{E}\exp(-itE)\mathcal{E}^{-1}.$$

More generally, one can define bounded functions of H, and these functions form an Abelian algebra. Whenever one can choose $m = 1$ in (4.16) (so that the (global) spectral multiplicity of H equals 1), this algebra is maximal Abelian: It cannot be enlarged without losing the commutativity property.

The spectral theorem has a generalization to several commuting s.a. operators $H_1, H_2, \ldots, H_N$. Here, commuting means by definition that the corresponding evolutions $\exp(-it_jH_j)$, $j = 1, \ldots, N$, commute. (For bounded $H_1, \ldots, H_N$, this is equivalent to $[H_k, H_l] = 0$. But for unbounded operators, the commutator need not be densely defined, and even if it is and vanishes, this does not entail that the evolutions commute.) Again, it is often simpler to transform to a representation where $H_1, \ldots, H_N$ become nonlinear functions of several variables (instead of multiplication by E_1, $\ldots$, E_N, respectively). The free AΔOs (4.8) are a case in point; cf. (4.9), (4.10). Note that they all have spectrum $[0,\infty)$ with infinite multiplicity. The Abelian algebra generated by the bounded functions of $S_1, \ldots, S_N$ (say) is, however, almost multiplicity-free (maximal): Its commutant is generated by the permutation operators

$$(P_\sigma\psi)(x) \equiv \psi(\sigma^{-1}(x)), \quad \sigma \in S_N, \psi \in L^2(\mathbb{R}^N, dx), \tag{4.17}$$

where $\tau(x)$ stands for $(x_{\tau(1)}, \ldots, x_{\tau(N)})$.

4.2   The Nonrelativistic Case: Commuting PDOs

The quantization prescription (4.1) applied to the Hamiltonian (2.13) yields the PDO

$$H = \frac{1}{2} \sum_{j=1}^{N} \hat{p}_j^2 + g^2 \sum_{\substack{j,k=1 \\ j<k}}^{N} V(x_j - x_k), \quad g > 0. \tag{4.18}$$

Specializing $V(x)$ to the type I–IV cases (2.18)–(2.21), the first question to answer is now, whether there exist N independent PDOs (including H) that commute pairwise. Equivalently, one should try and establish whether H is an *integrable* PDO; cf. the first few paragraphs of Section 4.1.

There are two sets of obvious candidates for these commuting PDOs. Clearly, the quantization $P \equiv \sum_j \hat{p}_j$ of the first power trace I_1/symmetric function S_1 of the respective Lax matrices commutes with H; also, H is the unambiguous quantization of I_2 (or equivalently $\frac{1}{2}S_1^2 - S_2$) up to an (irrelevant) λ-dependent constant for type IV. For $N > 2$ one can now try to continue either with $I_k(x,\hat{p})$, $k = 3, \ldots, N$, or with $S_k(x,\hat{p})$, $k = 3, \ldots, N$. But for the power traces ordering problems arise: in summands contributing to $I_k(x,p)$ for $k > 2$ the quantities x_j and p_j can occur simultaneously, so the ordering is ambiguous. By contrast, whenever p_j occurs in a monomial contributing to some principal minor of $L(x,p)$, the position x_j cannot occur, so that the expression $S_k(x,\hat{p})$ yields an unambiguous PDO for all $k \in \{1, \ldots, N\}$. Thus one need only show that $S_1(\hat{p}), \ldots, S_N(x,\hat{p})$ commute. (If so, one can define $I_k(x,\hat{p})$, $k > 2$, unambiguously via the Newton identities (2.34).)

We have no doubt that these PDOs indeed commute. But as we have learned, there appears to be no complete proof in the published literature. In Ref. 2, Olshanetsky and Perelomov do present a proof, but upon scrutiny this proof turns out to be incomplete on two counts—as we shall detail shortly.

Fortunately, there are complete proofs that H (4.18) is *integrable* for all of the systems of type I–VI; only the connection between the higher-order commuting PDOs and the quantized $S_k(L)$ has not been completely clarified yet. As we see it now, the first complete proof for systems of type I–III was given by Heckman [15], Opdam [46, 47], and Heckman and Opdam [48], and for systems of type IV by Oshima and H. Sekiguchi [49]. In our survey [5] we presented independent proofs of integrability for the nonrelativistic quantum systems of type I, II, III, V, and VI—as a corollary of integrability at the relativistic level, proved first in Ref. 4. We shall return to the latter strategy in Section 4.3.

We proceed by discussing the partial proof in Ref. 2 and the complete proof of Ref. 49. The starting point of Ref. 2 is a *classical* Hamiltonian of the form

$$J_N(x,p) \equiv \exp\left(\sum_{\substack{j,k=1 \\ j<k}}^{N} h(x_j - x_k)\partial_{p_j}\partial_{p_k} \right) p_1 \cdots p_N. \tag{4.19}$$

Expanding the exponential, this can be written

$$J_N = \sum_{l=0}^{[N/2]} \frac{1}{2^l l! (N-2l)!}$$
$$\times \sum_{\sigma \in S_N} \sigma(h(x_1-x_2) \cdots h(x_{2l-1}-x_{2l}) p_{2l+1} \cdots p_N) \quad (4.20)$$

Next, defining recursively

$$J_{k-1} \equiv \frac{\left\{ \sum_{j=1}^N x_j, J_k \right\}}{N-k+1}, \quad k = N, N-1, \ldots, 1, \quad (4.21)$$

and using (4.20), one obtains

$$J_k = \frac{1}{(N-k)!} \sum_{l=0}^{[k/2]} \frac{1}{2^l l! (k-2l)!}$$
$$\times \sum_{\sigma \in S_N} \sigma(h(x_1-x_2) \cdots h(x_{2l-1}-x_{2l}) p_{2l+1} \cdots p_k). \quad (4.22)$$

In particular, this yields

$$J_0 = 1, \quad J_1 = \sum_j p_j, \quad J_2 = \sum_{j<k} p_j p_k + \sum_{j<k} h(x_j - x_k),$$
$$J_3 = \sum_{i<j<k} p_i p_j p_k + \sum_{\substack{j<k \\ i \neq j,k}} h(x_j - x_k) p_i,$$

so that

$$\tfrac{1}{2} J_1^2 - J_2 = \frac{1}{2} \sum_j p_j^2 - \sum_{j<k} h(x_j - x_k) \equiv H. \quad (4.23)$$

Hence, the choice

$$h(x) = -g^2 \mathcal{P}(x) - C \quad (4.24)$$

yields the classical type IV Hamiltonian (2.13), (2.21) when one takes $C = 0$. More generally, one easily checks

$$J_k(h+C) = \frac{1}{(N-k)!} \sum_{j=0}^{[k/2]} \frac{C^j}{2^j j!} (N-k+2j)! J_{k-2j}(h).$$

In words, shifting h by a constant amounts to a linear reshuffling of J_1, ..., J_N.

Evidently, all of these quantities have unambiguous quantizations, and the proof of Ref. 2 is concerned with the commutativity of $J_1, \ldots, J_N$ given by the quantizations of (4.19), (4.21). Their proof uses classical input, which we now sketch. The Hamiltonian $J_N(x,p)$ (4.19) was introduced by Sawada and Kotera [50] for the type I case, where $h(x) = -g^2/x^2$. They showed that J_N Poisson commutes with H (4.23) and observed that this entails that the Hamiltonians J_k defined via (4.21) commute with H, as well. (Indeed, this easily follows from the Jacobi identity [50].) Subsequently, Wojciechowski [51] generalized J_N (4.19) to the type IV case and showed that when J_k is defined via the recurrence relation (4.21), then $J_1, \ldots, J_N$ are in involution.

Now Olshanetsky and Perelomov do prove that the quantum versions of H and J_N commute as well, by showing that the additional terms in the commutator (compared to the terms in the Poisson bracket) sum to zero; cf. Ref. 2, p. 336. Then it follows just as in the classical case that H also commutes with $J_{N-1}, \ldots, J_1$. But in Ref. 2 it is not proved (or even made plausible) that the additional terms in *arbitrary* commutators $[J_k, J_l]$ sum to zero.

The latter gap is closed in the work by Oshima and Sekiguchi [49]. Specifically, they prove (among other things) that the quantizations of J_k (4.22) with (4.24) in force commute without using any classical input (so that involutivity of $J_1(x,p), \ldots, J_N(x,p)$ follows as a corollary). They also prove a most useful uniqueness result, which we shall have occasion to invoke in Section 4.3. More precisely, we only need a slightly weaker version of Theorem 5.2 in Ref. [49], which we now state.

Suppose $G_k(x,\hat{p})$, $k = 1, \ldots, N$, $x \in \mathbb{R}^N$, are N commuting PDOs with meromorphic dependence on $x_1, \ldots, x_N$. Suppose that the PDOs are invariant under arbitrary permutations in S_N and that they are of the form

$$G_1 = \sum_j \hat{p}_j, \quad G_2 = \sum_{j<k} \hat{p}_j \hat{p}_k - g_2(x), \tag{4.25}$$

$$G_k = \sum_{i_1 < \cdots < i_k} \hat{p}_{i_1} \cdots \hat{p}_{i_k} - g_k(x,\hat{p}), \quad k = 3, \ldots, N, \tag{4.26}$$

with g_k of degree $< k - 1$ in $\hat{p}$. Then the commutative PDO algebra generated by $G_1, \ldots, G_N$ coincides with the algebra generated by the above $J_1, \ldots, J_N$. In particular, one must have

$$g_2(x) = C_1 \sum_{j<k} \mathcal{P}(x_j - x_k; \omega, \omega') + C_2. \tag{4.27}$$

Thus far, we have restricted ourselves to the type I–IV systems. Let us now turn to the Toda systems. Choosing

$$h(x) = -g^2 \left(\mathcal{P}(x) - \mathcal{P}(\lambda) \right) = -g^2 \frac{s(\lambda + x) s(\lambda - x)}{s^2(\lambda) s^2(x)}, \quad \omega' \equiv \frac{i\pi}{\mu}, \tag{4.28}$$

with $\lambda \equiv \omega + i\pi/\mu$, and substituting (2.66) and (2.68) in J_N (4.19), it follows from previous calculations that the limit $\omega \to \infty$ exists and yields (with the convention (3.15) in effect)

$$J_N = \exp\left(-a^2 \sum_{j=1}^N \exp\mu(x_j - x_{j-1})\partial_{p_j}\partial_{p_{j-1}}\right)p_1\cdots p_N. \qquad \text{(V) (4.29)}$$

The limit $\omega \to \infty$ can also be taken in the recurrence relation (4.21), and so one winds up with commuting PDOs $J_1(x,\hat{p}), \ldots, J_N(x,\hat{p})$. In particular, one gets in this way

$$J_1 = \sum_j \hat{p}_j, \quad J_2 = \sum_{j<k} \hat{p}_j\hat{p}_k - a^2 \sum_{j=1}^N \exp\mu(x_j - x_{j-1}),$$

so that

$$\tfrac{1}{2}J_1^2 - J_2 = \sum_{j=1}^N \left(\tfrac{1}{2}\hat{p}_j^2 + a^2 \exp\mu(x_j - x_{j-1})\right) = H(\text{V}).$$

Substituting (2.60) and taking $a \to 0$ now yields the nonperiodic Toda analogs, in particular

$$J_N = \exp\left(-\sum_{j=2}^N \exp\mu(x_j - x_{j-1})\partial_{p_j}\partial_{p_{j-1}}\right)p_1\ldots p_N. \qquad \text{(VI) (4.30)}$$

Alternatively, one can obtain the commuting type VI PDOs from the hyperbolic PDOs by substituting (2.69) and (2.70) in $J_1, \ldots, J_N$ with

$$h(x) = -\frac{g^2\mu^2}{4\,\mathrm{sh}^2(\mu x/2)} \qquad \text{(II) (4.31)}$$

and taking $\epsilon \to 0$. To complete the connection diagram (2.62) at the quantum nonrelativistic level, it remains to specify the direct IV $\to$ VI transition. To this end we use once more (4.28), but now with $\lambda \equiv \omega$. Substituting then (2.72) and (2.73) in J_N (4.19), it follows that J_N converges to (4.30); taking successively $k = N-1, \ldots, 1$, one deduces that J_k converges to the type VI J_k.

To conclude this subsection, we elaborate on the relation between J_1, $\ldots, J_N$ and the symmetric functions $S_1, \ldots, S_N$ of the type II Lax matrix (2.35) (and its type I and III versions), its λ-dependent type IV generalization (2.51), and the type V and VI Lax matrices (2.57), (2.59), respectively. First, it should be noted that whenever J_N and S_N are equal, equality of J_k and S_k for all $k \in \{1,\ldots,N\}$ follows. Indeed, the functions $S_1, \ldots, S_N$ also satisfy the recurrence relation (4.21). (The coefficient of p_j in S_k equals the sum of all principal minors of order $k-1$ not containing the index j. Each such minor does not contain $N-k+1$ indices, so it occurs $N-k+1$ times in the Poisson bracket at the right-hand side.)

Consequently, if $J_N(x,p)$ were equal to $S_N(x,p)$, the connection diagram (2.62) for the quantum versions would be immediate from the classical connections detailed in Section 2.3. Now when we choose $h(x)$ equal to (4.28), we must choose the same λ in (2.51) to obtain equality for $N = 2$. But then we do *not* get equality for $N = 3$, unless

$$\lambda \equiv \omega_i, \quad i = 1, 2, 3, \quad \omega_1 \equiv \omega, \quad \omega_2 \equiv -\omega - \omega', \quad \omega_3 \equiv \omega' \quad (4.32)$$

(modulo the period lattice), and even for these three choices, it is quite likely that one does not get equality for *arbitrary N*.

Turning now to the assertions of Ref. 1, it is claimed on p. 326 that their J_N (eq. (4.8) in l.c.) equals the determinant of Lax matrices specified on pp. 322–323. However, no proof of this claim is presented. Instead, the reader is referred to the papers by Sawada and Kotera [50] and Wojciechowski [51] already mentioned. But in these papers equality of J_N to a determinant is neither proved nor claimed to be valid.

Translated into our notation, the above claim says in particular that when one takes $h(x)$ in J_N (4.19) equal to (4.31), then J_N equals $S_N(L)$, with L given by (2.35). Undoubtedly this is true, and we have checked equality for $N \leq 4$. However, we are not aware of a complete proof. Changing $1/\mathrm{sh}$ to coth in (4.31) and in (2.35), Ref. 1 claims once more equality of J_N and S_N. (This choice amounts to the $\lambda = i\pi/\mu$ specialization of the λ-dependent type II Lax matrix, i.e., (2.51) with $\omega = \infty$.) This claim is *false*: J_4 and S_4 differ by a constant, so J_N and S_N differ for all $N \geq 5$, too.

In the elliptic case Ref. 1 allows three choices for h and the Lax matrix. These choices amount to the above choices (4.32). Again, the claim that J_N equals S_N is incorrect in general: J_4 and S_4 differ by a constant that depends on the choice, so that nontrivial differences for $N \geq 5$ result. We conjecture that the correct result for arbitrary $\lambda \in \mathbb{C}$ in (2.51) reads

$$S_N(\lambda) = J_N(h) + \sum_{k=0}^{N-2} c_{k,N}(\lambda) J_k(h), \quad h(x) = -g^2 \mathcal{P}(x) \quad (?) \ (4.33)$$

where $J_1(h), \ldots, J_N(h)$ are given by (4.22). (We have checked (4.33) for $N \leq 3$.) For the periodic Toda case we conjecture that

$$S_N(w) = J_N + (ia)^N (w^{-1} + (-)^N w) \quad (?)$$

with $S_N(w)$ the determinant of (2.57) and J_N given by (4.29). If so, equality of S_k and J_k for $k < N$ and equality of the type VI S_k and J_k for $k \leq N$ would follow, of course.

4.3   The Relativistic Case: Commuting AΔOs

As already discussed in Section 4.1, we run into ordering problems when we perform the canonical quantization (4.1) in the Poisson commuting Hamiltonians $S_k(x,p)$ (3.19), $\pm k \in \{1, \ldots, N-1\}$. More precisely, in writing down $S_k(x,p)$, we have automatically opted for an ordering, but for this ordering the prescription (4.1) does not necessarily yield commuting AΔOs.

As a model for the ambiguity at hand, let us look at a Hamiltonian of the form $h(x,p) \equiv e^p f(x)$. Writing it as $f(x)e^p$ yields a different AΔO upon quantization (unless $f(x)$ happens to have period $i\hbar$). As it turns out, both of these orderings in (3.19) give rise to noncommuting AΔOs. (Taking $N > 2$ and generic f, f_T.) Now for these two choices the resulting AΔOs are not even formally symmetric, so one can try next the orderings symbolized by $f(x)^{1/2}e^p f(x)^{1/2}$ and $e^{p/2}f(x)e^{p/2}$ (which do yield formally symmetric AΔOs). Again, these choices spoil commutativity, though. At this point it should be emphasized that no general results are known from which an ordering choice preserving commutativity would follow.

Such a choice does exist, however. Specializing first to the type I case, where $f(x) = (1 + \beta^2 g^2/x^2)^{1/2}$ (with $f(x) > 0$ for $x \in \mathbb{R}^*$), it can be symbolized by writing the model Hamiltonian $h(x,p)$ as $f_-(x)e^p f_+(x)$, where $f_\pm(x) \equiv (1 \pm i\beta g/x)^{1/2}$. Recalling $\beta, g > 0$, the square-root branches may and will be fixed by requiring $f_\pm(x) \to 1$ for $g \downarrow 0$. (To be quite precise, we require this for $x \in \mathbb{R}^*$. Since the x-shifts involved are in the imaginary direction, the branch points off the real axis are not encountered.)

We now turn to the corresponding ordering choice for the type IV Hamiltonians: It is obtained by choosing

$$f_\pm(x) \equiv \left(\frac{s(x \pm i\beta g)}{s(x)} \right)^{1/2}, \quad g \downarrow 0 \implies f_\pm(x) \to 1. \tag{4.34}$$

Written out, the corresponding AΔOs read

$$S_{\pm k}(x, \hat{p}) \equiv \sum_{\substack{I \subset \{1,\ldots,N\} \\ |I|=k}} \prod_{\substack{i \in I \\ j \notin I}} f_\mp(x_i - x_j)$$
$$\times \exp\left(\pm \beta \sum_{i \in I} \hat{p}_i \right) \prod_{\substack{i \in I \\ j \notin I}} f_\pm(x_i - x_j). \quad \text{(I − IV)} \tag{4.35}$$

Thus, their classical symbols yield the same functions as (3.19); moreover, (3.21) holds true again. (To check this, use $f_\delta(-x) = f_{-\delta}(x)$.) Using (3.18), we also get AΔOs $H(x,\hat{p})$ and $P(x,\hat{p})$ belonging to the algebra generated by the AΔOs (4.35). It can be shown that the commutativity of this algebra boils down to the functional equations

$$\sum_{\substack{I \subset \{1,\ldots,N\} \\ |I|=k}} \left(\prod_{\substack{i \in I \\ j \notin I}} \frac{s(x_i - x_j - \gamma)s(x_i - x_j + \gamma - \rho)}{s(x_i - x_j)s(x_i - x_j - \rho)} - (x \to -x) \right) = 0 \tag{4.36}$$

which hold true for all $N > 1$, $k \in \{1, \ldots, N\}$, $x \in \mathbb{C}^N, \gamma, \rho \in \mathbb{C}$. These results can be found in Ref. 4, with $s(x)$ replaced by the Weierstrass σ-function; in view of the relation (2.50), this difference is inconsequential.

The functional equations (4.36) encapsulate the integrability of all of the models considered in these lectures. We continue by elaborating on this assertion. First, dividing by ρ and sending ρ to 0, one obtains the functional equations

$$\sum_{\substack{I\subset\{1,\ldots,N\}\\|I|=k}} \sum_{i\in I} \partial_i \prod_{\substack{i\in I\\j\notin I}} \frac{s(x_i - x_j - \gamma)s(x_i - x_j + \gamma)}{s(x_i - x_j)^2} = 0.$$

These amount to the functional equations (3.20) that express involutivity of the classical Hamiltonians $S_1, \ldots, S_N$ [3]. Now we have shown in Section 3 that the latter commutativity result entails integrability for all of the relativistic and nonrelativistic *classical* systems of type I–Vi. Thus, it remains to explain how (4.36) entails integrability for their quantum versions.

We begin with the relativistic level. Of course, then we need only consider the Toda systems, since the type I–III systems are included in type IV via (2.65). In fact, we are going to detail the transitions in the diagram (2.62) at the quantum relativistic level. To this end we adapt the reasoning followed at the classical level to the AΔOs $S_1, \ldots, S_N$, rewritten as

$$S_l = \sum_{\substack{I\subset\{1,\ldots,N\}\\|I|=l}} \prod_{\substack{j\in I\\k\notin I}} f_-(x_j - x_k)f_+(x_j - x_k - i\hbar\beta)\exp(\beta\hat{p}_I); \qquad (4.37)$$

cf. (4.35). (The corresponding transitions for $S_{-1}, \ldots, S_{-N}$ can be dealt with by taking $\beta \to -\beta$.)

First, we consider the transition II → VI. Substituting (3.24) and (2.69) in $f_-(x_j - x_k)f_+(x_j - x_k - i\hbar\beta)$, we obtain a function f_{jk} with limits

$$f_{jk} \to \begin{cases} 1, & |j - k| > 1, \\ (1 + \beta^2 \exp[\mu(x_j - i\hbar\beta - x_{j-1})])^{1/2}, & k = j - 1, \\ (1 + \beta^2 \exp[\mu(x_{j+1} - x_j)])^{1/2}, & k = j + 1, \end{cases}$$

for $\epsilon \to 0$. Hence, the limit of S_l (4.37) exists and can be rewritten

$$S_k = \sum_{\substack{I\subset\{1,\ldots,N\}\\|I|=k}} \prod_{\substack{j\in I\\j+1\notin I}} f_T(x_{j+1} - x_j)\exp(\beta\hat{p}_I) \times \prod_{\substack{j\in I\\j-1\notin I}} f_T(x_j - x_{j-1}) \quad \text{(Toda)} \quad (4.38)$$

with the convention (3.16) in effect.

Second, we handle the IV → V limit, starting again from (4.37). To this end we use the analog of the factorization (3.27) for the functions f_+ and f_-, and mimic the reasoning in the classical case. Thus we introduce

$$S_{r,k} \equiv \frac{S_k}{\rho(g)^{k(N-k)}}, \quad \rho(g) \equiv \exp\left(\frac{\mu\beta^2 g(g - \hbar)}{4\omega}\right),$$

and substitute (2.66) and (3.29). Then the limit $\omega \to \infty$ exists and can be rewritten as (4.38), but now with (3.15) in force. (Just as in the classical case, the infinite products supply the extra boundary terms, as compared to the nonperiodic case.)

Third, the IV $\to$ VI transition can be made by substituting (2.72) and (3.31) in the AΔOs $S_{r,k}$ and taking $\omega \to \infty$. To complete the diagram (2.62) it remains to specify the V $\to$ VI limit. As before, it suffices to substitute (2.60) in the type V AΔOs S_k (4.38) and take $a \to 0$ to obtain the type VI S_k.

It can also be shown directly that the Toda AΔOs (4.38) commute. This can be reduced to the functional equations

$$
\sum_{\substack{I \subset \{1,\ldots,N\} \\ |I|=k}} \prod_{\substack{i \in I \\ i-1 \notin I}} f_T^2(x_{i-1} - x_i) \prod_{\substack{i \in I \\ i+1 \notin I}} f_T^2(x_i - x_{i+1} + \lambda)
$$

$$
= \sum_{\substack{I \subset \{1,\ldots,N\} \\ |I|=k}} \prod_{\substack{i \in I \\ i-1 \notin I}} f_T^2(x_{i-1} - x_i + \lambda) \prod_{\substack{i \in I \\ i+1 \notin I}} f_T^2(x_i - x_{i+1}) \quad (4.39)
$$

which hold true for all $N > 1$, $k \in \{1,\ldots,N\}$, $x \in \mathbb{C}^N$, $\lambda \in \mathbb{C}$ [37]. (Note this yields the classical functional equations (3.20) when one divides by λ and takes λ to 0.)

We now consider the nonrelativistic limit $\beta \to 0$, handling first the elliptic case. We start from commuting AΔOs

$$
A_k(\beta) \equiv \sum_{l=0}^{k} (-)^{k+l} \binom{N-l}{N-k} c_l(\lambda, i\beta g) S_l, \quad k = 1,\ldots,N, \quad (4.40)
$$

where

$$
c_k(\lambda, \alpha) \equiv s(\lambda)^{-k} s(\lambda - \alpha)^{k-1} s(\lambda + (k-1)\alpha), \quad k = 1,\ldots,N.
$$

The point of this definition is that the classical versions of these AΔOs, after division by β^k, converge to the Hamiltonians $S_k(L_{nr})$ for $\beta \to 0$, where L_{nr} is the type IV Lax matrix (2.51). (To see this, recall (3.38), (3.39), and (3.45)–(3.48).)

Expanding c_l and S_l in powers of β, we now write

$$
A_k(\beta) = \sum_{m=0}^{\infty} A_{k,m} \beta^m. \quad (4.41)
$$

(This expansion is meant in the sense of formal power series.) Then we calculate

$$
A_{1,0} = 0, \quad A_{1,1} = \sum_j \hat{p}_j \quad (4.42)
$$

$$A_{2,0} = A_{2,1} = 0,$$

$$A_{2,2} = \sum_{j<k} \left(\hat{p}_j \hat{p}_k - g(g-\hbar) \left(\mathcal{P}(x_j - x_k) + \frac{\eta}{\omega} \right) \right)$$
$$+ \frac{g^2}{2} N(N-1) \left(\mathcal{P}(\lambda) + \frac{\eta}{\omega} \right). \quad (4.43)$$

It is important to observe that this result differs from the classical expansion: In $A_{2,2}$ one has a coefficient $g(g-\hbar)$ and not g^2, since the partials in (4.35) do not commute with the f_+-factors. (Equivalently, the β-dependence of the coefficients in (4.37) differs from that of their classical versions.)

Next, we note that the integer

$$n_k \equiv \min\{l \,|\, A_{k,l} \neq 0\}, \quad k = 1, \ldots, N,$$

equals k in the classical case, and must be $\leq k$ in the quantum case, since we clearly have

$$A_{k,k} = \sum_{i_1 < \cdots < i_k} \hat{p}_{i_1} \cdots \hat{p}_{i_k}, \quad g = 0.$$

Taking for granted that $n_k = k$ in the quantum case as well, it would follow that

$$[A_{k,k}, A_{l,l}] = 0, \quad k, l = 1, \ldots, N,$$

and since we have (using obvious notation)

$$H_{\mathrm{nr}}\left(g^2 \to g(g-\hbar)\right) = \tfrac{1}{2}(A_{1,1})^2 - A_{2,2} + \text{constant}$$

integrability of the nonrelativistic Hamiltonian would follow.

Unfortunately, a direct proof that n_k equals k appears intractable. The problem is that one gets an avalanche of additional terms arising when partials act on f_+-factors. A priori, these extra terms might lower n_k as compared to the classical case, and yield PDOs A_{k,n_k} that are not independent for $k = 1, \ldots, N$. Note also that the connection of $A_{k,k}$ to the quantized symmetric functions of L_{nr} is quite opaque due to the extra terms.

We shall now show that n_k indeed equals k, by making a detour involving a function $w(\beta; x)$ that is also an important ingredient for obtaining relativistic eigenfunctions (as we shall see in Section 6). In the process, we shall obtain a rather explicit formula for $A_{k,k}$.

The function w is introduced in Ref. 31. It is a solution to the first-order analytic difference equation

$$w(x - i\hbar\beta) = \frac{f_-^2(x)}{f_+^2(x - i\hbar\beta)} w(x) \quad (4.44)$$

that is meromorphic, even and 2ω-periodic in x. Moreover, it has no poles for real x and is positive for $x \in (0, 2\omega)$, and it satisfies

$$\lim_{\beta \to 0} w(\beta; x) = C \exp\left(2\frac{g}{\hbar} \ln s(x)\right), \quad x \in (0, 2\omega). \tag{4.45}$$

Here, the positive constant C is irrelevant for what follows, and the logarithm is chosen real. (Note that (2.63) entails positivity of $s(x)$ on $(0, 2\omega)$.) Setting

$$\Delta(\beta; x) \equiv C' \prod_{j<k} w(\beta; x_j - x_k), \quad C' > 0,$$

and using (4.37), evenness of w and (4.44), we obtain transformed AΔOs

$$S_l^t \equiv \Delta^{-1/2} S_l \Delta^{1/2} = \sum_{\substack{I \subset \{1,\ldots,N\} \\ |I|=l}} \prod_{\substack{j \in I \\ k \notin I}} f_-^2(x_j - x_k) \exp(\beta \hat{p}_I). \tag{4.46}$$

Next, we replace S_l by S_l^t in (4.40), obtaining an AΔO $A_k^t(\beta)$. In view of (4.45), the formal power series expansion of this AΔO has as its lowest coefficient the PDO

$$A_{k,n_k}^t = \Pi(x)^{-1} A_{k,n_k} \Pi(x) \tag{4.47}$$

with

$$\Pi(x) \equiv \prod_{j<k} |s(x_j - x_k)|^{g/\hbar}. \tag{4.48}$$

To prove $n_k = k$, it therefore suffices to show that $A_{k,l}^t$ vanishes for $l < k$.

The crux is now, that we need only show this for the *classical* version $A_k^t(x, p)$. Indeed, in S_k^t (4.46) all partials occur to the right of $\hbar$-*independent* coefficients, so the PDO $A_{k,l}^t$ is obtained by substituting (4.1) in the normally ordered expansion coefficient $A_{k,l}^t(x, p)$. Here, *normal ordering* denotes the procedure of putting x-dependent coefficients to the left of monomials in $p_1, \ldots, p_N$ (classical case) or $\hat{p}_1, \ldots, \hat{p}_N$ (quantum case); we shall symbolize normal ordering by double dots in the sequel. In particular, we can now rewrite (4.47) as

$$A_{k,n_k} = \Pi(x) : A_{k,n_k}^t(x, \hat{p}) : \Pi(x)^{-1}.$$

We proceed by proving that $A_{k,l}^t(x, p)$ vanishes for $l < k$. To this end we first notice that the function $c_l(\lambda, i\beta g) S_l^t(x, p)$ is the lth symmetric function of the matrix

$$\tilde{L}_{jk} \equiv \tilde{d}_j D_{jk}$$

with D given by (3.47), and

$$\tilde{d}_j \equiv \exp(\beta p_j) \prod_{l \neq j} f_-^2(x_j - x_l).$$

(To verify this, one need only repeat the calculation leading to (3.48), making the pertinent changes in (3.37); note that $f_-^2(x)f_-^2(-x) = f^2(x)$.)
   Next, we observe that

$$\tilde{L} = 1_N + \beta(L_{\mathrm{nr}} + E) + O(\beta^2), \quad \beta \to 0,$$

where the extra matrix is given by

$$E \equiv \mathrm{diag}\big(z_1(x), \ldots, z_N(x)\big)$$

with

$$z_j(x) \equiv -ig \sum_{l \neq j} \frac{s'(x_j - x_l)}{s(x_j - x_l)}, \quad j = 1, \ldots, N. \tag{4.49}$$

Now this entails

$$\lim_{\beta \to 0} \beta^{-k} A_k^t(\beta) = S_k(L_{\mathrm{nr}} + E).$$

From this we deduce not only that $A_{k,l}^t(x,p)$ vanishes for $l < k$ (as desired), but also—returning to the quantum level—that

$$A_{k,k} = \Pi(x){:}S_k\big(L_{\mathrm{nr}}(x,\hat{p}) + E(x)\big){:}\Pi(x)^{-1}. \tag{4.50}$$

   The upshot is that we have now derived integrability of the nonrelativistic type IV quantum system from that of its relativistic version. Moreover, we have obtained rather explicit formulas for the commuting PDOs $A_{k,k}$. To exploit the formula (4.50), it is important to observe (recall (4.48) and (4.49))

$$\Pi(x)\hat{p}_j\Pi(x)^{-1} = \hat{p}_j - z_j(x), \tag{4.51}$$

$$\hat{p}_j(z_i) = -g\hbar\left(\mathcal{P}(x_j - x_i) + \frac{\eta}{\omega}\right) \equiv -\tilde{h}(x_j - x_i), \quad j \neq i. \tag{4.52}$$

Let us call the latter formula a *contraction*. Expanding (4.50) and using (4.51), we may move all $\hat{p}_j$'s to the right, picking up contractions along the way. Then we wind up with an expression that can be written

$$A_{k,k} = S_k\big(L_{\mathrm{nr}}(x,\hat{p})\big) + R_k(x,\hat{p}) \tag{4.53}$$

where $R_k(x,p)$ is normally ordered and consists of all terms involving at least one contraction. As such, $R_k(x,p)$ has degree $< k - 1$ in p. From this one easily deduces that $A_{k,k}(x,\hat{p})$ has the form (4.25), (4.26). (For $k = 1$ and $k = 2$ we reobtain (4.42) and (4.43), respectively, as should be the case, of course.) Therefore, we are now in the position to invoke the uniqueness result of Ref. 49 detailed in the paragraph containing (4.27): The commutative algebra generated by $A_{k,k}$, $k = 1, \ldots, N$, coincides with the algebra generated by $J_1(h), \ldots, J_N(h)$, with

$$h(x) \equiv -g(g - \hbar)\mathcal{P}(x). \tag{4.54}$$

It is not hard to see that $A_{k,k}$ is actually a *linear* combination of J_1, $\ldots$, J_k: Any monomial in $A_{k,k}(x,p)$ involves a given p_j only once, so no products of the J_l can occur. Notice also that the contractions are responsible for changing g^2 to $g(g - \hbar)$ in (4.54). In fact, we conjecture that one has

$$\Pi(x) \sum_{|I|=l} :\prod_{i\in I}(\hat{p}_i + z_i(x)):\Pi(x)^{-1} = J_l(\tilde{h}), \quad l = 1, \ldots, N, \quad (?) \tag{4.55}$$

with $\tilde{h}$ given by (4.52). (We have checked this for $l \leq 4$.) Clearly, (4.55) would be useful to render the formula (4.53) for $A_{k,k}$ even more explicit.

Of course, the above holds true for the type I–III systems, too. If one could prove (4.55), it would easily follow from the IV → V and II → VI transitions that one has

$$A_{T,k,k} = S_k\big(L_{\mathrm{nr}}(x,\hat{p})\big) \qquad \text{(Toda)} \quad (?) \tag{4.56}$$

in the Toda case. (Here, $A_{T,k,k}$ denotes the kth expansion coefficient of the AΔO $A_{T,k}(\beta)$ that is defined via (4.40) with $c_l S_l$ replaced by S_l (4.38) for type VI and by Σ_l (3.52) (with S_l given by (4.38)) for type V.) In particular, this would entail that the quantized symmetric functions of the Lax matrices (2.57) and (2.59) commute.

Independent of the validity of (4.56), it can be shown that $n_k = k$ in the Toda case [5]; the reasoning presented in Ref. 5 applies to type I–III as well, but leaves open the type IV case. (The statements in Ref. 5 concerning the relation of $A_{k,k}$ and quantized symmetric functions should be ignored, though; this is because these rely on the unproven assertions in Refs. 1 and 2 discussed above.)

5   Action-Angle Transforms

5.1   Introductory Examples

At the end of Section 2.1 we have introduced action-angle maps, a notion associated with an arbitrary Liouville integrable system

$$\mathcal{S} = \langle \Omega, \omega, I_1, \ldots, I_N \rangle. \tag{5.1}$$

As we have seen in Sections 2 and 3, the Calogero-Moser and Toda systems and their relativistic generalizations *are* Liouville integrable, so one is led to the problem of constructing *explicit* action-angle maps for these systems. Now the Liouville-Arnol'd theorem is of little help in that enterprise, since it is merely concerned with existence and general structure under some quite subtle assumptions, whose direct verification is often intractable. Provided the joint orbits (2.14) contained in Ω_i are compact (so that $k_i = N$ in (2.16)), there exist integral representations for the action and angle variables. (These are actually the reason for the term *integrable* system, historically speaking.) Even so, one would really like to express these integrals in terms of known functions, or obtain at least more information on the range of variation of the actions, the functional dependence of $I_1, \ldots, I_N$ on the actions, etc. More generally, the decomposition into invariant submanifolds $\Omega_1, \Omega_2, \ldots$ should be made explicit.

In order to provide more perspective for this circle of problems, we shall consider some quite elementary examples. Our starting point is the Hamiltonian

$$H(x, p) = \frac{p^2}{2} + V(x), \quad V \in C^\infty(\mathbb{R}), \tag{5.2}$$

defined on $\Omega = \mathbb{R}^2$ equipped with its standard form $\omega = dx \wedge dp$. Then we have $dH(x_0, p_0) \neq 0$ unless both $p_0 = 0$ and $V'(x_0) = 0$; such points yield equilibrium solutions $(x(t), p(t)) = (x_0, p_0)$, $\forall t \in \mathbb{R}$. Since $dH \neq 0$ on an open dense set, H is integrable.

The H flow is not complete without further restrictions on V, however. For instance, taking $V = -x^4/2$, one gets a solution $(1/(1-t), 1/(1-t)^2)$ to Hamilton's equations (2.1) that is defined only for $t < 1$. To ensure completeness (and hence Liouville integrability), let us henceforth assume that V is bounded below. (By energy conservation this entails an upper bound on $|p|$; since one has $\dot{x} = p$, this prevents escape to infinity in finite time.)

Of course, the simplest external field with this property is the constant field

$$V(x) = V_0.$$

Discarding the line of equilibria $\{p = 0\}$, one is left with two open, connected and invariant submanifolds Ω_1 and Ω_2 on which $p > 0$ and $p < 0$, respectively. These are of the form $\mathbb{R} \times A_i$ with A_i open and connected, and the flow $(x(t), p(t)) = (x_0 + p_0 t, p_0)$ is obviously linear in time. Hence one can take $\widehat{\Omega}_i = \Omega_i$ and Φ_i equal to the identity, $i = 1, 2$.

Next, we consider the harmonic field

$$V(x) = \frac{x^2}{2}. \tag{5.3}$$

In this case one clearly obtains the flow

$$e^{tH}(x_0, p_0) = (x_0 \cos t + p_0 \sin t, p_0 \cos t - x_0 \sin t).$$

This flow is nonlinear in t, but away from the equilibrium $(0,0)$, it can be linearized by a symplectic map

$$\Phi: \langle \Omega_1, dx \wedge dp \rangle \to \langle \widehat{\Omega}, d\hat{x} \wedge d\hat{p} \rangle, \quad (x,p) \mapsto (\hat{x}, \hat{p}), \qquad (5.4)$$

where

$$\Omega_1 \equiv \Omega \setminus \{(0,0)\}, \qquad (5.5)$$
$$\widehat{\Omega} \equiv \mathbb{T}^1 \times (0, \infty).$$

As announced below (2.17), the torus $\mathbb{T}^1$ is viewed as $\mathbb{R}/2\pi\mathbb{Z}$ and coordinatized by $\hat{x} \in (-\pi, \pi]$; explicitly, Φ reads

$$\hat{x} \equiv \arctan(x/p), \quad \hat{p} \equiv (p^2 + x^2)/2 = H(x,p), \qquad (5.6)$$

so that the inverse $\mathcal{E}$ is given by

$$x = (2\hat{p})^{1/2} \sin \hat{x}, \quad p = (2\hat{p})^{1/2} \cos \hat{x},$$

and one has

$$\exp(t\widehat{H})(\hat{x}_0, \hat{p}_0) = (\hat{x}_0 + t, \hat{p}_0). \qquad (5.7)$$

In this example all orbits are periodic, but in general a Hamiltonian of the form (5.2) has both periodic and nonperiodic orbits. Indeed, taking as a third example

$$V(x) = -\exp(-x^2) \qquad (5.8)$$

the level set $H(x,p) = E$ yields an equilibrium for $E = -1$, a periodic orbit for $E \in (-1,0)$, and two nonperiodic orbits for $E \geq 0$. Discarding the equilibrium at $(0,0)$ and the level set $E = 0$ separating bound and unbound orbits (*separatrix*), one obtains three open, connected and invariant submanifolds $\Omega_1, \Omega_2, \Omega_3$, corresponding to $E \in (-1,0)$, and to $E > 0$ with $p > 0$ and $p < 0$, respectively.

The action variable $\hat{p}(x,p)$ on Ω_1 is now determined (uniquely up to a constant) as the function $f(H(x,p))$ that is such that all of its orbits have primitive period 2π. It can be seen that the function

$$f(E) = \frac{1}{2\pi} A(E) = \frac{2}{\pi} \int_0^{V^{-1}(E)} p(x,E)\, dx, \quad E \in (-1,0), \qquad (5.9)$$

$$p(x,E) \equiv \left(\frac{E - V(x)}{2} \right)^{1/2}, \quad V(x) \leq E, \qquad (5.10)$$

has this property. Here, $A(E)$ denotes the phase space area enclosed by the level curve $H(x, p) = E$. Note this yields $\hat{p} = H$ for the special case (5.3), in agreement with (5.6). For (5.8) or any other potential with the same shape (such as $-1/\operatorname{ch}^2 x$), the oscillation period is nonconstant on Ω_1, so the transformed Hamiltonian must be a nonlinear function of $\hat{p}$. (In this regard the above harmonic oscillator example is highly nongeneric.)

There is also an integral representation for the canonically conjugate angle variable $\hat{x} \in (-\pi, \pi]$ (cf. Ref. 9, p. 281); this variable is uniquely determined up to addition (mod 2π) of an arbitrary function of $\hat{p}$. Since we have $H = f^{-1} \circ \hat{p}$, the H flow increases the angle by 2π after a time $2\pi \partial f / \partial E \equiv T(E)$. Using (5.9) and (5.10) to calculate $T(E)$, one now verifies that this yields the correct oscillation period.

The integral trajectories in Ω_2 and Ω_3 are scattering orbits, and so one has far more freedom in the choice of action coordinate. Indeed, one can take $\hat{p} = f(H)$ for *any* function $f \in C^\infty((0, \infty))$ with positive or negative derivative on $(0, \infty)$, and obtain a canonically conjugate angle variable $\hat{x}(x, p)$ taking values in $\mathbb{R}$. There is, however, a special choice that is singled out by a physical interpretation: one can choose $\hat{p}(x, p)$ equal to the limit of $p(t)$ for $t \to \infty$, so that $\hat{p} = (2H)^{1/2}$ on Ω_2 and $\hat{p} = -(2H)^{1/2}$ on Ω_3. As canonically conjugate position one can choose the asymptotic position x^+ determined by

$$\big(x(t), p(t)\big) \sim (x^+ + t\hat{p}, \hat{p}), \quad t \to \infty.$$

Alternatively, one can work with the $t \to -\infty$ asymptotics, yielding action-angle variables $(x^-, \hat{p})$. Restricting attention to Ω_2, canonicity of the maps

$$U_\pm : \mathbb{R} \times (0, \infty) \equiv \widehat{\Omega}_2 \to \Omega_2, \quad (x^\pm, \hat{p}) \mapsto (x, p), \qquad (5.11)$$

is then a consequence of scattering theory. Indeed, they are just the wave maps

$$U_\pm = \lim_{t \to \pm\infty} \exp(-tH) \circ \exp(t\widehat{H}),$$

$$\widehat{H}(x^\pm, \hat{p}) \equiv \frac{\hat{p}^2}{2} = U_\pm^* H, \qquad (5.12)$$

which are related via the scattering map

$$S \equiv U_+^{-1} U_- : \widehat{\Omega}_2 \to \widehat{\Omega}_2, \quad (x^-, \hat{p}) \mapsto (x^+, \hat{p}).$$

(Writing

$$x^+ = x^- + \delta(\hat{p})$$

the scattering is encoded in the function $\delta(\hat{p})$: it registers how much the particle has advanced as compared to a freely moving particle with velocity

$\hat{p}$.) Summarizing, one can choose the action-angle map Φ_2 equal to U_+^{-1} or U_-^{-1}.

As a fourth example, let us take

$$V(x) = x^2 \exp(-x^2).$$

This potential has a well around the origin and two maxima $V(\pm 1) = e^{-1}$. Thus the level set $E = 0$ yields an equilibrium $(0,0)$, whereas the level set $E \in (0, e^{-1})$ splits up into three orbits: Two scattering orbits (reflection at the bumps) and one periodic orbit (oscillation around the origin). For $E > e^{-1}$ the level set has two connected components, corresponding to scattering orbits (transmission from left and right). The separatrix level set $E = e^{-1}$ is connected, but not diffeomorphic to $\mathbb{R}$ or $\mathbb{T}^1$. Indeed, this set splits up into eight distinct orbits: Discarding the two equilibria at $(\pm 1, 0)$, one is left with six connected components, each of which yields an orbit. (Draw the phase diagram to see this.) Deleting the origin and the separatrix from $\Omega = \mathbb{R}^2$, one obtains a set with five connected components, viz., one bound state submanifold Ω_1, two reflection submanifolds Ω_2, Ω_3 and two transmission submanifolds Ω_4, Ω_5. On each of these the flow can be linearized by a canonical map Φ_i, involving the 1-torus and the integral (5.9) on Ω_1, and scattering theory objects on $\Omega_2, \ldots, \Omega_5$.

Turning now to the case $N > 1$, we note that there is a trivial, yet instructive way of manufacturing Liouville integrable systems: One can take

$$I_j(x, p) \equiv \frac{p_j^2}{2} + V_j(x_j), \quad j = 1, \ldots, N, \ V_j \in C^\infty(\mathbb{R}), \ V_j \geq 0,$$

to get N commuting Hamiltonians on $\Omega = \mathbb{R}^{2N}$ with its standard form (2.3). The submanifolds $\Omega_1, \Omega_2, \ldots$ and action-angle maps $\Phi_1, \Phi_2, \ldots$ then have a product structure, and the complement of $\Omega_1 \cup \Omega_2 \cup \cdots$ will consist of a union of sets that are built up from equilibria and separatrices for each of the Hamiltonians $I_1, \ldots, I_N$. The dimension of the latter excluded sets can vary from $2N - 1$ (take, e.g., (x_1, p_1) on a separatrix for I_1) to 0 (take (x_j, p_j) to be an equilibrium for I_j, $j = 1, \ldots, N$). Moreover, the integer k_i in (2.16) takes all allowed values $0, \ldots, N$ when each of $I_1, \ldots, I_N$ has both bound and scattering orbits.

From these examples one already gleans that the $N > 1$ situation can be quite complicated. An additional complication for $N > 1$ cannot be easily illustrated, since it is absent for the rather artificial product situation just discussed: To obtain invariant submanifolds of the form (2.15), (2.16) one may have to discard additional separatrices whose location is a matter of choice (just as branch cuts ensuring one-valuedness can be chosen at will). In geometric parlance, this phenomenon amounts to the occurrence of a nontrivial fiber bundle structure (with the fiber equal to (2.16)). We shall encounter an explicit example of this situation toward the end of

Section 5.3. Since the general theory should cover all possibilities at once, it is clear that one must make rather intricate technical assumptions in order to obtain invariant submanifolds that are symplectically diffeomorphic to manifolds of the form (2.15), (2.16). A detailed account can be found in Ref. 11; see also Ref. 52 for general information on the way in which the invariant tori can bifurcate and degenerate, and on various related matters.

5.2 Wave Maps and Pure Soliton Systems

In view of the general picture sketched in the previous subsection one should be prepared to encounter considerable complications in constructing explicit action-angle transforms for a given Liouville integrable system. There exists however a physically important class of integrable systems for which neither equilibria nor separatrices occur, so that one needs to consider only one invariant submanifold, namely, all of Ω. These integrable systems are defined by the Hamiltonian H (2.13) on the phase space Ω (2.23), with $V(x)$ a pair potential having the salient features of the type II potential $1/\operatorname{sh}^2 \nu x$. Specifically, $V(x)$ is a strictly monotone decreasing function on $(0, \infty)$ with a divergence for $x \downarrow 0$ preventing collisions, and rapid decay to 0 at ∞. Let us denote the class of such repulsive potentials by $\mathcal{R}$, and fix $V \in \mathcal{R}$. We proceed by describing how the inverses of the wave maps from scattering theory can now be used as action-angle maps, in much the same way as for the above example; cf. (5.11)–(5.12). (In the example, however, the potential is attractive, so that transmission occurs.)

First of all, any point in Ω yields an orbit with asymptotics

$$\big(x(t), p(t)\big) \sim (x^\pm + tp^\pm, p^\pm), \quad t \to \pm\infty,$$

where

$$p_1^+ > \cdots > p_N^+, \quad p_1^- < \cdots < p_N^-. \tag{5.13}$$

Introducing incoming and outgoing phase spaces

$$\Omega^+ \equiv \{(x^+, p^+) \in \mathbb{R}^{2N} \mid p_N^+ < \cdots < p_1^+\}, \tag{5.14}$$
$$\Omega^- \equiv \{(x^-, p^-) \in \mathbb{R}^{2N} \mid p_N^- > \cdots > p_1^-\},$$

equipped with their canonical forms

$$\omega^\delta \equiv \sum_{j=1}^{N} dx_j^\delta \wedge dp_j^\delta, \quad \delta = +, -,$$

and Hamiltonians

$$H^\delta(x^\delta, p^\delta) \equiv \frac{1}{2}\sum_{j=1}^{N}(p_j^\delta)^2, \quad \delta = +, -, \tag{5.15}$$

the wave maps are given by

$$U_\delta \equiv \lim_{t \to \delta\infty} \exp(-tH) \circ \exp(tH^\delta), \quad \delta = +, -. \tag{5.16}$$

Observe that the composition makes sense for $t \to \delta\infty$: the restrictions (5.13) ensure that the point $(x^\delta + tp^\delta, p^\delta)$ belongs to Ω (2.23) for δt large enough. Since the Hamiltonian flows at the right-hand side of (5.16) are canonical, the wave maps are symplectic maps

$$U_\delta : \Omega^\delta \to \Omega, \quad (x^\delta, p^\delta) \mapsto (x, p).$$

(To prove this rigorously is not easy, though.) The intertwining relations

$$U_\delta \circ \exp(tH^\delta) = \exp(tH) \circ U_\delta, \quad \delta = +, -, \tag{5.17}$$

show that U_δ^{-1} can indeed be viewed as a linearizing canonical transformation $\Phi_\delta : \Omega \to \Omega^\delta$; cf. also (2.11), (2.12).

Liouville integrability of H is now easily established. Indeed, defining

$$H_k^\delta(x^\delta, p^\delta) \equiv \frac{1}{k} \sum_{j=1}^{N} (p_j^\delta)^k, \quad k = 1, \ldots, N,$$

the Hamiltonians $H_1^\delta, \ldots, H_N^\delta$ are obviously in involution. Moreover, they are independent; their gradients are in fact linearly independent on *all* of Ω^δ, since the relevant determinant is a Vandermonde determinant that has no zeros on Ω^δ. Setting

$$\check{H}_k^\delta \equiv H_k^\delta \circ U_\delta^{-1}, \quad \delta = +, -, k = 1, \ldots, N, \tag{5.18}$$

it follows that $\check{H}_1^\delta, \ldots, \check{H}_N^\delta$ are in involution, and that their gradients are linearly independent on all of Ω. Since we have

$$H_2^\delta = H^\delta, \quad \check{H}_2^\delta = H, \quad \delta = +, -,$$

in view of (5.15) and the intertwining relations (5.17), it follows that H is integrable. Moreover, because the map U_δ is canonical, (5.18) entails that it intertwines the flows $\exp(tH_k^\delta)$ and $\exp(t\check{H}_k^\delta)$. Since the former flows are manifestly complete, the latter are, too. Hence, H is Liouville integrable, as claimed.

Now for a general $V \in \mathcal{R}$ it is quite unlikely that the commutative subalgebras of $\mathcal{I}_H$ generated by $\check{H}_1^+, \ldots, \check{H}_N^+$ and $\check{H}_1^-, \ldots, \check{H}_N^-$ are equal, and even more unlikely that $\check{H}_k^+$ and $\check{H}_k^-$ are equal for $k > 2$. (By translation invariance one does have equality for $k = 1$. Indeed, in that case one obtains the total momentum $\sum_{j=1}^{N} p_j$.) Assuming however that they *are* equal, it follows that

$$\sum_{j=1}^{N} p_j^+(x, p)^k = \sum_{j=1}^{N} p_j^-(x, p)^k, \quad k = 1, \ldots, N, \tag{5.19}$$

and so we deduce that the scattering map

$$S = U_+^{-1} U_- : \Omega^- \to \Omega^+, \quad (x^-, p^-) \mapsto (x^+, p^+)$$

conserves momenta. More specifically, from (5.13) we deduce (2.47).

Conversely, whenever S conserves momenta, one has the equalities (5.19) and so the pullback Hamiltonians $\check{H}_k^+$ and $\check{H}_k^-$ are equal for $k = 1, \dots, N$. A priori, these Hamiltonians need not have a polynomial dependence on the momenta $p_1, \dots, p_N$, however. But for the only potentials in $\mathcal{R}$ for which conservation of momenta is known to hold true, this turns out to be the case. Indeed, these potentials are the type II potentials $V(x) = 1/\operatorname{sh}^2 \nu x$, whose long-time asymptotics has already been discussed in Section 2.2; cf. (2.42)–(2.48). In view of (2.46) and (2.33) one has

$$\check{H}_k^\delta = H_k, \quad \delta = +, -, k = 1, \dots, N,$$

and H_k is indeed a polynomial in $p_1, \dots, p_N$ with x-dependent coefficients; cf. (2.41).

It is not known whether any other $V \in \mathcal{R}$ exist for which S conserves momenta, but this seems extremely unlikely. At any rate, it makes sense to single out the integrable systems (5.1) for which not only each initial state is a scattering state, but also the momenta are conserved under the scattering. We shall call such systems *pure soliton systems*. As we have seen in Sections 2 and 3, the systems of type I, II, and VI *are* pure soliton systems, not only at the nonrelativistic, but also at the relativistic level. In the next subsection we shall obtain detailed information on the type I and II wave and scattering maps, but for brevity we do not consider the type VI case. (A study of the latter case can be found in Ref. 37.)

5.3 Systems of Type I, II, and III

As explained in the previous subsection, one can choose one of the two inverse wave maps U_+^{-1}, U_-^{-1} as an action-angle map for the systems of type I and Ii. Indeed, the action-angle map Φ we are going to construct will turn out to be equal to U_+^{-1} for the type I systems, but for type II it is slightly different, and the difference will be crucial. In all four cases the map is written

$$\Phi : \Omega \to \widehat{\Omega}, \quad (x, p) \mapsto (\hat{x}, \hat{p}), \tag{5.20}$$

with Ω the type I and II phase space (2.23), and $\widehat{\Omega}$ the outgoing phase space Ω^+ (5.14). Hence, Ω and $\widehat{\Omega}$ are related by

$$\widehat{\Omega} = I(\Omega), \quad I(x, y) \equiv (y, x), \quad x, y \in \mathbb{R}^N. \tag{5.21}$$

For the type III systems we should consider three distinct phase spaces; cf. (2.24)–(2.30). As it happens, it suffices to delete a codimension-2 variety containing partial equilibria from each of these to obtain an invariant submanifold on which an action-angle map can be defined. Thus, no separatrices occur in this case, too. (Recall that a manifold remains connected after discarding a subset whose codimension is greater than one.)

The key tool in the construction of the action-angle map for all of the systems of type I–III is a commutation relation of the Lax matrix defined in Sections 2.2 and 3.1 with a diagonal matrix $A(x)$ given by

$$A(x) = \mathrm{diag}\big(d(x_1), \ldots, d(x_N)\big), \quad d(y) \equiv \begin{cases} y, & \text{(I)} \\ \exp(\mu y), & \text{(II)} \\ \exp(i\mu y). & \text{(III)} \end{cases} \qquad (5.22)$$

Specializing first to the type I and II cases, the symmetric functions $\check{D}_k(x)$ of $A(x)$ evidently give rise to an integrable system on Ω, so the Hamiltonians

$$D_k \equiv \check{D}_k \circ \mathcal{E}, \quad \mathcal{E} \equiv \Phi^{-1}, \qquad (5.23)$$

yield an integrable system on the action-angle phase space $\widehat{\Omega}$. Now this would not be of much interest by itself; in fact, this observation applies to *any* diagonal matrix of the form (5.22) with $d'(x) > 0$, say. The crux is, however, that the integrable systems thus obtained are of physical interest: they are type I or II systems!

Specifically, denoting the *dual systems* just defined by a caret, they are given by

$$\begin{aligned} \hat{\mathrm{I}}_{\mathrm{nr}} &\simeq \mathrm{I}_{\mathrm{nr}}, \quad \hat{\mathrm{I}}_{\mathrm{rel}} \simeq \mathrm{II}_{\mathrm{nr}}, \\ \widehat{\mathrm{II}}_{\mathrm{nr}} &\simeq \mathrm{I}_{\mathrm{rel}}, \quad \widehat{\mathrm{II}}_{\mathrm{rel}} \simeq \mathrm{II}_{\mathrm{rel}}. \end{aligned} \qquad (5.24)$$

Moreover, the inverse $\mathcal{E}$ of Φ serves as an action-angle map for these systems, as can already be seen from (5.23).

For the type III case the function $d(y)$ (5.22) is not real-valued, but now one can for example consider the symmetric functions of the matrix $A(x) + A(x)^*$. As it turns out, the dual systems arising from the latter Hamiltonians are once more pure soliton systems (in a slightly more general sense than described in the previous subsection), and the action-angle maps are intimately related to the wave maps for these pure soliton systems. We shall return to the dual type III systems later on.

We continue by supplying the details of the construction of Φ for the relativistic type II systems. The above-mentioned commutation relation reads

$$\tfrac{1}{2}\coth(z)[A, L] = e \otimes e - \tfrac{1}{2}(AL + LA) \qquad (5.25)$$

where we take

$$z \equiv \frac{i\beta\mu g}{2} \in i(0, \pi). \qquad (5.26)$$

Also, e is the vector-valued function given by (3.36), L is the Lax matrix (3.35), and A the diagonal matrix (5.22). (Note that (5.25) is invariant under taking $z, A, L \to -z, L, A$; this symmetry property will eventually lead to the self-duality of the $\mathrm{II}_{\mathrm{rel}}$ system.) Since L is self-adjoint, there exists a unitary U such that

$$\hat{L} \equiv U^* L U = \mathrm{diag}(\lambda_1, \dots, \lambda_N), \quad \lambda_i \in \mathbb{R}.$$

Transforming (5.25) with U and setting

$$\hat{A} \equiv U^* A U, \quad \hat{e} \equiv U^* e,$$

one readily obtains

$$\tfrac{1}{2}\hat{A}_{jk}[\coth(z)(\lambda_k - \lambda_j) + \lambda_k + \lambda_j] = \hat{e}_j \bar{\hat{e}}_k. \tag{5.27}$$

Now A has positive spectrum, and since U is unitary, it follows that $\hat{A}_{jj} > 0$. Moreover, we have

$$\prod_{j=1}^{N} \lambda_j = |\hat{L}| = |L| = \exp\big(\beta(p_1 + \cdots + p_N)\big) > 0$$

so that $\lambda_j \neq 0$, $j = 1, \dots, N$. Taking $j = k$ in (5.27) we deduce $\hat{e}_j \neq 0$ and $\lambda_j > 0$. Consequently, we may define a vector $\hat{p} \in \mathbb{R}^N$ by setting

$$\lambda_j \equiv \exp(\beta \hat{p}_j), \quad j = 1, \dots, N.$$

Next, we rewrite (5.27) as

$$\hat{A}_{jk} = \hat{e}_j \bar{\hat{e}}_k \exp\left(-\frac{\beta(\hat{p}_j + \hat{p}_k)}{2}\right) \frac{\mathrm{sh}\, z}{\mathrm{sh}(\beta(\hat{p}_k - \hat{p}_j)/2 + z)}. \tag{5.28}$$

Recalling (3.33), (3.34), and (3.23) we deduce

$$|\hat{A}| = \prod_j (|\hat{e}_j|^2 \exp(-\beta \hat{p}_j)) \prod_{j<k} \left(\frac{\mathrm{sh}^2(\beta(\hat{p}_j - \hat{p}_k)/2)}{\mathrm{sh}^2(\beta(\hat{p}_j - \hat{p}_k)/2) + \sin^2(\beta \mu g/2)}\right)$$

Now we have $|\hat{A}| = |A| > 0$ and $\sin(\beta \mu g/2) > 0$ (cf. (5.26)), so that

$$\hat{p}_j \neq \hat{p}_k, \quad 1 \leq j < k \leq N.$$

The upshot is, that L has positive and nondegenerate spectrum. The gauge ambiguity in the diagonalizing unitary is therefore given by the product of a permutation matrix and a diagonal phase matrix. To render U unique, we get rid of this gauge freedom: We require

$$\hat{L} = \mathrm{diag}\big(\exp(\beta \hat{p}_1), \dots, \exp(\beta \hat{p}_N)\big), \quad \hat{p}_N < \cdots < \hat{p}_1, \tag{5.29}$$

to fix the permutation matrix, and

$$\hat{e}_j > 0, \quad j = 1, \ldots, N, \tag{5.30}$$

to fix the phase matrix. (Recall $\hat{e} \equiv U^* e$; also, recall we have already shown $\hat{e}_j \neq 0$, so (5.30) makes sense.)

Since U is now uniquely determined, the vector $\hat{e}$ is uniquely determined, too. Therefore, we can now introduce a vector $\hat{x} \in \mathbb{R}^N$ by parametrizing $\hat{e}_j$ as

$$\hat{e}_j = \exp\left(\frac{\beta\hat{p}_j + \mu\hat{x}_j}{2}\right) \prod_{l \neq j}\left(1 + \frac{\sin^2(\beta\mu g/2)}{\operatorname{sh}^2(\beta(\hat{p}_j - \hat{p}_l)/2)}\right)^{1/4}. \tag{5.31}$$

This ensures that the map Φ (5.20) satisfies

$$\hat{e}(\hat{x}, \hat{p}) = e(\mu, \beta, g; \hat{p}, \hat{x}) \tag{5.32}$$

where $e(\beta, \mu, g; x, p)$ is given by (3.36), (3.23); recalling (5.28), this entails

$$\hat{A}(\hat{x}, \hat{p}) = L(\mu, \beta, g; \hat{p}, \hat{x})^t \tag{5.33}$$

where $L(\beta, \mu, g; x, p)$ is given by (3.35).

The relation (5.33) amounts to the self-duality announced previously (cf. (5.24)). Indeed, it entails that the symmetric functions of $\hat{A}(\hat{x}, \hat{p})$ equal the previous Hamiltonians S_k, with β and μ interchanged and $\hat{p}_j$ and $\hat{x}_j$ playing the role of positions and momenta, respectively.

To show that $\Phi(\beta, \mu, g; x, p)$ is actually a bijection onto $\widehat{\Omega}$ (5.21), it suffices to construct a map $\mathcal{E}(\beta, \mu, g; \hat{x}, \hat{p}) \colon \widehat{\Omega} \to \Omega$ that satisfies

$$\mathcal{E} \circ \Phi = \operatorname{id}(\Omega), \quad \Phi \circ \mathcal{E} = \operatorname{id}(\widehat{\Omega}).$$

In view of the self-duality relations (5.32), (5.33) this is quite straightforward: One need only run the construction backwards to obtain a map $\mathcal{E}$ with these two properties. Then it follows that $\mathcal{E}$ is the inverse of the bijection Φ and, in addition, one infers

$$\mathcal{E}(\beta, \mu, g; \hat{x}, \hat{p}) = I \circ \Phi(\mu, \beta, g; \hat{p}, \hat{x})$$

where I is the flip map (5.21). (The transpose in (5.33) can be traded for a sign flip of g, and it is not hard to see that Φ is even in g.)

A complete proof that the bijection Φ is in fact a canonical diffeomorphism involves more work. A crucial ingredient of the proof that can be found in Ref. 53 is the scattering theory associated to the S_1 flow, which was already sketched in Section 3.1. We shall return to the wave and scattering maps shortly, but it is convenient to obtain first a crucial finite-time result, which follows rather easily when one takes for granted that Φ is a symplectic map.

We begin by observing that due to its spectral properties L admits a logarithm. Specifically, we have

$$\ln L = U \operatorname{diag}(\beta\hat{p}_1, \ldots, \beta\hat{p}_N)U^*;$$

cf. (5.29). For $h \in C^\infty(\mathbb{R})$ we can then define

$$h(\beta^{-1}\ln L) \equiv U \operatorname{diag}(h(\hat{p}_1), \ldots, h(\hat{p}_N))U^*. \tag{5.34}$$

We now study Hamiltonians of the form

$$H_h \equiv \operatorname{Tr} h(\beta^{-1}\ln L), \quad h \in C^\infty(\mathbb{R}). \tag{5.35}$$

These include the Hamiltonians $S_{\pm 1}$, H and P from Section 3.1, since we have

$$h(y) = \left\{\begin{matrix} \exp(\pm\beta y) \\ \beta^{-2}\operatorname{ch}\beta y \\ \beta^{-1}\operatorname{sh}\beta y \end{matrix} \quad \begin{matrix} S_{\pm 1} \\ H \\ P \end{matrix} \right\} = H_h. \tag{5.36}$$

Similarly, all power traces of L are included (take $h(y) = \exp(k\beta y)/k$, $k = 1, 2, \ldots$).

Let $t \in \mathbb{R}$ and $Q \equiv (x, p) \in \Omega$, and define the matrix

$$A_h(t, Q) \equiv A(Q)\exp\left(t\mu h'\left(\beta^{-1}\ln L(Q)\right)\right).$$

For $t = 0$ this matrix reads $\operatorname{diag}(\exp(\mu x_1), \ldots, \exp(\mu x_N))$, so it has manifestly positive and simple spectrum. We claim that it actually has positive and simple spectrum for *all* $t \in \mathbb{R}$. Furthermore, we claim that the H_h flow is complete and that the configuration space projection of the integral curve $\exp(tH_h)(Q)$ is given by

$$x_j(t) = \mu^{-1}\ln\alpha_j(t), \quad j = 1, \ldots, N, 0 < \alpha_N(t) < \cdots < \alpha_1(t),$$

where $\alpha_1, \ldots, \alpha_N$ are the ordered eigenvalues of A_h.

Exploiting the above map Φ and its canonicity property, the proof of these claims is quite short. Indeed, setting

$$\hat{Q} \equiv \Phi(Q) = (\hat{x}, \hat{p}), \quad \hat{H}_h \equiv H_h \circ \Phi^{-1}, \tag{5.37}$$

we have

$$\hat{H}_h(\hat{Q}) = \operatorname{Tr} U \operatorname{diag}(h(\hat{p}_1), \ldots, h(\hat{p}_N))U^* = \sum_{j=1}^{N} h(\hat{p}_j).$$

Therefore, the $\hat{H}_h$ flow on $\hat{\Omega}$ reads

$$\exp(t\hat{H}_h)(\hat{x}, \hat{p}) = (\hat{x}_1 + th'(\hat{p}_1), \ldots, \hat{x}_N + th'(\hat{p}_N), \hat{p}) \tag{5.38}$$

so it is manifestly complete. Since Φ is canonical, it follows from (5.37) that

$$\exp(tH_h) = \mathcal{E} \circ \exp(t\widehat{H}_h) \circ \Phi. \tag{5.39}$$

Hence the H_h flow is complete, too, as claimed. Finally, denoting similarity by $\sim$, we obtain

$$
\begin{aligned}
A_h(t, Q) &\sim \hat{A}(\widehat{Q}) \exp\left(t\mu h'\left(\beta^{-1} \ln \hat{L}(\widehat{Q})\right)\right) \\
&= \hat{A}(\hat{x}, \hat{p}) \operatorname{diag}\left(\exp(t\mu h'(\hat{p}_1)), \dots, \exp(t\mu h'(\hat{p}_N))\right) \\
&\sim \hat{A}(\hat{x}_1 + th'(\hat{p}_1), \dots, \hat{x}_N + th'(\hat{p}_N), \hat{p}) \\
&= \hat{A}\left(\exp(t\widehat{H}_h)(\widehat{Q})\right) \\
&\sim A\left(\exp(tH_h)(Q)\right)
\end{aligned} \tag{5.40}
$$

where we used $\hat{A} = U^* A U$ and (5.34) in the first and second steps, (5.28) and (5.31) in the third, (5.38) in the fourth, and (5.39) in the last step. The remaining claims are now evident from (5.40).

The results just proved yield a quite explicit description of the position part of the integral curve, which is especially useful to study its long-time asymptotics. Indeed, (5.40) entails that we can express this asymptotics in terms of the spectral asymptotics for $t \to \pm\infty$ of the matrix

$$\hat{A}(\hat{x}, \hat{p}) \operatorname{diag}\left(\exp(td_1), \dots, \exp(td_N)\right), \quad d_j \equiv \mu h'(\hat{p}_j). \tag{5.41}$$

Let us now assume that $h''(y) > 0$, so that $h'(y)$ is strictly increasing. (This holds true for S_1 and H, for instance; cf. (5.36).) Since $\hat{A}$ is a positive matrix, we should determine the spectral asymptotics of matrices of the form

$$E(t) = Me^{tD}, \quad M > 0, D = \operatorname{diag}(d_1, \dots, d_N), d_N < \cdots < d_1.$$

For $t \to \infty$ ($t \to -\infty$), this can be expressed in terms of upper (lower) corner principal minors of M. Specifically, let us set

$$m_1^+ \equiv M(1), \quad m_2^+ \equiv \frac{M(1,2)}{M(1)}, \quad m_3^+ \equiv \frac{M(1,2,3)}{M(1,2)}, \dots,$$

$$m_1^- \equiv M(N), \quad m_2^- \equiv \frac{M(N-1, N)}{M(N)}, \dots,$$

so that $m_1^+ = M_{11}$, $m_2^+ = (M_{11}M_{22} - M_{12}M_{21})/M_{11}$, etc.. Then it is proved in Appendix A of Ref. 53 that the (ordered) eigenvalues $\alpha_1(t), \dots, \alpha_N(t)$ of $E(t)$ satisfy

$$\exp(-td_j)\alpha_{j \atop N-j+1}(t) - m_{j \atop N-j+1}^{\pm} = O\left(\exp(\mp tR)\right), \quad t \to \pm\infty,$$

where

$$R \equiv \min(d_1 - d_2, \ldots, d_{N-1} - d_N).$$

(Notice that these formulas are trivially true when M is diagonal.)

To apply this to the concrete matrix (5.41), we need the relevant principal minors of $\hat{A}$. But these are easily calculated explicitly by using (5.33) and recalling (3.37). Proceeding in this way we find

$$m^{\pm}_{\substack{j \\ N-j+1}} \left(\hat{A}(\hat{x}, \hat{p})\right) = \exp\left(\mu \hat{x}_j \mp \frac{\mu}{2} \Delta_j(\hat{p})\right)$$

where

$$\Delta_j(p) = \left(\sum_{k<j} - \sum_{k>j}\right) \delta(p_j - p_k), \tag{5.42}$$

$$\delta(p) = \frac{1}{\mu} \ln\left(1 + \frac{\sin^2(\beta\mu g/2)}{\operatorname{sh}^2(\beta p/2)}\right). \tag{5.43}$$

Putting the pieces together, we obtain

$$x_{\substack{j \\ N-j+1}}(t) = \hat{x}_j \mp \tfrac{1}{2}\Delta_j(\hat{p}) + th'(\hat{p}_j) + O\left(\exp(\mp tR)\right), \quad t \to \pm\infty$$

and using isospectrality of $L\left(\exp(tH_h)(Q)\right)$ we deduce

$$p_{\substack{j \\ N-j+1}}(t) = \hat{p}_j + O\left(\exp(\mp tR)\right), \quad t \to \pm\infty. \tag{5.44}$$

Now for the special case $h(y) = \exp(\beta y)$ we have $H_h = S_1$, so (5.42)–(5.44) render the asymptotics (3.22), (2.43) and (2.48) explicit: One has

$$x^+_j = \hat{x}_j - \tfrac{1}{2}\Delta_j(\hat{p}), \quad p^+_j = \hat{p}_j, \qquad j = 1, \ldots, N, \tag{5.45}$$

$$x^-_{N-j+1} = \hat{x}_j + \tfrac{1}{2}\Delta_j(\hat{p}), \quad p^-_{N-j+1} = \hat{p}_j, \quad j = 1, \ldots, N,$$

so that the scattering map reads

$$x^+_j = x^-_{N-j+1} + \Delta_{N-j+1}(p^-), \quad p^+_j = p^-_{N-j+1}, \quad j = 1, \ldots, N. \tag{5.46}$$

In particular, $\delta(p)$ (5.43) is the shift incurred in the two-particle interaction (as compared to a billiard ball collision). However, the results just obtained are far more general: They entail that the wave and scattering maps are shared by a vast class of dynamics, containing in particular all power traces of L. Note that the map $U_+^{-1} \circ \Phi^{-1}$ given by (5.45) differs from the identity solely by the shifts $-\Delta_j(\hat{p})/2$.

Let us now indicate how the above can be specialized to the $\mathrm{II_{nr}}$, $\mathrm{I_{rel}}$, and $\mathrm{I_{nr}}$ systems. First, taking $\beta \to 0$, the matrix $(L - 1_N)/\beta$ converges to the II_{nr} Lax matrix (2.35), whereas the dual Lax matrix (5.33) converges to the (transpose of the) type I_{rel} Lax matrix, with β replaced by μ, and $\hat{p}_j$ and $\hat{x}_j$ playing the role of x_j and p_j, respectively. Second, taking $\mu \to 0$, the matrix L becomes the I_{rel} Lax matrix, whereas $(\hat{A} - 1_N)/\mu$ converges to the II_{nr} Lax matrix, with $\mu, x, p \to \beta, \hat{p}, \hat{x}$. Third, taking μ to 0 after the first limit, or β to 0 after the second, one obtains the I_{nr} matrices $L(x,p)$ and $L(\hat{p}, \hat{x})^t$ in a way that will be clear by now. Thus the duality properties (5.24) follow. Note that the shifts (5.42) and (5.43) vanish for the I_{rel} and I_{nr} cases, whereas the II_{nr} case yields a pair shift

$$\delta(p) = \frac{1}{\mu} \ln\left(1 + \frac{\mu^2 g^2}{p^2}\right).$$

We continue by sketching how action-angle maps for the type III systems can be constructed, starting once more from the commutation relation (5.25). Of course, we should replace μ by $i\mu$ so as to obtain the relativistic type III Lax matrix. As it happens, the construction of the maps involves a lot more work than the construction of the II_{rel} map Φ detailed above. Therefore, we only mention some key points, referring for the details to Ref. 40.

First, the commutation relation can be once more exploited to derive crucial spectral information: Fixing (x,p) in the maximal type III phase space $\tilde{\Omega}$ (2.24), the Lax matrix $L(x,p)$ satisfies

$$L(x,p) \sim \text{diag}\big(\exp(\beta \hat{p}_1), \ldots, \exp(\beta \hat{p}_N)\big),$$
$$\hat{p}_j - \hat{p}_{j+1} \geq \mu g, \quad j = 1, \ldots, N-1 \quad (5.47)$$

In words, its spectrum is not only positive and simple, but also has gaps. Points in $\tilde{\Omega}$ for which one or more gaps are minimal correspond to partial equilibria. In particular, the set of points where all gaps are minimal is given by

$$E = \left\{ (x,p) \in \tilde{\Omega} \, \middle| \, \begin{array}{l} x_j - x_{j+1} = 2\pi/N\mu, j = 1, \ldots, N-1, \\ p_k = c, k = 1, \ldots, N \end{array} \right\}. \quad (5.48)$$

For initial values in E all of the commuting flows are of the form $\big(x(t), p(t)\big)$ $= (x_0 + c_0(t, \ldots, t), p_0)$; thus, all particles move uniformly along the line and no internal motion occurs. (Physically speaking, the classical molecule is in its ground state.)

A similar picture applies to the quotient phase spaces Ω' and Ω, since they may be coordinatized as subsets of $\tilde{\Omega}$. Of course, the interpretation of the flows on (the quotients of) E is now different: The N particles rotate around the ring, keeping a fixed angular distance $2\pi/N$ to nearest neighbors.

Deleting the points in the three phase spaces for which L has one or more minimal spectral gaps, one is left with restricted phase spaces that serve as the definition domains of the respective action-angle maps. We continue by specializing to the "minimal' phase space Ω and its restriction Ω_r. Then the action-angle map is of the form

$$\Phi \colon \Omega_r \to \widehat{\Omega} = \mathbb{T}^N \times A_N, \quad (x, p) \mapsto (\hat{x}, \hat{p}),$$

where

$$A_N \equiv \{\hat{p} \in \mathbb{R}^N \mid \hat{p}_j - \hat{p}_{j+1} > \mu g, j = 1, \ldots, N-1\}. \tag{5.49}$$

The simplest dual dynamics is obtained by transforming

$$\check{D}(x) = \tfrac{1}{2} \operatorname{Tr}(A(x) + A(x)^*) = \sum_{j=1}^{N} \cos(\mu x_j). \tag{5.50}$$

It reads

$$D(\hat{x}, \hat{p}) = \sum_{j=1}^{N} \cos(\mu \hat{x}_j) \prod_{k \neq j} \left(1 - \frac{\operatorname{sh}^2(\beta \mu g/2)}{\operatorname{sh}^2(\beta(\hat{p}_j - \hat{p}_k)/2)}\right)^{1/2}. \quad (\text{IÎI}_{\text{rel}}) \tag{5.51}$$

From a physical point of view, the dual dynamics $D(p, x)$ describes N particles on a line, whose distances $x_j - x_{j+1}$ are bounded below by μg, whereas their momenta p_j vary over the first Brillouin zone $(-\pi/\mu, \pi/\mu]$. (Of course, the 1-tori might also be coordinatized by the interval $(-\pi, \pi]$, but then undesirable scale factors would crop up in the dual quantities.)

The flow generated by D and the higher-power trace flows are not complete on $\widehat{\Omega}$. This is obvious from the fact that the corresponding flows on Ω (which *are* complete, of course) do not leave Ω_r invariant. However, it can be shown that the map Φ admits an extension to a symplectic map $\Phi^\sharp$ from $\langle \Omega, \omega \rangle$ onto a symplectic manifold $\langle \widehat{\Omega}^\sharp, \hat{\omega}^\sharp \rangle$ in which $\widehat{\Omega}$ is densely embedded, and on $\widehat{\Omega}^\sharp$ the flows are complete. We have dubbed this extension the *harmonic oscillator transform*, since it extends Φ in much the same way as the identity map in the example (5.3)–(5.7) extends the map (5.4) (when one views Φ as supplying new coordinates for the dense submanifold (5.5)). Thus, in the canonical variables coordinatizing $\widehat{\Omega}^\sharp$, the commuting flows have a trigonometric dependence on time (for the internal variables), and equilibria are no longer excluded.

As already mentioned, the dual flows on $\widehat{\Omega}^\sharp$ have a solitonic long-time asymptotics, but now this holds true only on an open dense unequal velocity subset that depends on the flow one selects. For instance, it is clear from (5.50) that the flow

$$\exp(t\check{D})(x, p) = (x, p_1 + \mu t \sin \mu x_1, \ldots, p_N + \mu t \sin \mu x_N)$$

does not yield $|p_j(t) - p_k(t)| \to \infty$ for $j \neq k$ unless $\sin \mu x_j \neq \sin \mu x_k$, and when $x_k = \pi/\mu - x_j \pmod{2\pi/\mu}$ this condition is violated. A related complication is that, after deleting the equal velocity subvariety, one winds up with several connected components on which the ordering of the velocities is not the same.

The dynamics-dependent separatrices just described are reminiscent of the phenomenon mentioned in the last paragraph of Section 5.1. Indeed, the above setting yields a simple example for this phenomenon. The action-angle phase space Ω for the commuting Hamiltonians on $\widehat{\Omega}^{\sharp}$ is not of the product form (2.15), (2.16); only after discarding a suitable codimension-one subvariety it takes this form. For instance, when one deletes the variety $\{x_1 = \pi/\mu\}$, one is left with a manifold that is the product of $\mathbb{R}^N$ and the interior of the set F_N (2.28). (Geometrically, Ω can be viewed as a nontrivial fiber bundle: The base manifold equals G/Γ_x, where G is the configuration space (2.24) and Γ_x the restriction of the map Γ (2.25) to G, the fiber equals $\mathbb{R}^N$, and the transition functions are cyclic permutations on the fiber.)

For the III$_b$ system described at the end of Section 3.2, a harmonic oscillator transform on the extended phase space $\mathbb{R}^2 \times \mathbb{P}^{N-1}$ has been constructed as well [40]. This transform is self-dual, in the same sense as the II$_{rel}$ transform. Especially in the III$_b$ context, a new interpretation of the (analog of the) flip map I (5.21) suggests itself. This reinterpretation is, however, also useful for the systems of type I–II already treated, and we proceed to describe it in that setting.

First, recall that the action variables play the role of positions for the dual systems. If we now agree to interchange the order of the factors in (2.15) and its various concretizations encountered above, then we no longer have any need for the flip map. (Though notational problems do remain: How should one choose notation making clear that the limits of $p(t)$ in (2.46) (for instance) are to be viewed as *positions* without creating confusion?) In particular, we are then free to *identify* Ω and $\widehat{\Omega}$ for the type I and II systems. Now this is acceptable, but it should be realized that there is a price to pay. One has

$$\Phi^* \omega = -\omega, \tag{5.52}$$

the minus sign being caused by the flip. In words, Φ becomes an antisymplectomorphism of $\langle \Omega, \omega \rangle$.

Next, we define the involutory antisymplectomorphism

$$\mathcal{T}(x, p) \equiv (x, -p)$$

(time reversal) and the symplectomorphism

$$\mathcal{F} \equiv \mathcal{T} \circ \Phi. \tag{5.53}$$

Now it is far from easy to see, but true, that one has

$$\Phi \circ \mathcal{C} = \mathcal{T} \circ \Phi, \quad \Phi \circ \mathcal{T} = \mathcal{C} \circ \Phi, \tag{5.54}$$

where $\mathcal{C}$ is the involutory antisymplectomorphism

$$\mathcal{C}(x_1,\ldots,x_N,p_1,\ldots,p_N) \equiv (-x_N,\ldots,-x_1,p_N,\ldots,p_1).$$

(This follows by analytic continuation from results obtained in Ref. 40; cf. especially l.c. (2.118), (2.119).)

The point of the above is, that the symplectomorphism $\mathcal{F}$ on Ω transforms the flows generated by the Hamiltonians H_h (5.35) into new flows on Ω that are linear in time. Specifically, using (5.38) one easily verifies

$$(\mathcal{F} \circ \exp(tH_h) \circ \mathcal{F}^{-1})(x,p)$$
$$= (x, p_1 - th'(-x_1), \ldots, p_N - th'(-x_N)). \quad (5.55)$$

The equations (5.52)–(5.55) hold true for each of the four type I and II systems, with (5.35) replaced by

$$H_h \equiv \operatorname{Tr} h(L), \quad h \in C^\infty(\mathbb{R}), \qquad\qquad (\mathrm{I}_{\mathrm{nr}}, \mathrm{II}_{\mathrm{nr}})$$

in the nonrelativistic case (where $\beta = 0$).

Specializing now to the $\mathrm{II}_{\mathrm{rel}}$ case, and taking $\beta = \mu$ so as to avoid scalings, we have additional information: Self-duality translates into Φ being an involution. From (5.53) and (5.54) we then deduce the relations

$$\mathcal{F}^{-1} = \mathcal{P}\mathcal{F} = \mathcal{F}\mathcal{P} = \mathcal{C}\mathcal{F}\mathcal{C} = \mathcal{T}\mathcal{F}\mathcal{T}, \quad \mathcal{F}^4 = \operatorname{id}(\Omega) \quad (\mathrm{II}_{\mathrm{rel}}, \beta = \mu), \quad (5.56)$$

where $\mathcal{P}$ (parity) is the involutory symplectomorphism

$$\mathcal{P}(x_1,\ldots,x_N,p_1,\ldots,p_N) \equiv (-x_N,\ldots,-x_1,-p_N,\ldots,-p_1).$$

These relations prepare us for the state of affairs on the quantum level, to which we now turn.

6   Eigenfunction Transforms

6.1   Preliminaries

As we have seen in Sections 4.2 and 4.3, the Poisson commuting Hamiltonians of the type I–VI systems admit quantizations as commuting PDOs and AΔOs, respectively. This section is concerned with the eigenfunctions of these operators, especially inasmuch as these are relevant to the question whether the integrable PDOs and AΔOs are Hilbert integrable—a notion introduced in Section 4.1. Before elaborating on this notion, let us delineate the PDOs and AΔOs we intend to study.

First of all, we shall restrict attention to the systems of type I–IV. For information on joint eigenfunctions for Toda-type PDOs we refer to [54].

For the Toda-type AΔOs (4.38) no eigenfunctions are known, and it may well be that these AΔOs are not Hilbert integrable. (Note in this connection that they are not even formally symmetric. Nevertheless, it is conceivable that the enormous multiplier freedom for eigenfunctions of AΔOs can be exploited to construct a unitary eigenfunction transform.)

Secondly, for type I–III we discuss the $N = 2$ case in some detail, but we also present information on $N > 2$ transforms, particularly for type IIi. At the relativistic level our starting point is formed by the commuting AΔOs (4.35). Omitting the λ-dependent factor c_l in (4.40) and expanding the resulting AΔO $A_k(\beta)$ according to (4.41), we obtain commuting PDOs $A_{1,1}, \ldots, A_{N,N}$ that will be our starting point at the nonrelativistic level.

Taking $\hbar \equiv 1$ from now on, the latter PDOs have the form

$$A_{1,1} = \sum_j \hat{p}_j, \quad A_{2,2} = \sum_{j<k} (\hat{p}_j \hat{p}_k - g(g-1)V(x_j - x_k)), \tag{6.1}$$

$$A_{k,k} = \sum_{i_1 < \cdots < i_k} \hat{p}_{i_1} \cdots \hat{p}_{i_k} + \rho_k, \quad k > 2, \tag{6.2}$$

where ρ_k has order $< k - 1$ in $\hat{p}$. Also, $V(x)$ is given by (2.18)–(2.20), respectively. Recall that the change $g^2 \to g(g-1)$ is a natural consequence of the nonrelativistic limit. (As will be seen, the eigenfunctions would have a quite awkward dependence on g without this change.)

Thirdly, for type IV we consider again the AΔOs (4.35) at the relativistic level. Here, however, there is no obvious choice for the spectral parameter entering (4.40), and so we wind up with λ-dependent PDOs $A_{k,k}$, $k = 1$, $\ldots$, N, when we take the nonrelativistic limit. In particular, $A_{2,2}$ is given by (4.43). We shall in fact restrict attention to the $N = 2$ case. Indeed, eigenfunctions for $N > 2$ are known only for $g = 0, 1$; observe in this connection that the AΔOs are "free" for $g = 1$, too; cf. (4.37), (4.34) with $g = \hbar$. (There are also some preliminary results for $N > 2$ and $g = 2, 3$, $\ldots$ in the nonrelativistic case [55].)

We now turn to some general remarks on the problem of proving Hilbert integrability for the commuting PDOs and AΔOs. Let us notice first that this problem has a distinctly analytic flavor—as opposed to the questions associated with the weaker notion of integrability, which are of an algebraic character. This parallels the situation at the classical level, where the completeness assumption defining Liouville integrability belongs to global analysis—as opposed to the local, algebraic notion of involutivity. In the classical setting, however, the extra requirement is quite easily verified for the models of type I–VI (as we have seen in Sections 2 and 3). By contrast, Hilbert integrability can only be verified by actually constructing a unitary joint eigenfunction transform—the quantum analog of the (inverse of the) action-angle transform.

A natural question is, therefore, whether there exists a more easily verified criterion from which the *existence* of such a diagonalizing transform would follow. For instance, one might try and show first that the pertinent operators have a well-defined action on a common dense invariant domain in $L^2(G, dx)$. But even when such a domain would exist, and the operators would commute and would be essentially self-adjoint on it, it would not follow that the corresponding time evolutions commute—which is necessary if a joint eigenfunction transform (as defined in Section 4.1) is to exist. (In case the operators are Hilbert integrable, a domain with all of the above properties does exist, viz., the subspace $\mathcal{E}\big(C_0^\infty(\Lambda)\big)$. This readily follows from Nelson's analytic vector theorem; cf. Ref. 43, p. 202.)

To get some feeling for what is involved here, we continue with some simple examples. First, we recall that the Fourier transform (4.7) diagonalizes both constant coefficient PDOs and the AΔOs (4.8); cf. (4.9), (4.10). Specializing to the type I and II systems, we can use this transform to obtain free transforms of the form (4.11)–(4.13), as follows.

We begin by noting that wave functions belonging to the symmetric or antisymmetric subspaces $L_s^2(\mathbb{R}^N)$ and $L_a^2(\mathbb{R}^N)$, respectively, are uniquely determined by their restrictions to the wedge G (2.23) (Weyl chamber). Obviously, $\mathcal{E}$ (4.7) intertwines the permutation operators (4.17) and their counterparts on $L^2(\mathbb{R}^N, dp)$, so it gives rise to unitary transforms

$$\mathcal{E}_s, \mathcal{E}_a \colon L^2(G, dp) \to L^2(G, dx).$$

Explicitly, one obtains

$$(\mathcal{E}_s\phi)(x) = (2\pi)^{-N/2} \sum_{\sigma \in S_N} \int_G dp \, \exp\big(ip \cdot \sigma(x)\big)\phi(p), \qquad x \in G,$$

$$(\mathcal{E}_a\phi)(x) = (2\pi)^{-N/2} \sum_{\sigma \in S_N} (-)^\sigma \int_G dp \, \exp\big(ip \cdot \sigma(x)\big)\phi(p), \qquad x \in G.$$

In this way we get two examples for a unitary of the form (4.11)–(4.13): Λ equals G (2.23), μ equals Lebesgue measure, $M_k(p)$ in (4.13) is given by $e_k(p)$ (4.10), $M_1(p)$ in (4.11) is given by $\sum_j p_j^2/2$, and $M_k(p)$ in (4.11) depends, for $k > 1$, on the choice of commuting PDOs $I_2, \ldots, I_N$. The domains of the two distinct self-adjoint operators on $L^2(G, dx)$ obtained via $\mathcal{E}_s$ and $\mathcal{E}_a$ (both associated with the PDO $H_{\mathrm{nr}} = \sum_j \hat{p}_j^2/2$) then consist of restrictions to G of symmetric and antisymmetric functions (respectively) in so-called Sobolev spaces. Similarly, the domains of the two self-adjoint operators associated via $\mathcal{E}_s$ ($\mathcal{E}_a$) with the AΔO $H_{\mathrm{rel}} = \sum_j \mathrm{ch}(\beta\hat{p}_j)/\beta^2$ consist of restrictions to G of (anti)symmetric functions that admit analytic continuation to the polystrip $\big(\mathbb{R} + i(-\beta, \beta)\big)^N$ and that have some further L^2-properties.

Due to the singularities of the potentials on the walls of G, the interacting Hamiltonians do not have a well-defined action on either of the two pertinent domains, however. For the same reason, it is not at all clear that a dense domain exists on which the interacting H_{rel} is symmetric—as opposed to the interacting Hamiltonian H_{nr} (Schrödinger operator), which is symmetric on the dense subspace $C_0^\infty(G)$ (cf. the discussion at the end of Section 4.1).

Returning to the above free transforms, let us identify $L^2(G, dp)$ and $L^2(G, dx)$ via the antilinear map $f(p) \mapsto \overline{f(x)}$. This is the quantum version of the identification of Ω and $\widehat{\Omega}$ explained in the paragraph containing (5.52): It entails that $\mathcal{E}_s$ and $\mathcal{E}_a$ become *antiunitary* operators from $L^2(G, dx)$ onto itself. Introducing time reversal

$$(\mathcal{T}\psi)(x) \equiv \overline{\psi(x)}, \quad \psi \in L^2(G, dx),$$

parity

$$(\mathcal{P}\psi)(x_1, \ldots, x_N) \equiv \psi(-x_N, \ldots, -x_1), \quad \psi \in L^2(G, dx),$$

and the conjugation

$$\mathcal{C} \equiv \mathcal{T}\mathcal{P},$$

it is then clear that one has

$$\mathcal{F}^{-1} = \mathcal{P}\mathcal{F} = \mathcal{F}\mathcal{P} = \mathcal{C}\mathcal{F}\mathcal{C} = \mathcal{T}\mathcal{F}\mathcal{T},$$
$$\mathcal{F}^4 = 1, \quad \mathcal{F} \equiv \mathcal{T}\mathcal{E}_s^{-1}, \mathcal{T}\mathcal{E}_a^{-1}. \tag{6.3}$$

This should be compared to (5.56).

We conjecture that for suitably restricted $g \geq 0$ there exists a unitary $\mathrm{II}_{\mathrm{rel}}$ eigenfunction transform $\mathcal{E}_g$ given by (4.12), (4.13), with $\Lambda = G$, $\mu = dp$, $M_k = e_k$ (4.10), and $\mathcal{E}_0 = \mathcal{E}_s$, $\mathcal{E}_1 = \mathcal{E}_a$; with the above identification in effect, the unitary $\mathcal{F} \equiv \mathcal{T}\mathcal{E}_g^{-1}$ on $L^2(G, dx)$ obeys (6.3), provided $\beta = \mu$. To date, we have only shown this for $N = 2$ and

$$g \in \left[0, 1 + \frac{2\pi}{\beta\mu}\right].$$

(We have also obtained some fragmentary results on eigenfunctions for $N > 2$, but here the unitarity region is still unclear.) We shall present the pertinent $N = 2$ eigenfunctions in Section 6.3. We intend to return to the corresponding transforms elsewhere.

6.2   Type III Eigenfunctions for Arbitrary N

We proceed by discussing eigenfunction transforms for the type III systems. Recall from Section 2.2 that one can distinguish three phase spaces for these systems, the choice depending on the physical interpretation. Accordingly, there are three state spaces at the quantum level, namely, the L^2-spaces over the sets G, F_N' and F_N; cf. (2.24)–(2.30).

We shall focus attention on the latter choice, which corresponds to indistinguishable particles on a ring. (Once the transforms on $L^2(F_N)$ are known, one can obtain transforms on $L^2(F'_N)$ and $L^2(G)$ via suitable co-ordinate changes.) To ease the notation we take $\mu \equiv 1$ throughout this subsection.

Consider first the free case. Then we can exploit the well-known Fourier series orthonormal base for $L^2(\mathbb{T}^N) \simeq L^2((-\pi, \pi]^N, dx)$, viz., the functions

$$K(x, n) \equiv (2\pi)^{-N/2} \exp(ix \cdot n), \quad x \in (-\pi, \pi]^N, n \in \mathbb{Z}^N. \tag{6.4}$$

Specifically, the transform

$$\mathcal{E} \colon l^2(\mathbb{Z}^N) \to L^2(\mathbb{T}^N), \quad \phi(n) \mapsto \sum_{n \in \mathbb{Z}^N} K(x, n)\phi(n),$$

is a unitary operator that intertwines the permutation operators on the two Hilbert spaces involved, so it gives rise to two unitaries $\mathcal{E}_s$ and $\mathcal{E}_a$ mapping the symmetric and antisymmetric subspaces (respectively) of $l^2(\mathbb{Z}^N)$ onto those of $L^2(\mathbb{T}^N)$. Now the symmetric and antisymmetric functions in $L^2(\mathbb{T}^N)$ are uniquely determined by their restrictions to F_N; indeed, one has

$$F_N \simeq \mathbb{T}^N_{\neq}/S_N, \quad \mathbb{T}^N_{\neq} \equiv \{x \in (-\pi, \pi]^N \mid x_i \neq x_j, 1 \leq i < j \leq N\}.$$

That is, F_N can be viewed as a fundamental set for the action of S_N on $\mathbb{T}^N_{\neq}$. Likewise,

$$\mathbb{Z}^N_+ \equiv \{n \in \mathbb{Z}^N \mid n_N \leq \cdots \leq n_1\}$$

is a fundamental set for the S_N-action on $\mathbb{Z}^N$.

Specializing to the symmetric case first, we therefore obtain an orthonormal base

$$M_n(x) \equiv r_n \sum_{\sigma \in S_N} \exp(in \cdot \sigma(x)), \quad x \in F_N, n \in \mathbb{Z}^N_+,$$

for $L^2(F_N, dx)$, provided the normalization constant $r_n > 0$ is suitably chosen. Equivalently, we obtain a unitary

$$\mathcal{E}_s \colon l^2(\mathbb{Z}^N_+) \to L^2(F_N, dx), \quad \phi(n) \mapsto \sum_{n \in \mathbb{Z}^N_+} M_n(x)\phi(n)$$

which obviously diagonalizes the free AΔOs and PDOs.

Similarly, the antisymmetrized base functions $K(x, n)$ can be used. Specifically, we need here

$$\mathbb{Z}^N_{+, \neq} \equiv \{n \in \mathbb{Z}^N \mid n_N < \cdots < n_1\}$$

and the functions

$$A_n(x) \equiv r \sum_{\sigma \in S_N} (-)^\sigma \exp(in \cdot \sigma(x)), \quad x \in F_N, n \in \mathbb{Z}^N_{+, \neq},$$

to obtain an orthonormal base for $L^2(F_N, dx)$ (for a suitable constant $r > 0$). Hence we get a unitary

$$\mathcal{E}_a\colon l^2(\mathbb{Z}^N_{+,\neq}) \to L^2(F_N, dx), \quad \phi(n) \mapsto \sum_{n \in \mathbb{Z}^N_{+,\neq}} A_n(x)\phi(n)$$

which also diagonalizes the free AΔOs and PDOs.

With these elementary examples and the associated notation at our disposal, we can now turn to a description of the $g \geq 0$ type III transforms $\mathcal{E}_g$. The pertinent eigenfunctions were introduced (in somewhat different guises) in Refs. 14 and 15 for $\beta = 0$ and in Ref. 16 for $\beta > 0$; cf. also Ref. 17. Not surprisingly, one has $\mathcal{E}_0 = \mathcal{E}_s$ for all N, but $\mathcal{E}_1$ is only equal to $\mathcal{E}_a$ for N odd. Of course, $\mathcal{E}_1$ is a free transform for N even, too, but in this case it is the antisymmetric reduction of the Fourier base for $L^2(\mathbb{T}^N)$ with *antiperiodic* boundary conditions, i.e., the functions (6.4) with $n \in (\mathbb{Z} + 1/2)^N$.

To handle the general case, it is convenient to introduce the vector

$$\rho(g) \equiv \frac{g}{2}(N-1, N-3, \ldots, -N+1)$$

and the set

$$\Lambda_g \equiv \{p \in \mathbb{R}^N \mid p - \rho(g) \in \mathbb{Z}^N_+\}. \tag{6.5}$$

Then the transforms are unitaries of the form

$$\mathcal{E}_g\colon l^2(\Lambda_g) \to L^2(F_N, dx), \quad \phi(p) \mapsto \sum_{p \in \Lambda_g} E(x, p)\phi(p).$$

Here, the β-dependence is suppressed; in the relativistic case $\beta > 0$ the kernel $E(x, p)$ is a joint eigenfunction of the AΔOs (4.35), satisfying

$$S_{\pm k}E(x, p) = \sum_{|I|=k} \exp(\pm \beta p_I)E(x, p), \quad k = 1, \ldots, N, \tag{6.6}$$

and taking $\beta \to 0$ one then deduces that for the PDOs (6.1), (6.2) one has

$$A_{k,k}E(x, p) = \sum_{i_1 < \cdots < i_k} p_{i_1} \cdots p_{i_k} E(x, p). \tag{6.7}$$

As a consequence, the type III AΔOs and PDOs are Hilbert integrable: (4.11)–(4.13) hold true, with μ the counting measure having support on Λ_g.

The eigenfunctions $E(x, p)$ can be written

$$E(x, p) = \Delta(x)^{1/2} P_{p-\rho(g)}(x), \quad x \in F_N, p \in \Lambda_g. \tag{6.8}$$

Using a tilde from now on to denote equality up to a positive constant, $\Delta(x)$ is of the form

$$\Delta(x) \sim \prod_{\substack{j,k=1 \\ j<k}}^{N} w(x_j - x_k). \tag{6.9}$$

Furthermore, $P_n(x)$, $n \in \Lambda_0$, is a finite linear combination of the above polynomials $M_n(x)$. We continue by providing more details on the weight function $w(x)$ and the linear combination involved.

For $\beta > 0$ the w-function is proportional to an infinite product, namely,

$$w(x) \sim \prod_{l=0}^{\infty} \frac{(1 - \exp[ix - l\beta])(x \to -x)}{(1 - \exp[ix - (l+g)\beta])(x \to -x)} \tag{6.10}$$

while for $\beta = 0$ one has

$$w(x) \sim \left| \sin \frac{x}{2} \right|^{2g}. \tag{$\beta = 0$} \tag{6.11}$$

The w-function (6.10) is the trigonometric degeneration of the elliptic w-function from Ref. [31], which was used in Section 4.3; cf. (4.44)–(4.46). As such, the function $\Delta(x)$ (6.9) gives rise to AΔOs S_k^t of the form (4.46). (This is easily verified directly.) On account of (6.6) and (6.8), these AΔOs satisfy

$$S_k^t P_n(x) = \sum_{|I|=k} \exp\left(\beta \sum_{i \in I} (n_i + \rho(g)_i) \right) P_n(x), \quad n \in \Lambda_0, \tag{6.12}$$

i.e., they admit multivariable polynomials as joint eigenfunctions. Taking $\beta \to 0$, this holds true for the correspondingly transformed PDOs $A_{k,k}^t$, too (recall (4.45), (4.47), and (4.48) in this connection).

Let us now describe the structure of the polynomials $P_n(x)$ in more detail. To this end we need a partial order on the set $\Lambda_0 = \mathbb{Z}_+^N$:

$$n \leq m$$
$$\iff \sum_{j=1}^{N}(n_j - m_j) = 0, \quad \sum_{j=1}^{k}(n_j - m_j) \leq 0, k = 1, \ldots, N - 1. \tag{6.13}$$

Thus, for a fixed $m \in \Lambda_0$ there are only finitely many $n \in \Lambda_0$ satisfying $n \leq m$; in particular, for m_j of the form $(j, \ldots, j)$ there is only one such n, namely $n = m_j$. The polynomials are now of the *triangular* form

$$P_m(x) = \sum_{n \leq m} c_{mn} M_n(x), \quad c_{mm} > 0 \tag{6.14}$$

with coefficients c_{mn} depending on β and g.

Before explaining how the coefficients c_{mn} are determined, we insert a remark on the (in)significance of fractional statistics for the eigenfunctions $E(x,p)$ (6.8) at hand. Obviously, the polynomials $P_m(x)$, $x \in F_N$, extend to functions that are entire and symmetric in $x_1, \ldots, x_N$, both in the nonrelativistic ($\beta = 0$) and the relativistic ($\beta > 0$) case. Now consider the factor (cf. (6.9), (6.11))

$$\Delta(x)^{1/2} \sim \prod_{j<k} \sin \tfrac{1}{2}(x_j - x_k)^g, \quad x \in F_N. \qquad (\beta = 0) \quad (6.15)$$

We have omitted the bars in (6.11), since all of the sines occurring in (6.15) are positive on F_N (2.28). Clearly, the function at the right-hand side (and hence $E(x,p)$ as well) extends to a function that is entire and symmetric (antisymmetric) in $x_1, \ldots, x_N$ for g even (odd), whereas it has logarithmic branch points on the walls of the Weyl alcove F_N for noninteger g. This is the so-called fractional statistics phenomenon: The particles are viewed as bosons (fermions) for g even (odd), and as anyons for noninteger g.

We would like to point out that this state of affairs is an artefact of the nonrelativistic limit. For $\beta > 0$ and $g > 0$ one gets from (6.10) (by splitting off the $l = 0$ numerator factor at the right-hand side)

$$\Delta(x)^{1/2} \sim \prod_{j<k} \sin \tfrac{1}{2}(x_j - x_k) w_r(x_j - x_k)^{1/2}, \quad x \in F_N, \qquad (\beta > 0)$$

where $w_r(x)$ extends to a meromorphic, *even* function that is *positive* on $\mathbb{R}$. (In fact, from (6.10) one reads off that $w_r(x)$ has no zeros and poles in the strip $|\operatorname{Im} x| < \min(\beta, g)$.) As a consequence, no trace of fractional statistics remains: For all $g > 0$ the particles *are* fermions. (Of course, for $g = 0$ the function $E(x,p)$, $x \in F_N$, extends to a *symmetric* function; there are noncommuting operations involved here.)

A similar change of anyons to fermions occurs for the $\beta > 0$ eigenfunctions of type II—inasmuch as these are known. It is caused by the same phenomenon as for type III: fixing $g \notin \mathbb{N}$, the (meromorphic) w-function has an infinite number of poles and zeros coalescing on the imaginary axis for $\beta \downarrow 0$ [31].

We continue by explaining how the coefficients in (6.14) are determined. To this end we introduce the renormalized polynomials

$$Q_m(x) \equiv \frac{P_m(x)}{c_{mm}} = M_m(x) - R_m(x),$$

$$R_m(x) = \sum_{n<m} d_{mn} M_n(x), \quad d_{mn} \equiv -\frac{c_{mn}}{c_{mm}},$$

the point being that $R_m(x)$ equals the orthogonal projection of $M_m(x)$ onto the subspace

$$\mathcal{H}_m \equiv \left\{ \sum_{n<m} c_n M_n \,\bigg|\, c_n \in \mathbb{C} \right\} \qquad (6.16)$$

in the Hilbert space $L^2(F_N, \Delta(x)dx)$. This determines the coefficients d_{mn} uniquely, and then c_{mm} is determined by requiring that $P_m(x)$ be a unit vector in the latter Hilbert space. In particular, for $g = 0$ the function $\Delta(x)$ reduces to a positive constant (cf. (6.9)–(6.11)), and so $R_m(x)$ vanishes. Hence one obtains $\mathcal{E}_0 = \mathcal{E}_s$, as announced above.

It is obvious from the previous paragraph that $\Delta(x)$ uniquely determines the polynomials $Q_m(x)$, and that one has

$$\int_{F_N} dx \, \Delta(x)\overline{Q}_m(x)Q_n(x) = 0, \quad n < m. \tag{6.17}$$

However, this much is true for a quite arbitrary weight function. Indeed, for $N = 2$ and $\Delta(x) \equiv f(x_1 - x_2)$ with $f(x)$ any positive, continuous and even function on $[-2\pi, 2\pi]$, the above characterization yields an orthonormal base. (It amounts to the well-known Gram-Schmidt procedure.) But for $N > 2$ the vectors M_n in the subspace $\mathcal{H}_m$ (6.16) are not totally ordered. (For example, taking

$$m = (5, 4, 0), \quad n_1 = (5, 2, 2), \quad n_2 = (4, 4, 1),$$

one has $n_1 < m$ and $n_2 < m$, yet neither $n_1 < n_2$ nor $n_2 < n_1$ holds true.) The above weight functions have the (very restrictive) property that (6.17) holds not only for $n < m$ but for all pairs $n \neq m$.

We proceed by sketching how the (transformed) AΔOs and PDOs can be used to prove this orthogonality property; in the process it will become clear why Q_m is a joint eigenfunction. Let us denote a fixed operator among the AΔOs or PDOs by A. Each of the operators has two key properties. First, it satisfies

$$AM_m = E_m M_m + \sum_{n<m} \alpha_{mn} M_n. \tag{6.18}$$

That is, it is triangular w.r.t. the partial order (6.13) on the free boson eigenstates M_m. Second, it is symmetric on the dense subspace of $L^2(F_N, \Delta(x)dx)$ spanned by these functions; this amounts to

$$(AM_m, M_n) = (M_m, AM_n), \quad \forall m, n \in \Lambda_0, \tag{6.19}$$

where $(\cdot, \cdot)$ denotes the scalar product in $L^2(F_N, \Delta(x)dx)$.

These two features, combined with the above definition of the polynomials Q_m, entail that Q_m is an eigenfunction of A with eigenvalue E_m. Indeed, by virtue of triangularity of A and Q_m one obtains

$$AQ_m = E_m Q_m + \sum_{n<m} a_{mn} Q_n \equiv E_m Q_m + L_m$$

so it suffices to show $L_m = 0$. Now (6.17) entails $(Q_m, L_m) = 0$ and so one infers (using also symmetry of A)

$$(L_m, L_m) = (AQ_m, L_m) = (Q_m, AL_m) = \sum_{n<m} a_{mn}(Q_m, AQ_n)$$

$$= \sum_{n<m} a_{mn} \sum_{j\leq n} a_{nj}(Q_m, Q_j) = 0$$

since $j \leq n < m$. Therefore, L_m vanishes.

The upshot is that the eigenfunction property of the polynomials Q_m has been reduced to the triangularity and symmetry relations (6.18) and (6.19). Taking these relations for granted—their proofs are not difficult, but would take us too far afield—it is not hard to see that the eigenvalues E_m are in fact given by (6.12) in the AΔO case and, therefore, by (6.7) in the PDO case. Fixing a pair $m \neq n$, we shall now show how this eigenvalue structure entails the announced orthogonality $(Q_m, Q_n) = 0$.

Clearly, we need only prove that at least one of the AΔOs or PDOs A_k, $k = 1, \ldots, N$, yields eigenvalues $E_m^{(k)} \neq E_n^{(k)}$ on Q_m and Q_n, respectively. (Indeed, orthogonality then follows from symmetry in a well-known way.) Let us assume all eigenvalues are equal. Now the eigenvalues are the symmetric functions of a matrix of the form $\text{diag}(\lambda_1, \ldots, \lambda_N)$ with $\lambda_N < \cdots < \lambda_1$. But then equality of all eigenvalues entails that the two vectors $\lambda(m), \lambda(n)$ involved are equal, so that $m = n$, a contradiction.

With orthogonality established, one need only normalize to obtain the orthonormal base $\{P_n(x)\}_{n \in \Lambda_0}$ of $L^2(F_N, \Delta(x)dx)$ and the corresponding orthonormal base $\{E(x, p)\}_{p \in \Lambda_g}$ of $L^2(F_N, dx)$. Note that the argument in the previous paragraph not only proves orthogonality, but also—using standard quantum-mechanical parlance—completeness of the set of observables $\{A_1, \ldots, A_N\}$. Rephrased in functional analytic language, the von Neumann algebra, generated by the bounded functions of the self-adjoint operators on $L^2(F_N, dx)$ associated with $A_1, \ldots, A_N$ via the unitary $\mathcal{E}_g$, is maximal Abelian.

The crucial triangularity property (6.18) was first observed and exploited by Sutherland [14, 36], who used the PDO $A = H_{nr}^t = (A_{1,1}^t)^2/2 - A_{2,2}^t$. For $\beta > 0$ the AΔOs S_k^t and the polynomials Q_m were first introduced by Macdonald [16, 17]. To be more specific, he considers several root systems at once and accordingly works with the center-of-mass versions of the S_k^t and Q_m; for the detailed relation we refer to Section 5.2 of Ref. 6.

We conclude this subsection by pointing out some illuminating relations between the classical and quantum diagonalizing transforms. First, denoting the closure of the range of variation A_N (5.49) of the actions by A_N^{cl}, the set Λ_g (6.5) is a lattice-type subset of A_N^{cl}, whose boundary points (vectors p for which $p_j - p_{j+1} = g$ for some j) are also boundary points of A_N^{cl} (partial equilibria). Moreover, the eigenvalues of the quantum Hamiltonians on a joint eigenfunction $E(x, p)$ are obtained by evaluating their classical counterparts in points of Ω whose action vector equals $p \in \Lambda_g \subset A_N^{cl}$. (Semiclassical quantization is exact.)

Second, in agreement with the correspondence just sketched, the quantum ground state is obtained by choosing $p = \rho(g)$; cf. (6.8). (This choice yields the minimal eigenvalue for the defining Hamiltonians, as is readily verified.) In fact, the triangular structure (6.14) entails

$$P_{m_0}(x) = c_0 > 0, \quad m_0 \equiv (0, \ldots, 0),$$

so that the ground state reads explicitly

$$E(x, \rho(g)) = c_0 \Delta(x)^{1/2}. \tag{6.20}$$

Third, the joint eigenfunction property of P_{m_0} can be translated into a set of N functional equations. Indeed, recalling (4.46) and the definition of $f_-(x)^2$ (cf. (4.34) and (2.65)), we see that (6.12) with $n = m_0$ amounts to

$$\sum_{\substack{I \subset \{1,\dots,N\} \\ |I|=k}} \prod_{\substack{j \in I \\ l \notin I}} \frac{\sin(x_j - x_l - i\beta g)/2}{\sin(x_j - x_l)/2} = \sum_{\substack{I \subset \{1,\dots,N\} \\ |I|=k}} \exp\left(\frac{1}{2}\beta g \sum_{j \in I}(N + 1 - 2j)\right).$$

These functional equations can also be proved directly; cf. Lemma A.5 in Ref. 40. In the classical context they are exploited to prove that the type III Lax matrix has minimal spectral gaps on the set E (5.48). Consequently, they encode both the classical and the quantum ground-state properties; the quantum ground-state (6.20) corresponds to the classical equilibrium subset E_0 of E obtained by putting $c = 0$ at the right-hand side of (5.48).

6.3   Type II and IV Eigenfunctions for $N = 2$

From now on we specialize to the $N = 2$ case. It is convenient to separate off the center-of-mass motion by employing new coordinates

$$X \equiv \frac{x_1 + x_2}{2}, \qquad\qquad P \equiv p_1 + p_2,$$

$$x \equiv x_1 - x_2, \qquad\qquad p \equiv \frac{p_1 - p_2}{2}.$$

Then the dependence of the eigenfunctions on the center-of-mass coordinates X, P and internal coordinates x, p can be factorized:

$$E_2\big((x_1, x_2), (p_1, p_2)\big) = \exp(iXP)E(x, p). \tag{6.21}$$

Indeed, in this way we obtain an eigenfunction of the AΔOs $S_{\pm 2}$ and PDO $A_{1,1}$ with eigenvalues $\exp(\pm\beta P)$ and P, respectively, and it remains to consider the operators $S_{\pm 1}$ and $A_{2,2}$ (6.1).

Specializing first to the type II case, we restrict the choice of the spectral variables p_1, p_2 by insisting that for $\beta > 0$ the AΔOs $S_{\pm 1}$ yield eigenvalues

$$\exp(\pm\beta p_1) + \exp(\pm\beta p_2) = 2\exp\left(\pm\beta\frac{P}{2}\right)\mathrm{ch}(\beta p)$$

and that for $\beta = 0$ the PDO $A_{2,2}$ yields eigenvalue

$$p_1 p_2 = \frac{P^2}{4} - p^2.$$

on E_2 (6.21). Using the new parameter $\nu \equiv \mu/2$ (which minimizes the occurrence of numerical factors), it follows that the function $E(x,p)$ at the right-hand side of (6.21) should satisfy

$$\left(\frac{\operatorname{sh}\nu(x - i\beta g)\operatorname{sh}\nu(x - i\beta + i\beta g)}{\operatorname{sh}\nu x\operatorname{sh}\nu(x - i\beta)}\right)^{1/2} E(x - i\beta, p) + (\beta \to -\beta)$$

$$= 2\operatorname{ch}(\beta p)E(x,p),$$

$$-\partial_x^2 E(x,p) + g(g - 1)\frac{\nu^2}{\operatorname{sh}^2 \nu x}E(x,p) = p^2 E(x,p),$$

for $\beta > 0$ and $\beta = 0$, respectively.

We write the elliptic generalizations of the operators at the left-hand sides as

$$H_{\mathrm{rel}} = \left(\frac{s(x - i\beta g)}{s(x)}\right)^{1/2} T_{i\beta}^x \left(\frac{s(x + i\beta g)}{s(x)}\right)^{1/2} + (\beta \to -\beta), \tag{6.22}$$

$$H_{\mathrm{nr}} = -\frac{d^2}{dx^2} + g(g - 1)\left(\mathcal{P}(x) + \frac{\eta}{\omega}\right). \tag{6.23}$$

Here, we have introduced the shift

$$(T_\xi^x f)(x) \equiv f(x - \xi), \quad \xi \in \mathbb{C} \tag{6.24}$$

and we are discarding a λ-dependent constant in (4.43). In this case there is no obvious parametrization for eigenvalues, as will be seen below. Just as for the type III case considered in Section 6.2, it is in fact more convenient to work with the measure $w(x)dx$ instead of Lebesgue measure dx, where $w(x)$ is the w-function that already appeared in Section 4.3; cf. (4.44), (4.45). After the corresponding similarity transformation, one obtains from (6.22) and (6.23) the operators (recall (4.46)–(4.48))

$$H_{\mathrm{rel}}^t = \frac{s(x - i\beta g)}{s(x)}T_{i\beta}^x + (\beta \to -\beta), \tag{6.25}$$

$$H_{\mathrm{nr}}^t = -\frac{d^2}{dx^2} - 2g\frac{s'(x)}{s(x)}\frac{d}{dx} - g\frac{s''(x)}{s(x)} + g(g - 1)\left(\mathcal{P}(x) + \frac{\eta}{\omega} - \frac{s'(x)^2}{s(x)^2}\right).$$

For the hyperbolic specialization,

$$H_{\mathrm{rel}}^t = \frac{\operatorname{sh}\nu(x - i\beta g)}{\operatorname{sh}\nu x}T_{i\beta}^x + (\beta \to -\beta), \tag{6.26}$$

$$H_{\mathrm{nr}}^t = -\frac{d^2}{dx^2} - 2g\nu\coth\nu x\frac{d}{dx} - g^2\nu^2,$$

the dual operators expected from the classical level read (recall (5.24))

$$\widehat{H}_{\mathrm{rel}}^t = \frac{\operatorname{sh}\beta(p - i\nu g)}{\operatorname{sh}\beta p}T_{i\nu}^p + (\nu \to -\nu), \tag{6.27}$$

$$\widehat{H}_{\mathrm{nr}}^t = \frac{p - i\nu g}{p}T_{i\nu}^p + (\nu \to -\nu).$$

More specifically, writing

$$E(x,p) \equiv Cw(x)^{1/2}R(x,p)\hat{w}(p)^{1/2}$$

(where $C > 0$ is a normalization constant), one should have

$$H^t_{\mathrm{rel}}R(x,p) = 2\,\mathrm{ch}(\beta p)R(x,p), \quad \widehat{H}^t_{\mathrm{rel}}R(x,p) = 2\,\mathrm{ch}(\nu x)R(x,p), \quad (6.28)$$
$$H^t_{\mathrm{nr}}R(x,p) = p^2 R(x,p), \quad \widehat{H}^t_{\mathrm{nr}}R(x,p) = 2\,\mathrm{ch}(\nu x)R(x,p). \quad (6.29)$$

In our survey [5] we have already pointed out that this expectation is satisfied in the nonrelativistic case. Indeed, here one has

$$R(x,p) = {}_2F_1\left(\frac{1}{2}\left(g+i\frac{p}{\nu}\right), \frac{1}{2}\left(g-i\frac{p}{\nu}\right), g+\frac{1}{2}; -\mathrm{sh}^2\,\nu x\right);$$

cf. l.c. (3.35), and the dual equation follows from the contiguous relations for the Gauss hypergeometric function ${}_2F_1$. In l.c. (3.37)–(3.42), we have also presented solutions to (6.28) with all of the expected properties, taking, however, $g \in \mathbb{N}$.

The general solution (which we have obtained in recent years) reduces to this special case for $g \in \mathbb{N}$, but it has a quite different appearance. As mentioned in the introduction to this chapter, its structure is actually such as to admit a straightforward generalization to a simultaneous eigenfunction for four Askey-Wilson–type hyperbolic AΔOs, each of which depends on four coupling constants. We shall describe these AΔOs and the general solution first, using notation that is convenient to bring out the symmetry properties, and then detail the pertinent specializations. We write the four couplings involved as

$$\mathbf{c} = (c_0, c_1, c_2, c_3).$$

It is convenient to introduce matrices

$$I \equiv \begin{pmatrix} 1 & 0 & 0 & 0 \\ 0 & 0 & 1 & 0 \\ 0 & 1 & 0 & 0 \\ 0 & 0 & 0 & 1 \end{pmatrix}, \quad J \equiv \frac{1}{2}\begin{pmatrix} 1 & 1 & 1 & 1 \\ 1 & 1 & -1 & -1 \\ 1 & -1 & 1 & -1 \\ 1 & -1 & -1 & 1 \end{pmatrix},$$

satisfying

$$IJ = JI, \quad K = K^* = K^{-1}, \quad K = I, J,$$

and dual couplings given by

$$\hat{\mathbf{c}} \equiv J\mathbf{c} = (\hat{c}_0, \hat{c}_1, \hat{c}_2, \hat{c}_3).$$

To define the four AΔOs, we use the functions

$$s_\delta(z) \equiv \mathrm{sh}\left(\frac{\pi z}{a_\delta}\right), \quad c_\delta(z) \equiv \mathrm{ch}\left(\frac{\pi z}{a_\delta}\right), \quad \delta = +, -, a_+, a_- > 0,$$

and the shift (6.24). Now we introduce the AΔO

$$A_\delta(\mathbf{c}; z) \equiv C_\delta(z)(T^z_{ia_{-\delta}} - 1) + C_\delta(-z)(T^z_{-ia_{-\delta}} - 1) + 2c_\delta(2i\hat{c}_0), \quad \delta = +, -,$$

where

$$C_\delta(z) \equiv \frac{s_\delta(z - ic_0)}{s_\delta(z)} \frac{c_\delta(z - ic_1)}{c_\delta(z)} \frac{s_\delta(z - ic_2 - ia_{-\delta}/2)}{s_\delta(z - ia_{-\delta}/2)} \frac{c_\delta(z - ic_3 - ia_{-\delta}/2)}{c_\delta(z - ia_{-\delta}/2)}.$$

The function R we are about to introduce is a joint eigenfunction of the AΔOs

$$A_+(\mathbf{c}; v), \quad A_-(I\mathbf{c}; v), \quad A_+(\hat{\mathbf{c}}; \hat{v}), \quad A_-(I\hat{\mathbf{c}}; \hat{v}), \tag{6.30}$$

with eigenvalues

$$2c_+(2\hat{v}), \quad 2c_-(2\hat{v}), \quad 2c_+(2v), \quad 2c_-(2v), \tag{6.31}$$

respectively. (Note in this connection that the two AΔOs acting on v or on $\hat{v}$ commute; cf. also the paragraph containing (4.14).) It is given by an integral involving products of the function

$$G(a_+, a_-; z) \equiv \exp\left(i \int_0^\infty \frac{dy}{y} \left(\frac{\sin 2yz}{2\,\mathrm{sh}(a_+ y)\,\mathrm{sh}(a_- y)} - \frac{z}{a_+ a_- y} \right) \right)$$

where $|\operatorname{Im} z| < \frac{1}{2}(a_+ + a_-)$. This building block is studied in detail in Section 3.1 of Ref. 31. The G-function extends to a meromorphic function that satisfies three elementary first-order analytic difference equations. It may be viewed as a generalization of the gamma function, and the integral representation for the eigenfunction R may be viewed as a generalization of the Barnes representation for the hypergeometric function.

The G-function is manifestly symmetric in a_+, a_-; we shall suppress these parameters. Taking from now on

$$v, \; \hat{v}, \; c_0, \; \hat{c}_0 > 0, \; c_0 + c_1 < \tfrac{1}{2}a_+ + a_-,$$
$$c_0 + c_2 < a_+ + \tfrac{1}{2}a_-, \; c_0 + c_3 < \tfrac{1}{2}(a_+ + a_-)$$

(this restriction is imposed to ease the exposition), we define

$$R(a_+, a_-, \mathbf{c}; v, \hat{v}) \equiv (a_+ a_-)^{-1/2} \int_\Gamma dz\, I(a_+, a_-, \mathbf{c}; v, \hat{v}, z)$$

where Γ is a contour along the real axis, indented downward near the origin so as to avoid a simple pole. The integrand is given by

$$I \equiv F(c_0; v, z) K(a_+, a_-, \mathbf{c}; z) F(\hat{c}_0; \hat{v}, z).$$

The function F is symmetric in a_+, a_-:

$$F(c; y, z) \equiv \left(\frac{G(y + z + ic - i(a_+ + a_-)/2)}{G(y + ic - i(a_+ + a_-)/2)} \right)(y \to -y).$$

By contrast, the kernel function K is not symmetric:

$$K(a_+, a_-, \mathbf{c}; z) \equiv \frac{1}{G(z + i(a_+ + a_-)/2)} \frac{G(i(c_0 + c_1) - ia_-/2)}{G(z + i(c_0 + c_1) - ia_-/2)}$$
$$\cdot \frac{G(i(c_0 + c_2) - ia_+/2)}{G(z + i(c_0 + c_2) - ia_+/2)} \frac{G(i(c_0 + c_3))}{G(z + i(c_0 + c_3))}.$$

However, it satisfies

$$K(a_+, a_-, \mathbf{c}; z) = K(a_-, a_+, I\mathbf{c}; z)$$

which entails

$$R(a_+, a_-, \mathbf{c}; v, \hat{v}) = R(a_-, a_+, I\mathbf{c}; v, \hat{v}).$$

It is also clear from the above that one has

$$R(a_+, a_-, \mathbf{c}; v, \hat{v}) = R(a_+, a_-, \hat{\mathbf{c}}; \hat{v}, v), \tag{6.32}$$
$$R(a_+, a_-, \mathbf{c}; v, \hat{v}) = R(a_-, a_+, I\hat{\mathbf{c}}; \hat{v}, v).$$

From these symmetry properties it is immediate that the AΔOs (6.30) have eigenvalues (6.31) on R iff one of the AΔOs yields the respective eigenvalue.

Even a sketch of the proof of the eigenfunction property would carry us too far afield—among other things, it involves various properties of the G-function [31]. To proceed, we detail one simple way to obtain the relativistic type II eigenfunction $R(x, p)$: One can take

$$R(x, p) \equiv R(2\pi, \beta v, (\beta v g, \beta v g, 0, 0); vx, \beta p).$$

Indeed, with this choice of variables the AΔOs $A_+(\mathbf{c}; v)$ and $A_+(\hat{\mathbf{c}}; \hat{v})$ turn into H_{rel}^t (6.26) and $\widehat{H}_{\mathrm{rel}}^t$ (6.27), respectively, and (6.28) and self-duality (symmetry under $v, \beta, x, p \to \beta, v, p, x$) follow by specialization from (6.30), (6.31), and (6.32), respectively.

As our final topic, we turn to relativistic type IV eigenfunctions, taking

$$g = M + 1 = 2, 3, 4, \ldots. \tag{6.33}$$

The w-function corresponding to this choice reads [31]

$$w_M(x) = Cs^2(x) \prod_{j=1}^{M} s(x + ij\beta)s(x - ij\beta), \quad C > 0,$$

and the corresponding AΔO H_{rel}^t (6.25) will be written A_M. Anticipating the outcome of an Ansatz to be detailed shortly, we find eigenfunctions

$$A_M R_M(x, \lambda) = E_M(\lambda) R_M(x, \lambda) \tag{6.34}$$

depending on a complex parameter λ that plays the role of p for the previous case. (It does not reduce to p for the hyperbolic specialization, however.) For an infinite sequence of λ-values the function $R_M(x, \lambda)$ is 2ω-periodic or 2ω-antiperiodic; taking $\beta(M+1) \in (0, -2i\omega')$ (for simplicity), this countable infinity of functions spans a subspace $\mathcal{D}$ in the Hilbert space $\mathcal{H} \equiv L^2([0, 2\omega], w_M(x)dx)$ on which the AΔO A_M is symmetric; in particular, the eigenvalues of A_M on the sequence of functions are real. (Probably, $\mathcal{D}$ is dense in $\mathcal{H}$; if so, A_M is Hilbert integrable.)

The function R_M is of the form

$$R_M(x, \lambda) = F_M(x, \lambda) + F_M(-x, \lambda)$$

where $F_M(\pm x, \lambda)$ are linearly independent eigenfunctions with eigenvalue $E_M(\lambda)$. Since A_M is parity-invariant, we need only discuss $F_M(x, \lambda)$. This eigenfunction is found via an Ansatz of the form

$$F_M(x, \lambda) = \left(\prod_{k=-M}^{M} s(x + i\beta k) \right)^{-1} \prod_{j=1}^{M} s(x - z_j) \\ \times \exp\left(\frac{ix}{2\beta} \ln\left(\frac{s(z_j - i\beta)}{s(z_j + i\beta)} \right) \right) \tag{6.35}$$

(This Ansatz was inspired by our previous results for the hyperbolic $g \in \mathbb{N}$ case [5].) Requiring that (6.34) with $R \to F$ hold true, one readily obtains

$$E_M(\lambda) = C_M(x) \prod_{j=1}^{M} \left(s(i\beta + z_j) s(i\beta - z_j) \right)^{-1/2}$$

where

$$C_M(x) \equiv i^M \frac{s(x + i\beta M)}{s(x)} \prod_{j=1}^{M} \frac{s(x - i\beta - z_j)}{s(x - z_j)} s(i\beta - z_j) + (\beta \to -\beta).$$

Therefore, we obtain an eigenfunction whenever $C_M(x)$ does not depend on x.

We proceed by studying if and when the zeros $z_1, \ldots, z_M$ can be chosen such that this happens. We begin with the simplest case by far, namely, $M = 1$. We take z_1 incongruent to 0 (modulo the period lattice), so that the poles at $x \equiv 0$ and $x \equiv z_1$ of the two summands of $C_1(x)$ are simple. Now $C_1(x)$ is elliptic in x, and since the residues at $x = 0$ and $x = z_1$

cancel, $C_1(x)$ is constant. Thus we may put $x = i\beta$ to obtain (taking λ equal to z_1 from now on)

$$E_1(\lambda) = \frac{is(2i\beta)}{s(i\beta)^2} \left(\frac{1}{\mathcal{P}(\lambda) - \mathcal{P}(i\beta)} \right)^{1/2}. \tag{6.36}$$

Summarizing, $F_1(x, \lambda)$ (6.35) is an eigenfunction of A_1 with eigenvalue $E_1(\lambda)$ (6.36). Letting the zero $z_1 = \lambda$ vary over the line segment between $i\beta$ and ω', the eigenvalue $E_1(\lambda)$ is positive and decreases monotonically from ∞ to a finite limit.

Next, we take $M > 1$. Again, $C_M(x)$ is readily seen to be elliptic, but now $C_M(x)$ is not constant in general. However, we can proceed as follows. First, choose $z_1, \ldots, z_M$ pairwise incongruent, and incongruent to 0. Then the two summands of $C_M(x)$ have only simple poles, and the residues at $x \equiv 0$ cancel. Requiring now that the residue sum at the poles $x \equiv z_k$ vanish, too, we obtain

$$s(z_k + i\beta M) \prod_{\substack{j=1 \\ j \neq k}}^{M} s(z_k - z_j - i\beta) \prod_{j=1}^{M} s(z_j - i\beta) - (\beta \to -\beta) = 0,$$
$$k = 1, \ldots, M. \tag{6.37}$$

This constraint system of M equations for M unknowns $Z \equiv (z_1, \ldots, z_M)$ admits the solution

$$Z_0 \equiv (i\beta, 2i\beta, \ldots, Mi\beta)$$

as is readily verified. An application of the implicit function theorem now shows that there exists a solution curve $\lambda \mapsto Z(\lambda)$ near Z_0, with λ equal to z_1. Hence $C_M(x)$ is constant on the curve, and so we obtain eigenfunctions and eigenvalues depending on the curve parameter λ.

We intend to elaborate on the above assertions in Ref. 28. We conclude our sketch of type IV results by detailing the relation to the Lamé functions that solve the eigenfunction problem for the operator H_{nr} (6.23), with the restriction (6.33) in force.

First, taking β to 0 in (6.35) yields

$$F_M(x, \lambda) \to s(x)^{-2M-1} \prod_{j=1}^{M} s(x - z_j) \exp\left(\frac{xs'(z_j)}{s(z_j)} \right). \tag{6.38}$$

Second, dividing the constraints (6.37) by β and letting $\beta \to 0$, we obtain

$$M \frac{s'(z_k)}{s(z_k)} - \sum_{\substack{j=1 \\ j \neq k}}^{M} \frac{s'(z_k - z_j)}{s(z_k - z_j)} - \sum_{j=1}^{M} \frac{s'(z_j)}{s(z_j)} = 0, \quad k = 1, \ldots, M. \tag{6.39}$$

Third, recalling (2.50), we have

$$\frac{s'(x)}{s(x)} = \zeta(x) - \frac{\eta x}{\omega}$$

where ζ is the Weierstrass ζ-function. Therefore, the limit functions (6.38) and constraints (6.39) amount to the Lamé functions and associated constraints that are specified by Whittaker and Watson; cf. p. 572 and p. 574, respectively, of Ref. 12. Quite recently, Etingof and Kirillov have tied in the latter functions with the representation theory of affine Lie algebras [20].

7   REFERENCES

1. M. A. Olshanetsky and A. M. Perelomov. Classical integrable finite-dimensional systems related to Lie algebras. *Phys. Rep.*, 71 (5): 313–400, 1981.

2. M. A. Olshanetsky and A. M. Perelomov. Quantum integrable systems related to Lie algebras. *Phys. Rep.*, 94 (6): 313–404, 1983.

3. S. N. M. Ruijsenaars and H. Schneider. A new class of integrable systems and its relation to solitons. *Ann. Phys.*, 170 (2): 370–405, 1986.

4. S. N. M. Ruijsenaars. Complete integrability of relativistic Calogero-Moser systems and elliptic function identities. *Commun. Math. Phys.*, 110 (2): 191–213, 1987.

5. S. N. M. Ruijsenaars. Finite-dimensional soliton systems. In B. Kupershmidt, ed., *Integrable and Superintegrable Systems*. World Scientific, Teaneck, NJ, pages 165–206, 1990.

6. J. F. van Diejen. Commuting difference operators with polynomial eigenfunctions. *Comp. Math.*, 95 (2): 183–233, 1995.

7. J. F. van Diejen. Integrability of difference Calogero-Moser systems. *J. Math. Phys.*, 35 (6): 2983–3004, 1994.

8. J. F. van Diejen. Difference Calogero-Moser systems and finite Toda chains. *J. Math. Phys.*, 36 (3): 1299–1323, 1995.

9. V. I. Arnol'd. *Mathematical Methods of Classical Mechanics*, volume 60 of *Graduate Texts in Mathematics*. Springer-Verlag, New York, 1978.

10. W. Thirring. *Lehrbuch der mathematischen Physik. I. Klassische dynamische Systeme.* Springer-Verlag, Vienna, 1977.

11. R. Abraham and J. E. Marsden. *Foundations of Mechanics*, 2nd edition. Benjamin-Cummings, Reading, MA, 1978.

12. E. T. Whittaker and G. N. Watson. *A Course of Modern Analysis*. Cambridge Univ. Press, Cambridge, 1973.

13. M. Reed and B. Simon. *Methods of Modern Mathematical Physics III. Scattering Theory*. Academic, New York, 1979.

14. B. Sutherland. Exact results for a quantum many-body problem in one dimension II. *Phys. Rev.*, A5: 1372–1376, 1972.

15. G. J. Heckman. Root systems and hypergeometric functions II. *Comp. Math.*, 64 (3): 353–373, 1987.

16. I. G. Macdonald. Orthogonal polynomials associated with root systems. preprint, 1988.

17. I. G. Macdonald. Orthogonal polynomials associated with root systems. In P. Nevai, ed., *Orthogonal Polynomials*, (Columbus, OH, 1989), volume C294 of *NATO ASI*, 1990. Kluwer, Dordrecht, pages 311–318.

18. M. F. E. de Jeu. The Dunkl transform. *Invent. Math.*, 113 (1): 147–162, 1993.

19. P. I. Etingof and A. A. Kirillov, Jr. Macdonald's polynomials and representations of quantum groups. *Math. Res. Lett.*, 1 (3): 279–296, 1994.

20. P. I. Etingof and A. A. Kirillov, Jr. Representations of affine Lie algebras, parabolic differential equations, and Lamé functions. *Duke Math. J.*, 74 (3): 585–614, 1994.

21. M. Noumi. Macdonald's symmetric polynomials as zonal spherical functions on some quantum homogeneous spaces. *Adv. Math.*, to appear.

22. I. Cherednik. Difference-elliptic operators and root systems. *Internat. Math. Res. Notices*, 1995 (1): 43–58, 1995.

23. R. Askey and J. Wilson. *Some Basic Hypergeometric Orthogonal Polynomials that Generalize Jacobi Polynomials*, volume 319 of *Mem. Amer. Math. Soc.* Amer. Math. Soc., Providence, RI, 1985.

24. R. Floreanini and L. Vinet. Quantum algebras and q-special functions. *Ann. Phys.*, 221 (1): 53–70, 1993.

25. T. H. Koornwinder. Orthogonal polynomials in connection with quantum groups. In P. Nevai, ed., *Orthogonal Polynomials*, (Columbus, OH, 1989), volume C294 of *NATO ASI*, 1990. Kluwer, Dordrecht, pages 257–292.

26. T. H. Koornwinder. Askey-Wilson polynomials as zonal spherical functions on the SU(2) quantum group. *SIAM J. Math. Anal.*, 24 (3): 795–813, 1993.

27. N. J. Vilenkin and A. U. Klimyk. *Representation of Lie Groups and Special Functions III. Classical and Quantum Groups and Special Functions*, volume 75 of *Mathematics and Its Applications (Soviet Series)*. Kluwer, Dordrecht, 1992.

28. S. N. M. Ruijsenaars. Generalized Lamé functions. *J. Math. Phys.*, to appear.

29. S. N. M. Ruijsenaars. Relativistic Calogero-Moser systems and solitons. In M. Ablowitz, B. Fuchssteiner, and M. Kruskal, eds., *Topics in Soliton Theory and Exactly Solvable Nonlinear Equations*, (Oberwolfach, 1986), 1987. World Scientific, Singapore, pages 182–190.

30. S. N. M. Ruijsenaars. Action-angle maps and scattering theory for some finite-dimensional integrable systems II. Solitons, antisolitons, and their bound states. *Publ. RIMS, Kyoto Univ.*, 30 (6): 865–1008, 1994.

31. S. N. M. Ruijsenaars. First-order analytic difference equations and integrable quantum systems. *J. Math. Phys.*, 38: 1069–1146, 1997.

32. G. Felder and A. P. Veselov. Shift operators for the quantum Calogero-Sutherland problems via the Knizhnik-Zamolodchikov equation. *Commun. Math. Phys.*, 160 (2): 259–273, 1994.

33. J. Moser. Three integrable Hamiltonian systems connected with isospectral deformations. *Adv. Math.*, 16: 197–220, 1975.

34. I. M. Krichever. Elliptic solutions of the Kadomtsev-Petviashvili equation and integrable systems of particles. *Func. Anal. Appl.*, 14 (4): 282–290, 1980.

35. V. I. Inozemtsev. The finite Toda lattices. *Commun. Math. Phys.*, 121 (4): 629–638, 1989.

36. B. Sutherland. An introduction to the Bethe ansatz. In B. S. Shastry, S. S. Jha, and V. Singh, eds., *Exactly Solvable Problems in Condensed Matter and Relativistic Field Theory*, (Panchgani, 1985), volume 242 of *Lecture Notes in Physics*, 1985. Springer-Verlag, Berlin, pages 1–95.

37. S. N. M. Ruijsenaars. Relativistic Toda systems. *Commun. Math. Phys.*, 133 (2): 217–247, 1990.

38. G. Frobenius. Über die elliptischen functionen zweiter art. *J. Reine und Angew. Math.*, 93: 53–68, 1882.

39. A. K. Raina. An algebraic geometry study of the *b-c* system with arbitrary twist fields and arbitrary statistics. *Commun. Math. Phys.*, 140 (2): 373–397, 1991.

40. S. N. M. Ruijsenaars. Action-angle maps and scattering theory for some finite-dimensional integrable systems III. Sutherland type systems and their duals. *Publ. RIMS, Kyoto Univ.*, 31 (2): 247–353, 1995.

41. J. von Neumann. *Mathematical Foundations of Quantum Mechanics.* Princeton Univ. Press, Princeton, NJ, 1955.

42. M. Reed and B. Simon. *Methods of Modern Mathematical Physics* I. *Functional Analysis.* Academic, New York, 1972.

43. M. Reed and B. Simon. *Methods of Modern Mathematical Physics* II. *Fourier Analysis, Self-Adjointness.* Academic, New York, 1975.

44. M. Reed and B. Simon. *Methods of Modern Mathematical Physics* IV. *Analysis of Operators.* Academic, New York, 1978.

45. W. Thirring. *Lehrbuch der mathematischen Physik.* III. *Quantenmechanik von Atomen und Molekülen.* Springer-Verlag, Vienna, 1979.

46. E. M. Opdam. Root systems and hypergeometric functions III. *Comp. Math.*, 67 (1): 21–49, 1988.

47. E. M. Opdam. Root systems and hypergeometric functions IV. *Comp. Math.*, 67 (2): 191–209, 1988.

48. G. J. Heckman and E. M. Opdam. Root systems and hypergeometric functions I. *Comp. Math.*, 64 (3): 329–352, 1987.

49. T. Oshima and H. Sekiguchi. Commuting families of differential operators invariant under the action of a Weyl group. *J. Math. Sci. Univ. Tokyo*, 2 (1): 1–75, 1995.

50. K. Sawada and T. Kotera. Integrability and a solution for the one-dimensional N-particle system with inversely quadratic pair potentials. *J. Phys. Soc. Japan*, 39 (6): 1614–1618, 1975.

51. S. Wojciechowski. Involutive set of integrals for completely integrable many-body problems with pair interaction. *Lett. Nuovo Cim.*, 18: 103–107, 1977.

52. A. T. Fomenko. *Integrability and Nonintegrability in Geometry and Mechanics*, volume 31 of *Mathematics and Its Applications (Soviet Series)*. Kluwer, Dordrecht, 1988.

53. S. N. M. Ruijsenaars. Action-angle maps and scattering theory for some finite-dimensional integrable systems I. The pure soliton case. *Commun. Math. Phys.*, 115 (1): 127–165, 1988.

54. R. Goodman and N. R. Wallach. Classical and quantum-mechanical systems of Toda lattice-type III. Joint eigenfunctions of the quantized systems. *Commun. Math. Phys.*, 105 (3): 473–509, 1986.

55. J. Dittrich and V. I. Inozemtsev. On the structure of eigenvectors of the multidimensional Lamé operator. *J. Phys A: Math. Gen.*, 26 (16): L753–L756, 1993.

8

Discrete Gauge Theories

Mark de Wild Propitius
F. Alexander Bais

ABSTRACT We present a self-contained treatment of planar gauge theories broken down to some finite residual gauge group H via the Higgs
mechanism. The main focus is on the discrete H gauge theory describing
the long-distance physics of such a model. The spectrum features global H
charges, magnetic vortices, and dyonic combinations. Due to the Aharonov-
Bohm effect, these particles exhibit topological interactions. Among other
things, we review the Hopf algebra related to this discrete H gauge theory, which provides a unified description of the spin, braid, and fusion
properties of the particles in this model. Exotic phenomena such as flux
metamorphosis, Alice fluxes, Cheshire charge, (non-)Abelian braid statistics, the generalized spin-statistics connection, and non-Abelian Aharonov-
Bohm scattering are explained and illustrated by representative examples.

1   Broken Symmetry Revisited

Symmetry has become one of the major guiding principles in physics during
the twentieth century. Over the past century, we have gradually progressed
from external to internal, from global to local, from finite to infinite, from
ordinary to supersymmetry, and quite recently arrived at the notion of
Hopf algebras or quantum groups.

In general, a physical system consists of a finite or infinite number of
degrees of freedom which may or may not interact. The dynamics is pre-
scribed by a set of evolution equations which follow from varying the action
with respect to the different degrees of freedom. A symmetry then corresponds to a group of transformations on the space-time coordinates and/or
the degrees of freedom that leave the action and therefore also the evolution
equations invariant. External symmetries have to do with invariances (e.g.,
Lorentz invariance) under transformations on the space-time coordinates.
Symmetries not related to transformations of space-time coordinates are
called internal symmetries. We also discriminate between global symmetries and local symmetries. A global or rigid symmetry transformation is
the same throughout space-time and usually leads to a conserved quantity.
Turning a global symmetry into a local symmetry, i.e., allowing the symmetry transformations to vary continuously from one point in space time to

another, requires the introduction of additional *gauge* degrees of freedom mediating a force. It is this so-called gauge principle that has eventually led to the extremely successful standard model of the strong and electroweak interactions between the elementary particles based on the local gauge group $SU(3) \times SU(2) \times U(1)$.

The use of symmetry considerations has been extended significantly by the observation that a symmetry of the action is *not* automatically a symmetry of the ground state of a physical system. If the action is invariant under some symmetry group G and the ground state only under a subgroup H of G, the symmetry group G is said to be spontaneously broken down to H. The symmetry is not completely lost, though, for the broken generators of G transform one ground state into another.

The physics of a broken global symmetry is quite different from a broken local (gauge) symmetry. The signature of a broken continuous *global* symmetry group G in a physical system is the occurrence of massless scalar degrees of freedom, the so-called Goldstone bosons. Specifically, each broken generator of G gives rise to a massless Goldstone boson field. Well-known realizations of Goldstone bosons are the long-range spin waves in a ferromagnet, in which the rotational symmetry is broken below the Curie temperature through the appearance of spontaneous magnetization. An application in particle physics is the low-energy physics of the strong interactions, where the spontaneous breakdown of (approximate) chiral symmetry leads to (approximately) massless pseudoscalar particles such as the pions.

In the case of a broken *local* (gauge) symmetry, by way of contrast, the would be massless Goldstone bosons conspire with the massless gauge fields to form massive vector fields. This celebrated phenomenon is known as the Higgs mechanism. The canonical example in condensed matter physics is the ordinary superconductor. In the phase transition from the normal to the superconducting phase, the $U(1)$ gauge symmetry is spontaneously broken to the finite cyclic group $\mathbb{Z}_2$ by a condensate of Cooper pairs. This leads to a mass M_A for the photon field in the superconducting medium as witnessed by the Meissner effect: magnetic fields are expelled from a superconducting region and have a characteristic penetration depth which in proper units is just the inverse of the photon mass M_A. Moreover, the Coulomb interactions among external electric charges in a superconductor are of finite range $\sim 1/M_A$. The Higgs mechanism also plays a key role in the unified theory of weak and electromagnetic interactions, that is, the Glashow-Weinberg-Salam model where the product gauge group $SU(2) \times U(1)$ is broken to the $U(1)$ subgroup of electromagnetism. In this context, the massive vector particles correspond to the W and Z bosons mediating the short-range weak interactions. More speculative applications of the Higgs mechanism are those where the standard model of the strong, weak, and electromagnetic interactions is embedded in a grand unified model with a large simple gauge group. The most ambitious attempts invoke supersymmetry as well.

In addition to the aforementioned characteristics in the spectrum of fundamental excitations, there are in general other fingerprints of a broken symmetry in a physical system. These are usually called *topological excitations* or just *defects* and correspond to collective degrees of freedom carrying "charges" or quantum numbers which are conserved for topological reasons, not related to a manifest symmetry of the action. (See, for example, Refs. 1–4 for reviews). It is exactly the appearance of these topological charges which renders the corresponding collective excitations stable. Topological excitations may manifest themselves as particlelike, stringlike or planarlike objects (solitons), or have to be interpreted as quantum-mechanical tunneling processes (instantons). Depending on the model in which they occur, these excitations carry evocative names like kinks, domain walls, vortices, cosmic strings, Alice strings, monopoles, skyrmions, texture, sphalerons, and so on. Defects are crucial for a full understanding of the physics of systems with a broken symmetry and lead to a host of rather unexpected and exotic phenomena which are in general of a nonperturbative nature.

The prototypical example of a topological defect is the Abrikosov-Nielsen-Olesen flux tube in the type II superconductor with broken U(1) gauge symmetry [5, 6]. The topologically conserved quantum number characterizing these defects is the magnetic flux, which indeed can only take discrete values. A beautiful but unfortunately not yet observed example in particle physics is the 't Hooft-Polyakov monopole [7, 8] occurring in any grand unified model in which a simple gauge group G is broken to a subgroup H containing the electromagnetic U(1) factor. Here, it is the quantized magnetic charge carried by these monopoles that is conserved for topological reasons. In fact, the discovery that these models support magnetic monopoles reconciled the two well-known arguments for the quantization of electric charge, namely, Dirac's argument based on the existence of a magnetic monopole [9] and the obvious fact that the U(1) generator should be compact as it belongs to a larger compact gauge group.

An example of a model with a broken global symmetry supporting topological excitations is the effective sigma model describing the low-energy strong interactions for the mesons—that is, the phase with broken chiral symmetry mentioned before. One may add a topological term and a stabilizing term to the action and obtain a theory that features topological particlelike objects called skyrmions, which have exactly the properties of the baryons. See Refs. 10–12. So, upon extending the effective model for the Goldstone bosons, we recover the complete spectrum of the underlying strong interaction model (quantum chromodynamics) and its low-energy dynamics. Indeed, this picture leads to an attractive phenomenological model for baryons.

Another area of physics where defects may play a fundamental role is cosmology. See, for instance, Ref. 13 for a recent review. According to the standard cosmological hot Big Bang scenario, the universe cooled down through a sequence of local and/or global symmetry breaking phase transitions in a very early stage. The question of the actual formation of defects in these phase transitions is of prime importance. It has been argued, for instance, that magnetic monopoles might have been produced copiously. As they tend to dominate the mass in the universe, however, magnetic monopoles are notoriously hard to accommodate and if indeed formed, they have to be "inflated away." Phase transitions that see the production of (local or global) cosmic strings, on the other hand, are much more interesting. In contrast with magnetic monopoles, the presence of cosmic strings does not lead to cosmological disasters and according to an attractive but still speculative theory cosmic strings may even have acted as seeds for the formation of galaxies and other large-scale structures in the present-day universe.

Similar symmetry breaking phase transitions are extensively studied in condensed matter physics. We have already mentioned the transition from the normal to the superconducting phase in superconducting materials of type II, which may give rise to the formation of magnetic flux tubes. In the field of low-temperature physics, there also exists a great body of both theoretical and experimental work on the transitions from the normal to the many superfluid phases of helium-3 in which line and point defects arise in a great variety, e.g., [14]. Furthermore, in uniaxial nematic liquid crystals, point defects, line defects and texture arise in the transition from the disordered to the ordered phase in which the rotational global symmetry group $SO(3)$ is broken down to the semidirect product group $U(1) \times_{\text{s.d.}} \mathbb{Z}_2$. Biaxial nematic crystals, in turn, exhibit a phase transition in which the global rotational symmetry group is broken down to the product group $\mathbb{Z}_2 \times \mathbb{Z}_2$ yielding line defects labeled by the elements of the (non-Abelian) quaternion group $\overline{D}_2$, e.g., [2]. Nematic crystals are cheap materials and as compared to helium-3, for instance, relatively easy to work with in the laboratory. The symmetry-breaking phase transitions typically appear at temperatures that can be reached by a standard kitchen oven, whereas the size of the defects occurring is such that these can be seen by means of a simple microscope. Hence, these materials form an easily accessible experimental playground for the investigation of defect-producing phase transitions and as such may partly mimic the physics of the early universe in the laboratory. For some recent ingenious experimental studies on the formation and the dynamics of topological defects in nematic crystals making use of high-speed film cameras, the interested reader is referred to Refs. 15 and 16.

From a theoretical point of view, many aspects of topological defects have been studied and understood. At the classical level, one may roughly sketch the following programme. One first uses simple topological arguments, usually of the homotopy type, to see whether a given model does exhibit topological charges. Subsequently, one may try to prove the existence of the corresponding classical solutions by functional analysis methods or just by explicit construction of particular solutions. On the other hand, one may in many cases determine the dimension of the solution or moduli space and its dependence on the topological charge using index theory. Finally, one may attempt to determine the general solution space more or less explicitly. In this respect, one has been successful in varying degree. In particular, the self-dual instanton solutions of Yang-Mills theory on S^4 have been obtained completely.

The physical properties of topological defects can be probed by their interactions with the ordinary particles or excitations in the model. This amounts to investigating (quantum) processes in the background of the defect. In particular, one may calculate the one-loop corrections to the various quantities characterizing the defect, which involves studying the fluctuation operator. Here, one naturally has to distinguish the modes with zero eigenvalue from those with nonzero eigenvalues. The nonzero modes generically give rise to the usual renormalization effects, such as mass and coupling constant renormalization. The zero modes, which often arise as a consequence of the global symmetries in the theory, lead to collective coordinates. Their quantization yields a semiclassical description of the spectrum of the theory in a given topological sector, including the external quantum numbers of the soliton such as its energy and momentum and its internal quantum numbers such as its electric charge, e.g., [1, 3].

In situations where the residual gauge group H is non-Abelian, the analysis outlined in the previous paragraph is rather subtle. For instance, the naive expectation that a soliton can carry internal electric charges that form representations of the complete unbroken group H is wrong. As only the subgroup of H that commutes with the topological charge can be globally implemented, these internal charges form representations of this so-called centralizer subgroup. (See Refs. 17–19 for the case of magnetic monopoles and Refs. 20 and 21 for the case of magnetic vortices). This makes the full spectrum of topological and ordinary quantum numbers in such a broken phase rather intricate.

Also, an important effect on the spectrum and the interactions of a theory with a broken gauge group is caused by the introduction of additional topological terms in the action, such as a nonvanishing θ angle in $(3+1)$-dimensional space-time and the Chern-Simons term in $(2+1)$ dimensions. It has been shown by Witten that in case of a nonvanishing θ angle, for example, magnetic monopoles carry electric charges which are shifted by an amount proportional to $\theta/2\pi$ and their magnetic charge [22].

Other results are even more surprising. A broken gauge theory only containing bosonic fields may support topological excitations (dyons), which on the quantum level carry half-integral spin and are fermions, thereby realizing the counterintuitive possibility to make fermions out of bosons [23, 24]. It has subsequently been argued by Wilczek [25] that in $(2+1)$-dimensional space-time one can even have topological excitations, namely, flux/charge composites, which behave as anyons [26], i.e., particles with fractional spin and quantum statistics interpolating between bosons and fermions. The possibility of anyons in two spatial dimensions is not merely of academic interest, as many systems in condensed matter physics, for example, are effectively described by (2+1)-dimensional models. Indeed, anyons are known to be realized as quasiparticles in fractional quantum Hall systems [27, 28]. Further, it has been been shown that an ideal gas of electrically charged anyons is superconducting [29–32]. At present it is unclear whether this new and rather exotic type of superconductivity is actually realized in nature.

Furthermore, remarkable calculations by 't Hooft revealed a nonperturbative mechanism for baryon decay in the standard model through instantons and sphalerons [33]. Afterwards, Rubakov [34] and Callan [35, 36] discovered the phenomenon of baryon decay catalysis induced by grand unified monopoles. Baryon number violating processes also occur in the vicinity of grand unified cosmic strings as has been established by Alford et al. [37].

So far, we have given a (rather incomplete) enumeration of properties and processes that involve the interactions between topological and ordinary excitations. However, the interactions between defects themselves can also be highly nontrivial. Here, one should not only think of ordinary interactions corresponding to the exchange of field quanta. Consider, for instance, the case of Alice electrodynamics which occurs if some non-Abelian gauge group (e.g., SO(3)) is broken to the non-Abelian subgroup U(1) $\times_{\text{s.d.}}$ $\mathbb{Z}_2$, that is, the semidirect product of the electromagnetic group U(1) and the additional cyclic group $\mathbb{Z}_2$ whose nontrivial element reverses the sign of the electromagnetic fields [38]. This model features magnetic monopoles and in addition a magnetic $\mathbb{Z}_2$ string (the so-called Alice string) with the miraculous property that if a monopole (or an electric charge for that matter) is transported around the string, its charge will change sign. In other words, a particle is converted into its own antiparticle. This non-Abelian analogue of the celebrated Aharonov-Bohm effect [39] is of a topological nature. That is, it only depends on the number of times the particle winds around the string and is independent of the distance between the particle and the string.

Similar phenomena occur in models in which a continuous gauge group is spontaneously broken down to some *finite* subgroup H. The topological defects supported by such a model are stringlike in three spatial dimensions and carry a magnetic flux corresponding to an element h of the residual gauge group H. As these stringlike objects trivialize one spatial dimension, we may just as well descend to the plane, for convenience. In this arena, these defects become magnetic vortices, i.e., particlelike objects of characteristic size $1/M_H$ with M_H the symmetry breaking scale. Besides these topological particles, the broken phase features matter charges labeled by the unitary irreducible representations Γ of the residual gauge group H. Since all gauge fields are massive, there are no ordinary long-range interactions among these particles. The remaining long-range interactions are topological Aharonov-Bohm interactions. If the residual gauge group H is non-Abelian, for instance, the non-Abelian fluxes $h \in H$ carried by the vortices exhibit flux metamorphosis [40]. In the process of circumnavigating one vortex with another vortex their fluxes may change. Moreover, if a charge corresponding to some representation Γ of H is transported around a vortex carrying the magnetic flux $h \in H$, it returns transformed by the matrix $\Gamma(h)$ assigned to the element h in the representation Γ of H.

The spontaneously broken (2+1)-dimensional models just mentioned will be the subject of these lecture notes. One of our aims is to show that the long-distance physics of such a model is, in fact, governed by a Hopf algebra or quantum group based on the residual finite gauge group H [20, 41–43]. This algebraic framework manifestly unifies the topological and nontopological quantum numbers as dual aspects of a single symmetry concept. The results presented here strongly suggests that revisiting the symmetry breaking concept in general will reveal similar underlying algebraic structures.

The outline of these notes, which are intended to be accessible to a reader with a minimal background in field theory, quantum mechanics, and finite group theory, is as follows. In Section 2, we start with a review of the basic physical properties of a planar gauge theory broken down to a finite gauge group via the Higgs mechanism. The main focus will be on the discrete H gauge theory describing the long distance of such a model. We argue that in addition to the aforementioned magnetic vortices and global H charges the complete spectrum also consists of dyonic combinations of the two and establish the basic topological interactions among these particles. In Section 3, we then turn to the Hopf algebra related to this discrete H gauge theory and elaborate on the unified description this framework gives of the spin, braid, and fusion properties of the particles. Finally, the general formalism developed in the foregoing sections is illustrated by an explicit non-Abelian example in Section 4, namely, a planar gauge theory spontaneously broken down to the double dihedral gauge group $\bar{D}_2$. Among other things, exotic phenomena like Cheshire charge, Alice fluxes, non-Abelian braid statistics, and non-Abelian Aharonov-Bohm scattering are explained there.

Let us conclude this preface with some remarks on conventions. Throughout these notes units in which $\hbar = c = 1$ are employed. Latin indices take the values 1, 2. Greek indices run from 0 to 2. Further, x^1 and x^2 denote spatial coordinates and $x^0 = t$ the time coordinate. The signature of the three-dimensional metric is taken as $(+, -, -)$. Unless stated otherwise, we adopt Einstein's summation convention.

2   Basics

2.1   Introduction

The planar gauge theories we will study in these notes are given by an action of the general form

$$S = S_{\text{YMH}} + S_{\text{matter}}. \tag{2.1}$$

The continuous gauge group G of this model is assumed to be broken down to some finite subgroup H of G by means of the Higgs mechanism. That is, the Yang-Mills Higgs part S_{YMH} of the action features a Higgs field whose nonvanishing vacuum expectation values are only invariant under the action of H. Further, the matter part S_{matter} describes matter fields covariantly coupled to the gauge fields. A discussion of the implications of adding a Chern-Simons term to the spontaneously broken planar gauge theory (2.1) is beyond the scope of these notes. For this, the interested reader is referred to Refs. 41–44.

Since the unbroken gauge group H is finite, all gauge fields are massive and it seems that the low-energy or equivalently the long-distance physics of the model (2.1) is trivial. This is not the case though. It is the occurrence of topological defects and the persistence of the Aharonov-Bohm effect that renders the long-distance physics nontrivial. Specifically, the defects supported by these models are (particlelike) vortices of characteristic size $1/M_H$, with M_H the symmetry breaking scale. These vortices carry magnetic fluxes labeled by the elements h of the residual gauge group H.[1] In other words, the vortices introduce nontrivial holonomies in the locally flat gauge fields. Consequently, if the residual gauge group H is non-Abelian, these fluxes exhibit nontrivial topological interactions: in the process in which one vortex circumnavigates another, the associated magnetic fluxes feel each other's holonomies and affect each other through conjugation. This is in a nutshell the long-distance physics described by the Yang-Mills Higgs part S_{YMH} of the action. The matter fields, covariantly coupled to the gauge fields in the matter part S_{matter} of the action, form multiplets which transform irreducibly under the broken gauge group G. In the broken phase, these branch to irreducible representations of the residual gauge group H. So, the matter fields introduce point charges in the broken phase labeled by the unitary irreducible representations Γ of H. When such a charge encircles a magnetic flux $h \in H$, it undergoes an Aharonov-Bohm effect: it returns transformed by the matrix $\Gamma(h)$ assigned to the group element h in the representation Γ of H.

---

[1] Here, we tacitly assume that the broken gauge group G is simply connected. If G is not simply connected and the model does not contain Dirac monopoles/instantons, then the vortices carry fluxes labeled by the elements of the lift $\overline{H}$ of H into the universal covering group $\overline{G}$ of G. See Section 2.4.1 in this connection.

In this section, we establish the complete spectrum of the discrete H gauge theory describing the long-distance physics of the spontaneously broken model (2.1), which besides the aforementioned matter charges and magnetic vortices also consists of dyons obtained by composing these charges and vortices. In addition, we elaborate on the basic topological interactions between these particles. The discussion is organized as follows. In Section 2.2, we start by briefly recalling that particle interchanges in the plane are organized by braid groups. Section 2.3 then contains an analysis of a planar Abelian Higgs model in which the U(1) gauge group is spontaneously broken to the cyclic subgroup $\mathbb{Z}_N$. The main emphasis will be on the $\mathbb{Z}_N$ gauge theory that describes the long-distance physics of this model. Among other things, we show that the spectrum indeed consists of $\mathbb{Z}_N$ fluxes, $\mathbb{Z}_N$ charges and dyonic combinations of the two, establish the quantum-mechanical Aharonov-Bohm interactions between these particles and argue that as a result the wave functions of the multiparticle configurations in this model realize nontrivial Abelian representations of the related braid group. Finally, the subtleties involved in the generalization to models in which a non-Abelian gauge group G is broken to a non-Abelian finite group H are dealt with in Section 2.4.

2.2 Braid Groups

Let us consider a system of n indistinguishable particles moving on a manifold M, which is assumed to be connected and path connected for convenience. The classical configuration space of this system is given by

$$\mathcal{C}_n(M) = (M^n - D)/S_n, \tag{2.2}$$

where the action of the permutation group S_n on the particle positions is divided out to account for the indistinguishability of the particles. Moreover, the singular configurations D in which two or more particles coincide are excluded. The configuration space (2.2) is in general multiply connected. This means that there are different kinematical options to quantize this multiparticle system. To be precise, there is a quantization associated to each unitary irreducible representation (UIR) of the fundamental group $\pi_1\big(\mathcal{C}_n(M)\big)$. See, for instance, Refs. 45–48.

It is easily verified that for manifolds M with dimension larger than 2, we have the isomorphism $\pi_1\big(\mathcal{C}_n(M)\big) \simeq S_n$. Hence, the inequivalent quantizations of multiparticle systems moving on such manifolds are labeled by the UIRs of the permutation group S_n. There are two one-dimensional UIRs of S_n. The trivial representation naturally corresponds with Bose statistics. In this case, the system is quantized by a (scalar) wave function, which is symmetric under all permutations of the particles. The antisymmetric representation, on the other hand, corresponds with Fermi statistics, i.e., we are dealing with a wave function which acquires a minus sign under odd

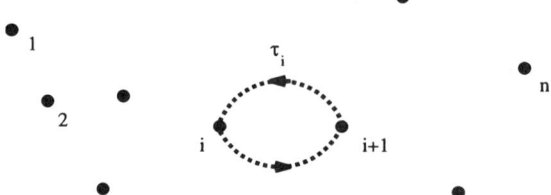

FIGURE 1. The braid operator τ_i establishes a counterclockwise interchange of the particles i and $i+1$ in a set of n numbered indistinguishable particles in the plane.

permutations of the particles. Finally, parastatistics is also conceivable. In this case, the system is quantized by a multicomponent wave function which transforms as a higher-dimensional UIR of S_n.

It has been known for some time that quantum statistics for identical particles moving in the plane ($M = \mathbb{R}^2$) can be much more exotic then in three or more dimensions [26, 49]. The point is that the fundamental group of the associated configuration space $\mathcal{C}_n(\mathbb{R}^2)$ is not given by the permutation group, but rather by the so-called braid group $B_n(\mathbb{R}^2)$ [50]. In contrast with the permutation group S_n, the braid group $B_n(\mathbb{R}^2)$ is a non-Abelian group of *infinite* order. It is generated by $n-1$ elements $\tau_1, \ldots, \tau_{n-1}$, where τ_i establishes a counterclockwise interchange of the particles i and $i+1$ as depicted in Figure 1. These generators are subject to the relations

$$\begin{aligned}
\tau_i\tau_{i+1}\tau_i &= \tau_{i+1}\tau_i\tau_{i+1}, & i &= 1,\ldots,n-2, \\
\tau_i\tau_j &= \tau_j\tau_i, & |i-j| &\geq 2,
\end{aligned} \tag{2.3}$$

which can be presented graphically as in Figures 2 and 3, respectively. In fact, the permutation group S_n ruling the particle exchanges in three or more dimensions is given by the same set of generators with relations (2.3) *and* the additional relations $\tau_i^2 = e$ for all $i \in 1, \ldots, n-1$. These last relations are absent for $\pi_1\big(\mathcal{C}_n(\mathbb{R}^2)\big) \simeq B_n(\mathbb{R}^2)$, since in the plane a counterclockwise particle interchange τ_i ceases to be homotopic to the clockwise interchange τ_i^{-1}.

The one-dimensional UIRs of the braid group $B_n(\mathbb{R}^2)$ are labeled by an angular parameter $\Theta \in [0, 2\pi)$ and are defined by assigning the same phase factor to all generators. That is,

$$\tau_i \mapsto \exp(i\Theta), \tag{2.4}$$

for all $i \in 1, \ldots, n-1$. The quantization of a system of n identical particles in the plane corresponding to an arbitrary but fixed $\Theta \in [0, 2\pi)$ is then given by a multivalued (scalar) wave function that generates the quantum-statistical phase $\exp(i\Theta)$ upon a counterclockwise interchange of two adjacent particles. For $\Theta = 0$ and $\Theta = \pi$, we are dealing with bosons and

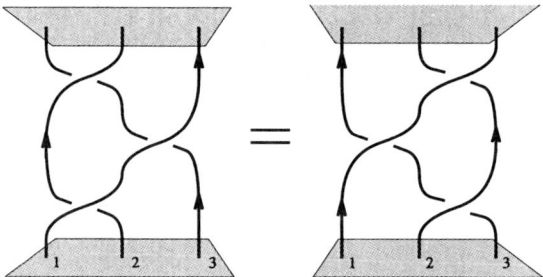

FIGURE 2. Pictorial presentation of the braid relation $\tau_1\tau_2\tau_1 = \tau_2\tau_1\tau_2$. The particle trajectories corresponding to the composition of exchanges $\tau_1\tau_2\tau_1$ (diagram at the l.h.s.) can be continuously deformed into the trajectories associated with the composition of exchanges $\tau_2\tau_1\tau_2$ (r.h.s. diagram).

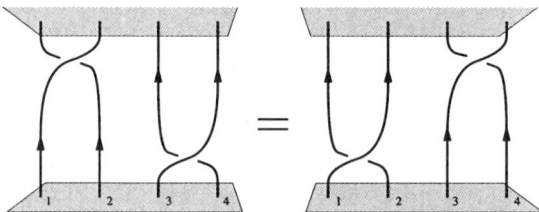

FIGURE 3. The braid relation $\tau_1\tau_3 = \tau_3\tau_1$ expresses the fact that the particle trajectories displayed in the l.h.s. diagram can be continuously deformed into the trajectories in the r.h.s. diagram.

fermions respectively. The particle species related to other values of Θ have been called anyons [49]. Quantum statistics deviating from conventional permutation statistics is known under various names in the literature, e.g., fractional statistics, anyon statistics, and exotic statistics. We adopt the following nomenclature. An identical particle system described by a (multivalued) wave function that transforms as an one-dimensional (Abelian) UIR of the braid group $B_n(\mathbb{R}^2)$ ($\Theta \neq 0, \pi$) is said to realize Abelian braid statistics. If an identical particle system is described by a multicomponent wave function carrying an higher-dimensional UIR of the braid group, then the particles are said to obey non-Abelian braid statistics.

A system of n distinguishable particles moving in the plane, in turn, is described by the nonsimply connected configuration space

$$\mathcal{Q}_n(\mathbb{R}^2) = (\mathbb{R}^2)^n - D.$$

The fundamental group of this configuration space is the so-called colored braid group $P_n(\mathbb{R}^2)$, also known as the pure braid group. The colored braid group $P_n(\mathbb{R}^2)$ is the subgroup of the ordinary braid group $B_n(\mathbb{R}^2)$ generated by the monodromy operators

$$\gamma_{ij} := \tau_i \cdots \tau_{j-2} \tau_{j-1}^2 \tau_{j-2}^{-1} \cdots \tau_i^{-1}, \quad \text{with } 1 \leq i < j \leq n. \tag{2.5}$$

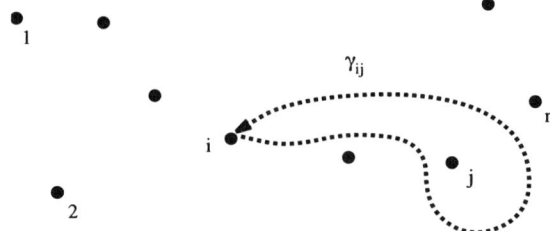

FIGURE 4. The monodromy operator γ_{ij} takes particle i counterclockwise around particle j.

Here, the τ_i's are the generators of $B_n(\mathbb{R}^2)$ acting on the set of n numbered distinguishable particles as displayed in Figure 1. It then follows from the definition (2.5) that the monodromy operator γ_{ij} takes particle i counterclockwise around particle j as depicted in Figure 4. The different UIRs of $P_n(\mathbb{R}^2)$ now label the inequivalent ways to quantize a system of n distinguishable particles in the plane. Finally, a planar system that consists of a subsystem of identical particles of one type, a subsystem of identical particles of another type and so on, is of course also conceivable. The fundamental group of the configuration space of such a system is known as a partially colored braid group. Let the total number of particles of this system again be n, then the associated partially colored braid group is the subgroup of the ordinary braid group $B_n(\mathbb{R}^2)$ generated by the braid operators that interchange identical particles and the monodromy operators acting on distinguishable particles. See, for example, Refs. 51 and 52.

To conclude, the fundamental excitations in planar discrete gauge theories, namely, magnetic vortices and matter charges, are in principle bosons. As will be argued in the next sections, in the first quantized description, these particles acquire braid statistics through the Aharonov-Bohm effect. Hence, depending on whether we are dealing with a system of identical particles, a system of distinguishable particles or a mixture, the associated multiparticle wave function generically transforms as a nontrivial representation of the ordinary braid group, colored braid group, or partially colored braid group, respectively.

2.3 $\mathbb{Z}_N$ Gauge Theory

The simplest example of a broken gauge theory is an U(1) gauge theory spontaneously broken down to the cyclic subgroup $\mathbb{Z}_N$. This symmetry breaking scheme occurs in an Abelian Higgs model in which the field that condenses carries charge Ne, with e the fundamental charge unit [53]. The case $N = 2$ is in fact realized in the ordinary BCS superconductor, as the field that condenses in the BCS superconductor is that associated with the Cooper pair carrying charge $2e$.

This subsection is devoted to a discussion of such an Abelian Higgs model in $(2 + 1)$-dimensional space-time. We focus on the $\mathbb{Z}_N$ gauge theory describing the long-range physics. The organization is as follows. In Section 2.3.1, we will start with a brief review of the screening mechanism for the electromagnetic fields of external matter charges q in the Higgs phase. We will argue that the external matter charges, which are multiples of the fundamental charge e rather then multiples of the Higgs charge Ne, are surrounded by screening charges provided by the Higgs condensate. These screening charges screen the electromagnetic fields around the external charges. Thus, no long-range Coulomb interactions persist among the external charges. The main point of Section 2.3.2 will be, however, that the screening charges do *not* screen the Aharonov-Bohm interactions between the external charges and the magnetic vortices, which also feature in this model. *As a consequence, long-range Aharonov-Bohm interactions persist between the vortices and the external matter charges in the Higgs phase.* Upon circumnavigating a magnetic vortex (carrying a flux ϕ which is a multiple of the fundamental flux unit $2\pi/(Ne)$ in this case) with an external charge q (being a multiple of the fundamental charge unit e) the wave function of the system picks up the Aharonov-Bohm phase $\exp(iq\phi)$. These Aharonov-Bohm phases lead to observable low-energy scattering effects from which we conclude that the physically distinct superselection sectors in the Higgs phase can be labeled as (a, n), where a stands for the number of fundamental flux units $2\pi/(Ne)$ and n for the number of fundamental charge units e. In other words, the spectrum of the $\mathbb{Z}_N$ gauge theory in the Higgs phase consists of pure charges n, pure fluxes a, and dyonic combinations. Given the remaining long-range Aharonov-Bohm interactions, these charge and flux quantum numbers are defined modulo N. Having identified the spectrum and the long-range interactions as the topological Aharonov-Bohm effect, we proceed with a closer examination of this $\mathbb{Z}_N$ gauge theory in Section 2.3.3. It will be argued that multiparticle systems in general satisfy Abelian braid statistics. That is, the wave functions realize one-dimensional representations of the associated braid group. In particular, identical dyons behave as anyons. We will also discuss the composition rules for the charge/flux quantum numbers when two particles are brought together. A key result of this subsection is a topological proof of the spin-statistics connection for the particles in the spectrum. This proof is of a general nature and applies to all the theories that will be discussed in these notes.

2.3.1   Coulomb Screening

The planar Abelian Higgs model which we will study is governed by the following action

$$S = \int d^3x \, (\mathcal{L}_{\text{YMH}} + \mathcal{L}_{\text{matter}}), \tag{2.6}$$

$$\mathcal{L}_{\mathrm{YMH}} = -\frac{1}{4} F^{\kappa\nu} F_{\kappa\nu} + (\mathcal{D}^{\kappa}\Phi)^{*}\mathcal{D}_{\kappa}\Phi - V(|\Phi|), \qquad (2.7)$$

$$\mathcal{L}_{\mathrm{matter}} = -j^{\kappa} A_{\kappa}. \qquad (2.8)$$

The Higgs field Φ is assumed to carry the charge Ne w.r.t. the compact U(1) gauge symmetry. In the conventions we will adopt in these notes, this means that the covariant derivative reads $\mathcal{D}_{\rho}\Phi = (\partial_{\rho} + iNeA_{\rho})\Phi$. Furthermore, the potential

$$V(|\Phi|) = \frac{\lambda}{4}(|\Phi|^2 - v^2)^2, \quad \lambda, v > 0, \qquad (2.9)$$

endows the Higgs field with a nonvanishing vacuum expectation value $|\langle\Phi\rangle| = v$, which implies that the global continuous U(1) symmetry is spontaneously broken. However, in this particular model the symmetry is not completely broken. Under global symmetry transformations $\Lambda(\alpha)$, with $\alpha \in [0, 2\pi)$ being the U(1) parameter, the ground states transform as

$$\Lambda(\alpha)\langle\Phi\rangle = e^{iN\alpha}\langle\Phi\rangle,$$

since the Higgs field is assumed to carry the charge Ne. Clearly, the residual symmetry group of the ground states is the finite cyclic group $\mathbb{Z}_N$ corresponding to the elements $\alpha = 2\pi k/N$ with $k \in 0, 1, \dots, N-1$.

Further, the field equations following from variation of the action (2.6) w.r.t. the vector potential A_{κ} and the Higgs field Φ are simply inferred as

$$\partial_{\nu} F^{\nu\kappa} = j^{\kappa} + j_{H}^{\kappa},$$

$$\mathcal{D}_{\kappa}\mathcal{D}^{\kappa}\Phi^{*} = -\frac{\partial V}{\partial\Phi},$$

where

$$j_{H}^{\kappa} = iNe(\Phi^{*}\mathcal{D}^{\kappa}\Phi - (\mathcal{D}^{\kappa}\Phi)^{*}\Phi), \qquad (2.10)$$

denotes the Higgs current.

In this section, we will only be concerned with the Higgs screening mechanism for the electromagnetic fields induced by the matter charges described by the conserved matter current j^{κ} in (2.8). For convenience, we discard the dynamics of the fields that are associated with this current and simply treat j^{κ} as being external. In fact, for our purposes the only important feature of the current j^{κ} is that it allows us to introduce global U(1) charges q in the Higgs medium, which are multiples of the fundamental charge e rather then multiples of the Higgs charge Ne, so that all conceivable charge sectors can be discussed.

Let us first recall some of the basic dynamical features of this model. First of all, the complex Higgs field

$$\Phi(x) = \rho(x)\exp(i\sigma(x))$$

describes two physical degrees of freedom: the charged Goldstone boson field $\sigma(x)$ and the physical field $\rho(x) - v$ with mass $M_H = v\sqrt{2\lambda}$ corresponding to the charged neutral Higgs particles. The Higgs mass M_H sets the characteristic energy scale of this model. At energies larger then M_H, the massive Higgs particles can be excited. At energies smaller then M_H on the other hand, the massive Higgs particles can not be excited. For simplicity we will restrict ourselves to the latter low-energy regime. In that case, the Higgs field is completely condensed, i.e., it acquires ground-state values everywhere

$$\Phi(x) \mapsto \langle\Phi(x)\rangle = v\exp(i\sigma(x)). \tag{2.11}$$

The condensation of the Higgs field implies that in the low-energy regime, the Higgs model is governed by the effective action obtained from the action (2.6) by the following simplification

$$\mathcal{L}_{\text{YMH}} \mapsto -\frac{1}{4}F^{\kappa\nu}F_{\kappa\nu} + \frac{M_A^2}{2}\tilde{A}^\kappa\tilde{A}_\kappa, \tag{2.12}$$

$$\tilde{A}_\kappa := A_\kappa + \frac{1}{Ne}\partial_\kappa\sigma, \tag{2.13}$$

$$M_A := Nev\sqrt{2}. \tag{2.14}$$

Thus, the dynamics of the Higgs medium arising here is described by the effective field equations inferred from varying the effective action w.r.t. the gauge field A_κ and the Goldstone boson σ, respectively,

$$\partial_\nu F^{\nu\kappa} = j^\kappa + j_{\text{scr}}^\kappa, \tag{2.15}$$

$$\partial_\kappa j_{\text{scr}}^\kappa = 0, \tag{2.16}$$

with

$$j_{\text{scr}}^\kappa = -M_A^2\tilde{A}^\kappa, \tag{2.17}$$

the simple form the Higgs current (2.10) takes in the low-energy regime.

It is easily verified that the field equations (2.15) and (2.16) can be cast in the following form

$$(\partial_\nu\partial^\nu + M_A^2)\tilde{A}^\kappa = j^\kappa, \tag{2.18}$$

$$\partial_\kappa\tilde{A}^\kappa = 0, \tag{2.19}$$

which clearly indicates that the gauge-invariant vector field $\tilde{A}_\kappa$ has become massive. More specifically, in this $(2+1)$-dimensional setting it describes a two-component massive photon field carrying the mass M_A defined in (2.14). Consequently, the electromagnetic fields around sources in the Higgs medium decay exponentially with mass M_A. Of course, the number of degrees of freedom is conserved. We started with an unbroken theory with two physical degrees of freedom, $\rho - v$ and σ, for the Higgs field and one for the massless gauge field, A_κ. After spontaneous symmetry breaking the Goldstone boson σ conspires with the gauge field A_κ to form a massive vector field $\tilde{A}_\kappa$ with two degrees of freedom, while the real scalar field ρ decouples in the low-energy regime.

Let us finally turn to the response of the Higgs medium to the external point charges $q = ne$ (with $n \in \mathbb{Z}$) introduced by the matter current j^κ in (2.8). From (2.18), we infer that the gauge-invariant combined field $\tilde{A}_\kappa$ around this current drops off exponentially with mass M_A. Hence, the gauge field A_κ necessarily becomes pure gauge at distances much larger then $1/M_A$ from these point charges, and the electromagnetic fields generated by this current vanish accordingly. In other words, the electromagnetic fields generated by the external matter charges q are completely screened by the Higgs medium. From the field equations (2.15) and (2.16), it is clear how the Higgs screening mechanism works. The external matter current j^κ induces a screening current (2.17) in the Higgs medium proportional to the vector field $\tilde{A}_\kappa$. This becomes most transparent upon considering Gauss's law in this case,

$$Q = \int d^2x \nabla \cdot \mathbf{E} = q + q_{\text{scr}} = 0,$$

which shows that the external point charge q is surrounded by a cloud of screening charge density j^0_{scr} with support of characteristic size $1/M_A$. The contribution of the screening charge $q_{\text{scr}} = \int d^2x \, j^0_{\text{scr}} = -M_A^2 \int d^2x \, \tilde{A}^0 = -q$ to the long-range Coulomb fields completely cancels the contribution of the external charge q. Thus, we arrive at the well-known result that long-range Coulomb interactions between external matter charges vanish in the Higgs phase.

It has long been believed that with the vanishing of the Coulomb interactions, there are no long-range interactions left for the external charges in the Higgs phase. However, it was indicated by Krauss, Wilczek and Preskill [53, 54] that this is not the case. They noted that when the U(1) gauge group is not completely broken, but instead we are left with a finite cyclic manifest gauge group $\mathbb{Z}_N$ in the Higgs phase, the external matter charges may still have long-range Aharonov-Bohm interactions with the magnetic vortices also featuring in this model. These interactions are of a purely quantum-mechanical nature with no classical analogue. The physical mechanism behind the survival of Aharonov-Bohm interactions was subsequently uncovered in [44]: the induced screening charges q_{scr} accompanying the matter charges only couple to the Coulomb interactions and not to the Aharonov-Bohm interactions. As a result, the screening charges only screen the long-range Coulomb interactions among the external matter charges, but not the aforementioned long-range Aharonov-Bohm interactions between the matter charges and the magnetic vortices. We will discuss this phenomenon in further detail in the next section.

2.3.2 Survival of the Aharonov-Bohm Effect

A distinguishing feature of the Abelian Higgs model (2.7) is that it supports stable vortices carrying magnetic flux [5, 6]. These are static classical solutions of the field equations with finite energy and correspond to topological defects in the Higgs condensate, which are pointlike in our $(2 + 1)$-dimensional setting. Here, we will briefly review the basic properties of these magnetic vortices and subsequently elaborate on their long-range Aharonov-Bohm interactions with the screened external charges.

The energy density following from the action (2.7) for time-independent field configurations reads

$$\mathcal{E} = \tfrac{1}{2}(E^i E^i + B^2) + (NeA_0)^2|\Phi|^2 + \mathcal{D}_i\Phi(\mathcal{D}_i\Phi)^* + V(|\Phi|). \qquad (2.20)$$

All the terms occurring here are obviously positive definite. For field configurations of finite energy these terms should therefore vanish separately at spatial infinity. The potential (2.9) vanishes for ground states only. Thus, the Higgs field is necessarily condensed (2.11) at spatial infinity. Of course, the Higgs condensate can still make a nontrivial winding in the manifold of ground states. Such a winding at spatial infinity corresponds to a nontrivial holonomy in the Goldstone boson field

$$\sigma(\theta + 2\pi) - \sigma(\theta) = 2\pi a, \qquad (2.21)$$

where a is required to be an integer in order to leave the Higgs condensate (2.11) itself single valued, while θ denotes the polar angle. Requiring the fourth term in (2.20) to be integrable translates into the condition

$$\mathcal{D}_i\Phi(r \to \infty) \sim \tilde{A}_i(r \to \infty) = 0, \qquad (2.22)$$

with $\tilde{A}_i$ the gauge-invariant combination of the Goldstone boson and the gauge field defined in (2.13). Consequently, the nontrivial holonomy in the Goldstone boson field has to be compensated by an holonomy in the gauge fields and the vortices carry magnetic flux ϕ quantized as

$$\phi = \oint dl^i \, A^i = \frac{1}{Ne} \oint dl^i \, \partial_i\sigma = \frac{2\pi a}{Ne} \quad \text{with } a \in \mathbb{Z}. \qquad (2.23)$$

To proceed, the third term in the energy density (2.20) disappears at spatial infinity if and only if $A_0(r \to \infty) = 0$, and all in all we see that the gauge field A_κ is pure gauge at spatial infinity, so the first two terms vanish automatically. To end up with a regular field configuration corresponding to a nontrivial winding (2.21) of the Higgs condensate at spatial infinity, the Higgs field Φ should obviously become zero somewhere in the plane. Thus the Higgs phase is necessarily destroyed in some finite region in the plane. A closer evaluation of the energy density (2.20) shows that the Higgs field grows monotonically from its zero value to its asymptotic ground-state value (2.11) at the distance $1/M_H$, the so-called core size [5, 6]. Outside the core we are in the Higgs phase, and the physics is described by the effective Lagrangian (2.12), while inside the core the U(1) symmetry is restored. The magnetic field associated with the flux (2.23) of the vortex reaches its maximum inside the core where the gauge fields are massless. Outside the core the gauge fields become massive and the magnetic field drops off exponentially with the mass M_A. The core size $1/M_H$ and the penetration depth $1/M_A$ of the magnetic field are the two length scales characterizing the magnetic vortex. The formation of magnetic vortices depends on the ratio of these two scales. An evaluation of the free energy (see, for instance, Ref. 55) yields that magnetic vortices can be formed iff $M_H/M_A = \sqrt{\lambda}/Ne \geq 1$. We will always assume that this inequality is satisfied, so that magnetic vortices may indeed appear in the Higgs medium. In other words, we assume that we are dealing with a superconductor of type II.

To summarize, there are two dually charged types of sources in the Higgs medium. On the one hand, we have the vortices ϕ being sources for screened magnetic fields, and on the other hand the external charges q being sources for screened electric fields. The magnetic fields of the vortices are localized within regions of length scale $1/M_H$ dropping off with mass M_A at larger distances. The external charges are point particles with Coulomb fields completely screened at distances $> 1/M_A$. Henceforth, we will restrict our considerations to the low-energy regime (or alternatively send the Higgs mass M_H and the mass M_A of the gauge field to infinity by sending the symmetry-breaking scale to infinity). This means that the distances between the sources remain much larger then the Higgs length scale $1/M_H$. In other words, the electromagnetic fields associated with the magnetic- and electric sources never overlap and the Coulomb interactions between these sources vanish in the low-energy regime. Thus, from a classical point of view there are no long-range interactions left between the sources. From a quantum-mechanical perspective, however, it is known that in ordinary electromagnetism shielded localized magnetic fluxes can affect electric charges even though their mutual electromagnetic fields do not interfere. When an electric charge q encircles a localized magnetic flux ϕ, it notices the nontrivial holonomy in the locally flat gauge fields around the flux and in this process the wave function picks up a quantum phase $\exp(iq\phi)$ in the first quantized description. This is the celebrated Aharonov-Bohm effect [39], which is a purely quantum-mechanical effect with no classical analogue. These long-range Aharonov-Bohm interactions are of a topological nature, i.e., as long as the charge never enters the region where the flux is localized, the Aharonov-Bohm interactions only depend on the number of windings of the charge around the flux and not on the distance between the charge and the flux. Due to a remarkable cancellation in the effective action (2.12), the screening charges q_{scr} accompanying the external charges do *not* exhibit the Aharonov-Bohm effect. As a result the

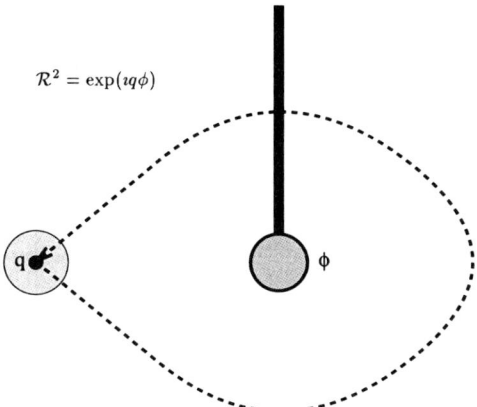

$$\mathcal{R}^2 = \exp(\imath q\phi)$$

FIGURE 5. Taking a screened external charge q around a magnetic vortex ϕ in the Higgs medium generates the Aharonov-Bohm phase $\exp(iq\phi)$. We have emphasized the extended structure of these sources, although this structure will not be probed in the low-energy regime to which we confine ourselves here. The shaded region around the external point charge q represents the cloud of screening charge of characteristic size $1/M_A$. The flux of the vortex is confined to the shaded circle bounded by the core at the distance $1/M_H$ from its center. The string attached to the core represents the Dirac string of the flux, i.e., the strip in which the nontrivial parallel transport in the gauge fields takes place.

long-range Aharonov-Bohm effect *persists* between the external charges q and the magnetic vortices ϕ in the Higgs phase. We will argue this in further detail.

Consider the system depicted in Figure 5 consisting of an external charge q and a magnetic vortex ϕ in the Higgs medium well separated from each other. We have depicted these sources as extended objects, but in the low-energy regime their extended structure will never be probed and it is legitimate to describe these sources as point particles moving in the plane. The magnetic vortex introduces a nontrivial holonomy (2.23) in the gauge fields to which the external charge couples through the matter coupling (2.8)

$$- \int d^2 x\, j^\kappa A_\kappa = \frac{q\phi}{2\pi}\dot{\chi}_\phi\big(\mathbf{y}(t) - \mathbf{z}(t)\big). \qquad (2.24)$$

Here, $\mathbf{y}(t)$ and $\mathbf{z}(t)$ respectively denote the worldlines of the external charge q and magnetic vortex ϕ in the plane. In the conventions we will use throughout these notes, the nontrivial parallel transport in the gauge fields around the magnetic vortices takes place in a thin strip (hereafter simply called the Dirac string) attached to the core of the vortex going off to spatial infinity in the direction of the positive vertical axis. This situation can always be reached by a smooth gauge transformation, and simplifies the bookkeeping for the braid processes involving more than two particles. The multivalued function $\chi_\phi(\mathbf{x})$ with support in the aforementioned strip of parallel transport is a direct translation of this convention. It increases from 0 to 2π if the strip is passed from right to left. Thus, when the external charge q moves through this strip once in the counterclockwise fashion indicated in Figure 5, the topological interaction Lagrangian (2.24) generates the action $q\phi$. In the same process the screening charge $q_{scr} = -q$ accompanying the external charge q also moves through this strip of parallel transport. Since the screening charge has a sign opposite to the sign of the external charge, it seems, at first sight, that the total topological action associated with encircling a flux by a screened external charge vanishes. This is not the case, though. The screening charge q_{scr} not only couples to the holonomy in the gauge field A_κ around the vortex but also to the holonomy in the Goldstone boson field σ. This follows directly from the effective low-energy Lagrangian (2.12). Let j^κ_{scr} be the screening current (2.17) associated with the screening charge q_{scr}. The interaction term in (2.12) couples this current to the massive gauge invariant field $\tilde{A}_\kappa$ around the vortex: $-j^\kappa_{scr}\tilde{A}_\kappa$. As we have seen in (2.22), the holonomies in the gauge field and the Goldstone boson field are related at large distances from the core of the vortex, such that $\tilde{A}_\kappa$ strictly vanishes. As a consequence, the interaction term $-j^\kappa_{scr}\tilde{A}_\kappa$ vanishes and indeed the matter coupling (2.24) summarizes all the remaining long-range interactions in the low-energy regime [44].

Being a total time derivative, the topological interaction term (2.24) does not appear in the equations of motion and has no effect at the classical level. In the first quantized description, however, the appearance of this term has far-reaching consequences. This is most easily seen using the path integral method for quantization. In the path integral formalism, the transition amplitude or propagator from one point in the configuration space at some time to another point at some later time, is given by a weighed sum over all the paths connecting the two points. In this sum, the paths are weighed by their action $\exp(iS)$. If we apply this prescription to our charge/flux system, we see that the Lagrangian (2.24) assigns amplitudes differing by $\exp(iq\phi)$ to paths differing by an encircling of the external charge q around the flux ϕ. Thus nontrivial interference takes place between paths associated with different winding numbers of the charge around the flux. This is the Aharonov-Bohm effect which becomes observable in quantum interference experiments [39], such as low-energy scattering experiments of external charges from the magnetic vortices. The cross sections measured in these Aharonov-Bohm scattering experiments can be found in Section 4.4.

There are two equivalent ways to present the appearance of the Aharonov-Bohm interactions. In the above discussion of the path integral formalism we kept the topological Aharonov-Bohm interactions in the Lagrangian for this otherwise free charge/flux system. In this description we work with single-valued wave functions on the configuration space for a given time slice

$$\Psi_{q\phi}(\mathbf{y}, \mathbf{z}, t) = \Psi_q(\mathbf{y}, t)\Psi_\phi(\mathbf{z}, t) \quad \text{with } \mathbf{y} \neq \mathbf{z}.$$

The factorization of the wave functions follows because there are no inter-actions between the external charge and the magnetic flux other than the topological one (2.24). The time evolution of these wave functions is given by the propagator associated with the two-particle Lagrangian

$$L = \frac{1}{2}m_q\dot{\mathbf{y}}^2 + \frac{1}{2}m_\phi\dot{\mathbf{z}}^2 + \frac{q\phi}{2\pi}\dot{\chi}_\phi\big(\mathbf{y}(t) - \mathbf{z}(t)\big).$$

Equivalently, we may absorb the topological interaction (2.24) in the bound-ary condition of the wave functions and work with multivalued wave func-tions

$$\widetilde{\Psi}_{q\phi}(\mathbf{y}, \mathbf{z}, t) := e^{iq\phi\chi_\phi(\mathbf{y}-\mathbf{z})/2\pi}\Psi_q(\mathbf{y}, t)\Psi_\phi(\mathbf{z}, t),$$

which propagate with a completely free two-particle Lagrangian [50] (see also Ref. 56)

$$\tilde{L} = \tfrac{1}{2}m_q\dot{\mathbf{y}}^2 + \tfrac{1}{2}m_\phi\dot{\mathbf{z}}^2.$$

We cling to the latter description from now on. That is, we will always absorb the topological interaction terms in the boundary condition of the wave functions. For later use and convenience, we set some more conven-tions. We will adopt a compact Dirac notation emphasizing the internal charge/flux quantum numbers of the particles. In this notation, the quan-tum state describing a charge or flux localized at some position $\mathbf{x}$ in the plane is presented as

$$|\text{charge/flux}\rangle := |\text{charge/flux}, \mathbf{x}\rangle = |\text{charge/flux}\rangle|\mathbf{x}\rangle.$$

To proceed, the charges $q = ne$ will be abbreviated by the number n of fundamental charge units e and the fluxes ϕ by the number a of fundamen-tal flux units $2\pi/Ne$. With the two-particle quantum state $|n\rangle|a\rangle$ we then indicate the multivalued wave function

$$|n\rangle|a\rangle := e^{ina\chi_a(\mathbf{x}-\mathbf{y})/N}|n, \mathbf{x}\rangle|a, \mathbf{y}\rangle,$$

where by convention the particle that is located most left in the plane (in this case the external charge $q = ne$), appears most left in the tensor product. The process of transporting the charge adiabatically around the flux in a counterclockwise fashion as depicted in Figure 5 is now summarized by the action of the monodromy operator on this two-particle state

$$\mathcal{R}^2|n\rangle|a\rangle = e^{2\pi ina/N}|n\rangle|a\rangle, \tag{2.25}$$

which boils down to a residual global $\mathbb{Z}_N$ transformation by the flux a of the vortex on the charge n.

Given the remaining long-range Aharonov-Bohm interactions (2.25) in the Higgs phase, the labeling of the charges and the fluxes by integers is, of course, highly redundant. Charges n differing by a multiple of N can not be distinguished. The same holds for the fluxes a. Hence, the charge and flux quantum numbers are defined modulo N in the residual manifest $\mathbb{Z}_N$ gauge theory describing the long-distance physics of the model (2.6). Besides these pure $\mathbb{Z}_N$ charges and fluxes the full spectrum naturally consists of charge/flux composites or dyons produced by fusing the charges and fluxes. We return to a detailed discussion of this spectrum and the topological interactions it exhibits in the next section.

Let us recapitulate our results from a more conceptual point of view (see also Refs. 54, 57, and 58 in this connection). In unbroken (compact) quantum electrodynamics, the quantized matter charges $q = ne$ (with $n \in \mathbb{Z}$), corresponding to the different unitary irreducible representations (UIRs) of the global symmetry group $U(1)$, carry long-range Coulomb fields. In other words, the Hilbert space of this theory decomposes into a direct sum of orthogonal charge superselection sectors that can be distinguished by measuring the associated Coulomb fields at spatial infinity. Local observables preserve this decomposition, since they cannot affect these long-range properties of the charges. The charge sectors can alternatively be distinguished by their response to global $U(1)$ transformations, since these are related to physical measurements of the Coulomb fields at spatial infinity through Gauss's law. Let us emphasize that the states in the Hilbert space are of course invariant under local gauge transformations, i.e., gauge transformations with finite support, which become trivial at spatial infinity.

Here, we touch upon the important distinction between global symmetry transformations and local gauge transformations. Although both leave the action of the model invariant, their physical meaning is rather different. A global symmetry (independent of the coordinates) is a true symmetry of the theory and in particular leads to a conserved Noether current. Local gauge transformations, on the other hand, correspond to a redundancy in the variables describing a given model and should therefore be modded out in the construction of the physical Hilbert space. In the $U(1)$ gauge theory under consideration the fields that transform nontrivially under the global $U(1)$ symmetry are the matter fields. The associated Noether current j^κ shows up in the Maxwell equations. More specifically, the conserved Noether charge $q = \int d^2x\, j^0$, being the generator of the global symmetry, is identified with the Coulomb charge $Q = \int d^2x\, \nabla \cdot \mathbf{E}$ through Gauss's law. This is the aforementioned relation between the global symmetry transformations and physical Coulomb charge measurements at spatial infinity.

Although the long-range Coulomb fields vanish when this $U(1)$ gauge theory is spontaneously broken down to a finite cyclic group $\mathbb{Z}_N$, we are still able to detect $\mathbb{Z}_N$ charge at arbitrary long distances through the Aharonov-Bohm effect. In other words, there remains a relation between residual global symmetry transformations and physical charge measurements at spatial infinity. The point is that we are left with a *gauged* $\mathbb{Z}_N$ symmetry in the Higgs phase, as witnessed by the appearance of stable magnetic fluxes in the spectrum. The magnetic fluxes introduce holonomies in the (locally flat) gauge fields, which take values in the residual manifest gauge group $\mathbb{Z}_N$ to leave the Higgs condensate single valued. To be specific, the holonomy of a given flux is classified by the group element picked up by the Wilson loop operator

$$W(\mathcal{C}, \mathbf{x}_0) = P \exp\left(ie \oint A^i \, dl^i \right) \in \mathbb{Z}_N, \tag{2.26}$$

where $\mathcal{C}$ denotes a loop enclosing the flux starting and ending at some fixed base point $\mathbf{x}_0$ at spatial infinity. The path ordering indicated by P is trivial in this Abelian case. These fluxes can be used for charge measurements in the Higgs phase by means of the Aharonov-Bohm effect (2.25). This purely quantum-mechanical effect, boiling down to a global $\mathbb{Z}_N$ gauge transformation on the charge by the group element (2.26), is topological. It persists at arbitrary long-ranges and therefore distinguishes the nontrivial $\mathbb{Z}_N$ charge sectors in the Higgs phase. Thus the result of the Higgs mechanism for the charge sectors can be summarized as follows: the charge superselection sectors of the original U(1) gauge theory, which were in one-to-one correspondence with the UIRs of the global symmetry group U(1), branch to UIRs of the residual (gauged) symmetry group $\mathbb{Z}_N$ in the Higgs phase.

An important conclusion from the foregoing discussion is that a spontaneously broken U(1) gauge theory in general can have distinct Higgs phases corresponding to different manifest cyclic gauge groups $\mathbb{Z}_N$. The simplest example is a U(1) gauge theory with two Higgs fields; one carrying a charge Ne and the other a charge e. There are in principle two possible Higgs phases in this particular theory, depending on whether the $\mathbb{Z}_N$ gauge symmetry remains manifest or not. In the first case only the Higgs field with charge Ne is condensed and we are left with nontrivial $\mathbb{Z}_N$ charge sectors. In the second case the Higgs field carrying the fundamental charge e is condensed. No charge sectors survive in this completely broken phase. These two Higgs phases, separated by a phase transition, can clearly be distinguished by probing the existence of $\mathbb{Z}_N$ charge sectors. This is exactly the content of the nonlocal order parameter constructed by Preskill and Krauss [54] (see also Refs. 58–62 in this context). In contrast with the Wilson loop operator and the 't Hooft loop operator distinguishing the Higgs and confining phase of a given gauge theory through the dynamics of electric and magnetic flux tubes [63, 64], this order parameter is of a topological nature. To be specific, in this $(2 + 1)$-dimensional setting it amounts to evaluating the expectation value of a closed electric flux tube linked with a closed magnetic flux loop corresponding to the worldlines of a minimal $\mathbb{Z}_N$ charge/anticharge pair linked with the worldlines of a minimal $\mathbb{Z}_N$ magnetic flux/antiflux pair. If the $\mathbb{Z}_N$ gauge symmetry is manifest, this order parameter gives rise to the Aharonov-Bohm phase (2.25), whereas it becomes trivial in the completely broken phase with minimal stable flux $2\pi/e$.

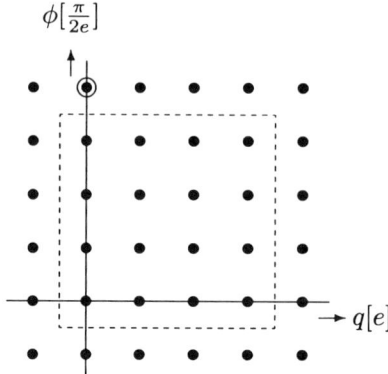

$\phi[\frac{\pi}{2e}]$

$q[e]$

FIGURE 6. The spectrum of a Higgs phase featuring a residual manifest gauge group $\mathbb{Z}_4$ compactifies to the particles inside the dashed box. The particles outside the box are identified with the ones inside by means of modulo 4 calculus along the charge and flux axes. The modulo 4 calculus for the fluxes corresponds to Dirac monopoles/instantons, if these are present. That is, the minimal monopole tunnels the encircled flux into the vacuum.

2.3.3   Braid and Fusion Properties of the Spectrum

We proceed with a more thorough discussion of the topological interactions described by the residual $\mathbb{Z}_N$ gauge theory featuring in the Higgs phase of the model (2.6). As we have argued in the previous section, the complete spectrum consists of pure $\mathbb{Z}_N$ charges labeled by n, pure $\mathbb{Z}_N$ fluxes labeled by a and dyons produced by fusing these charges and fluxes:

$$|a\rangle \times |n\rangle = |a, n\rangle \quad \text{with } a, n \in 0, 1, \ldots, N-1. \tag{2.27}$$

We have depicted this spectrum for a $\mathbb{Z}_4$ gauge theory in Figure 6.

The topological interactions described by a $\mathbb{Z}_N$ gauge theory are completely governed by the Aharonov-Bohm effect (2.25) and can simply be summarized as follows

$$\mathcal{R}^2 |a, n\rangle |a', n'\rangle = e^{2\pi i(na' + n'a)/N} |a, n\rangle |a', n'\rangle, \tag{2.28}$$

$$\mathcal{R} |a, n\rangle |a, n\rangle = e^{2\pi i na/N} |a, n\rangle |a, n\rangle, \tag{2.29}$$

$$|a, n\rangle \times |a', n'\rangle = |[a + a'], [n + n']\rangle, \tag{2.30}$$

$$\mathcal{C} |a, n\rangle = |[-a], [-n]\rangle, \tag{2.31}$$

$$\mathcal{T} |a, n\rangle = e^{2\pi i na/N} |a, n\rangle. \tag{2.32}$$

The expressions (2.28) and (2.29) sum up the braid properties of the particles in the spectrum (2.27). These realize Abelian representations of the braid groups discussed in Section 2.2. Of course, for distinguishable particles only the monodromies, as contained in the pure or colored braid groups are relevant. (See the discussion concerning relation (2.5) for the definition of colored braid groups). In the present context, particles carrying different charge and magnetic flux are distinguishable. When a given particle $|a, n\rangle$ located at some position in the plane is adiabatically transported around another remote particle $|a', n'\rangle$ in the counterclockwise fashion depicted in Figure 4, the total multivalued wave function of the system picks up the Aharonov-Bohm phase displayed in (2.28). In this process, the charge n of the first particle moves through the Dirac string attached to the flux a' of the second particle, while the charge n' of the second particle moves through the Dirac string of the flux a of the first particle. In short, the total Aharonov-Bohm effect for this monodromy is the composition of a global $\mathbb{Z}_N$ symmetry transformation on the charge n by the flux a' and a global transformation on the charge n' by the flux a. We confined ourselves to the case of two particles so far. The generalization to systems containing more then two particles is straightforward. The quantum states describing these systems are tensor products of localized single-particle states $|a, n, \mathbf{x}\rangle$, where we cling to the convention that the particle that appears most left in the plane appears most left in the tensor product. These multivalued wave functions carry Abelian representations of the colored braid group: the action of the monodromy operators (2.5) on these wave functions boils down to the quantum phase in expression (2.28).

For identical particles, i.e., particles carrying the same charge n and flux a, the braid operation depicted in Figure 1 becomes meaningful. In this braid process, in which two adjacent identical particles $|a, n\rangle$ located at different positions in the plane are exchanged in a counterclockwise way, the charge of the particle that moves "behind" the other dyon encounters the Dirac string attached to the flux of the latter. The result of this exchange in the multivalued wave function is the quantum-statistical phase factor (see expression (2.4) of Section 2.2) presented in (2.29). In other words, the dyons in the spectrum of this $\mathbb{Z}_N$ theory are anyons. In fact, these charge/flux composites are very close to Wilczek's original proposal for anyons [25].

An important aspect of this theory is that the particles in the spectrum (2.27) satisfy the canonical spin-statistics connection. The proof of this connection is of a topological nature and applies in general to all the models that will be considered in these notes. The fusion rules play a role in this proof and we will discuss these first.

Fusion and braiding are intimately related. Bringing two particles together is essentially a local process. As such, it can never affect global properties of the system. Hence, the single-particle state that arises after fusion should exhibit the same global properties as the two-particle state we started with. In this topological theory, the global properties of a given configuration are determined by its braid properties with the different particles in the spectrum (2.27). In the previous section, we had already established that the charges and fluxes become $\mathbb{Z}_N$ quantum numbers under these braid properties. Therefore, the complete set of fusion rules, determining the way the charges and fluxes of a two-particle state compose into the charge and flux of a single-particle state when the pair is brought together, can be summarized as (2.30). The rectangular brackets denote modulo N calculus such that the sum always lies in the range $0, 1, \ldots, N - 1$.

It is worthwhile to digress a little on the dynamical mechanism underlying the modulo N calculus compactifying the flux part of the spectrum. This modulo calculus is induced by magnetic monopoles, when these are present. The presence of magnetic monopoles can be accounted for by assuming that the compact U(1) gauge theory (2.6) arises from a spontaneously broken SO(3) gauge theory. The monopoles we obtain in this particular model are the regular 't Hooft-Polyakov monopoles [7, 8]. Let us, alternatively, assume that we have singular Dirac monopoles [9] in this compact U(1) gauge theory. In three spatial dimensions, these are point particles carrying magnetic charges g quantized as $2\pi/e$. In the present $(2 + 1)$-dimensional Minkowski setting, they become instantons describing flux tunneling events $|\Delta\phi| = 2\pi/e$. As has been shown by Polyakov [65], the presence of these instantons in unbroken U(1) gauge theory has a striking dynamical effect. It leads to linear confinement of electric charge. In the broken version of these theories, in which we are interested, electric charge is screened and the presence of instantons in the Higgs phase merely implies that the magnetic flux (2.23) of the vortices is conserved modulo N

$$\text{instanton} : a \mapsto a - N. \tag{2.33}$$

In other words, a flux N moving in the plane (or N minimal fluxes for that matter) can disappear by ending on an instanton. The fact that the instantons tunnel between states that cannot be distinguished by the braidings in this theory is nothing but the $(2 + 1)$-dimensional space-time translation of the unobservability of the Dirac string in three spatial dimensions.

We turn to the connection between spin and statistics. There are in principle two approaches to prove this deep relation, both having their own merits. One approach, originally due to Wightman [66], involves the axioms of local relativistic quantum field theory, and leads to the observation that integral spin fields commute, while half integral spin fields anticommute. The topological approach that we will take here was first proposed by Finkelstein and Rubinstein [11]. It does not rely upon the heavy framework of local relativistic quantum field theory and among other things applies to the topological defects considered in this thesis. The original formulation of Finkelstein and Rubinstein was in the $(3 + 1)$-dimensional context, but it naturally extends to $(2 + 1)$-dimensional space-time as we will discuss now [67, 68]. See also Refs. 69 and 70 for an algebraic approach.

The crucial ingredient in the topological proof of the spin-statistics connection for a given model is the existence of an antiparticle for every particle in the spectrum, such that the pair can annihilate into the vacuum after fusion. Consider the process depicted at the l.h.s. of the equality

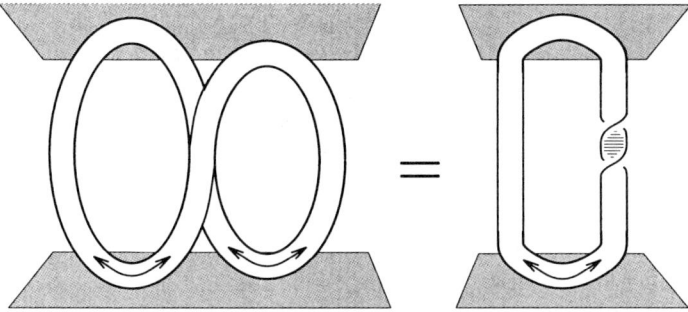

FIGURE 7. Canonical spin-statistics connection. The trajectories describing a counterclockwise interchange of two particles in separate particle/antiparticle pairs (the 8 laying on its back) can be continuously deformed into a single pair in which the particle undergoes a counterclockwise rotation over an angle of 2π around its own center (the 0 with a twisted leg).

sign in Figure 7. It describes the creation of two separate identical particle/antiparticle pairs from the vacuum, a subsequent counterclockwise exchange of the particles of the two pairs and finally annihilation of the pairs. To keep track of the writhing of the particle trajectories we depict them as ribbons with a white and a dark side. It is easily verified now that the closed ribbon associated with the process just explained can be continuously deformed into the ribbon at the r.h.s., which corresponds to a counterclockwise rotation of the particle over an angle of 2π around its own center. In other words, the effect of interchanging two identical particles in a consistent quantum description should be the same as the effect of rotating one particle over an angle of 2π around its center. The effect of this rotation in the wave function is the spin factor $\exp(2\pi i s)$ with s the spin of the particle, which in contrast with three spatial dimensions may be any real number in two spatial dimensions. Therefore, the result of exchanging the two identical particles necessarily boils down to a quantum-statistical phase factor $\exp(i\Theta)$ in the wave function being the same as the spin factor

$$\exp(i\Theta) = \exp(2\pi i s). \tag{2.34}$$

This is the canonical spin-statistics connection. Actually, a further consistency condition can be inferred from this ribbon argument. The writhing in the particle trajectory can be continuously deformed to a writhing with the same orientation in the antiparticle trajectory. Hence, the antiparticle necessarily carries the same spin and statistics as the particle.

   Sure enough the topological proof of the canonical spin-statistics connection applies to the $\mathbb{Z}_N$ gauge theory at hand. First of all, we can naturally assign an antiparticle to every particle in the spectrum (2.27) through the charge conjugation operator (2.31). Under charge conjugation the charge and flux of the particles in the spectrum reverse sign and amalgamating a particle with its charge conjugated partner yields the quantum numbers of the vacuum as follows from the fusion rules (2.30). Thus the basic assertion for the above ribbon argument is satisfied. From the quantum-statistical phase factor (2.29) assigned to the particles and (2.34), we then conclude that the particles carry spin. Specifically, under rotation over 2π the single particle states should give rise to the spin factors displayed in (2.32). In fact, these spin factors can be interpreted as the Aharonov-Bohm phase generated when the charge of a given dyon rotates around its own flux. Of course, a small separation between the charge and the flux of the dyon is required for this interpretation. Also, note that the particles and their antiparticles indeed carry the same spin and statistics, as follows immediately from the invariance of the Aharonov-Bohm effect under charge conjugation.

Having established a complete classification of the topological interactions described by a $\mathbb{Z}_N$ gauge theory, we conclude with some remarks on the Aharonov-Bohm scattering experiments by which these interactions can be probed. (A concise discussion of these purely quantum mechanical experiments can be found in Section 4.4). It is the monodromy effect (2.28) that is measured in these two-particle elastic scattering experiments. To be explicit, the symmetric cross section for scattering a particle $|a, n\rangle$ from a particle $|a', n'\rangle$ is given by

$$\frac{d\sigma}{d\theta} = \frac{\sin^2(\pi(na' + n'a)/N)}{2\pi p \sin^2(\theta/2)},$$

with p the relative momentum of the two particles and θ the scattering angle. A subtlety arises in scattering experiments involving two identical particles, however. Quantum statistics enters the scene: exchange processes between the scatterer and the projectile have to be taken into account [32, 71]. This leads to the following cross section for Aharonov-Bohm scattering of two identical particles $|a, n\rangle$

$$\frac{d\sigma}{d\theta} = \frac{\sin^2(2\pi na/N)}{2\pi p \sin^2(\theta/2)} + \frac{\sin^2(2\pi na/N)}{2\pi p \cos^2(\theta/2)},$$

where the second term summarizes the effect of the extra exchange contribution to the direct scattering amplitude.

2.4   Non-Abelian Discrete Gauge Theories

The generalization of the foregoing analysis to spontaneously broken models in which we are left with a *non-Abelian* finite gauge group H involves some essentially new features. In this introductory section, we will establish the complete flux/charge spectrum of such a non-Abelian discrete H gauge theory and discuss the basic topological interactions among the different flux/charge composites. The outline is as follows. Section 2.4.1 contains a general discussion on the topological classification of stable magnetic vortices and the subtle role magnetic monopoles play in this classification. In Section 2.4.2, we subsequently review the properties of the non-Abelian magnetic vortices that occur when the residual symmetry group H is non-Abelian—the most important one being that these vortices exhibit a non-Abelian Aharonov-Bohm effect. To be specific, the fluxes of the vortices, which are labeled by the group elements of H, affect each other through conjugation when they move around each other [40]. Under the residual global symmetry group H the magnetic fluxes transform by conjugation as well, and the conclusion is that the vortices are organized in degenerate multiplets, corresponding to the different conjugacy classes of H. These classical properties will then be elevated into the first quantized description in which the magnetic vortices are treated as point particles moving in the plane. In Section 2.4.3, we finally turn to the matter charges that may occur in these Higgs phases and their Aharonov-Bohm interactions with the magnetic vortices. As has been pointed out in Refs. 54 and 72, these matter charges are labeled by the different UIRs Γ of the residual global symmetry group H and when such a charge encircles a non-Abelian vortex it picks up a global symmetry transformation by the matrix $\Gamma(h)$ associated with the flux h of the vortex in the representation Γ. To conclude, we elaborate on the subtleties [20] involved in the description of dyonic combinations of the non-Abelian magnetic fluxes and the matter charges Γ.

2.4.1   Classification of Stable Magnetic Vortices

Let us start by briefly specifying the spontaneously broken gauge theories in which we are left with a non-Abelian discrete gauge theory. In this case, we are dealing with a model governed by a Yang-Mills Higgs action of the form

$$S_{\text{YMH}} = \int d^3x \left(-\frac{1}{4} F^{a\kappa\nu} F^a_{\kappa\nu} + (\mathcal{D}^\kappa \Phi)^\dagger \cdot \mathcal{D}_\kappa \Phi - V(\Phi) \right). \qquad (2.35)$$

Here, the Higgs field Φ transforms according to some higher-dimensional representation of a continuous non-Abelian gauge group G, the superscript a naturally labels the generators of the Lie algebra of G and the potential $V(\Phi)$ gives rise to a degenerate set of ground states $\langle \Phi \rangle \neq 0$ which are only invariant under the action of a finite non-Abelian subgroup H of G. For simplicity, we make two assumptions. First of all, we assume that this Higgs potential is normalized such that $V(\Phi) \geq 0$ and equals zero for the ground states $\langle \Phi \rangle$. More importantly, we assume that all ground states can be reached from any given one by global G transformations. This last assumption implies that the ground-state manifold becomes isomorphic to the coset G/H. (Renormalizable examples of potentials doing the job for $G \simeq \text{SO}(3)$ and H some of its point groups can be found in Ref. 73). In the following, we will only be concerned with the low-energy regime of this theory, so that the massive gauge bosons can be ignored.

The topologically stable vortices that can be formed in the spontaneously broken gauge theory (2.35) correspond to noncontractible maps from the circle at spatial infinity (starting and ending at a fixed base point x_0) into the ground-state manifold G/H. Different vortices are related to noncontractible maps that cannot be continuously deformed into each other. In short, the different vortices are labeled by the elements of the fundamental group π_1 of G/H based at the particular ground state $\langle \Phi_0 \rangle$ the Higgs field takes at the base point x_0 in the plane. (Standard references on the use of homotopy groups in the classification of topological defects are Refs. 1–3 and 74. See also Ref. 75 for an early discussion on the occurrence of non-Abelian fundamental groups in models with a spontaneously broken *global* symmetry).

The content of the fundamental group $\pi_1(G/H)$ of the ground-state manifold for a specific spontaneously broken model (2.35) can be inferred from the exact sequence

$$0 \simeq \pi_1(H) \to \pi_1(G) \to \pi_1(G/H) \to \pi_0(H) \to \pi_0(G) \simeq 0, \qquad (2.36)$$

where the first isomorphism follows from the fact that H is discrete. For convenience, we restrict our considerations to continuous Lie groups G that are path connected, which accounts for the last isomorphism. If G is simply connected as well, i.e., $\pi_1(G) \simeq 0$, then the exact sequence (2.36) yields the isomorphism

$$\pi_1(G/H) \simeq H, \qquad (2.37)$$

where we used the result $\pi_0(H) \simeq H$, which holds for finite H. Thus, the different magnetic vortices in this case are in one-to-one correspondence with the group elements h of the residual symmetry group H. When G is *not* simply connected, however, this is not a complete classification. This can be seen by the following simple argument. Let $\overline{G}$ denote the universal covering group of G and $\overline{H}$ the corresponding lift of H into $\overline{G}$. We then have $G/H = \overline{G}/\overline{H}$ and in particular $\pi_1(G/H) \simeq \pi_1(\overline{G}/\overline{H})$. Since the universal covering group of G is by definition simply connected, that is, $\pi_1(\overline{G}) \simeq 0$, we obtain the following isomorphism from the exact sequence (2.36) for the lifted groups $\overline{G}$ and $\overline{H}$

$$\pi_1(G/H) \simeq \pi_1(\overline{G}/\overline{H}) \simeq \overline{H}. \qquad (2.38)$$

Hence, for a nonsimply connected broken gauge group G, the different stable magnetic vortices are labeled by the elements of $\overline{H}$ rather then H itself.

It should be emphasized that the extension (2.38) of the magnetic vortex spectrum is based on the tacit assumption that there are no Dirac monopoles featuring in this model. In any theory with a nonsimply connected gauge group G, however, we have the freedom to introduce singular Dirac monopoles "by hand" [1, 76]. The magnetic charges of these monopoles are characterized by the elements of the fundamental group $\pi_1(G)$, which is Abelian for continuous Lie groups G. The exact sequence (2.36) for the present spontaneously broken model now implies the identification

$$\pi_1(G) \simeq \mathrm{Ker}(\pi_1(G/H) \to \pi_0(H)) \simeq \mathrm{Ker}(\overline{H} \to H).$$

In other words, the magnetic charges of the Dirac monopoles are in one-to-one correspondence with the nontrivial elements of $\pi_1(G/H) \simeq \overline{H}$ associated with the trivial element in $\pi_0(H) \simeq H$. The physical interpretation of this formula is as follows. In the $(2 + 1)$-dimensional Minkowsky setting, in which we are interested, the Dirac monopoles become instantons describing tunneling events between magnetic vortices $\bar{h} \in \overline{H}$ differing by the elements of $\pi_1(G)$. Here, the decay or tunneling time will naturally depend exponentially on the actual mass of the monopoles. The important conclusion is that in the presence of these Dirac monopoles the magnetic fluxes $\bar{h} \in \overline{H}$ are conserved modulo the elements of $\pi_1(G)$ and the proper labeling of the stable magnetic vortices boils down to the elements of the residual symmetry group H itself

$$\overline{H}/\pi_1(G) \simeq H. \tag{2.39}$$

To proceed, the introduction of Dirac monopoles has a bearing on the matter content of the model as well. The only matter fields allowed in the theory with monopoles are those that transform according to an ordinary representation of G. Matter fields carrying a faithful representation of the universal covering group $\overline{G}$ are excluded. This means that the matter charges appearing in the broken phase correspond to ordinary representations of H, while faithful representations of the lift $\overline{H}$ do not occur. As a result, the fluxes $\bar{h} \in \overline{H}$ related by tunneling events induced by the Dirac monopoles cannot be distinguished through long-range Aharonov-Bohm experiments with the available matter charges, which is consistent with the fact that the stable magnetic fluxes are labeled by elements of H rather then $\overline{H}$ in this case.

The whole discussion can now be summarized as follows. First of all, if a simply connected gauge group G is spontaneously broken down to a finite subgroup H, we are left with a discrete H gauge theory in the low-energy regime. The magnetic fluxes are labeled by the elements of H, whereas the different electric charges correspond to the full set of UIRs of H. When we are dealing with a nonsimply connected gauge group G broken down to a finite subgroup H, there are two possibilities depending on whether we allow for Dirac monopoles/instantons in the theory or not. In case Dirac monopoles are ruled out, we obtain a discrete $\overline{H}$ gauge theory. The stable fluxes are labeled by the elements of $\overline{H}$ and the different charges by the UIRs of $\overline{H}$. If the model features singular Dirac monopoles, on the other hand, then the stable fluxes simply correspond to the elements of the group H itself, while the allowed matter charges constitute UIRs of H. In other words, we are left with a discrete H gauge theory under these circumstances.

Let us illustrate these general considerations by some explicit examples. First we return to the model discussed in Section 2.3, in which the non-simply connected gauge group $G \simeq U(1)$ is spontaneously broken down to the finite cyclic group $H \simeq \mathbb{Z}_N$. The topological classification (2.38) for this particular model gives

$$\pi_1(U(1)/\mathbb{Z}_N) \simeq \pi_1(\mathbb{R}/\mathbb{Z}_N \times \mathbb{Z}) \simeq \mathbb{Z}_N \times \mathbb{Z} \simeq \mathbb{Z}.$$

Thus, in the absence of Dirac monopoles, the different stable vortices are labeled by the integers in accordance with (2.23), where we found that the magnetic fluxes associated with these vortices are quantized as $\phi = 2\pi a/(Ne)$ with $a \in \mathbb{Z}$. In principle, we are dealing with a discrete $\mathbb{Z}$ gauge theory now and the complete magnetic flux spectrum could be distinguished by means of long-range Aharonov-Bohm experiments with electric charges q being fractions of the fundamental unit e, which correspond to the UIRs of $\mathbb{Z}$. Of course, this observation is rather academic in this context, since free charges carrying fractions of the fundamental charge unit e have never been observed. With matter charges q being multiples of e, the low-energy theory then boils down to a $\mathbb{Z}_N$ gauge theory, although the topologically stable magnetic vortices in the broken phase are labeled by the integers a. The Dirac monopoles/instantons that can be introduced in this theory correspond to the elements of $\pi_1(U(1)) \simeq \mathbb{Z}$. The presence of these monopoles, which carry magnetic charge $g = 2\pi m/e$ with $m \in \mathbb{Z}$, imply that the magnetic flux a of the vortices is conserved modulo N, as we have seen explicitly in (2.33). In other words, the proper labeling of the stable magnetic fluxes is by the elements of $\mathbb{Z}_N \times \mathbb{Z}/\mathbb{Z} \simeq \mathbb{Z}_N$, as indicated by (2.39). Moreover, electric charge is necessarily quantized in multiples of the fundamental charge unit e now, so that the tunneling events induced by the instantons are unobservable at long distances. The unavoidable conclusion then becomes that in the presence of Dirac monopoles, we are left with a $\mathbb{Z}_N$ gauge theory in the low-energy regime of this spontaneously broken model, in complete accordance with the general discussion of the foregoing paragraphs.

When a gauge theory at some intermediate stage of symmetry breaking exhibits regular 't Hooft-Polyakov monopoles, their effect on the stable magnetic vortex classification is automatically taken care of, as it should be because the monopoles cannot be left out in such a theory. Consider, for example, a model in which the nonsimply connected gauge group $G \simeq SO(3)$ is initially broken down to $H_1 \simeq U(1)$ and subsequently to $H_2 \simeq \mathbb{Z}_N$

$$SO(3) \rightarrow U(1) \rightarrow \mathbb{Z}_N. \tag{2.40}$$

The first stage of symmetry breaking is accompanied by the appearance of regular 't Hooft-Polyakov monopoles [7, 8] carrying magnetic charges characterized by the elements of the second homotopy group $\pi_2(SO(3)/U(1)) \simeq \mathbb{Z}$. A simple exact sequence argument shows

$$\pi_2(SO(3)/U(1)) \simeq \mathrm{Ker}\Big(\pi_1(U(1)) \to \pi_1(SO(3))\Big) \simeq \mathrm{Ker}(\mathbb{Z} \to \mathbb{Z}_2).$$

Hence, the magnetic charges of the regular monopoles correspond to the elements of $\pi_1(U(1))$ associated with the trivial element of $\pi_1(SO(3))$, that is, the even elements of $\pi_1(U(1))$. In short, the regular monopoles carry magnetic charge $g = 4\pi m/e$ with $m \in \mathbb{Z}$. To proceed, the residual topologically stable magnetic vortices emerging after the second symmetry breaking are labeled by the elements of $\overline{H}_2 \simeq \mathbb{Z}_{2N}$, which follows from (2.38)

$$\pi_1(SO(3)/\mathbb{Z}_N) \simeq \pi_1(SU(2)/\mathbb{Z}_{2N}) \simeq \mathbb{Z}_{2N}.$$

As in the previous example, the magnetic fluxes carried by these vortices are quantized as $\phi = 2\pi a/(Ne)$, while the presence of the regular 't Hooft-Polyakov monopoles now causes the fluxes a to be conserved modulo $2N$. The tunneling or decay time will depend on the mass of the regular monopoles, that is, the energy scale associated with the first symmetry breaking in the hierarchy (2.40). Here it is assumed that the original $SO(3)$ gauge theory does not feature Dirac monopoles ($g = 2\pi m/e$, with $m = 0, 1$) corresponding to the elements of $\pi_1(SO(3)) \simeq \mathbb{Z}_2$. This means that additional matter fields carrying faithful (half integral spin) representations of the universal covering group $SU(2)$ are allowed in this model, which leads to half integral charges $q = ne/2$ with $n \in \mathbb{Z}$ in the $U(1)$ phase. In the final Higgs phase, the half integral charges q and the quantized magnetic fluxes ϕ then span the complete spectrum of the associated discrete $\mathbb{Z}_{2N}$ gauge theory.

Let us now, instead, suppose that the original $SO(3)$ gauge theory contains Dirac monopoles. The complete monopole spectrum arising after the first symmetry breaking in (2.40) then consists of the magnetic charges $g = 2\pi m/e$ with $m \in \mathbb{Z}$, which implies that magnetic flux a is conserved modulo N in the final Higgs phase. This observation is in complete agreement with (2.39), which states that the proper magnetic flux labeling is by the elements of $\mathbb{Z}_{2N}/\mathbb{Z}_2 \simeq \mathbb{Z}_N$ under these circumstances. In addition, the incorporation of Dirac monopoles rules out matter fields which carry faithful representations of the universal covering group $SU(2)$. Hence, only integral electric charges are conceivable ($q = ne$ with $n \in \mathbb{Z}$) and all in all we end up with a discrete $\mathbb{Z}_N$ gauge theory in the Higgs phase. This last situation can alternatively be implemented by embedding this spontaneously broken $SO(3)$ gauge theory in an $SU(3)$ gauge theory. In other words, the symmetry breaking hierarchy is extended to

$$SU(3) \to SO(3) \to U(1) \to \mathbb{Z}_N. \tag{2.41}$$

The singular Dirac monopoles in the $SO(3)$ phase then turn into regular 't Hooft-Polyakov monopoles

$$\pi_2(SU(3)/SO(3)) \simeq \pi_1(SO(3)) \simeq \mathbb{Z}_2.$$

The unavoidable presence of these monopoles automatically implies that the magnetic flux a of the vortices in the final Higgs phase is conserved modulo N. To be specific, a magnetic flux $a = N$ can decay by ending on a regular monopole in this model, where the decay time will again depend on the mass of the monopole or equivalently on the energy scale associated with the first symmetry breaking in (2.41). The existence of such a dynamical decay process is implicitly taken care of in the classification (2.37), which indicates that the stable magnetic fluxes are indeed labeled by the elements of $\pi_1(SU(3)/\mathbb{Z}_N) \simeq \mathbb{Z}_N$.

To conclude, in the above examples we restricted ourselves to the case where we are left with an Abelian finite gauge group in the Higgs phase. Of course, the discussion extends to non-Abelian finite groups as well. The more general picture then becomes as follows. If the nonsimply connected gauge group $G \simeq SO(3)$ is spontaneously broken to some (possibly non-Abelian) finite subgroup $H \subset SO(3)$, then the topologically stable magnetic fluxes correspond to the elements of the lift $\overline{H} \subset SU(2) \simeq \overline{G}$. In the Higgs phase, we are then left with a discrete $\overline{H}$ gauge theory. If we have embedded $SO(3)$ in $SU(3)$ (or alternatively introduced the conceivable $\mathbb{Z}_2$ Dirac monopoles), on the other hand, then the topologically stable magnetic fluxes correspond to the elements of H itself and we end up with a discrete H gauge theory.

2.4.2 Flux Metamorphosis

Henceforth, we assume that the spontaneously broken gauge group G in our model (2.35) is simply connected, for convenience. Hence, the stable magnetic vortices are labeled by the elements of the non-Abelian residual symmetry group H, as indicated by the isomorphism (2.37).

We start with a discussion of the classical field configuration associated with a single static non-Abelian vortex in the plane. In principle, we are dealing with an extended object with a finite core size proportional to the inverse of the symmetry breaking scale M_H. In the low-energy regime, however, we can neglect this finite core size and we will idealize the vortex as a point singularity in the plane. For finite energy, the associated static classical field configuration then satisfies the equations $V(\Phi) = 0$, $F^{\kappa\nu} = 0$, $\mathcal{D}_i\Phi = 0$ and $A_0 = 0$ outside the core. These equations imply that the Higgs field takes ground state values $\langle\Phi\rangle$ and the Lie algebra valued vector potential A_κ is pure gauge so that all nontrivial curvature $F^{\kappa\nu}$ is localized inside the core. To be explicit, a path (and gauge) dependent solution w.r.t. an arbitrary but fixed ground state $\langle\Phi_0\rangle$ at an arbitrary but fixed base point

$\mathbf{x}_0$ can be presented as

$$\langle \Phi(\mathbf{x}) \rangle = W(\mathbf{x}, \mathbf{x}_0, \gamma)\langle \Phi_0 \rangle,$$

where the untraced path ordered Wilson line integral

$$W(\mathbf{x}, \mathbf{x}_0, \gamma) = P \exp\left(ie \int_{\mathbf{x}_0}^{\mathbf{x}} A^i \, dl^i \right),$$

is evaluated along an oriented path γ (avoiding the singularity where the vortex is located) from the base point to some other point $\mathbf{x}$ in the plane. Here, we merely used the fact that the relation $\mathcal{D}_i\langle \Phi \rangle = 0$ identifies the parallel transport in the Goldstone boson fields with that in the gauge fields, as we have argued in full detail for the Abelian case in Section 2.3.2. Now in order to keep the Higgs field single valued, the magnetic flux of the vortex, picked up by the Wilson line integral along a counterclockwise closed loop $\mathcal{C}$, which starts and ends at the base point and encloses the core, necessarily takes values in the subgroup H_0 of G that leaves the ground state $\langle \Phi_0 \rangle$ at the base point invariant, i.e.,

$$W(\mathcal{C}, \mathbf{x}_0) = P \exp\left(ie \oint A^i \, dl^i \right) = h \in H_0. \tag{2.42}$$

The untraced Wilson loop operator (2.42) completely classifies the long-range properties of the vortex solution. It is invariant under a continuous deformation of the loop $\mathcal{C}$ that keeps the base point fixed and avoids the core of the vortex. Moreover, it is invariant under continuous gauge transformations that leave the ground state $\langle \Phi_0 \rangle$ at the base point invariant. As in the Abelian case, we fix this residual gauge freedom by sending all nontrivial parallel transport into a narrow wedge or Dirac string from the core of the vortex to spatial infinity as depicted in Figure 8. It should be emphasized that our gauge-fixing procedure for these vortex solutions involves two physically irrelevant choices. First of all, we have chosen a fixed ground state $\langle \Phi_0 \rangle$ at the base point $\mathbf{x}_0$. This choice merely determines the embedding of the residual symmetry group in G to be the stability group H_0 of $\langle \Phi_0 \rangle$. A different choice for this ground state gives rise to a different embedding of the residual symmetry group, but will eventually lead to an unitarily equivalent quantum description of the discrete H gauge theory in the Higgs phase. For convenience, we subsequently fix the remaining gauge freedom by sending all nontrivial transport around the vortices to a small wedge. Of course, physical phenomena will not depend on this choice. In fact, an equivalent formulation of the low-energy theory, without fixing this residual gauge freedom for the vortices, can also be given; see, for example, Ref. 77.

In the gauge-fixed prescription described above, we are still able to perform global symmetry transformations $g \in H_0$ on the vortex solutions that

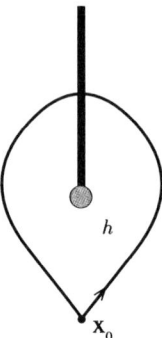

FIGURE 8. Single vortex solution. We have fixed the gauge freedom by sending all nontrivial parallel transport around the core in the Dirac string attached to the core. Thus, outside the core, the Higgs field takes the same ground-state value $\langle \Phi_0 \rangle$ everywhere except for the region where the Dirac string is localized. Here it makes a noncontractible winding in the ground state manifold. This winding corresponds to a holonomy in the gauge field classified by the result of the un-traced Wilson loop operator $W(\mathcal{C}, \mathbf{x}_0) = h \in H_0$, which picks up the non-Abelian magnetic flux located inside the core.

leave the ground state $\langle \Phi_0 \rangle$ invariant. These transformations affect the field configuration of the vortex in the following way

$$\Phi(\mathbf{x}) \mapsto g\Phi(\mathbf{x}),$$
$$A_\kappa(\mathbf{x}) \mapsto gA_\kappa(\mathbf{x})g^{-1}.$$

As an immediate consequence, we then obtain

$$W(\mathcal{C}, \mathbf{x}_0) \mapsto gW(\mathcal{C}, \mathbf{x}_0)g^{-1},$$

which shows that the flux of the vortex becomes conjugated $h \mapsto ghg^{-1}$ under a residual global symmetry transformation $g \in H_0$. The conclusion is that the non-Abelian vortex solutions are in fact organized in degenerate multiplets under the residual global symmetry transformations H_0, namely, the different conjugacy classes of H_0 denoted as $^A C$, where A labels a particular conjugacy class. For convenience, we will refer to the stability group of $\langle \Phi_0 \rangle$ as H from now on.

The different vortex solutions in a given conjugacy class $^A C$ of H, be-ing related by internal global symmetry transformations that leave the action (2.35) invariant, clearly carry the same external quantum numbers, that is, the total energy of the configuration, the core size, etc. These so-lutions only differ by their internal magnetic flux quantum number. This internal degeneracy becomes relevant in adiabatic interchange processes of remote vortices in the plane. Consider, for instance, the configuration of two remote vortices as presented in Figure 9. In the depicted adiabatic

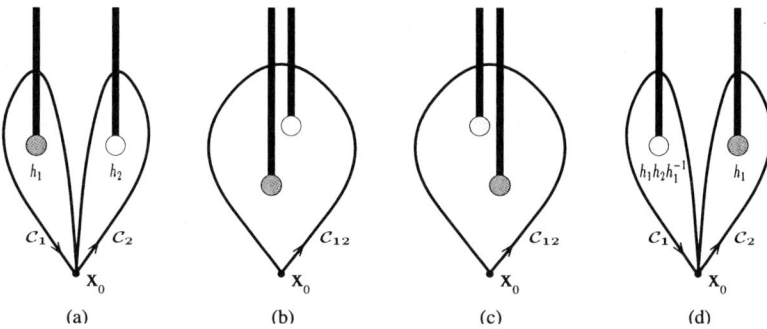

FIGURE 9. Flux metamorphosis. We start off with a classical configuration of two patched vortex solutions, as visualized in Figure (a). The vortices are initially assumed to carry the fluxes $W(\mathcal{C}_1, \mathbf{x}_0) = h_1$ and $W(\mathcal{C}_2, \mathbf{x}_0) = h_2$. The total flux of this configuration is picked up by the Wilson line integral along the loop $\mathcal{C}_{12}$ encircling both vortices as depicted in Figure (b): $W(\mathcal{C}_{12}, \mathbf{x}_0) = W(\mathcal{C}_1 \circ \mathcal{C}_2, \mathbf{x}_0) = W(\mathcal{C}_1, \mathbf{x}_0) \cdot W(\mathcal{C}_2, \mathbf{x}_0) = h_1 h_2$. Now suppose that the two vortices are interchanged in the counterclockwise fashion depicted in figures (b)–(d). In this process vortex 2 moves through the Dirac string attached to vortex 1 and as a result its flux will be affected $h_2 \mapsto h_2'$. Vortex 1, on the other hand, never meets any nontrivial parallel transport in the gauge fields and its flux remains the same. Since this local braid process should not be able to change the global properties of this system, i.e., the total flux, we have $h_1 h_2 = W(\mathcal{C}_{12}, \mathbf{x}_0) = W(\mathcal{C}_1, \mathbf{x}_0) \cdot W(\mathcal{C}_2, \mathbf{x}_0) = h_2' h_1$. Thus the flux of vortex 2 becomes conjugated $h_2' = h_1 h_2 h_1^{-1}$ by the flux of vortex 1 in this braid process.

counterclockwise interchange of these vortices, the vortex initially carrying the magnetic flux h_2 moves through the Dirac string attached to the other vortex. As a result, its flux picks up a global symmetry transformation by the flux h_1 of the latter, i.e., $h_2 \mapsto h_1 h_2 h_1^{-1}$, such that the total flux of the configuration is conserved. This classical non-Abelian Aharonov-Bohm effect appearing for noncommuting fluxes, which has been called flux metamorphosis [40], leads to physical observable phenomena. Suppose, for example, that the magnetic flux h_2 was a member of a flux/antiflux pair (h_2, h_2^{-1}) created from the vacuum. When h_2 encircles h_1, it returns as the flux $h_1 h_2 h_1^{-1}$ and will not be able to annihilate the flux h_2^{-1} anymore. Upon rejoining the pair we now obtain the stable flux $h_1 h_2 h_1^{-1} h_2$. Moreover, at the quantum level, flux metamorphosis leads to nontrivial Aharonov-Bohm scattering between non-Abelian vortices as we will argue in more detail later on.

Residual global symmetry transformations naturally leave the aforementioned observable Aharonov-Bohm effect for non-Abelian vortices invariant. This simply follows from the fact that these transformations commute with this non-Abelian Aharonov-Bohm effect. To be precise, a residual global symmetry transformation $g \in H$ on the two vortex configuration in Figure 9, for example, affects the flux of both vortices through conjugation

by the group element g, and it is easily verified that it makes no difference whether such a transformation is performed before the interchange is
started or after the interchange is completed. The extension of these classical considerations to configurations of more then two vortices in the plane
is straightforward. Braid processes, in which the fluxes of the vortices affect each other by conjugation, conserve the total flux of the configuration.
The residual global symmetry transformations $g \in H$ of the low-energy
regime, which act by an overall conjugation of the fluxes of the vortices in
the configuration by g, commute with these braid processes.

As in the Abelian case discussed in the previous sections, we wish to treat
these non-Abelian vortices as point particles in the first quantized description. The degeneracy of these vortices under the residual global symmetry
group H then indicates that we have to assign a finite-dimensional internal
Hilbert space V^A to these particles, which is spanned by the different fluxes
in a given conjugacy class $^A C$ of H and endowed with the standard inner
product [20]

$$\langle h' \mid h \rangle = \delta_{h',h}, \quad \forall h, h' \in {}^A C.$$

Under the residual global symmetry transformations the flux eigenstates
in this internal Hilbert space V^A are affected through conjugation

$$g \in H : \quad |h\rangle \mapsto |ghg^{-1}\rangle. \tag{2.43}$$

In general, the particle can be in a normalized linear combination of the different flux eigenstates in the internal Hilbert space V^A. The residual global
symmetry transformations (2.43) act linearly on such states. Of course, the
conjugated action of the residual symmetry group is in general reducible
and, at first sight, it seems that we have to decompose this internal Hilbert
space into the different irreducible components. This is not the case as
we will see in more detail later on (see the discussion concerning relation (2.50)). The point is that we can independently perform physical flux
measurements by means of quantum interference experiments with electric charges. These measurements project out a particular flux eigenstate.
Clearly, these flux measurements do not commute with the residual global
symmetry transformations and under their combined action the internal
Hilbert spaces V^A associated with the different conjugacy classes $^A C$ form
irreducible representations.

The complete quantum state of these particles consists of an internal
flux part and an external part. The quantum state describing a single par-
ticle in the flux eigenstate $|h_1\rangle \in V^{A_1}$ at a fixed position $\mathbf{y}$ in the plane,
for instance, is the formal tensor product $|h_1, \mathbf{y}\rangle = |h_1\rangle|\mathbf{y}\rangle$. To proceed,
the initial configuration depicted in Figure 9 is described by the multival-
ued two-particle quantum state $|h_1, \mathbf{y}\rangle|h_2, \mathbf{z}\rangle$, where again by convention
the particle located most left in the plane appears most left in the tensor
product. The result of an adiabatic counterclockwise interchange of the two

particles can now be summarized by the action of the braid operator

$$\mathcal{R}|h_1, \mathbf{y}\rangle|h_2, \mathbf{z}\rangle = |h_1 h_2 h_1^{-1}, \mathbf{y}\rangle|h_1, \mathbf{z}\rangle, \tag{2.44}$$

which acts linearly on linear combinations of these flux eigenstates. What we usually measure in quantum interference experiments, however, is the effect in the internal wave function of a monodromy of the two particles

$$\mathcal{R}^2|h_1, \mathbf{y}\rangle|h_2, \mathbf{z}\rangle = |(h_1 h_2)h_1(h_1 h_2)^{-1}, \mathbf{y}\rangle|h_1 h_2 h_1^{-1}, \mathbf{z}\rangle. \tag{2.45}$$

This non-Abelian Aharonov-Bohm effect can be probed either through a double slit experiment [71, 78] or through an Aharonov-Bohm scattering experiment as discussed in Section 4.4. In the first case, we keep one particle fixed between the two slits, whereas the other particle comes in as a plane wave. The geometry of the Aharonov-Bohm scattering experiment, depicted in Figure 13 is more or less similar. The interference pattern in both experiments is determined by the internal transition amplitude

$$\langle u_2|\langle u_1|\mathcal{R}^2|u_1\rangle|u_2\rangle, \tag{2.46}$$

where $|u_1\rangle$ and $|u_2\rangle$ respectively denote the properly normalized internal flux states of the two particles, which are generally linear combinations of the flux eigenstates in the corresponding internal Hilbert spaces V^{A_1} and V^{A_2}. The topological interference amplitudes (2.46) summarize all the physical observables for vortex configurations in the low-energy regime to which we confine ourselves here. As we have argued before, the residual global symmetry transformations affect internal multivortex states through an overall conjugation

$$g \in H: \quad |h_1\rangle|h_2\rangle \mapsto |gh_1 g^{-1}\rangle|gh_2 g^{-1}\rangle, \tag{2.47}$$

which commutes with the braid operator and therefore leave the interference amplitudes (2.46) invariant.

2.4.3 Including Matter

Let us now suppose that the total model is of the actual form

$$S = S_{\text{YMH}} + S_{\text{matter}}, \tag{2.48}$$

where S_{YMH} denotes the action for the non-Abelian Higgs model given in (2.35) and the action S_{matter} describes additional matter fields minimally coupled to the gauge fields. In principle, these matter fields correspond to multiplets which transform irreducibly under the spontaneously broken symmetry group G. Under the residual symmetry group H in the Higgs phase, however, these representations will become reducible and branch to UIRs Γ of H. Henceforth, it is assumed that the matter content of the model is such that all UIRs Γ of H are indeed realized. We will treat the different charges Γ, appearing in the Higgs phase in this way [54, 72], as point particles. In the first quantized description, these point charges then carry an internal Hilbert space, namely, the representation space associated with Γ. Let us now consider a configuration of a non-Abelian vortex in a flux eigenstate $|h\rangle$ at some fixed position in the plane and a remote charge Γ in a normalized internal charge state $|v\rangle$ fixed at another position. When the charge encircles the vortex in a counterclockwise fashion, it meets the Dirac string and picks up a global symmetry transformation by the flux of the vortex

$$\mathcal{R}^2|h,\mathbf{y}\rangle|v,\mathbf{z}\rangle = |h,\mathbf{y}\rangle|\Gamma(h)v,\mathbf{z}\rangle. \tag{2.49}$$

Here, $\Gamma(h)$ is the matrix assigned to the group element h in the representation Γ. Note, that this Aharonov-Bohm effect boils down to the Abelian one given in (2.25) in case the residual gauge group $H \simeq \mathbb{Z}_N$. Further, the residual global symmetry transformations on the two-particle configuration

$$g \in H: \quad |h,\mathbf{y}\rangle|v,\mathbf{z}\rangle \mapsto |ghg^{-1},\mathbf{y}\rangle|\Gamma(g)v,\mathbf{z}\rangle,$$

again commutes with the monodromy operation (2.49). Thus, the interference amplitudes

$$\langle v|\langle h|\mathcal{R}^2|h\rangle|v\rangle = \langle h \mid h\rangle\langle v \mid \Gamma(h)v\rangle = \langle v \mid \Gamma(h)v\rangle, \tag{2.50}$$

measured in either double-slit or Aharonov-Bohm scattering experiments involving these particles are invariant under the residual global symmetry transformations. As alluded to before, these interference experiments can be used to measure the flux of a given vortex [58, 71, 78, 79]. To that end, we place the vortex between the two slits (or alternatively use it as the scatterer in an Aharonov-Bohm scattering experiment) and evaluate the interference pattern for an incident beam of charges Γ in the same internal state $|v\rangle$. In this way, we determine the interference amplitude (2.50). Upon repeating this experiment a couple of times with different internal states for the incident charge Γ, we can determine all matrix elements of $\Gamma(h)$ and hence, iff Γ corresponds to a faithful UIR of H, the group element h itself. In a similar fashion, we may determine the charge Γ of a given particle and, moreover, its internal quantum state $|v\rangle$. In this case, we put the unknown charge between the double slit (or use it as the scatterer in an Aharonov-Bohm scattering experiment), measure the interference pattern for an incident beam of vortices in the same flux eigenstate $|h\rangle$ and again repeat this experiment for all $h \in H$.

At this point, we have established the purely magnetic flux and the purely electric charge superselection sectors of the discrete H gauge theory describing the long-distance physics of the model (2.48). The different magnetic sectors are labeled by the conjugacy classes $^A C$ of the residual gauge group H, whereas the different electric charge sectors correspond to the different UIRs Γ of H. The complete spectrum of this discrete gauge theory also contains dyonic combinations of these sectors. The relevant remark in this context is that we have not yet completely exhausted the action of the residual global symmetry transformations on the internal magnetic flux quantum numbers. As we have seen in (2.43), the residual global H transformations affect the magnetic fluxes through conjugation. The transformations that slip through this conjugation may in principle be implemented on an additional internal charge degree of freedom assigned to these fluxes [20]. More specifically, the global symmetry transformations that leave a given flux $|h\rangle$ invariant are those that commute with this flux, i.e., the group elements in the centralizer $^hN \subset H$. The internal charges that we can assign to this flux correspond to the different UIRs α of the group ^hN. Hence, the inequivalent dyons that can be formed in the composition of a global H charge Γ with a magnetic flux $|h\rangle$ correspond to the different irreducible components of the subgroup ^hN of H contained in the representation Γ. Two remarks are pertinent now. First of all, the centralizers of different fluxes in a given conjugacy class ^AC are isomorphic. Secondly, the full set of the residual global H symmetry transformations relate the fluxes in a given conjugacy class carrying unitary equivalent centralizer charge representations. In other words, the different dyonic sectors are labeled by $(^AC, \alpha)$, where ^AC runs over the different conjugacy classes of H and α over the different nontrivial UIRs of the associated centralizer. The explicit transformation properties of these dyons under the full global symmetry group H involve some conventions, which will be discussed in in the next chapter, where we will identify the Hopf algebra related to a discrete H gauge theory.

The physical observation behind the formal construction of the dyonic sectors [20] described above, is that we can, in fact, only measure the transformation properties of the charge of a given flux/charge composite under the centralizer of the flux of this composite; see also Ref. 71. A similar phenomenon occurs in the $(3 + 1)$-dimensional setting for monopoles carrying a non-Abelian magnetic charge, where it is known as the global color problem [17–19]. To illustrate this phenomenon, we suppose that we have a composite of a pure flux $|h\rangle$ and a pure global H charge Γ in some internal state $|v\rangle$. Thus, the complete internal state of the composite becomes $|h, v\rangle$. As we have argued before, the charge of a given object can be determined through double-slit or Aharonov-Bohm scattering experiments involving beams of vortices in the same internal flux state $|h'\rangle$ and repeating these experiments for all $h' \in H$. The interference amplitudes measured in this particular case are of the form

$$\langle h, v | \langle h' | \mathcal{R}^2 | h' \rangle | h, v \rangle = \langle h, v \mid h'hh'^{-1}, \Gamma(h')v \rangle \langle h' \mid (h'h)h'(h'h)^{-1} \rangle$$
$$= \langle v \mid \Gamma(h')v \rangle \delta_{h, h'hh'^{-1}},$$

where we used (2.45) and (2.49). As a result of the flux metamorphosis (2.45), the interference term is only nonzero for experiments involving fluxes h' that commute with the flux of the composite, i.e., $h' \in {}^h N$. Hence, we are only able to detect the response of the charge Γ of the composite to global symmetry transformations in ${}^h N$. This topological obstruction is usually summarized with the statement [21, 38, 54, 80] that in the background of a single vortex h, the only "realizable" global symmetry transformations are those taking values in the centralizer ${}^h N$.

Let us close this subsection with a summary of the main conclusions. First of all, the complete spectrum of the non-Abelian discrete H gauge theory describing the long-distance physics of the spontaneously broken model (2.48) can be presented as

$$({}^A C, \alpha), \tag{2.51}$$

where ${}^A C$ runs over the conjugacy classes of H and α denotes the different UIRs of the centralizer associated to a specific conjugacy class ${}^A C$. The purely magnetic sectors correspond to trivial centralizer representations and are labeled by the different nontrivial conjugacy classes. The pure charge sectors, on the other hand, correspond to the trivial conjugacy class (with centralizer the full group H) and are labeled by the different nontrivial UIRs of the residual symmetry group H. The other sectors describe the dyons in this theory. Note that the sectors (2.51) boil down to the sectors of the spectrum (2.27) in case $H \simeq \mathbb{Z}_N$.

The residual long-range interactions between the particles in the spectrum (2.51) of a discrete H gauge theory are topological Aharonov-Bohm interactions. In a counterclockwise braid process involving two given particles, the internal quantum state of the particle that moves through the Dirac string attached to the flux of the other particle picks up a global symmetry transformation by this flux. This (in general non-Abelian) Aharonov-Bohm effect conserves the total flux of the system and moreover commutes with the residual global H transformations, which act simultaneously on the internal quantum states of all the particles in the system. The last property ensures that the physical observables for a given system, which are all related to this Aharonov-Bohm effect, are invariant under global H transformations.

An exhaustive treatment of the spin, braid, and fusion properties of the particles in the spectrum (2.51) of a (non-Abelian) discrete H gauge theory involves the Hopf algebra $D(H)$ related to a discrete H gauge theory, which will be discussed in the next chapter. For notational simplicity, we will omit explicit mentioning of the external degrees of freedom of the particles in the following. In our considerations, we usually work with position eigenstates for the particles unless we are discussing double-slit or Aharonov-Bohm scattering experiments in which the incoming projectiles are in momentum eigenstates.

3 Algebraic Structure

It is by now well established that there are deep connections between two-dimensional rational conformal field theory, three-dimensional topological field theory, and quantum groups or Hopf algebras. See for instance Refs. 81–83 and references therein. Discrete H gauge theories, being examples of three dimensional topological field theories, naturally fit in this general scheme. As has been argued in Ref. 20 (see also Refs. 41 and 42) the algebraic structure related to a discrete H gauge theory is the quasitriangular Hopf algebra $D(H)$ being the result of applying Drinfel'd's quantum double construction [84, 85] to the Abelian algebra $\mathcal{F}(H)$ of functions on the finite group H.[2] Considered as a vector space, we then have $D(H) = \mathcal{F}(H) \otimes \mathbf{C}[H]$, where $\mathbf{C}[H]$ denotes the group algebra over the complex numbers $\mathbf{C}$. Loosely speaking, the elements spanning the Hopf algebra $D(H)$ signal the flux of the particles (2.51) in the spectrum of the related discrete H gauge theory and implement the residual global symmetry transformations. Under this action the particles form irreducible representations. Moreover, the algebra $D(H)$ provides a unified description of the spin, braid, and fusion properties of the particles. Henceforth, we will simply refer to the algebra $D(H)$ as the quantum double. This name, inspired by its mathematical construction, also summarizes nicely the physical content of a Higgs phase with a residual finite gauge group H. The topological interactions between the particles are of a quantum-mechanical nature, whereas the spectrum (2.51) exhibits an electric/magnetic self-dual (or double) structure.

In fact, the quantum double $D(H)$ was first proposed by Dijkgraaf et al. [87]. They identified it as the Hopf algebra associated with certain holomorphic orbifolds of rational conformal field theories [88] and the related three-dimensional topological field theories with finite gauge group H as introduced by Dijkgraaf and Witten [89]. The new insight that emerged in [20, 41, 42] was that such a topological field theory finds a natural realization as the residual discrete H gauge theory describing the long-range physics of gauge theories in which some continuous gauge group G is spontaneously broken down to a finite subgroup H.

Here, we review the notion of the quantum double $D(H)$ and elaborate on the unified description this framework gives of the spin, braid, and fusion properties of the topological and ordinary particles in the spectrum of a discrete H gauge theory.

---

[2]For a thorough treatment of Hopf algebras in general and related issues, the interested reader is referred to the excellent book by Shnider and Sternberg [86].

3.1   Quantum Double

As has been argued in Section 2.4.3, we are basically left with two physical operations on the particles (2.51) in the spectrum of a discrete H gauge theory. We can independently measure their magnetic flux and their electric charge through quantum interference experiments. The magnetic flux of a particle is given by a group element $h \in H$, while the charge forms an unitary irreducible representation of the centralizer $^h N$ of the flux $h \in H$ carried by the particle. Flux measurements then correspond to operators P_h projecting out a particular flux h, while the charge of a given particle can be detected through its transformation properties under the residual global symmetry transformations $g \in {}^h N \subset H$ that commute with the flux h of the particle.

The operators P_h projecting out the flux $h \in H$ of a given quantum state naturally realize the projector algebra

$$P_h P_{h'} = \delta_{h,h'} P_h, \tag{3.1}$$

with $\delta_{h,h'}$ the Kronecker delta function for the group elements h, $h' \in H$. As we have seen in relation (2.43), global symmetry transformations $g \in H$ affect the fluxes through conjugation. This implies that the flux projection operators and global symmetry transformations for a non-Abelian finite gauge group H do not commute

$$g P_h = P_{ghg^{-1}} g. \tag{3.2}$$

The combination of global symmetry transformations followed by flux measurements

$$\{P_h g\}_{h,g \in H},$$

generate the quantum double $D(H) = \mathcal{F}(H) \otimes \mathbf{C}[H]$ and the multiplication (3.1) and (3.2) of these elements can be recapitulated as[3]

$$P_h g \cdot P_{h'} g' = \delta_{h,gh'g^{-1}} P_h g g'.$$

The different particles (2.51) in the spectrum of the associated discrete H gauge theory constitute the complete set of inequivalent irreducible representations of the quantum double $D(H)$. To make explicit the irreducible action of the quantum double on these particles, we have to develop some further notation. To start with, we will label the group elements in the different conjugacy classes of H as

$$^A C = \{^A h_1, {}^A h_2, \ldots, {}^A h_k\}.$$

---

[3] In [20, 41, 42, 87] the elements of the quantum double were denoted by $^h \llcorner_g$. For notational simplicity, we use the presentation $P_h g$ in these notes.

Let $^AN \subset H$ be the centralizer of $^Ah_1$ and $\{^Ax_1, {}^Ax_2, \ldots, {}^Ax_k\}$ a set of representatives for the equivalence classes of $H/^AN$, such that $^Ah_i = {}^Ax_i{}^Ah_1{}^Ax_i^{-1}$. For convenience, we will always take $^Ax_1 = e$, with e the unit element in H. To proceed, the basis vectors of the unitary irreducible representation α of the centralizer ^AN will be denoted by $^\alpha v_j$. With these conventions, the internal Hilbert space V_α^A is spanned by the quantum states

$$\{|^Ah_i, {}^\alpha v_j\rangle\}_{i=1,\ldots,k}^{j=1,\ldots,\dim \alpha}. \tag{3.3}$$

The combined action of a global symmetry transformation $g \in H$ followed by a flux projection operation P_h on these internal flux/charge eigenstates spanning the Hilbert space V_α^A can then be presented as [87]

$$\Pi_\alpha^A(P_h g)|^Ah_i, {}^\alpha v_j\rangle = \delta_{h, g^Ah_ig^{-1}}|g^Ah_ig^{-1}, \alpha(\tilde{g})_{mj} {}^\alpha v_m\rangle, \tag{3.4}$$

with

$$\tilde{g} := {}^Ax_k^{-1} g^Ax_i, \tag{3.5}$$

and ^Ax_k defined through $^Ah_k := g^Ah_ig^{-1}$. It is easily verified that this element $\tilde{g}$ constructed from g and the flux ^Ah_i indeed commutes with $^Ah_1$ and therefore can be implemented on the centralizer charge. Two remarks are pertinent now. First of all, there is of course arbitrariness involved in the ordering of the elements in the conjugacy classes and the choice of the representatives ^Ax_k for the equivalence classes of the coset $H/^AN$. However, different choices lead to unitarily equivalent representations of the quantum double. Secondly, note that (3.4) is exactly the action anticipated in Section 2.4. The flux ^Ah_i of the associated particle is conjugated by the global symmetry transformation $g \in H$, while the part of g that slips through this conjugation is implemented on the centralizer charge of the particle. The operator P_h subsequently projects out the flux h.

We will now argue that the flux/charge eigenstates (3.3) spanning the internal Hilbert space V_α^A carry the same spin, i.e., a counterclockwise rotation over an angle of 2π gives rise to the same spin factor for all quantum states in V_α^A. As in our discussion of the (Abelian) $\mathbb{Z}_N$ gauge theory in Section 2.3.3, we assume a small seperation between the centralizer charge and the flux of the particles. In the aforementioned rotation, the centralizer charge of the particle then moves through the Dirac string attached to its flux and as a result picks up a transformation by this flux. The element in the quantum double that implements this effect on the internal quantum states (3.3) is the central element

$$\sum_h P_h h. \tag{3.6}$$

It signals the flux of the internal quantum state and implements this flux on the centralizer charge

$$\Pi_\alpha^A \left(\sum_h P_h h \right) |^A h_i, {}^\alpha v_j \rangle = |^A h_i, \alpha(^A h_1)_{mj} {}^\alpha v_m \rangle,$$

which boils down to the same matrix $\alpha(^A h_1)$ for all fluxes $^A h_i$ in $^A C$. Here, we used (3.4) and (3.5). Since $^A h_1$ by definition commutes with all the elements in the centralizer $^A N$, it follows from Schur's lemma that it is proportional to the unit matrix in the irreducible representation α

$$\alpha(^A h_1) = e^{2\pi i s(A,\alpha)} \mathbf{1}_\alpha. \tag{3.7}$$

This proves our claim. The conclusion is that there is an overall spin value $s_{(A,\alpha)}$ assigned to the sector $(^A C, \alpha)$. Note that the only sectors carrying a nontrivial spin are the dyonic sectors corresponding to nontrivial conjugacy classes paired with nontrivial centralizer charges.

The internal Hilbert space describing a system of two particles $(^A C, \alpha)$ and $(^B C, \beta)$ is the tensor product $V_\alpha^A \otimes V_\beta^B$. The extension of the action of the quantum double $D(H)$ on the single-particle states (3.4) to the two-particle states in $V_\alpha^A \otimes V_\beta^B$ is given by the comultiplication

$$\Delta(P_h g) = \sum_{h' \cdot h'' = h} P_{h'} g \otimes P_{h''} g, \tag{3.8}$$

which is an algebra morphism from $D(H)$ to $D(H) \otimes D(H)$. To be concrete, the tensor product representation of $D(H)$ carried by the two-particle internal Hilbert space $V_\alpha^A \otimes V_\beta^B$ is defined as $\Pi_\alpha^A \otimes \Pi_\beta^B(\Delta(P_h g))$. The action (3.8) of the quantum double on the internal two-particle quantum states in $V_\alpha^A \otimes V_\beta^B$ can be summarized as follows. In accordance with our observations in Sections 2.4.2 and 2.4.3, the residual global symmetry transformations $g \in H$ affect the internal quantum states of the two particles separately. The projection operator P_h subsequently projects out the total flux of the two-particle quantum state, i.e., the product of the two fluxes. Hence, the action (3.8) of the quantum double determines the global properties of a given two-particle quantum state, which are conserved under the local process of fusing the two particles. It should be mentioned now that the tensor product representation $(\Pi_\alpha^A \otimes \Pi_\beta^B, V_\alpha^A \otimes V_\beta^B)$ of $D(H)$ is in general reducible and can be decomposed into a direct sum of irreducible representations $(\Pi_\gamma^C, V_\gamma^C)$. The different single-particle states that can be obtained by the aforementioned fusion process are the states in the different internal Hilbert spaces V_γ^C that occur in this decomposition. We will return to an elaborate discussion of the fusion rules in Section 3.3.

An important property of the comultiplication (3.8) is that it is coassociative, i.e.,

$$(\mathrm{id} \otimes \Delta)\Delta(P_h g) = (\Delta \otimes \mathrm{id})\Delta(P_h g) = \sum_{h' \cdot h'' \cdot h''' = h} P_{h'} g \otimes P_{h''} g \otimes P_{h'''} g.$$

It means that the action of the quantum double $D(H)$ on the three-particle internal Hilbert space $V_\alpha^A \otimes V_\beta^B \otimes V_\gamma^C$ defined either through $(\text{id} \otimes \Delta)\Delta$ or through $(\Delta \otimes \text{id})\Delta$ is the same. Extending the action of the quantum double to systems containing an arbitrary number of particles is now straightforward: the global symmetry transformations $g \in H$ are implemented on all the particles separately, while the operator P_h projects out the total flux of the system.

The braid operation is formally implemented by the universal R-matrix, which is an element of $D(H) \otimes D(H)$

$$R = \sum_{h,g} P_g \otimes P_h g.$$

The R-matrix acts on a two-particle state as a global symmetry transformation on the second particle by the flux of the first particle. The physical braid operator $\mathcal{R}$ that effectuates a counterclockwise interchange of the two particles is defined as the action of this R matrix followed by a permutation σ of the two particles

$$\mathcal{R}_{\alpha\beta}^{AB} := \sigma \circ (\Pi_\alpha^A \otimes \Pi_\beta^B)(R).$$

To be explicit, on the two-particle charge flux eigenstate $|^A h_i, {}^\alpha v_j\rangle |^B h_m, {}^\beta v_n\rangle \in V_\alpha^A \otimes V_\beta^B$, we have

$$\mathcal{R}|^A h_i, {}^\alpha v_j\rangle |^B h_m, {}^\beta v_n\rangle = |^A h_i {}^B h_m {}^A h_i^{-1}, \beta({}^A \tilde{h}_i)_{ln} {}^\beta v_l\rangle |^A h_i, {}^\alpha v_j\rangle, \qquad (3.9)$$

where the element $^A \tilde{h}_i$ is defined as in (3.5). Note that the expression (3.9), which summarizes the braid operation on all conceivable two-particle states in this theory contains the braid effects established in Sections 2.4.2 and 2.4.3, namely, flux metamorphosis for two pure magnetic fluxes (2.44) and the Aharonov-Bohm effect for a pure magnetic flux with a pure charge (2.49).

It is now easily verified that the braid operator defined in (3.9) and the comultiplication given by (3.8) satisfy the quasitriangularity conditions

$$\mathcal{R}\Delta(P_h g) = \Delta(P_h g)\mathcal{R}, \qquad (3.10)$$
$$(\text{id} \otimes \Delta)(\mathcal{R}) = \mathcal{R}_2 \mathcal{R}_1, \qquad (3.11)$$
$$(\Delta \otimes \text{id})(\mathcal{R}) = \mathcal{R}_1 \mathcal{R}_2. \qquad (3.12)$$

Here, the braid operators $\mathcal{R}_1$ and $\mathcal{R}_2$, respectively, act as $\mathcal{R} \otimes \mathbf{1}$ and $\mathbf{1} \otimes \mathcal{R}$ on three-particle states in the internal Hilbert space $V_\alpha^A \otimes V_\beta^B \otimes V_\gamma^C$. The relation (3.10) expresses the fact that the braid operator commutes with the global symmetry transformations $g \in H$ and conserves the total magnetic flux of the configuration as measured by P_h. In addition, the quasitriangularity conditions (3.11) and (3.12), which can be presented graphically as in Figure 10, imply consistency between braiding and fusing.

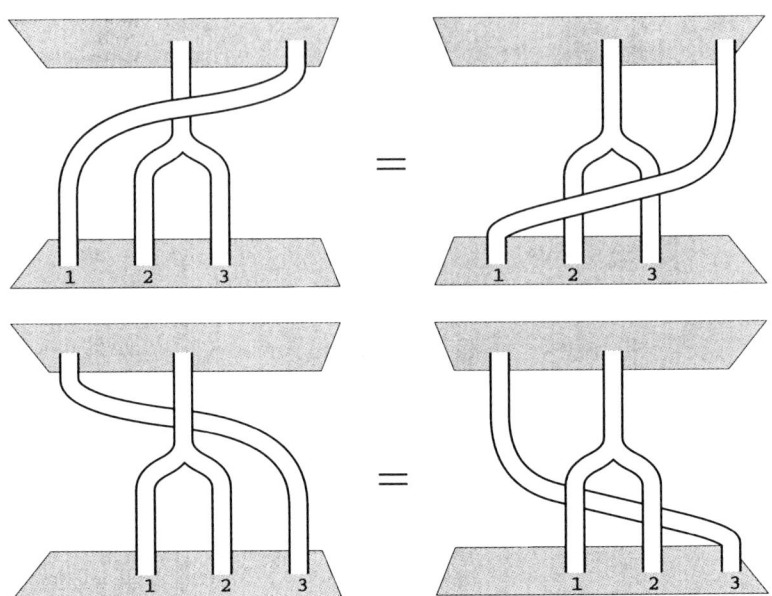

FIGURE 10. Compatibility of fusion and braiding as expressed by the quasitri-angularity conditions. It makes no difference whether a third particle braids with two particles separately or with the composite that arises after fusing these two particles. The ribbons represent the trajectories of the particles.

From the complete set of quasitriangularity conditions, it follows that the braid operator satisfies the Yang-Baxter equation

$$\mathcal{R}_1 \mathcal{R}_2 \mathcal{R}_1 = \mathcal{R}_2 \mathcal{R}_1 \mathcal{R}_2. \tag{3.13}$$

Thus the braid operators (3.9) define representations of the braid groups discussed in Section 2.2. These unitary representations are in general reducible. So the internal Hilbert space describing a multiparticle system in general splits up into a direct sum of irreducible subspaces under the action of the braid group. The braid properties of the system depend on the particular irreducible subspace. If the dimension of the irreducible representation is one, we are dealing with Abelian braid statistics or ordinary anyons. If the dimension is larger then one, we are dealing with non-Abelian braid statistics, i.e., the non-Abelian generalization of anyons. Note that the latter higher-dimensional irreducible representations only occur for systems consisting of more than two particles, because the braid group for two particles is Abelian.

To conclude, the internal Hilbert space describing a multiparticle system in a discrete H gauge theory carries a representation of the internal symmetry algebra $D(H)$ and a braid group representation. Both representations are in general reducible. The quasitriangularity condition (3.10) implies (see, for instance, Refs. 81 and 82) that the action of the associated braid operators commutes with the action of the elements of $D(H)$. Hence, the multidyon internal Hilbert space can in fact be decomposed into a direct sum of irreducible subspaces under the direct product action of $D(H)$ and the braid group. We discuss this in further detail in the next two sections. We first introduce the notion of truncated braid groups.

3.2 Truncated Braid Groups

We turn to a closer examination of the braid group representations that occur in discrete H gauge theories. An important observation in this respect is that the braid operator (3.9) is of finite order:

$$\mathcal{R}^m = 1 \otimes 1, \tag{3.14}$$

with 1 the identity operator and m some integer depending on the specific particles on which the braid operator acts. In other words, we can assign a finite number m to any two-particle internal Hilbert space $V_\alpha^A \otimes V_\beta^B$, such that the effect of m braidings is trivial for all states in this internal Hilbert space. This result, which can be traced back directly to the finite order of H, implies that the multiparticle configurations appearing in a discrete H gauge theory actually realize representations of factor groups of the braid groups discussed in Section 2.2. Consider, for instance, a system consisting of n indistinguishable particles. Hence, all particles carry the same internal Hilbert space V_α^A and the n particle internal Hilbert space describing this system is the tensor product space $(V_\alpha^A)^{\otimes n}$. The abstract generator τ_i, which establishes a counterclockwise interchange of the two adjacent particles i and $i+1$, acts on this internal Hilbert space by means of the operator

$$\tau_i \mapsto \mathcal{R}_i, \tag{3.15}$$

with

$$\mathcal{R}_i := 1^{\otimes(i-1)} \otimes \mathcal{R} \otimes 1^{\otimes(n-i-1)}. \tag{3.16}$$

That is, the generator τ_i acts as (3.9) on the ith and $(i+1)^{\text{th}}$ entry in the tensor product space $(V_\alpha^A)^{\otimes n}$. As follows from (3.13) and (3.14), the homomorphism (3.15) furnishes a representation of the braid group

$$\tau_i \tau_{i+1} \tau_i = \tau_{i+1} \tau_i \tau_{i+1}, \quad i = 1, \ldots, n-2$$
$$\tau_i \tau_j = \tau_j \tau_i, \qquad |i-j| \geq 2, \tag{3.17}$$

with the *extra* relation

$$\tau_i^m = e, \quad i = 1, \ldots, n-1, \tag{3.18}$$

where e denotes the unit element or trivial braid. For obvious reasons, we will call the factor groups with defining relations (3.17) and the additional relation (3.18) *truncated* braid groups $B(n, m)$. Here n naturally stands for the number of particles and m for the order of the generators τ_i.

The observation of the previous paragraph naturally extends to a system containing n distinguishable particles, i.e., the particles carry different internal Hilbert spaces or "colors" now. The group that governs the monodromy properties of such a system is the truncated version $P(n, m)$ of the colored braid group $P_n(\mathbb{R}^2)$ defined in Section 2.2. To be specific, the truncated colored braid group $P(n, m)$ is the subgroup of $B(n, m)$ generated by the elements

$$\gamma_{ij} = \tau_i \cdots \tau_{j-2} \tau_{j-1}^2 \tau_{j-2}^{-1} \cdots \tau_i^{-1}, \quad 1 \leq i < j \leq n,$$

with the extra relation (3.18) incorporated. Thus the generators of the pure braid group satisfy

$$\gamma_{ij}^{m/2} = e,$$

from which it is clear that the colored braid group $P(n, m)$ is, in fact, only defined for even m. The representation of the colored braid group $P(n, m)$ realized by a system of n different particles in a discrete H gauge theory then becomes

$$\gamma_{ij} \mapsto \mathcal{R}_i \cdots \mathcal{R}_{j-1} \mathcal{R}_j^2 \mathcal{R}_{j-1}^{-1} \cdots \mathcal{R}_i^{-1},$$

where the operators $\mathcal{R}_i$ defined by expression (3.16) now act on the tensor product space $V_{\alpha_1}^{A_1} \otimes \cdots \otimes V_{\alpha_n}^{A_n}$ of n different internal Hilbert spaces $V_{\alpha_l}^{A_l}$ with $l \in 1, 2, \ldots, n$.

Finally, a "mixture" of the above systems is of course also possible, that is, a system containing a subsystem consisting of n_1 particles with "color" $V_{\alpha_1}^{A_1}$, a subsystem of n_2 particles carrying the different "color" $V_{\alpha_2}^{A_2}$, and so on. Such a system realizes a representation of a truncated partially colored braid group (see Section 2.2 and the references given there for the definition of ordinary partially colored braid groups). Let $n = n_1 + n_2 + \cdots$ again be the total number of particles in the system. The truncated partially colored braid group associated with this system then becomes the subgroup of some truncated braid group $B(n, m)$, generated by the braid operations on particles with the same "color" and the monodromy operations on particles carrying different "color."

The appearance of truncated rather than ordinary braid groups facilitates the decomposition of a given multiparticle internal Hilbert space into irreducible subspaces under the braid/monodromy operations. The point is that the representation theory of ordinary braid groups is rather complicated due to their infinite order. The extra relation (3.18) for truncated braid groups $B(n, m)$, however, causes these to become finite for various values of the labels n and m, which leads to identifications with well-known groups of finite order [90]. It is instructive to consider some of these cases explicitly. The truncated braid group $B(2, m)$ for two indistinguishable particles, for instance, has only one generator τ, which satisfies $\tau^m = e$. Thus, we obtain the isomorphism

$$B(2, m) \simeq \mathbb{Z}_m.$$

For $m = 2$, the relations (3.17) and (3.18) are the defining relations of the permutation group S_n on n strands

$$B(n, 2) \simeq S_n.$$

A less trivial example is the non-Abelian truncated braid group $B(3, 3)$ for three indistinguishable particles. By explicit construction from the defining relations (3.17) and (3.18), we arrive at the identification

$$B(3, 3) \simeq \bar{T},$$

with $\bar{T}$ the lift of the tetrahedral group into SU(2). The structure of the truncated braid group $B(3, 4)$ and its subgroup $P(3, 4)$, which for example occur in a $\bar{D}_2$ gauge theory (see Section 4.3), can be found in Section 4.5.

To our knowledge, truncated braid groups have not been studied in the literature so far and a complete classification is not available. An interesting group-theoretical question in this context is whether the truncated braid groups are of finite order for all values of the labels n and m.

3.3 Fusion, Spin, Braid Statistics, and All That . . .

Let $(\Pi_\alpha^A, V_\alpha^A)$ and (Π_β^B, V_β^B) be two irreducible representations of the quantum double $D(H)$ as defined in (3.4). The tensor product representation $(\Pi_\alpha^A \otimes \Pi_\beta^B, V_\alpha^A \otimes V_\beta^B)$, constructed by means of the comultiplication (3.8), need not be irreducible. In general, it gives rise to a decomposition

$$\Pi_\alpha^A \otimes \Pi_\beta^B = \bigoplus_{C, \gamma} N_{\alpha\beta C}^{AB\gamma} \Pi_\gamma^C, \tag{3.19}$$

where $N_{\alpha\beta C}^{AB\gamma}$ stands for the multiplicity of the irreducible representation $(\Pi_\gamma^C, V_\gamma^C)$. From the orthogonality relation for the characters of the irreducible representations of $D(H)$, we infer [87]

$$N_{\alpha\beta C}^{AB\gamma} = \frac{1}{|H|} \sum_{h,g} \mathrm{tr}\left(\Pi_\alpha^A \otimes \Pi_\beta^B \left(\Delta(P_h g)\right)\right) \mathrm{tr}\left(\Pi_\gamma^C(P_h g)\right)^*, \tag{3.20}$$

where $|H|$ denotes the order of the group H and $*$ indicates complex conjugation. The fusion rule (3.19) now determines which particles $(^C C, \gamma)$ can be formed in the composition of two given particles $(^AC, \alpha)$ and $(^BC, \beta)$, or if read backwards, gives the decay channels of the particle $(^CC, \gamma)$.

The fusion algebra, spanned by the elements Π_α^A with multiplication rule (3.19), is commutative and associative and can therefore be diagonalized. The matrix implementing this diagonalization is the so-called modular S matrix [91]

$$
\begin{aligned}
S_{\alpha\beta}^{AB} &:= \frac{1}{|H|} \operatorname{tr} \mathcal{R}^{-2}{}_{\alpha\beta}^{AB} \\
&= \frac{1}{|H|} \sum_{\substack{^A h_i \in {}^A C \\ ^B h_j \in {}^B C \\ [^A h_i, {}^B h_j] = e}} \operatorname{tr}\big(\alpha(^A x_i^{-1}{}^B h_j{}^A x_i)\big)^* \operatorname{tr}\big(\beta(^B x_j^{-1}{}^A h_i{}^B x_j)\big)^*, \quad (3.21)
\end{aligned}
$$

The modular S matrix (3.21) contains all information concerning the fusion algebra defined in (3.19). In particular, the multiplicities (3.20) can be expressed in terms of the modular S matrix by means of Verlinde's formula [91]

$$
N_{\alpha\beta C}^{AB\gamma} = \sum_{D,\delta} \frac{S_{\alpha\delta}^{AD} S_{\beta\delta}^{BD} (S^*)_{\gamma\delta}^{CD}}{S_{0\delta}^{eD}}. \quad (3.22)
$$

Whereas the modular S matrix is determined through the monodromy operator following from (3.9), the modular matrix T contains the spin factors (3.7) assigned to the particles in the spectrum of a discrete H gauge theory

$$
T_{\alpha\beta}^{AB} := \delta_{\alpha,\beta}\delta^{A,B} \exp\big(2\pi i s_{(A,\alpha)}\big) = \delta_{\alpha,\beta}\delta^{A,B} \frac{1}{d_\alpha} \operatorname{tr}\big(\alpha(^A h_1)\big), \quad (3.23)
$$

where d_α stands for the dimension of the centralizer charge representation α of the particle $(^AC, \alpha)$. The matrices (3.21) and (3.23) now realize an unitary representation of the modular group $SL(2, \mathbb{Z})$ with the following relations [88]

$$
\mathcal{C} = (ST)^3 = S^2, \quad (3.24)
$$

$$
S^* = \mathcal{C}S = S^{-1}, \qquad S^t = S, \quad (3.25)
$$

$$
T^* = T^{-1}, \qquad\qquad T^t = T. \quad (3.26)
$$

The relations (3.25) and (3.26) express the fact that the matrices (3.21) and (3.23) are symmetric and unitary. To proceed, the matrix $\mathcal{C}$ defined in (3.24) represents the charge conjugation operator, which assigns an unique antipartner $\mathcal{C}(^AC, \alpha) = (^{\bar A}C, \bar\alpha)$ to each particle $(^AC, \alpha)$ in the spectrum, such that the vacuum channel occurs in the fusion rule (3.19) for the particle/antiparticle pairs. Also, note that the complete set of relations imply that the charge conjugation matrix C commutes with the modular matrix T, which implies that a given particle carries the same spin as its antipartner.

Having determined the fusion rules and the associated modular algebra, we turn to the issue of braid statistics and the fate of the spin statistics connection in non-Abelian discrete H gauge theories. Let us emphasize from the outset that much of what follows has been established elsewhere in a more general setting. See Refs. 81, 82, and 92 and the references therein for the conformal field theory point of view and Refs. 69, 70, and 83 for the related $(2+1)$-dimensional space-time perspective.

Let us first discuss a system consisting of two distinguishable particles $(^AC, \alpha)$ and $(^BC, \beta)$. The associated two-particle internal Hilbert space $V_\alpha^A \otimes V_\beta^B$ carries a representation of the Abelian truncated colored braid group $P(2, m)$ with $m/2 \in \mathbb{Z}$ the order of the monodromy matrix $\mathcal{R}^2$ for this particular two-particle system. This representation decomposes into a direct sum of one-dimensional irreducible subspaces, each being labeled by the associated eigenvalue of the monodromy matrix $\mathcal{R}^2$. Recall from Section 3.1, that the monodromy operation commutes with the action of the quantum double. This implies that the decomposition (3.19) simultaneously diagonalizes the monodromy matrix. To be specific, the two-particle total flux/charge eigenstates spanning a given fusion channel V_γ^C all carry the same monodromy eigenvalue, which in addition can be shown to satisfy the generalized spin-statistics connection [87]

$$K_{\alpha\beta\gamma}^{ABC} \mathcal{R}^2 = e^{2\pi i \left(s_{(C,\gamma)} - s_{(A,\alpha)} - s_{(B,\beta)} \right)} K_{\alpha\beta\gamma}^{ABC}. \qquad (3.27)$$

Here, $K_{\alpha\beta\gamma}^{ABC}$ stands for the projection on the irreducible component V_γ^C of $V_\alpha^A \otimes V_\beta^B$. In other words, the monodromy operation on a two-particle state in a given fusion channel is the same as a clockwise rotation over an angle of 2π of the two particles separately accompanied by a counterclockwise rotation over an angle of 2π of the single-particle state emerging after fusion. This is consistent with the observation that these two processes can be continuously deformed into each other, see the associated ribbon diagrams depicted in Figure 11. The discussion can now be summarized by the statement that the total internal Hilbert space $V_\alpha^A \otimes V_\beta^B$ decomposes into the following direct sum of irreducible representations of the direct product $D(H) \times P(2, m)$

$$\bigoplus_{C, \gamma} N_{\alpha\beta C}^{AB\gamma} (\Pi_\gamma^C, \Lambda_{C-A-B}),$$

where Λ_{C-A-B} denotes the one-dimensional irreducible representation of $P(2, m)$ in which the monodromy generator γ_{12} acts as (3.27).

The analysis for a configuration of two indistinguishable particles $(^AC, \alpha)$ is analogous. The total internal Hilbert space $V_\alpha^A \otimes V_\alpha^A$ decomposes into one-dimensional irreducible subspaces under the action of the truncated braid

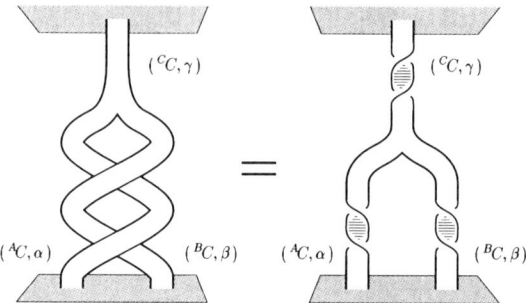

FIGURE 11. Generalized spin-statistics connection. The displayed ribbon diagrams are homotopic as can be checked with the pair of pants you are presently wearing. This means that a monodromy of two particles in a given fusion channel followed by fusion of the pair can be continuously deformed into the process describing a rotation over an angle of -2π of the two particles separately followed by fusion of the pair and a final rotation over an angle of 2π of the composite.

group $B(2, m)$ with m the order of the braid operator $\mathcal{R}$, which depends on the system under consideration. By the same argument as before, the two-particle total flux/charge eigenstates spanning a given fusion channel V_γ^C all carry the same one-dimensional representation of $B(2, m)$. The quantum-statistical parameter assigned to this channel now satisfies the square root version of the generalized spin-statistics connection (3.27)

$$K_{\alpha\alpha\gamma}^{AAC}\mathcal{R} = \epsilon e^{\pi i\left(s(C,\gamma)-2s(A,\alpha)\right)}K_{\alpha\alpha\gamma}^{AAC}, \qquad (3.28)$$

with ϵ a sign depending on whether the fusion channel V_γ^C appears in a symmetric or an antisymmetric fashion [81, 82]. In other words, the internal space Hilbert space for a system of two indistinguishable particles $(^AC, \alpha)$ breaks up into the following irreducible representations of the direct product $D(H) \times B(2, m)$

$$\bigoplus_{C,\gamma} N_{\alpha\alpha C}^{AA\gamma}(\Pi_\gamma^C, \Lambda_{C-2A}), \qquad (3.29)$$

with Λ_{C-2A} the one-dimensional representation of the truncated braid group $B(2, m)$ defined in (3.28).

The result (3.28) is actually rather surprising. It states that indistinguishable particle systems in a non-Abelian discrete H gauge theory quite generally violate the canonical spin-statistics connection (2.34). More accurately, in a non-Abelian discrete gauge theory we are dealing with the generalized connection (3.28), which incorporates the canonical one. In fact, the canonical spin-statistics connection is retrieved in some particular channels occurring in (3.29), as we will argue now. Let us first emphasize that the basic assertions for the ribbon proof depicted in Figure 7 are naturally satisfied in the non-Abelian setting as well. For every particle $(^AC, \alpha)$

in the spectrum there exists an antiparticle $(^{\bar{A}}C, \bar{\alpha})$ such that under the proper composition the pair acquires the quantum numbers of the vacuum and may decay. Moreover, every particle carries the same spin as its antipartner, as indicated by the fact that the charge conjugation operator $\mathcal{C}$ commutes with the modular matrix T. It should be noted now that the ribbon proof in Figure 7 actually *only* applies to states in which the particles that propagate along the exchanged ribbons are in strictly identical internal states. Otherwise the ribbons cannot be closed. Indeed, we find that the action (3.9) of the braid operator on two particles in identical internal flux/charge eigenstates

$$\mathcal{R}|^A h_i, {}^\alpha v_j\rangle|^A h_i, {}^\alpha v_j\rangle = |^A h_i, \alpha(^A h_1)_{mj}{}^\alpha v_j\rangle|^A h_i, {}^\alpha v_j\rangle, \qquad (3.30)$$

boils down to the diagonal matrix (3.7) and therefore to the same spin factor (3.31) for all i, j

$$\exp(i\Theta_{(A,\alpha)}) = \exp(2\pi i s_{(A,\alpha)}). \qquad (3.31)$$

The conclusion is that the canonical spin-statistics connection is restored in the fusion channels spanned by linear combinations of the states (3.30) in which the particles are in strictly identical internal flux/charge eigenstates. The quantum-statistical parameter (3.28) assigned to these channels reduces to the spin factor (3.31). Thus the effect of a counterclockwise interchange of the two particles in the states in these channels is the same as the effect of rotating one of the particles over an angle of 2π. To conclude, the closed ribbon proof does not apply to the other channels and we are left with the more involved connection (3.28) following from the open ribbon argument displayed in Figure 11.

Finally, higher-dimensional irreducible braid group representations are conceivable for a system that consists of more than two particles. The occurrence of such representations simply means that the generators of the braid group cannot be diagonalized simultaneously. What happens in this situation is that under the full set of braid operations, the system jumps between isotypical fusion channels, i.e., fusion channels of the same type or "color." Let us make this statement more precise. To keep the discussion general, we do not specify the nature of the particles in the system. Depending on whether the system consists of distinguishable particles, indistinguishable particles or some "mixture," we are dealing with a truncated braid group, a colored braid group, or a partially colored braid group, respectively. The internal Hilbert for such a system again decomposes into a direct sum of irreducible subspaces (or fusion channels) under the action of the quantum double $D(H)$. Given the fact that the action of the associated braid group commutes with that of the quantum double, we are left with two possibilities. First of all, there will in general be some fusion channels separately being invariant under the action of the full braid group. As in the two-particle systems discussed before, the total flux/charge eigenstates spanning such a fusion channel, say V_γ^C, carry the same one-dimensional irreducible representation Λ_{ab} of the braid group. That is, these states realize Abelian braid statistics with the same quantum-statistical parameter. The fusion channel V_γ^C then carries the irreducible representation $(\Pi_\gamma^C, \Lambda_{ab})$ of the direct product of the quantum double and the braid group. In addition, it is also feasible that states carrying the *same* total flux and charge in *different* (isotypical) fusion channels are mixed under the action of the full braid group. In that case, we are dealing with a higher-dimensional irreducible representation of the truncated braid group or non-Abelian braid statistics. Note that non-Abelian braid statistics is conceivable, if and only if some fusion channel, say V_δ^D, occurs more then once in the decomposition of the Hilbert space under the action of the quantum double. Only then there are some orthogonal states with the same total flux and charge available to span an higher-dimensional irreducible representation of the braid group. The number n of fusion channels V_δ^D related by the action of the braid operators now constitutes the dimension of the irreducible representation Λ_{nonab} of the braid group and the multiplicity of this representation is the dimension d of the fusion channel V_δ^D. To conclude, the direct sum of these n fusion channels V_δ^D carries an $(n \cdot d)$-dimensional irredicible representation $(\Pi_\delta^D, \Lambda_{nonab})$ of the direct product of the quantum double and the braid group.

4 $\overline{D}_2$ Gauge Theory

In this last section, we will illustrate the foregoing general considerations with one of the simplest non-Abelian discrete H gauge theories, namely, that with finite gauge group the double dihedral group $H \simeq \overline{D}_2$. See also Refs. 20, 41, and 42 in this connection. The plan is as follows. In Section 4.1, we establish the spectrum of a $\overline{D}_2$ gauge theory, the spin factors assigned to the particles and the fusion rules. Here, we also elaborate on a feature special for non-Abelian discrete H gauge theories: a pair of non-Abelian magnetic fluxes can carry charges that are not localized on any of the two fluxes nor anywhere else. Among other things, we will show that these so-called Cheshire charges a non-Abelian flux pair may carry, can be excited by monodromy processes with other particles in the spectrum. In Section 4.2, we treat the (non-Abelian) cross sections measured in Aharonov-Bohm scattering experiments involving the particles in a $\overline{D}_2$ gauge theory. Further, the issue of (non-Abelian) braid statistics realized by the multiparticle configurations in this theory will be dealt with in Section 4.3. We have also included two appendices. Section 4.4 contains a concise review of the Aharonov-Bohm scattering experiment focussing on the cross sections appearing in (non-)Abelian discrete H gauge theories. Finally, in Section 4.5, we give the group structure of two particular truncated braid groups which enter the analysis in Section 4.3.

4.1   Alice in Physics

A $\overline{D}_2$ gauge theory may, for instance, arise as "the long-distance remnant" of a Higgs model of the form (2.48) in which the gauge group $G \simeq \mathrm{SU}(2)$ is spontaneously broken down to the double dihedral group $H \simeq \overline{D}_2 \subset \mathrm{SU}(2)$. Since $\mathrm{SU}(2)$ is simply connected, the fundamental group $\pi_1(\mathrm{SU}(2)/\overline{D}_2)$ coincides with the residual symmetry group $\overline{D}_2$. Hence, the stable magnetic fluxes in this broken theory are indeed labeled by the group elements of $\overline{D}_2$. See the discussion concerning the isomorphism (2.37) in Section 2.4.1. In the following, we will not dwell any further on the explicit details of this or other possible embeddings in broken gauge theories and simply focus on the features of the $\overline{D}_2$ gauge theory itself. We start with a discussion of the spectrum.

The double dihedral group $\overline{D}_2$ is a group of order 8 with a nontrivial center of order 2. The magnetic fluxes associated with its group elements are organized in the conjugacy classes exhibited in Table 1. There are five conjugacy classes which we will denote as e, $\bar{e}$, X_1, X_2 and X_3. The conjugacy class e naturally corresponds to the trivial magnetic flux sector, while the conjugacy class $\bar{e}$ consists of the nontrivial center element. The conjugacy classes X_1, X_2, and X_3 all contain two commuting elements of order 4. In other words, a $\overline{D}_2$ gauge theory features four nontrivial purely magnetic flux sectors: one singlet flux $\bar{e}$ and three different doublet fluxes X_1, X_2, and X_3. The purely electric charge sectors, on the other hand, correspond to the UIRs of $\overline{D}_2$. From the character table displayed in Table 2, we infer that there are four nontrivial pure charges in the spectrum: three singlet charges J_1, J_2, J_3 and one doublet charge χ. The magnetic fluxes X_a and $\overline{X}_a$ (with $a \in 1, 2, 3$) act on the doublet charge χ as $i\sigma_a$ and $-i\sigma_a$, respectively, where the symbol σ_a denotes the Pauli matrices. Let us now turn to the dyonic sectors. These are constructed by assigning a nontrivial centralizer representation to the nontrivial fluxes. The centralizers associated with the different flux sectors can be found in Table 1. The flux $\bar{e}$ obviously commutes with the full group $\overline{D}_2$, while the centralizer of the other flux sectors is the cyclic group $\mathbb{Z}_4$. Hence, we arrive at thirteen

TABLE 1. Conjugacy classes of the double dihedral group $\overline{D}_2$ together with their centralizers.

Conjugacy class	Centralizer
$e = \{e\}$	$\overline{D}_2$
$\bar{e} = \{\bar{e}\}$	$\overline{D}_2$
$X_1 = \{X_1, \overline{X}_1\}$	$\mathbb{Z}_4 \simeq \{e, X_1, \bar{e}, \overline{X}_1\}$
$X_2 = \{X_2, \overline{X}_2\}$	$\mathbb{Z}_4 \simeq \{e, X_2, \bar{e}, \overline{X}_2\}$
$X_3 = \{X_3, \overline{X}_3\}$	$\mathbb{Z}_4 \simeq \{e, X_3, \bar{e}, \overline{X}_3\}$

TABLE 2. Character tables of $\overline{D}_2$ and $\mathbb{Z}_4$.

$\overline{D}_2$	e	$\bar{e}$	X_1	X_2	X_3
1	1	1	1	1	1
J_1	1	1	1	-1	-1
J_2	1	1	-1	1	-1
J_3	1	1	-1	-1	1
χ	2	-2	0	0	0

$\mathbb{Z}_4$	e	X_a	$\bar{e}$	$\overline{X}_a$
Γ^0	1	1	1	1
Γ^1	1	i	-1	$-i$
Γ^2	1	-1	1	-1
Γ^3	1	$-i$	-1	i

different dyons: three singlet dyons and one doublet dyon associated with the flux $\bar{e}$ and nine doublets dyons associated with the fluxes X_1, X_2, and X_3 paired with nontrivial $\mathbb{Z}_4$ representations. All in all, the spectrum of this theory features 22 particles, which will be labeled as

$$\begin{aligned}
1 &:= (e, 1), & \bar{1} &:= (\bar{e}, 1), \\
J_a &:= (e, J_a), & \bar{J}_a &:= (\bar{e}, J_a), \\
\chi &:= (e, \chi), & \bar{\chi} &:= (\bar{e}, \chi), \\
\sigma_a^+ &:= (X_a, \Gamma^0), & \sigma_a^- &:= (X_a, \Gamma^2), \\
\tau_a^+ &:= (X_a, \Gamma^1), & \tau_a^- &:= (X_a, \Gamma^3),
\end{aligned} \tag{4.1}$$

for convenience. Note that the square of the dimensions of the internal Hilbert spaces carried by these particles indeed add up to the order of the quantum double $D(\overline{D}_2)$: $8 \cdot 1^2 + 14 \cdot 2^2 = 8^2$.

As has been argued in Section 3.3, the topological interactions described by a discrete H gauge theory are encoded in the associated modular matrices S and T. The modular T matrix (3.23) contains the spin factors assigned to the different particles. With relation (3.7) and Table 2, we easily infer the following spin factors for the particles in the spectrum (4.1) of a $\overline{D}_2$ gauge theory

Particle	$\exp(2\pi i s)$
$1, J_a$	1
$\bar{1}, \bar{J}_a$	1
$\chi, \bar{\chi}$	$1, -1$
$\sigma_a^{\pm}$	± 1
$\tau_a^{\pm}$	$\pm i$.

$$\tag{4.2}$$

The modular S matrix (3.21), on the other hand, is determined by the monodromy matrix following from (3.9). A lengthy but straightforward calculation shows that the modular S matrix for a $\overline{D}_2$ gauge theory takes the form displayed in Table 3. We proceed by enumerating the fusion rules following from plugging this modular S matrix in Verlinde's formula (3.22).

The fusion rules for the purely electric charges are, of course, dictated

TABLE 3. Modular S-matrix of the quantum double $D(\bar{D}_2)$ up to an overall factor $\frac{1}{8}$. We defined $\epsilon_{ab} := 1$ iff $a = b$ and $\epsilon_{ab} := -1$ iff $a \neq b$.

S	1	$\bar{1}$	J_a	$\bar{J}_a$	χ	$\bar{\chi}$	σ_a^+	σ_a^-	τ_a^+	τ_a^-
1	1	1	1	1	2	2	2	2	2	2
$\bar{1}$	1	1	1	1	-2	-2	2	2	-2	-2
J_b	1	1	1	1	2	2	$2\epsilon_{ab}$	$2\epsilon_{ab}$	$2\epsilon_{ab}$	$2\epsilon_{ab}$
$\bar{J}_b$	1	1	1	1	-2	-2	$2\epsilon_{ab}$	$2\epsilon_{ab}$	$-2\epsilon_{ab}$	$-2\epsilon_{ab}$
χ	2	-2	2	-2	4	-4	0	0	0	0
$\bar{\chi}$	2	-2	2	-2	-4	4	0	0	0	0
σ_b^+	2	2	$2\epsilon_{ab}$	$2\epsilon_{ab}$	0	0	$4\delta_{ab}$	$-4\delta_{ab}$	0	0
σ_b^-	2	2	$2\epsilon_{ab}$	$2\epsilon_{ab}$	0	0	$-4\delta_{ab}$	$4\delta_{ab}$	0	0
τ_b^+	2	-2	$2\epsilon_{ab}$	$-2\epsilon_{ab}$	0	0	0	0	$-4\delta_{ab}$	$4\delta_{ab}$
τ_b^-	2	-2	$2\epsilon_{ab}$	$-2\epsilon_{ab}$	0	0	0	0	$4\delta_{ab}$	$-4\delta_{ab}$

by the representation ring of $\bar{D}_2$

$$J_a \times J_a = 1, \quad J_a \times J_b = J_c, \quad J_a \times \chi = \chi, \quad \chi \times \chi = 1 + \sum_a J_a. \quad (4.3)$$

Here, the subscripts a, b, and c take the values 1, 2, or 3 and by convention $a \neq b$, $a \neq c$, and $b \neq c$. The latter convention will be used throughout the following. To continue, the dyons associated with the flux $\bar{1}$ are obtained by simply composing this flux with the purely electric charges

$$J_a \times \bar{1} = \bar{J}_a, \quad \chi \times \bar{1} = \bar{\chi}. \quad (4.4)$$

In a similar fashion, we construct the other dyons

$$J_a \times \sigma_a^+ = \sigma_a^+, \quad J_b \times \sigma_a^+ = \sigma_a^-, \quad \chi \times \sigma_a^+ = \tau_a^+ + \tau_a^-. \quad (4.5)$$

We now have produced all the constituents of the spectrum as given in (4.1). Recall from Section 3.3 that the fusion algebra is commutative and associative. This implies that the full set of fusion rules is, in fact, completely determined by a minimal subset. Bearing this in mind, amalgamation involving the flux $\bar{1}$ is unambiguously prescribed by (4.4) and

$$\bar{1} \times \bar{1} = 1, \quad \bar{1} \times \sigma_a^\pm = \sigma_a^\pm, \quad \bar{1} \times \tau_a^\pm = \tau_a^\mp. \quad (4.6)$$

The complete set of fusion rules is then fixed by the previous ones together with

$$J_a \times \tau_a^\pm = \tau_a^\pm, \quad J_b \times \tau_a^\pm = \tau_a^\mp, \quad \chi \times \tau_a^\pm = \sigma_a^+ + \sigma_a^-,$$

and

$$\sigma_a^s \times \sigma_a^s = 1 + J_a + \bar{1} + \bar{J}_a \qquad (4.7)$$

$$\sigma_a^s \times \sigma_b^s = \sigma_c^+ + \sigma_c^- \qquad (4.8)$$

$$\sigma_a^s \times \tau_a^s = \chi + \bar{\chi} \qquad (4.9)$$

$$\sigma_a^s \times \tau_b^s = \tau_c^+ + \tau_c^- \qquad (4.10)$$

$$\tau_a^s \times \tau_a^s = 1 + J_a + \bar{J}_b + \bar{J}_c \qquad (4.11)$$

$$\tau_a^s \times \tau_b^s = \sigma_c^+ + \sigma_c^-, \qquad (4.12)$$

with $s \in +, -$.

A few remarks concerning this fusion algebra are pertinent. First of all, it is easily verified that the class algebra of $\bar{D}_2$ is respected as an overall selection rule. The class multiplication in the fusion rule (4.7), for instance, reads $X_a * X_a = 2e + 2\bar{e}$. The appearance of the class algebra naturally expresses magnetic flux conservation: in establishing the fusion rule, all fluxes in the consecutive conjugacy classes are multiplied out. Further, the modular S matrix as given in Table 3 is real and therefore equal to its inverse as follows directly from relation (3.25). Consequently, the charge conjugation operator $\mathcal{C}$ is trivial, i.e., it acts on the spectrum (4.1) as the unit matrix $\mathcal{C} = S^2 = 1$. Hence, all particles in this $\bar{D}_2$ gauge theory feature as their own antiparticle. Only two similar particles are able to annihilate, as witnessed by the occurrence of the vacuum channel 1 in the fusion rule for two similar particles.

At first sight, the message of the fusion rule (4.7) is actually rather remarkable. It seems that the fusion of two pure fluxes σ_a^+ may give rise to electric charge creation. One could start wondering about electric charge conservation at this point. Electric charge is conserved though. Before fusion this charge was present in the form of so-called nonlocalizable Cheshire charge [20, 42, 54, 80], i.e., the nontrivial representation of the global symmetry group $\bar{D}_2$ carried by the flux pair. This becomes clear upon writing the fusion rule (4.7) in terms of the two-particle flux states corresponding to the different channels:

$$\frac{1}{\sqrt{2}}\{|\bar{X}_a\rangle|X_a\rangle + ||X_a\rangle|\bar{X}_a\rangle\} \mapsto 1, \qquad (4.13)$$

$$\frac{1}{\sqrt{2}}\{|\bar{X}_a\rangle|X_a\rangle - |X_a\rangle|\bar{X}_a\rangle\} \mapsto J_a, \qquad (4.14)$$

$$\frac{1}{\sqrt{2}}\{|X_a\rangle|X_a\rangle + |\bar{X}_a\rangle|\bar{X}_a\rangle\} \mapsto \bar{1}, \qquad (4.15)$$

$$\frac{1}{\sqrt{2}}\{|X_a\rangle|X_a\rangle - |\bar{X}_a\rangle|\bar{X}_a\rangle\} \mapsto \bar{J}_a. \qquad (4.16)$$

The identification of the two-particle flux states at the l.h.s. with the single-particle states at the r.h.s. is established by the action (3.8) of the quantum double $D(\overline{D}_2)$ on the two-particle states. On the one hand, we can perform global $\overline{D}_2$ symmetry transformations from which we learn the total charge carried by the flux pair. As indicated by the comultiplication (3.8), the global $\overline{D}_2$ transformations act through overall conjugation. The total flux of the pair, on the other hand, is formally obtained by applying the flux projection operators (3.1) and is nothing but the product of the two fluxes of the pair. Since $X_a \cdot \overline{X}_a = e$, the total flux of the two-particle state in (4.13), for instance, vanishes. Moreover, it is easily verified that this state is invariant under global under global $\overline{D}_2$ transformations. Thus it corresponds to the vacuum channel 1. In a similar fashion, we obtain the identification of the other two-particle states with the single-particle states. Note that these two-particle quantum states describing the flux pairs are nonseparable. The two fluxes are correlated: by measuring the flux of one particle of the pair, we instantaneously fix the flux of the other. This is the famous Einstein-Podolsky-Rosen (EPR) paradox [93]. Just as in the notorious experiment with two spin-$\frac{1}{2}$ particles in the singlet state, it is no longer possible to make a flux measurement on one particle without affecting the other instantaneously. The Cheshire charge carried by the flux pair depends on the symmetry properties of these nonseparable quantum states. The symmetric quantum states correspond to the trivial charge 1, whereas the antisymmetric quantum states carry the nontrivial charge J_a. It is clear that the charge J_a cannot be localized on any of the fluxes nor anywhere else. It is a property of the pair and only becomes localized when the fluxes are brought together in a fusion process. It is this elusive nature, reminiscent of the smile of the Cheshire cat in *Alice's Adventures in Wonderland* [94], that formed the motivation to call such a charge a Cheshire charge [54, 78].

The Cheshire charge J_a of the flux pair can be excited by encircling one flux in the pair by the doublet charge χ [42, 54, 78]. Here, we draw on a further analogy with Alice's adventures. The magnetic fluxes X_a and $\overline{X}_a$ act by means of $i\sigma_a$ and $-i\sigma_a$, respectively, on the doublet charge χ, where the symbol σ_a denotes the Pauli matrices. This means that when a charge χ with its orientation down is adiabatically transported around, for example, the flux X_2, it returns with its orientation up:

$$\mathcal{R}^2 |X_2\rangle |\begin{pmatrix} 0 \\ 1 \end{pmatrix}\rangle = |X_2\rangle |\begin{pmatrix} 1 \\ 0 \end{pmatrix}\rangle, \tag{4.17}$$

as follows from (3.9). In terms of Alice's adventures: the charge has gone through the looking-glass. For this reason the flux X_2 has been called an Alice flux [38, 54, 80]. The other fluxes $X_a, \overline{X}_a$ affect the doublet charge χ in a similar way. Let us now consider the process depicted in Figure 12. We start with the creation of a charge/anticharge pair χ and a flux/antiflux pair σ_a^+ from the vacuum. Thus both pairs do not carry Cheshire charge at this stage. They are in the vacuum channel of the corresponding fusion rules (4.3) and (4.7). Next, one member of the charge pair encircles a flux in the flux pair. The flip of the charge orientation (4.17) leads to an exchange

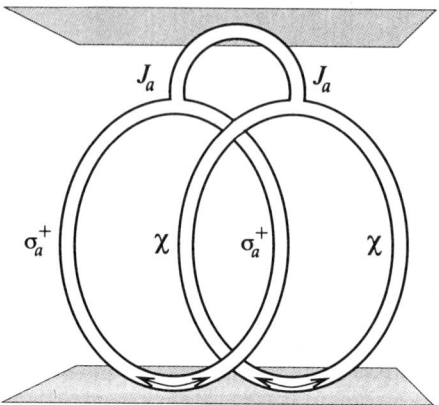

FIGURE 12. A charge/anticharge pair χ and a flux/antiflux pair σ_a^+ are created from the vacuum at a certain time slice. The ribbons represent the worldlines of the particles. After the charge χ has encircled the flux σ_a^+, both particle/antiparticle pairs carry the nonlocalizable Cheshire charge J_a. These Cheshire charges become localized upon rejoining the members of the pairs. Subsequently, the two charges J_a annihilate each other.

of the internal quantum numbers of the pairs: both pairs carry a Cheshire charge J_a after this process, i.e., both pairs are in the J_a channel of the associated fusion rules. The global charge of the configuration is conserved. Both charges J_a can be annihilated by bringing them together as follows from the fusion rule $J_a \times J_a = 1$ given in (4.3). These phenomena can be made explicit by writing this process in terms of the corresponding correlated internal quantum states:

$$
\begin{aligned}
1 &\longrightarrow \frac{1}{2}\{|\overline{X}_2\rangle|X_2\rangle + |X_2\rangle|\overline{X}_2\rangle\}\{|\left(\begin{smallmatrix}1\\0\end{smallmatrix}\right)\rangle|\left(\begin{smallmatrix}0\\1\end{smallmatrix}\right)\rangle - |\left(\begin{smallmatrix}0\\1\end{smallmatrix}\right)\rangle|\left(\begin{smallmatrix}1\\0\end{smallmatrix}\right)\rangle\} \\
&\xrightarrow{1\otimes\mathcal{R}^2\otimes 1} \frac{1}{2}\{|\overline{X}_2\rangle|X_2\rangle - |X_2\rangle|\overline{X}_2\rangle\}\{|\left(\begin{smallmatrix}0\\1\end{smallmatrix}\right)\rangle|\left(\begin{smallmatrix}0\\1\end{smallmatrix}\right)\rangle + |\left(\begin{smallmatrix}1\\0\end{smallmatrix}\right)\rangle|\left(\begin{smallmatrix}1\\0\end{smallmatrix}\right)\rangle\} \\
&\longrightarrow |J_2\rangle|J_2\rangle \\
&\longrightarrow 1. \tag{4.18}
\end{aligned}
$$

Here, we used (3.9) and the fact that the fluxes X_2 and $\overline{X}_2$ act by means of the matrices $i\sigma_2$ and $-i\sigma_2$ (with σ_2 the second Pauli matrix), respectively, on the charge χ. After the charge has encircled the flux, the flux pair is in the antisymmetric quantum state (4.14) with Cheshire charge J_2, while the same observation holds for the quantum state of the charge pair. Before fusion the charge pair was in the antisymmetric vacuum representation 1, while the state that emerges after the monodromy carries the Cheshire charge J_2. For convenience, we restricted ourselves to the flux pair σ_2^+ here. The argument for the other flux pairs is completely similar though.

The foregoing discussion naturally extends to the exchange of magnetic quantum numbers in monodromy processes involving noncommuting fluxes, that is, the occurence of flux metamorphosis (2.45). If we replace the doublet charge χ pair in Figure 12 by a flux pair σ_b^+ (with $a \neq b$) starting off in the vacuum channel (4.13), both flux pairs end up in the nontrivial flux channel (4.15) after the monodromy and both pairs now carry the total flux $\bar{1}$. Upon fusing the members of the pairs, these Cheshire fluxes become localized and subsequently annihilate each other according to their fusion rule given in (4.6).

Let us close this subsection by briefly summarizing the profound role the fusion rules play as overall selection rules in the flux/charge exchange processes among the particles. It is natural to confine our considerations to multiparticle systems which are overall in the vacuum sector denoted by 1, i.e., multiparticle systems for which the total flux and charge vanishes. Hence, in the $\bar{D}_2$ gauge theory under consideration the particles necessarily appear in pairs of the same species, as we have seen explicitly in the example of such a system in Figure 12. The fusion rules then classify the different total fluxes and Cheshire charges these pairs can carry and determine the flux/charge exchanges that may occur in monodromy processes involving particles in different pairs.

4.2   Scattering Doublet Charges Off Alice Fluxes

The Aharonov-Bohm interactions among the particles in the spectrum of a non-Abelian discrete H gauge theory roughly fall into two classes. First of all, there are the interactions in which no internal flux/charge quantum numbers are exchanged between the particles. In other words, we are dealing with two particles for which the monodromy matrix following by taking the square of the braid matrix (3.9) is diagonal in the two-particle flux/charge eigenbasis with possibly different Aharonov-Bohm phases as diagonal elements. The cross sections measured in Aharonov-Bohm scattering experiments with two such particles simply follow from the well-known cross section derived by Aharonov and Bohm [39]. See relation (4.49) of Section 4.4. The more interesting Aharonov-Bohm interactions are those in which internal flux/charge quantum numbers are exchanged between two particles when these encircle each other. In that case, we are dealing with two particles for which the monodromy matrix is off diagonal in the two-particle flux/charge eigenbasis. The cross sections [20, 71, 95] measured in Aharonov-Bohm scattering experiments involving two such particles are briefly reviewed in Section 4.4. In this section, we will focus on a nontrivial example in the $\bar{D}_2$ gauge theory at hand, namely, an Aharonov-Bohm scattering experiment in which a doublet charge χ scatters from an Alice flux σ_2^+.

The total internal Hilbert space associated with the two-particle system consisting of a pure doublet charge χ together with a pure doublet flux σ_2^+

is four-dimensional. We define the following natural flux/charge eigenbasis in this internal Hilbert space

$$
\begin{aligned}
e_1 &= |X_2\rangle|\left(\begin{smallmatrix}1\\0\end{smallmatrix}\right)\rangle := |\uparrow\rangle|\uparrow\rangle, \\
e_2 &= |X_2\rangle|\left(\begin{smallmatrix}0\\1\end{smallmatrix}\right)\rangle := |\uparrow\rangle|\downarrow\rangle, \\
e_3 &= |\overline{X}_2\rangle|\left(\begin{smallmatrix}1\\0\end{smallmatrix}\right)\rangle := |\downarrow\rangle|\uparrow\rangle, \\
e_4 &= |\overline{X}_2\rangle|\left(\begin{smallmatrix}0\\1\end{smallmatrix}\right)\rangle := |\downarrow\rangle|\downarrow\rangle.
\end{aligned}
\tag{4.19}
$$

As has been mentioned before, the fluxes X_2 and $\overline{X}_2$, respectively, are represented by the matrices $i\sigma_2$ and $-i\sigma_2$ (with σ_2 the second Pauli matrix) in the doublet charge representation χ. From (3.9), we then infer that the monodromy matrix takes the following block diagonal form in this basis

$$
\mathcal{R}^2 = \begin{pmatrix} 0 & 1 & 0 & 0 \\ -1 & 0 & 0 & 0 \\ 0 & 0 & 0 & -1 \\ 0 & 0 & 1 & 0 \end{pmatrix},
\tag{4.20}
$$

which reflects the phenomenon discussed in the previous section: the orientation of the charge χ is flipped, when it is transported either around the Alice flux X_2 or around $\overline{X}_2$.

Let us now consider the Aharonov-Bohm scattering experiment in which the doublet charge χ scatters from the Alice flux σ_2^+. We assume that we are measuring with a detector that only gives a signal when a scattered charge χ enters the device with a specific orientation (either $\uparrow$ or $\downarrow$). Here we may, for instance, think of an apparatus in which we have captured the associated antiparticle. This is the charge with opposite orientation, as we have seen in (4.18). If the orientation of the scattered charge entering the device matches that of the antiparticle, the pair annihilates and we assume that the apparatus somehow gives a signal when such an annihilation process occurs. The cross section measured with such a detector involves the matrix elements of the scattering matrix

$$
\begin{aligned}
&\mathcal{R}^{-\theta/\pi}(1 - \mathcal{R}^2) \\
&= \sqrt{2}e^{-i\theta/2}
\begin{pmatrix}
\cos\frac{\pi-\theta}{4} & \sin\frac{\pi-\theta}{4} & 0 & 0 \\
-\sin\frac{\pi-\theta}{4} & \cos\frac{\pi-\theta}{4} & 0 & 0 \\
0 & 0 & \cos\frac{\pi-\theta}{4} & -\sin\frac{\pi-\theta}{4} \\
0 & 0 & \sin\frac{\pi-\theta}{4} & \cos\frac{\pi-\theta}{4}
\end{pmatrix},
\end{aligned}
\tag{4.21}
$$

for the flux/charge eigenstates (4.19). This scattering matrix is determined using the prescription (4.52) in the monodromy eigenbasis in which the above monodromy matrix (4.20) is diagonal, and subsequently transforming back to the flux/charge eigenbasis (4.19). Now suppose that the scatterer is in a particular flux eigenstate, while the projectile that comes in is a charge with a specific orientation and the detector is only sensitive for scattered charges with this specific orientation. Under these circumstances, the two-particle in and out states are the same, $|\text{in}\rangle = |\text{out}\rangle$, and equal to one of the flux/charge eigenstates in (4.19). In other words, we are measuring the scattering amplitudes on the diagonal of the scattering matrix (4.21). Note that the formal sum of the out state |out⟩ over a complete basis of flux eigenstates for the scatterer, as indicated in Section 4.4, boils down to one term here, namely, the flux eigenstate of the scatterer in the in state |in⟩. The other flux eigenstate does *not* contribute. The corresponding matrix element vanishes, because the flux of the scatterer is not affected when it is encircled by the charge χ. From equation (4.51) of Section 4.5, we finally obtain the following exclusive cross section for this scattering experiment:

$$\frac{d\sigma_+}{d\theta} = \frac{1 + \sin(\theta/2)}{8\pi p \sin^2(\theta/2)}. \tag{4.22}$$

The charge flip cross section, in turn, is measured by a detector which only signals scattered charges with an orientation opposite to the orientation of the charge of the projectile. In that case, the state |in⟩ is again one of the flux/charge eigenstates in (4.19), while the |out⟩ state we measure is the same as the in state, but with the orientation of the charge flipped. Thus we are now measuring the off diagonal matrix elements of the scattering matrix (4.21). In a similar fashion as before, we find the following form for the charge flip cross section

$$\frac{d\sigma_-}{d\theta} = \frac{1 - \sin(\theta/2)}{8\pi p \sin^2(\theta/2)}. \tag{4.23}$$

The exclusive cross sections (4.22) and (4.23), which are the same as derived for scattering of electric charges from Alice fluxes in Alice electrodynamics by Lo and Preskill [71], are clearly multivalued

$$\frac{d\sigma_\pm}{d\theta}(\theta + 2\pi) = \frac{d\sigma_\mp}{d\theta}(\theta).$$

This merely reflects the fact that a detector only signalling charges χ with their orientation up becomes a detector only signalling charges with orientation down (and vice versa), when it is transported in a counterclockwise way over an angle 2π around the scatterer. Specifically, in this parallel transport the antiparticle in our detector feels the holonomy in the gauge fields associated with the flux of the scatterer and returns with its orientation flipped. As a consequence, the device becomes sensitive for the opposite charge orientation after this parallel transport.

Verlinde's detector does not suffer from this multivaluedness. It does not discriminate between the orientations of the scattered charge, and gives a signal whenever a charge χ enters the device. This detector measures the total or inclusive cross section, i.e., both branches of the multivalued cross section (4.22) (or (4.23) for that matter). To be specific, the exclusive cross sections (4.22) and (4.23) combine as follows

$$\frac{d\sigma}{d\theta} = \frac{d\sigma_-}{d\theta} + \frac{d\sigma_+}{d\theta} = \frac{1}{4\pi p \sin^2(\theta/2)}.$$

into Verlinde's single-valued inclusive cross section (4.50) for this scattering experiment.

To conclude, the above analysis is easily extended to Aharonov-Bohm scattering experiments involving other particles in the spectrum (4.1) of this $\bar{D}_2$ gauge theory. It should be stressed, however, that a crucial ingredient in the derivation of the *multivalued* exclusive cross sections (4.22) and (4.23) is that the monodromy matrix (4.20) is off diagonal and has imaginary eigenvalues $\pm i$. In the other cases, where the monodromy matrices are diagonal or off diagonal with eigenvalues ± 1, as it appears for scattering two noncommuting fluxes σ_a^+ and σ_b^+ from each other, we arrive at *single-valued* exclusive cross sections.

4.3   Non-Abelian Braid Statistics

We finally turn to the issue of non-Abelian braid statistics. As we have argued in Section 3.2, the braidings and monodromies for multiparticle configurations appearing in discrete H gauge theories are governed by truncated rather then ordinary braid groups. To be precise, the total internal Hilbert space for a given multiparticle system carries a representation of some truncated braid group, which in general decomposes into a direct sum of irreducible representations. In this section, we identify the truncated braid groups ruling in this particular $\bar{D}_2$ gauge theory and elaborate on the aforementioned decomposition. We first consider the indistinguishable particle configurations in this model.

It can easily be verified that the braid operators acting on a configuration which only consists of the pure singlet charges J_a (with $a \in 1,\ 2,\ 3$) are of order one. The same holds for the singlet dyons $\bar{1}$ and $\bar{J}_a$. In other words, these particles behave as ordinary bosons, in accordance with the trivial spin factors (4.2) assigned to them. To proceed, the braid operators acting on a system of n doublet charges χ are of order two and therefore realize a (higher-dimensional) representation of the permutation group S_n. The same observation appears for the doublet dyons $\bar{\chi}$ and $\sigma_a^\pm$. The total internal Hilbert spaces for these indistinguishable particle systems can then be decomposed into a direct sum of subspaces, each carrying an irreducible representation of the permutation group S_n. The one-dimensional representations that appear in this decomposition naturally correspond either to Bose or Fermi statistics, while the higher-dimensional representations describe parastatistics. Finally, braid statistics occur for a system consisting of n dyons $\tau_a^\pm$. The braid operators that act on such a system are of order four. Hence, the associated internal Hilbert space splits up into a direct sum of irreducible representations of the truncated braid group $B(n, 4)$. The one-dimensional representations that occur in this decomposition realize Abelian braid statistics (anyons), whereas the higher-dimensional representations correspond to non-Abelian braid statistics (non-Abelian anyons). We will illustrate these features with two representative examples. We first examine a system containing two dyons τ_1^+. The irreducible braid group representations available for this system are one-dimensional, since the truncated braid group B(2, 4) for two particles is Abelian. We then turn to the more interesting system consisting of three dyons τ_1^+. In that case, we are dealing with non-Abelian braid statistics. The associated total internal Hilbert space breaks up into four one-dimensional irreducible subspaces and two two-dimensional irreducible subspaces under the action of the non-Abelian truncated braid group B(3, 4).

We start by setting some conventions. First of all, the two fluxes in the conjugacy class associated with the dyon τ_1^+ are ordered as indicated in Table 1, i.e.,

$$^1h_1 = X_1, \quad ^1h_2 = \overline{X}_1,$$

while we take the following coset representatives appearing in the definition (3.4) of the centralizer charge

$$^1x_1 = e, \quad ^1x_2 = X_2.$$

To lighten the notation a bit, we furthermore use the following abbreviation for the internal flux/charge eigenstates of the dyon τ_1^+

$$|\uparrow\rangle := |X_1, {}^1v\rangle, \quad |\downarrow\rangle := |\overline{X}_1, {}^1v\rangle.$$

Let us now consider a system consisting of two dyons τ_1^+. Under the action of the quantum double $D(\overline{D}_2)$, the internal Hilbert space $V_{\tau_1^+} \otimes V_{\tau_1^+}$ associated with this system decomposes according to the fusion rule (4.11), which we repeat for convenience

$$\tau_1^+ \times \tau_1^+ = 1 + J_1 + \bar{J}_2 + \bar{J}_3. \tag{4.24}$$

The two-particle states corresponding to the different fusion channels carry a one-dimensional (irreducible) representation of the Abelian truncated braid group B(2, 4) = $\mathbb{Z}_4$. We first establish the different irreducible pieces contained in the B(2, 4) representation carried by the total internal Hilbert space $V_{\tau_1^+} \otimes V_{\tau_1^+}$. This can be done by calculating the traces of the elements $\{e, \tau, \tau^2, \tau^3\}$ of B(2, 4) in this representation using the standard diagrammatic techniques (see, for instance, Refs. 96 and 97). From the character vector obtained in this way, we learn that this representation breaks up as

$$\Lambda_{B(2,4)} = 3\Gamma^1 + \Gamma^3,$$

with Γ^1 and Γ^3 the irreducible $\mathbb{Z}_4$ representations displayed in the character Table 1, i.e., $\Gamma^1(\tau) := i$ and $\Gamma^3(\tau) := -i$. After some algebra, we then arrive at the following basis of mutual eigenstates under the combined action of the quantum double and the truncated braid group

$$
\begin{array}{cccc}
V_{\tau_1^+} \otimes V_{\tau_1^+} & D(\bar{D}_2) & B(2,4) & \\
\hline
\frac{1}{\sqrt{2}}\{|{\uparrow}\rangle|{\downarrow}\rangle - |{\downarrow}\rangle|{\uparrow}\rangle\} & 1 & \Gamma^1 & (4.25) \\
\frac{1}{\sqrt{2}}\{|{\uparrow}\rangle|{\downarrow}\rangle + |{\downarrow}\rangle|{\uparrow}\rangle\} & J_1 & \Gamma^3 & (4.26) \\
\frac{1}{\sqrt{2}}\{|{\uparrow}\rangle|{\uparrow}\rangle + |{\downarrow}\rangle|{\downarrow}\rangle\} & \bar{J}_2 & \Gamma^1 & (4.27) \\
\frac{1}{\sqrt{2}}\{|{\uparrow}\rangle|{\uparrow}\rangle - |{\downarrow}\rangle|{\downarrow}\rangle\} & \bar{J}_3 & \Gamma^1, & (4.28)
\end{array}
$$

from which we conclude that the two-particle internal Hilbert space $V_{\tau_1^+} \otimes V_{\tau_1^+}$ decomposes into the following direct sum of one-dimensional irreducible representations of the direct product $D(\bar{D}_2) \times B(2,4)$

$$(1, \Gamma^1) + (J_1, \Gamma^3) + (\bar{J}_2, \Gamma^1) + (\bar{J}_3, \Gamma^1).$$

In accordance with the general discussion concerning relation (3.30), the two-particle states contained in (4.27) and (4.28), which are given by a linear combination of two states in which both particles are in the same internal quantum state, satisfy the canonical spin-statistics connection (3.31). That is, $\exp(i\Theta) = \exp(2\pi i s_{\tau_1^+}) = i$. Accidentally, the same observation appears for the two-particle state (4.25). Finally, the two-particle state displayed in (4.26) satisfies the generalized spin-statistics connection (3.28) and describes semion statistics with quantum-statistical parameter $\exp(i\Theta) = -i$.

We now extend our analysis to a system containing three dyons τ_1^+. From the fusion rules (4.24) and (4.4)–(4.6), we infer that the decomposition of the total internal Hilbert space under the action of the quantum double becomes

$$\tau_1^+ \times \tau_1^+ \times \tau_1^+ = 4\tau_1^+. \tag{4.29}$$

The occurrence of four equivalent fusion channels indicates that non-Abelian braid statistics is conceivable and it turns out that higher-dimensional irreducible representations of the truncated braid group $B(3,4)$ indeed appear. The structure of this group and its irreducible representations are discussed in Section 4.5. A lengthy but straightforward diagrammatic calculation of the character vector associated with the $B(3,4)$ representation carried by the three particle internal Hilbert space $V_{\tau_1^+} \otimes V_{\tau_1^+} \otimes V_{\tau_1^+}$ reveals the following irreducible pieces

$$\Lambda_{B(3,4)} = 4\Lambda_1 + 2\Lambda_5, \tag{4.30}$$

with Λ_1 and Λ_5 the irreducible representations of $B(3,4)$ exhibited in the character Table 4 of Section 4.5. The one-dimensional representation Λ_1 describes Abelian semion statistics, while the two-dimensional representation Λ_5 corresponds to non-Abelian braid statistics. From (4.29) and (4.30), we can immediately conclude that this three-particle internal Hilbert space breaks up into the following direct sum of irreducible subspaces under the action of the direct product $D(\overline{D}_2) \times B(3,4)$

$$2(\tau_1^+, \Lambda_1) + (\tau_1^+, \Lambda_5),$$

where (τ_1^+, Λ_1) labels a two-dimensional and (τ_1^+, Λ_5) a four-dimensional representation. A basis adapted to this decomposition can be cast in the following form

$V_{\tau_1^+} \otimes V_{\tau_1^+} \otimes V_{\tau_1^+}$	$D(\overline{D}_2)$	$B(3,4)$	
$\lvert\downarrow\rangle\lvert\downarrow\rangle\lvert\downarrow\rangle$	$\lvert\uparrow\rangle_1$	Λ_1	(4.31)
$\lvert\uparrow\rangle\lvert\uparrow\rangle\lvert\uparrow\rangle$	$\lvert\downarrow\rangle_1$	Λ_1	(4.32)
$\frac{1}{\sqrt{3}}\{\lvert\uparrow\rangle\lvert\uparrow\rangle\lvert\downarrow\rangle - \lvert\uparrow\rangle\lvert\downarrow\rangle\lvert\uparrow\rangle + \lvert\downarrow\rangle\lvert\uparrow\rangle\lvert\uparrow\rangle\}$	$\lvert\uparrow\rangle_2$	Λ_1	(4.33)
$\frac{1}{\sqrt{3}}\{\lvert\downarrow\rangle\lvert\downarrow\rangle\lvert\uparrow\rangle - \lvert\downarrow\rangle\lvert\uparrow\rangle\lvert\downarrow\rangle + \lvert\uparrow\rangle\lvert\downarrow\rangle\lvert\downarrow\rangle\}$	$\lvert\downarrow\rangle_2$	Λ_1	(4.34)
$\frac{1}{2}\{2\lvert\uparrow\rangle\lvert\uparrow\rangle\lvert\downarrow\rangle + \lvert\uparrow\rangle\lvert\downarrow\rangle\lvert\uparrow\rangle - \lvert\downarrow\rangle\lvert\uparrow\rangle\lvert\uparrow\rangle\}$	$\lvert\uparrow\rangle_3$	Λ_5	(4.35)
$\frac{1}{2}\{2\lvert\downarrow\rangle\lvert\downarrow\rangle\lvert\uparrow\rangle + \lvert\downarrow\rangle\lvert\uparrow\rangle\lvert\downarrow\rangle - \lvert\uparrow\rangle\lvert\downarrow\rangle\lvert\downarrow\rangle\}$	$\lvert\downarrow\rangle_3$	Λ_5'	(4.36)
$\frac{1}{\sqrt{2}}\{\lvert\uparrow\rangle\lvert\downarrow\rangle\lvert\uparrow\rangle + \lvert\downarrow\rangle\lvert\uparrow\rangle\lvert\uparrow\rangle\}$	$\lvert\uparrow\rangle_4$	Λ_5	(4.37)
$\frac{1}{\sqrt{2}}\{\lvert\downarrow\rangle\lvert\uparrow\rangle\lvert\downarrow\rangle + \lvert\uparrow\rangle\lvert\downarrow\rangle\lvert\downarrow\rangle\}$	$\lvert\downarrow\rangle_4$	Λ_5',	(4.38)

The subscript attached to the single-particle states in the second column label the four fusion channels showing up in (4.29). In other words, these states summarize the global properties of the three-particle states in the first column, that is, the total flux and charge, which are conserved under braiding. Each of the three-particle states in the first four rows carry the one-dimensional representation Λ_1 of the truncated braid group $B(3,4)$. The particles in these states obey semion statistics with quantum-statistical parameter $\exp(i\Theta) = i$ and satisfy the canonical spin-statistics connection (3.31). Finally, the states in the last four rows constitute a basis for the representation (τ_1^+, Λ_5). To be specific, the states (4.35) and (4.37), carrying the same total flux and charge, form a basis for a two-dimensional irreducible representation Λ_5 of the truncated braid group. The same remark holds for the states (4.36) and (4.38). For convenience, we have distinguished these two irreducible representations by a prime. Note that we have chosen a basis which diagonalizes the braid operator $\mathcal{R}_1$ acting on the first two particles with eigenvalues either i or $-i$, whereas the braid operator $\mathcal{R}_2$ for the last two particles mixes the states in the different fusion channels. Of course, this choice is quite arbitrary. By another basis choice, we could have reversed this situation.

Let us also comment briefly on the distinguishable particle systems that may occur. The maximal order of the monodromy operator for distinguishable particles in this $\bar{D}_2$ gauge theory is four. Hence, the distinguishable particle systems in this theory are governed by the truncated colored braid groups $P(n, 8)$ and their subgroups. A system consisting of the three different particles σ_1^+, σ_2^+ and τ_3^+, for instance, realizes a representation of the colored braid group $P(3, 4) \subset P(3, 8)$. (The group structure of $P(3, 4)$ and a classification of its irreducible representations are given in Section 4.5). In a similar fashion as before, it is easily inferred that this represention of $P(3, 4)$ breaks up into the following irreducible pieces

$$\Lambda_{P(3,4)} = 2\Omega_8 + 2\Omega_9, \tag{4.39}$$

with Ω_8 and Ω_9 the two-dimensional irreducible representations displayed in the character Table 5 of Section 4.5. Thus, this system obeys non-Abelian "monodromy statistics": the three monodromy operators displayed in (4.53) can not be diagonalized simultaneously. Further, from the fusion rules (4.8) and (4.9) in combination with (4.3)–(4.5), we obtain that the internal Hilbert space of this three-particle particle system splits up into the following irreducible components under the action of the quantum double $D(\bar{D}_2)$

$$\sigma_1^+ \times \sigma_2^+ \times \tau_3^+ = (\sigma_3^+ + \sigma_3^-) \times \tau_3^+ = 2\chi + 2\bar{\chi}. \tag{4.40}$$

It is then readily checked that a basis adapted to the simultaneous decomposition of the three-particle internal Hilbert space $V_{\sigma_1^+} \otimes V_{\sigma_2^+} \otimes V_{\tau_3^+}$ into UIRs of $D(\bar{D}_2)$ and $P(3, 4)$ reads:

$V_{\sigma_1^+} \otimes V_{\sigma_2^+} \otimes V_{\tau_3^+}$	$D(\bar{D}_2)$	$P(3,4)$	
$\frac{1}{\sqrt{2}}\{\lvert X_1\rangle\lvert \overline{X}_2\rangle + \lvert \overline{X}_1\rangle\lvert X_2\rangle\}\lvert \overline{X}_3, {}^1v\rangle$	χ	Ω_9	(4.41)
$\frac{1}{\sqrt{2}}\{\lvert X_1\rangle\lvert X_2\rangle + \lvert \overline{X}_1\rangle\lvert \overline{X}_2\rangle\}\lvert X_3, {}^1v\rangle$	χ	Ω_9'	(4.42)
$\frac{1}{\sqrt{2}}\{\lvert X_1\rangle\lvert \overline{X}_2\rangle - \lvert \overline{X}_1\rangle\lvert X_2\rangle\}\lvert \overline{X}_3, {}^1v\rangle$	χ'	Ω_9'	(4.43)
$\frac{1}{\sqrt{2}}\{\lvert X_1\rangle\lvert X_2\rangle - \lvert \overline{X}_1\rangle\lvert \overline{X}_2\rangle\}\lvert X_3, {}^1v\rangle$	χ'	Ω_9	(4.44)
$\frac{1}{\sqrt{2}}\{\lvert X_1\rangle\lvert \overline{X}_2\rangle + \lvert \overline{X}_1\rangle\lvert X_2\rangle\}\lvert X_3, {}^1v\rangle$	$\bar{\chi}$	Ω_8	(4.45)

$$\frac{1}{\sqrt{2}}\{|X_1\rangle|X_2\rangle + |\overline{X}_1\rangle|\overline{X}_2\rangle\}|\overline{X}_3, {}^1v\rangle \qquad \overline{\chi} \qquad \Omega'_8 \qquad (4.46)$$

$$\frac{1}{\sqrt{2}}\{|X_1\rangle|\overline{X}_2\rangle - |\overline{X}_1\rangle|X_2\rangle\}|X_3, {}^1v\rangle \qquad \overline{\chi}' \qquad \Omega'_8 \qquad (4.47)$$

$$\frac{1}{\sqrt{2}}\{|X_1\rangle|X_2\rangle - |\overline{X}_1\rangle|\overline{X}_2\rangle\}|\overline{X}_3, {}^1v\rangle \qquad \overline{\chi}' \qquad \Omega_8. \qquad (4.48)$$

As before, the isotypical fusion channels and the equivalent UIRs of $P(3,4)$ are distinguished by a prime. So, the three-particle states in (4.41) and (4.42), for example, form a basis for one of the two fusion channels χ in (4.40), whereas the states in (4.43) and (4.44) constitute a basis for the other fusion channel χ. The three-particle states in (4.41) and (4.44) then span one of the two UIRs Ω_9 of $P(3,4)$ in (4.39) and the states in (4.42) and (4.43) the other. Finally, from (4.41)–(4.48), we learn that the two fusion channels χ in (4.40) combine with the two UIRs Ω_9 of $P(3,4)$ in (4.39) and the two fusion channels $\overline{\chi}$ with the two UIRs Ω_8. Hence, the three-particle internal Hilbert space $V_{\sigma_1^+} \otimes V_{\sigma_2^+} \otimes V_{\tau_3^+}$ breaks up into the following two four-dimensional irreducible representations

$$(\chi, \Omega_9) + (\overline{\chi}, \Omega_8),$$

of the direct product $D(\overline{D}_2) \times P(3,4)$.

As a last blow, we return to the process depicted in Figure (12). After the double pair creation, we are dealing with a four-particle system consisting of a subsystem of two indistinguishable particles σ_2^+ and a subsystem of two indistinguishable particles χ. Recall from the sequence (4.18) that the two-particle state for the fluxes σ_2^+ was initially bosonic, whereas the two-particle state for the charges χ was fermionic. After the monodromy has taken place, the situation is reversed. The two-particle state for the fluxes σ_2^+ has become fermionic and the two-particle state for the charges χ bosonic. In other words, the exchange of Cheshire charge is accompanied by an exchange of quantum statistics; see also Ref. 52 in this connection. The total four-particle system now realizes a two-dimensional irreducible representation of the associated truncated partially colored braid group. The two braid operators $\mathcal{R}_1$ and $\mathcal{R}_3$ for the indistinguisbale particle exchanges in the two subsystems act diagonally with eigenvalues ± 1 and ∓ 1, respectively. Furthermore, under the repeated action of the monodromy operator $\mathcal{R}_2^2$, the subsystems simultaneously jump back and forth between the fusion channels 1 and J_2 with their associated Cheshire charge and quantum statistics.

4.4 Aharonov-Bohm Scattering

The only experiments in which the particles in a discrete H gauge theory leave "long-range fingerprints" are of a quantum-mechanical nature,

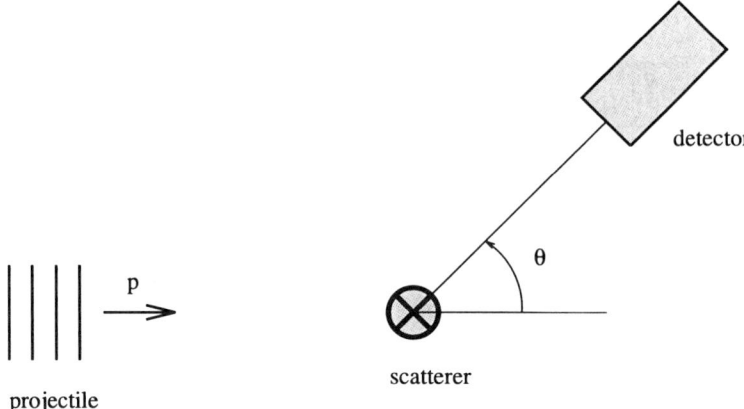

FIGURE 13. The geometry of the Aharonov-Bohm scattering experiment. The projectile comes in as a plane wave with momentum p and scatters elastically from a scatterer fixed at the origin. It is assumed that the projectile never enters the region where the scatterer is located. The cross section for the scattered projectile is measured by a detector placed at the scattering angle θ.

namely, quantum interference experiments, such as the double-slit experiment [71, 78] and the Aharonov-Bohm scattering experiment [39]. What we are measuring in these experiments is the way the particles affect their mutual internal flux/charge quantum numbers when they encircle each other. In other words, we are probing the content of the monodromy matrix $\mathcal{R}^2$ following from (3.9). In this subsection, we will give a concise discussion of two-particle Aharonov-Bohm scattering and provide the details entering the calculation of the cross sections in Section 4.2. For a recent review of the experimental status of the Aharonov-Bohm effect, the reader is referred to [98].

The geometry of the Aharonov-Bohm scattering experiment is depicted in Figure 13. It involves two particles, a projectile, and a scatterer fixed at the origin. The incoming external part of the total wave function is a plane wave for the projectile vanishing at the location of the scatterer. Nontrivial scattering takes place if and only if the monodromy matrix $\mathcal{R}^2$ acting on the internal part of the wave function is nontrivial.

In the Abelian discrete gauge theory discussed in Section 2.3, we only encountered the Abelian version. That is, the effect of a monodromy of the two particles in the internal wave function is just a phase

$$\mathcal{R}^2 = e^{2\pi i \alpha}.$$

The differential cross section for the quantum-mechanical scattering experiment involving such particles has been derived almost 40 years ago by Aharonov and Bohm [39]

$$\frac{d\sigma}{d\theta} = \frac{\sin^2(\pi\alpha)}{2\pi p \sin^2(\theta/2)}. \tag{4.49}$$

Here, θ denotes the scattering angle and p the momentum of the incoming plane wave of the projectile.

The particles appearing in a non-Abelian discrete gauge theory may in general exchange internal flux/charge quantum numbers when they encircle each other. This effect is described by nondiagonal monodromy *matrices* $\mathcal{R}^2$ acting on multicomponent internal wave functions. The cross section measured in Aharonov-Bohm scattering experiment involving these particles is a non-Abelian generalization of the Abelian one given in (4.49). An elegant closed formula for these non-Abelian cross sections has been derived by Erik Verlinde [95]. The crucial insight was that the monodromy matrix $\mathcal{R}^2$ for two particles can always be diagonalized, since the braid group for two particles is Abelian. In the monodromy eigenbasis in which the monodromy matrix $\mathcal{R}^2$ is diagonal, the non-Abelian problem then reduces to the Abelian one solved by Aharonov and Bohm. The solution can subsequently be cast in the basis independent form

$$\left.\frac{d\sigma}{d\theta}\right|_{\text{in}\rightarrow\text{all}} = \frac{1}{4\pi p \sin^2(\theta/2)}[1 - \text{Re}\langle\text{in}|\mathcal{R}^2|\text{in}\rangle], \tag{4.50}$$

with $|\text{in}\rangle$ the normalized two-particle incoming internal quantum state. Note that this cross section boils down to (4.49) for the Abelian case. We will always work in the natural two-particle flux/charge eigenbasis being the tensor product of the single particle internal basis states (3.3). In fact, in our applications the $|\text{in}\rangle$ state usually is a particular two-particle flux/charge eigenstate. The detector measuring the cross section (4.50) is a device that does not discriminate between the different internal disguises the scattered projectile can take. Specifically, in the scattering process discussed in Section 4.2, the Verlinde detector gives a signal, when the scattered pure doublet charge χ enters the apparatus with its charge orientation either up or down. In this sense, Verlinde's cross section (4.50) is inclusive.

Inspired by this work, Lo and Preskill subsequently introduced a finer detector [71]. Their device is able to distinguish between the different internal appearances of the projectile. In the scattering process studied in Section 4.2, for example, we can use a device, which only gives a signal if the projectile enters the device with its internal charge orientation up. The exclusive cross section measured with such a detector can be expressed as

$$\left.\frac{d\sigma}{d\theta}\right|_{\text{in}\rightarrow\text{out}} = \frac{1}{8\pi p \sin^2(\theta/2)}|\langle\text{out}|\mathcal{R}^{-\theta/\pi}(\mathbf{1} - \mathcal{R}^2)|\text{in}\rangle|^2, \tag{4.51}$$

where $|\text{in}\rangle$ and $|\text{out}\rangle$ denote normalized two-particle incoming- and outgoing internal quantum states. The outgoing state we observe depends on the detector we have installed, but since we only measure the projectile, so half of the out state, the state $|\text{out}\rangle$ in (4.51) should always be summed over a complete basis for the internal Hilbert space of the scatterer. The new ingredient in the exclusive cross section (4.51) is the matrix $\mathcal{R}^{-\theta/\pi}$. This matrix is defined as the diagonal matrix in the monodromy eigenbasis, which acts as

$$\mathcal{R}^{-\theta/\pi} := e^{-i\alpha\theta} \quad \text{with } \alpha \in [0, 1), \tag{4.52}$$

on a monodromy eigenstate characterized by the eigenvalue $\exp(2\pi i\alpha)$ under $\mathcal{R}^2$. By a basis transformation, we then find the matrix elements of $\mathcal{R}^{-\theta/\pi}$ in our favorite two-particle flux/charge eigenbasis.

A peculiar property of the exclusive cross section (4.51) is that it is in general multivalued. This is just a reflection of the fact that the detector can generally change its nature, when it is parallel transported around the scatterer. An apparatus that only detects projectiles with internal charge orientation up in the scattering process studied in Section 4.2, for example, becomes an apparatus, which only detects projectiles with charge orientation down, after a rotation over an angle of 2π around the scatterer. Verlinde's detector, giving a signal independent of the internal disguise of the projectile entering the device, obviously does *not* suffer from this multivaluedness. As a matter of fact, extending the aforementioned sum of the $|\text{out}\rangle$ state in (4.51) over a complete basis of the internal Hilbert space for the scatterer by a sum over a complete basis of the internal Hilbert space for the projectile and subsequently using the partition of unity, yields the single-valued inclusive cross section (4.50).

As a last remark, the cross sections for Aharonov-Bohm scattering experiments in which the projectile and the scatterer are indistinguishable particles contain an extra contribution due to conceivable exchange processes between the scatterer and the projectile [32, 71]. The incorporation of this exchange contribution merely amounts to diagonalizing the braid matrix $\mathcal{R}$ instead of the monodromy matrix $\mathcal{R}^2$.

4.5   B(3,4) and P(3,4)

In this last subsection, we give the structure of the truncated braid group B(3,4) and the truncated colored braid group P(3,4), which enter the discussion of the non-Abelian braid properties of certain three-particle configurations in a $\overline{D}_2$ gauge theory in Section 4.3.

According to the general definition (3.17)–(3.18) given in Section 3.2, the truncated braid group B(3,4) for three indistinguishable particles is generated by two elements τ_1 and τ_2 subject to the relations

$$\tau_1\tau_2\tau_1 = \tau_2\tau_1\tau_2, \quad \tau_1^4 = \tau_2^4 = e.$$

By explicit construction from these defining relations, which is a lengthy and not at all trivial job, it can be inferred that B(3,4) is a group of order 96, which splits up into the following conjugacy classes:

$$C_0^1 = \{e\},$$
$$C_0^2 = \{\tau_1\tau_2\tau_1\tau_2\tau_1\tau_2\},$$
$$C_0^3 = \{\tau_2^2\tau_1^2\tau_2^2\tau_1^2\},$$
$$C_0^4 = \{\tau_2^2\tau_1^3\tau_2^2\tau_1^3\},$$
$$C_1^1 = \{\tau_1, \tau_2, \tau_2\tau_1\tau_2^3, \tau_2^2\tau_1\tau_2^2, \tau_2^3\tau_1\tau_2, \tau_1^2\tau_2\tau_1^2\},$$
$$C_1^2 = \{\tau_1^3\tau_2\tau_1^2\tau_2, \tau_2^3\tau_1\tau_2^2\tau_1, \tau_2\tau_1^3\tau_2\tau_1^2, \tau_2^2\tau_1^2\tau_2^2\tau_1, \tau_1\tau_2^3\tau_1\tau_2^2, \tau_1^2\tau_2^2\tau_1^2\tau_2\},$$
$$C_1^3 = \{\tau_2\tau_1^3\tau_2\tau_1^3\tau_2, \tau_1^2\tau_2\tau_1^3\tau_2^2\tau_1, \tau_2^3\tau_1\tau_2^3\tau_1^2, \tau_1\tau_2^3\tau_1^2\tau_2^3\tau_1, \tau_2\tau_1\tau_2^3\tau_1^3\tau_2^2,$$
$$\tau_2\tau_1^2\tau_2^3\tau_1^2\tau_2\},$$
$$C_1^4 = \{\tau_2^2\tau_1^3\tau_2^2, \tau_1^2\tau_2^3\tau_1^2, \tau_2^3\tau_1^3\tau_2, \tau_1^3, \tau_2\tau_1^3\tau_2^3, \tau_2^3\},$$
$$C_2^1 = \{\tau_1\tau_2, \tau_2\tau_1, \tau_1^2\tau_2\tau_1^3, \tau_1^3\tau_2\tau_1^2, \tau_2\tau_1^2\tau_2^2\tau_1, \tau_2^2\tau_1\tau_2^3, \tau_2^3\tau_1\tau_2, \tau_1\tau_2^2\tau_1^2\tau_2\},$$
$$C_2^2 = \{\tau_1^2\tau_2\tau_1^3\tau_2\tau_1, \tau_1\tau_2\tau_1^3\tau_2\tau_1^2, \tau_2\tau_1^2\tau_2^2\tau_1^3, \tau_1\tau_2\tau_1^2\tau_2^2\tau_1^2,$$
$$\tau_2\tau_1^3\tau_2^2\tau_1^2, \tau_1\tau_2\tau_1^3\tau_2^2\tau_1, \tau_1^2\tau_2\tau_1^2\tau_2^2, \tau_1^2\tau_2^2\tau_1^3\tau_2\},$$
$$C_2^3 = \{\tau_1^3\tau_2\tau_1^3\tau_2^2\tau_1, \tau_1\tau_2^2\tau_1^3\tau_2\tau_1^3, \tau_2\tau_1^3\tau_2^2, \tau_2^2\tau_1^3\tau_2, \tau_2^3\tau_1^3, \tau_1\tau_2^3\tau_1^2, \tau_1^2\tau_2^3\tau_1, \tau_1^3\tau_2^3\},$$
$$C_2^4 = \{\tau_1^3\tau_2^3\tau_1^2, \tau_1^2\tau_2^3\tau_1^3, \tau_2^3\tau_1, \tau_1\tau_2^3, \tau_1\tau_2\tau_1\tau_2, \tau_1^3\tau_2, \tau_2\tau_1^3, \tau_2\tau_1\tau_2\tau_1\},$$
$$C_3^1 = \{\tau_1^2, \tau_2^2, \tau_1\tau_2^2\tau_1^3, \tau_2^2\tau_1^2\tau_2^2, \tau_1^2\tau_2^2\tau_1^2, \tau_1^3\tau_2^2\tau_1\},$$
$$C_3^2 = \{\tau_2\tau_1^2\tau_2, \tau_1\tau_2^2\tau_1, \tau_1^2\tau_2^2, \tau_2^3\tau_1^2\tau_2^3, \tau_1^3\tau_2^2\tau_1^3, \tau_2^2\tau_1^2\},$$
$$C_4^1 = \{\tau_1\tau_2\tau_1, \tau_1^2\tau_2, \tau_2^2\tau_1, \tau_2\tau_1^2, \tau_1\tau_2^2, \tau_1^3\tau_2\tau_1^3, \tau_1^3\tau_2\tau_1^3\tau_2^2\tau_1^2,$$
$$\tau_2\tau_1^3\tau_2^2\tau_1, \tau_1^2\tau_2^2\tau_1^3, \tau_1\tau_2^2\tau_1^3\tau_2, \tau_1^3\tau_2^2\tau_1^2, \tau_1\tau_2\tau_1^3\tau_2^2\},$$
$$C_4^2 = \{\tau_1\tau_2\tau_1\tau_2\tau_1\tau_2\tau_1\tau_2\tau_1, \tau_2\tau_1^2\tau_2^2, \tau_1\tau_2^2\tau_1^2, \tau_2^2\tau_1^2\tau_2, \tau_1^2\tau_2^2\tau_1, \tau_2\tau_1^3\tau_2,$$
$$\tau_2^3\tau_1^3\tau_2^3, \tau_2^3\tau_1^2, \tau_1^3\tau_2^2, \tau_2^2\tau_1^3, \tau_1^2\tau_2^3, \tau_1\tau_2^3\tau_1\}.$$

We organized the conjugacy classes such that $C_k^{i+1} = zC_k^i$, with $z = \tau_1\tau_2\tau_1\tau_2\tau_1\tau_2$ the generator of the center of $B(3,4)$. The character table of the truncated braid group $B(3,4)$ is displayed in Table 4.

The truncated colored braid group $P(3,4)$, which consists of the monodromy operations on a configuration of three distinguishable particles, is the subgroup of $B(3,4)$ generated by the elements

$$\gamma_{12} = \tau_1^2, \quad \gamma_{13} = \tau_1\tau_2^2\tau_1^{-1} = \tau_1\tau_2^2\tau_1^3, \quad \gamma_{23} = \tau_2^2, \tag{4.53}$$

which satisfy

$$\gamma_{12}^2 = \gamma_{13}^2 = \gamma_{23}^2 = e.$$

It can be verified that $P(3,4)$ is a group of order 16 which splits up into the following 10 conjugacy classes

$$C_0 = \{e\}, \qquad\qquad C_1 = \{\gamma_{13}\gamma_{12}\gamma_{23}\},$$

TABLE 4. Character table of the truncated braid group B(3, 4). We used $\eta :=i + 1$.

	C_0^1	C_0^2	C_0^3	C_0^4	C_1^1	C_1^2	C_1^3	C_1^4	C_2^1	C_2^2	C_2^3	C_2^4	C_3^1	C_3^2	C_4^1	C_4^2
Λ_0	1	1	1	1	1	1	1	1	1	1	1	1	1	1	1	1
Λ_1	1	−1	1	−1	i	$-i$	i	$-i$	−1	1	−1	1	−1	1	$-i$	i
Λ_2	1	1	1	1	−1	−1	−1	−1	1	1	1	1	1	1	−1	−1
Λ_3	1	−1	1	−1	$-i$	i	$-i$	i	−1	1	−1	1	−1	1	i	$-i$
Λ_4	2	2	2	2	0	0	0	0	−1	−1	−1	−1	2	2	0	0
Λ_5	2	−2	2	−2	0	0	0	0	1	−1	1	−1	−2	2	0	0
Λ_6	2	$2i$	−2	$-2i$	η	$-\eta^*$	$-\eta$	η^*	i	−1	$-i$	1	0	0	0	0
Λ_7	2	$2i$	−2	$-2i$	$-\eta$	η^*	η	$-\eta^*$	i	−1	$-i$	1	0	0	0	0
Λ_8	2	$-2i$	−2	$2i$	$-\eta^*$	η	η^*	$-\eta$	$-i$	−1	i	1	0	0	0	0
Λ_9	2	$-2i$	−2	$2i$	η^*	$-\eta$	$-\eta^*$	η	$-i$	−1	i	1	0	0	0	0
Λ_{10}	3	3	3	3	1	1	1	1	0	0	0	0	−1	−1	−1	−1
Λ_{11}	3	−3	3	−3	i	$-i$	i	$-i$	0	0	0	0	1	−1	i	$-i$
Λ_{12}	3	3	3	3	−1	−1	−1	−1	0	0	0	0	−1	−1	1	1
Λ_{13}	3	−3	3	−3	$-i$	i	$-i$	i	0	0	0	0	1	−1	$-i$	i
Λ_{14}	4	4	−4	−4	0	0	0	0	1	1	−1	−1	0	0	0	0
Λ_{15}	4	−4	−4	4	0	0	0	0	−1	1	1	−1	0	0	0	0

TABLE 5. Character table of the truncated colored braid group P(3, 4).

$P(3,4)$	C_0	C_1	C_2	C_3	C_4	C_5	C_6	C_7	C_8	C_9
Ω_0	1	1	1	1	1	1	1	1	1	1
Ω_1	1	1	1	1	−1	−1	1	−1	−1	1
Ω_2	1	1	1	1	−1	1	−1	1	−1	−1
Ω_3	1	1	1	1	1	−1	−1	−1	1	−1
Ω_4	1	−1	1	−1	−1	1	1	−1	1	−1
Ω_5	1	−1	1	−1	1	−1	1	1	−1	−1
Ω_6	1	−1	1	−1	1	1	−1	−1	−1	1
Ω_7	1	−1	1	−1	−1	−1	−1	1	1	1
Ω_8	2	$2i$	−2	$-2i$	0	0	0	0	0	0
Ω_9	2	$-2i$	−2	$2i$	0	0	0	0	0	0

$$C_2 = \{\gamma_{23}\gamma_{12}\gamma_{23}\gamma_{12}\}, \qquad C_3 = \{\gamma_{23}\gamma_{12}\gamma_{13}\},$$
$$C_4 = \{\gamma_{12}, \gamma_{23}\gamma_{12}\gamma_{23}\}, \qquad C_5 = \{\gamma_{23}, \gamma_{12}\gamma_{23}\gamma_{12}\}, \qquad (4.54)$$
$$C_6 = \{\gamma_{13}, \gamma_{12}\gamma_{13}\gamma_{12}\}, \qquad C_7 = \{\gamma_{13}\gamma_{12}, \gamma_{12}\gamma_{13}\},$$
$$C_8 = \{\gamma_{23}\gamma_{13}, \gamma_{13}\gamma_{23}\}, \qquad C_9 = \{\gamma_{12}\gamma_{23}, \gamma_{23}\gamma_{12}\}.$$

It turns out that the truncated colored braid group $P(3,4)$ is, in fact, isomorphic to the coxeter group denoted as $16/8$ in Ref. 99. Further, the center of $P(3,4)$ contained in the first four conjugacy classes in (4.54) is of order four and coincides with that of $B(3,4)$. Finally, the character table of $P(3,4)$ is given in Table 5.

5 Concluding Remarks and Outlook

We have given a thorough treatment of planar gauge theories in which some continuous gauge group G is broken down to a finite subgroup H by means of the Higgs mechanism. One of the main points has been that the long-distance physics of such a model is governed by a quantum group based on the residual finite gauge group H, namely, the quasitriangular Hopf algebra $D(H)$ being the result of Drinfel'd's quantum double construction applied to the Abelian algebra $\mathcal{F}(H)$ of functions on the finite group H. The different particles in the spectrum, i.e., magnetic vortices, global H charges, and dyonic combinations of the two, are in one-to-one correspondence with the inequivalent unitary irreducible representations of the quantum double $D(H)$. Moreover, the algebraic framework $D(H)$ provides an unified description of the spin, braid, and fusion properties of these particles.

The implications of adding a Chern-Simons term to these spontaneously broken models have been adressed in Refs. 41–44. A review was beyond the scope of these notes. Let us just briefly summarize the main results. The distinct Chern-Simons actions S_{CS} for a compact gauge group G are known to be in one-to-one correspondence with the elements of the cohomology group $H^4(BG, \mathbb{Z})$ of the classifying space BG with integer coefficients [89]. In particular, for finite groups H, this classification boils down to the cohomology group $H^3(H, \mathrm{U}(1))$ of the group H itself with coefficients in $\mathrm{U}(1)$. In other words, the different Chern-Simons theories with finite gauge group H, in fact, correspond to the independent algebraic 3-cocycles $\omega \in H^3(H, \mathrm{U}(1))$. Now suppose that we add a Chern-Simons term $S_{\mathrm{CS}} \in H^4(BG, \mathbb{Z})$ to a planar gauge theory of the form (2.48) in which the continuous gauge group G (assumed to be simply connected for convenience) is spontaneously broken down to a finite subgroup H. Hence, the total action of the model becomes

$$S = S_{\mathrm{YMH}} + S_{\mathrm{matter}} + S_{\mathrm{CS}}.$$

It can then be shown [43] that the long-distance physics of this model is described by a Chern-Simons theory with finite gauge group H and 3-

cocycle $\omega \in H^3\big(H, \mathrm{U}(1)\big)$ determined by the original Chern-Simons action $S_{\mathrm{CS}} \in H^4(BG, \mathbb{Z})$ for the broken gauge group G through the natural homomorphism $H^4(BG, \mathbb{Z}) \to H^3\big(H, \mathrm{U}(1)\big)$ induced by the inclusion $H \subset G$. The physical picture behind this natural homomorphism, also known as the restriction, is that the Chern-Simons term S_{CS} gives rise to additional Aharonov-Bohm interactions for the magnetic vortices. These additional topological interactions are summarized by a 3-cocycle ω for the residual gauge group H, as such being "the long-distance remnant" of the Chern-Simons term S_{CS} for the broken gauge group G. Accordingly, the quantum double $D(H)$ related to the discrete H gauge theory describing the long-distance physics in the absence of a Chern-Simons term is deformed into the quasi-quantum double $D^\omega(H)$ in the presence of a Chern-Simons term S_{CS}.

For convenience, we have restricted ourselves to $(2+1)$-dimensional Minkowski space-time in these notes. For a discussion of discrete H gauge theories on higher genus spatial surfaces, i.e., surfaces with handles, the reader is referred to Ref. 79. Further, most of our observations naturally extend to the $(3 + 1)$-dimensional setting in which the magnetic vortices become stringlike objects, that is, either closed or open magnetic flux tubes.

Also, we have treated the vortices, charges, and dyons featuring in these spontaneously broken models as point particles in the first quantized description. Rerunning the discussion in the framework of canonical quantization involves the construction of magnetic vortex creation operators and charge creation operators and an analysis of their nontrivial commutation relations [100].

An outstanding question is to what extent the emergence of the quantum double $D(H)$ is particular to the case of a *local* symmetry spontaneously broken down to a finite subgroup H. This point deserves further scrutiny, for the discrete residual symmetries which do arise in condensed matter systems available for experiments (such as nematic crystals and helium-3) are *global*. In this respect, it is noteworthy that it has recently been pointed out [101] that the spontaneous breakdown of a global symmetry to a finite subgroup can lead to particles that exhibit a phenomenon called "internal frame dragging" when they are adiabatically transported around a global string. As a consequence, these particles scatter with Aharonov-Bohm–like cross sections off a global string. Something similar happens in superfluid helium-3 [102]. See also Ref. 103 in this connection. Hence, it seems that just as in the local case, the spectrum of a model with a residual global discrete symmetry group H may feature H charges that can be detected at arbitrary large distances by Aharonov-Bohm experiments with the global strings. Furthermore, the global strings labeled by the group elements of H display the global analogue of flux metamorphosis. All this suggests that also in the global case the (semiclassical) long-distance physics is governed by the quantum double $D(H)$.

To conclude, another obvious next step is to generalize the quantum double construction for finite groups to continuous groups. The quantum double related to the semidirect product group $U(1) \times_{s.d.} \mathbb{Z}_2$, for instance, may be relevant in the discussion of Alice electrodynamics. Particularly interesting in this context is the case of $(2 + 1)$-dimensional gravity. As in a discrete gauge theory, the interactions between the massive and/or spinning particles in the spectrum of $(2+1)$-dimensional gravity are purely topological, e.g., Refs. 104–106. There are indications that the algebraic structure related to this topological field theory is the quantum double based on the $(2 + 1)$-dimensional homogeneous Lorentz group $SO(2, 1)$. These matters are currently under active investigation.

Acknowledgments: These notes are based on a chapter of the Ph.D. thesis [43] of one of the authors (M. de W.P.). F.A.B. would like to thank the organizers of the CRM-CAP Banff Summer School "Particles and Fields 1994" for a most stimulating research meeting in a terrific setting. We gratefully acknowledge helpful discussions with Danny Birmingham, Hoi-Kwong Lo, Nathalie Muller, Arjan van der Sijs, Peter van Driel, and Alain Verberkmoes. We would also like to mention that the concept of truncated braid groups was developed in collaboration with Peter van Driel [90].

6   References

1. S. Coleman. Classical lumps and their quantum descendents. In *Aspects of Symmetry*. Cambridge Univ. Press, Cambridge, pages 185–264, 1985.

2. N. D. Mermin. The topological theory of defects in ordered media. *Rev. Mod. Phys.*, 51 (3): 591–648, 1979.

3. J. Preskill. Vortices and monopoles. In P. Ramond and R. Stora, eds., *Architecture of the Fundamental Interactions at Short Distances*. North-Holland, Amsterdam, pages 235–338, 1987.

4. R. Rajaraman. *Solitons and Instantons*. North-Holland, Amsterdam, 1982.

5. A. Abrikosov. On the magnetic properties of superconductors of the second group. *Sov. Phys.-JETP*, 5 (6): 1174–1182, 1957.

6. H. B. Nielsen and P. Olesen. Vortex line models for dual strings. *Nucl. Phys.*, B61 (1): 45–61, 1973.

7. G. 't Hooft. Magnetic monopoles in unified gauge theories. *Nucl. Phys.*, B79: 276–284, 1974.

8. A. M. Polyakov. Particle spectrum in quantum field theory. *JETP Lett.*, 20 (6): 194–195, 1974.

9. P. A. M. Dirac. Quantised singularities in the electromagnetic field. *Proc. Roy. Soc. London*, A133: 60–72, 1931.

10. T. H. R. Skyrme. A nonlinear field theory. *Proc. Roy. Soc.*, A260: 127–138, 1961.

11. D. Finkelstein and J. Rubinstein. Connection between spin, statistics, and kinks. *J. Math. Phys.*, 9: 1762–1779, 1968. Reprinted in Ref. 32.

12. E. Witten. Current algebra, baryons, and quark confinement. *Nucl. Phys.*, B223 (2): 433–444, 1983.

13. R. H. Brandenberger. Topological defects and structure formation. *Int. J. Mod. Phys.*, 9 (13): 2117–2189, 1994.

14. G. E. Volovik. *Exotic Properties of Superfluid ^{3}He*, volume 1 of *Series in Modern Condensed Matter Physics*. World Scientific, Singapore, 1992.

15. M. Bowick, L. Chandar, E. A. Schiff, and A. Srivastava. The cosmological Kibble mechanism in the laboratory: string formation in liquid crystals. *Science*, 263: 943–945, 1994.

16. I. Chuang, R. Durrer, N. Turok, and B. Yurke. Cosmology in the laboratory: defect dynamics in liquid crystals. *Science*, 251: 1336–1342, 1991.

17. A. P. Balachandran, G. Marmo, N. Mukunda, J. S. Nilsson, E. C. G. Sudarshan, and F. Zaccaria. Monopole topology and the problem of color. *Phys. Rev. Lett.*, 50 (20): 1553–1555, 1983.

18. P. Nelson and A. Manohar. Global color is not always defined. *Phys. Rev. Lett.*, 50 (13): 943–945, 1983.

19. P. Nelson and S. Coleman. What becomes of global color. *Nucl. Phys.*, B237 (1): 1–31, 1984.

20. F. A. Bais, P. van Driel, and M. de Wild Propitius. Quantum symmetries in discrete gauge theories. *Phys. Lett.*, B280 (1-2): 63–70, 1992.

21. A. P. Balachandran, F. Lizzi, and V. G. Rodgers. Topological symmetry breakdown in cholesterics, nematics and ^{3}He. *Phys. Rev. Lett.*, 52 (20): 1818–1821, 1984.

22. E. Witten. Dyons of charge $e\theta/(2\pi)$. *Phys. Lett.*, B86 (3): 283–287, 1979.

23. P. Hasenfratz and G. 't Hooft. Fermion-boson puzzle in a gauge theory. *Phys. Rev. Lett.*, 36 (19): 1119–1122, 1976.

24. R. Jackiw and C. Rebbi. Spin from isospin in a gauge theory. *Phys. Rev. Lett.*, 36 (19): 1116–1119, 1976.

25. F. Wilczek. Magnetic flux, angular momentum and statistics. *Phys. Rev. Lett.*, 48: 1144–1146, 1982.

26. J. M. Leinaas and J. Myrheim. On the theory of identical particles. *Nuovo Cimento*, 37B: 1–23, 1977.

27. B. I. Halperin. Statistics of quasiparticles and the hierarchy of fractional quantized Hall states. *Phys. Rev. Lett.*, 52 (18): 1583–1586, 1984.

28. R. B. Laughlin. Anomalous quantum Hall effect: An incompressible quantum fluid with fractionally charged excitations. *Phys. Rev. Lett.*, 50 (18): 1395–1398, 1983.

29. Y.-H. Chen, F. Wilczek, E. Witten, and B. I. Halperin. On anyon superconductivity. *Int. J. Mod. Phys.*, B3 (7): 1001–1067, 1989.

30. A. L. Fetter, C. B. Hanna, and R. B. Laughlin. Random-phase approximation in the fractional-statistics gas. *Phys. Rev.*, B39 (13): 9679–9681, 1989.

31. R. B. Laughlin. Superconducting ground state of noninteracting particles obeying fractional statistics. *Phys. Rev. Lett.*, 60 (25): 2677–2680, 1988.

32. F. Wilczek, ed. *Fractional Statistics and Anyon Superconductivity.* World Scientific, Teaneck, NJ, 1990.

33. G. 't Hooft. Symmetry breaking through Bell-Jackiw anomalies. *Phys. Rev. Lett.*, 37 (1): 8–11, 1976; G. 't Hooft. Computation of the quantum effects due to a four-dimensional pseudoparticle. *Phys. Rev.*, D14 (12): 3432–3450, 1976.

34. V. A. Rubakov. Superheavy magnetic monopoles and proton decay. *Pis'ma Zh. Eksp. Teor. Fiz.*, 33 (12): 658–660, 1981; V. A. Rubakov. Superheavy magnetic monopoles and proton decay. *JETP Lett.*, 33 (12): 644–646, 1981; V. A. Rubakov. Adler-Bell-Jackiw anomaly and fermion number breaking in the presence of a magnetic monopole. *Nucl. Phys.*, B203 (2): 311–348, 1982.

35. C. Callan. Dyon-fermion dynamics. *Phys. Rev.*, D26 (8): 2058–2068, 1982.

36. C. Callan. Monopole catalysis of baryon decay. *Nucl. Phys.*, B212 (3): 391–400, 1983.

37. M. G. Alford, J. March-Russell, and F. Wilczek. Enhanced baryon number violation due to cosmic strings. *Nucl. Phys.*, B328 (1): 140–158, 1989.

38. A. S. Schwarz. Field theories with no local conservation of the electric charge. *Nucl. Phys.*, B208 (1): 141–158, 1982.

39. Y. Aharonov and D. Bohm. Significance of electromagnetic potential in the quantum theory. *Phys. Rev.*, 115: 485–491, 1959.

40. F. A. Bais. Flux metamorphosis. *Nucl. Phys.*, B170 (1, FS 1): 32–43, 1980.

41. F. A. Bais, P. van Driel, and M. de Wild Propitius. Anyons in discrete gauge theories with Chern-Simons terms. *Nucl. Phys.*, B393 (3): 547–570, 1993.

42. F. A. Bais and M. de Wild Propitius. Quantum groups in the Higgs phase. *Teoret. Mat. Fiz.*, 98 (3): 509–523, 1994.

43. M. de Wild Propitius. *Topological Interactions in Broken Gauge Theories*. Ph.D. thesis, Universiteit van Amsterdam, 1995.

44. F. A. Bais, A. Morozov, and M. de Wild Propitius. Charge screeing in the Higgs phase of Chern-Simons electrodynamics. *Phys. Rev. Lett.*, 71 (15): 2383–2386, 1993.

45. T. D. Imbo and J. March-Russell. Exotic statistics on surfaces. *Phys. Lett.*, B252 (1): 84–90, 1990.

46. M. G. G. Laidlaw and C. M. DeWitt. Feynman functional integrals for systems of indistinguishable particles. *Phys. Rev.*, D3 (6): 1375–1378, 1971.

47. L. S. Schulman. *Techniques and Applications of Path Integration*. Wiley, New York, 1981.

48. L. S. Schulman. Appoximate topologies. *J. Math. Phys.*, 12 (2): 304–314, 1971.

49. F. Wilczek. Quantum mechanics of fractional-spin particles. *Phys. Rev. Lett.*, 49 (14): 957–959, 1982.

50. Y.-S. Wu. General theory for quantum statistics in two dimensions. *Phys. Rev. Lett.*, 52 (24): 2103–2106, 1984.

51. L. Brekke, A. F. Falk, S. J. Hughes, and T. D. Imbo. Anyons from bosons. *Phys. Lett.*, B271 (1): 73–78, 1991.

52. L. Brekke, H. Dijkstra, A. F. Falk, and T. D. Imbo. Novel spin and statistical properties of non-Abelian vortices. *Phys. Lett.*, B304 (1-2): 127–133, 1993.

53. L. Krauss and F. Wilczek. Discrete gauge symmetry in continuum theories. *Phys. Rev. Lett.*, 62 (11): 1221–1223, 1989.

54. J. Preskill and L. Krauss. Local discrete symmetry and quantum-mechanical hair. *Nucl. Phys.*, B341 (1): 50–100, 1990.

55. P. G. de Gennes. *Superconductivity of Metals and Alloys*. Benjamin, New York, 1966.

56. S. Forte. Quantum mechanics and field theory with fractional spin and statistics. *Rev. Mod. Phys.*, 64 (1): 193–236, 1992.

57. K. Li. Remarks on local discrete symmetry. *Nucl. Phys.*, B361 (2): 437–450, 1991.

58. M. G. Alford and J. March-Russell. Discrete gauge theories. Fractional statistics in action. *Int. J. Mod. Phys.*, B5 (16-17): 2641–2673, 1991.

59. M. G. Alford, K.-M. Lee, J. March-Russell, and J. Preskill. Quantum field theory of non-Abelian strings and vortices. *Nucl. Phys.*, B384 (1-2): 251–317, 1992.

60. M. G. Alford and J. March-Russell. New order parameters for non-Abelian gauge theories. *Nucl. Phys.*, B369 (1-2): 276–298, 1992.

61. H.-K. Lo. Aharonov-Bohm order parameters for non-Abelian gauge theories. *Phys. Rev.*, D52 (12): 7247–7264, 1995; H.-K. Lo. Order parameters for non-Abelian gauge theories. Technical Report IASSNS-HEP-94/2, hep-th/9411133, Institute for Advanced Study, 1994; H.-K. Lo. Elusive order parameters for non-Abelian gauge theories. Technical Report IASSNS-HEP-95/4, hep-th/9502079, Institute for Advanced Study, 1995.

62. M. Polikarpov, U.-J. Wiese, and M. Zubkov. String representation of the Abelian Higgs theory and Aharonov-Bohm effect on the lattice. *Phys. Lett.*, B309: 133–138, 1993.

63. G. 't Hooft. On the phase transition towards permanent quark confinement. *Nucl. Phys.*, B138 (1): 1–25, 1978; G. 't Hooft. A property of electric and magnetic flux in non-Abelian gauge theories. *Nucl. Phys.*, B153 (1-2): 141–160, 1979.

64. K. Wilson. Confinement of quarks. *Phys. Rev.*, D10 (8): 2445–2459, 1974.

65. A. M. Polyakov. Quark confinement and topology of gauge groups. *Nucl. Phys.*, B120 (3): 429–458, 1977.

66. R. F. Streater and A. S. Wightman. *PCT, Spin, and Statistics and All That.* Benjamin, New York, 1964.

67. A. P. Balachandran, A. Daughton, Z.-C. Gu, G. Marmo, R. D. Sorkin, and A. M. Srivastava. A topological spin-statistics theorem or a use of the antiparticle. *Mod. Phys. Lett.*, A5 (20): 1575–1585, 1990.

68. A. P. Balachandran, R. D. Sorkin, W. D. McGlinn, L. O'Raifeartaigh, and S. Sen. The spin-statistics connection from homology groups of configuration space and an anyon Wess-Zumino term. *Int. J. Mod. Phys.*, A7 (27): 6887–6906, 1992.

69. J. Fröhlich and P.-A. Marchetti. Spin-statistics theorem and scattering in planar quantum field theories with braid statistics. *Nucl. Phys.*, B356 (3): 533–573, 1991.

70. J. Fröhlich, F. Gabbiani, and P.-A. Marchetti. Braid statistics in three-dimensional local quantum theory. In H.-C. Lee, ed., *Physics, Geometry, and Topology*, (Banff, 1989), volume 238 of *NATO ASI*, 1990. Plenum Press, New York, pages 15–79.

71. H.-K. Lo and J. Preskill. Non-Abelian vortices and non-Abelian statistics. *Phys. Rev.*, D48 (10): 4821–4834, 1993.

72. M. G. Alford, J. March-Russell, and F. Wilczek. Discrete quantum hair on black holes and the non-Abelian Aharonov-Bohm effect. *Nucl. Phys.*, B337 (3): 695–708, 1990.

73. B. A. Ovrut. Isotropy subgroups of SO(3) and Higgs potentials. *J. Math. Phys.*, 19 (2): 418–425, 1978.

74. H.-R. Trebin. The topology of nonuniform media in condensed matter physics. *Adv. Phys.*, 31 (3): 195–254, 1982.

75. V. Poénaru and G. Toulouse. The crossing of defects in ordered media and the topology of 3-manifolds. *J. Phys.*, 38 (8): 887–895, 1977.

76. F. A. Bais and R. Laterveer. Exact regular Z_N monopole solutions in gauge theories with nonadjoint Higgs representations. *Nucl. Phys.*, B307 (3): 487–511, 1988.

77. M. Bucher. The Aharonov-Bohm effect and exotic statistics for non-Abelian vortices. *Nucl. Phys.*, B350 (1-2): 163–178, 1991.

78. M. G. Alford, S. Coleman, and J. March-Russell. Disentangling non-Abelian discrete quantum hair. *Nucl. Phys.*, B351 (3): 735–748, 1991.

79. K.-M. Lee. Vortices on higher genus surfaces. *Phys. Rev.*, D49 (4): 2030–2040, 1994.

80. M. G. Alford, K. Benson, S. Coleman, J. March-Russell, and F. Wilczek. Interactions and excitations of non-Abelian vortices. *Phys. Rev. Lett.*, 64 (14): 1623–1635, 1990; M. G. Alford, K. Benson, S. Coleman, J. March-Russell, and F. Wilczek. Zero modes of non-Abelian vortices. *Nucl. Phys.*, B349 (2): 414–438, 1991.

81. L. Alvarez-Gaumé, C. Gomez, and G. Sierra. Hidden quantum symmetries in rational conformal field theories. *Nucl. Phys.*, B319 (1): 155–186, 1989.

82. L. Alvarez-Gaumé, C. Gomez, and G. Sierra. Duality and quantum groups. *Nucl. Phys.*, B330 (2-3): 347–398, 1990.

83. E. Witten. Quantum field theory and the Jones polynomials. *Commun. Math. Phys.*, 121 (3): 351–399, 1989.

84. V. G. Drinfel'd. Quantum groups. In *Proceedings of the International Congress of Mathematicians*, (Berkeley, 1986), 1987. Amer. Math. Soc. , Providence, RI, pages 798–820.

85. V. G. Drinfel'd. Quasi-Hopf algebras and Knizhnik-Zamolodchikov equations. In *Problems of Modern Quantum Field Theory*, (Alushta, 1989), 1989. Springer, Berlin, pages 1–13.

86. S. Shnider and S. Sternberg. *Quantum groups. From Coalgebras to Drinfel'd Algebras. A Guided Tour*, volume 2 of *Graduate Texts in Mathematical Physics*. International Press, Cambridge, MA, 1993.

87. R. Dijkgraaf, V. Pasquier, and P. Roche. Quasi Hopf algebras, group cohomology and orbifold models. In *Recent Advances in Field Theory*, (Annecy-le-Vieux, 1990), volume 18B of *Nuclear Phys. B. Proc. Suppl.*, 1991. North-Holland, Amsterdam, pages 60–72.

88. R. Dijkgraaf, C. Vafa, E. Verlinde, and H. Verlinde. The operator algebra of orbifold models. *Commun. Math. Phys.*, 123 (3): 485–526, 1989.

89. R. Dijkgraaf and E. Witten. Topological gauge theories and group cohomology. *Commun. Math. Phys.*, 129 (2): 393–429, 1990.

90. P. van Driel and M. de Wild Propitius. Truncated braid groups. unpublished, 1990.

91. E. Verlinde. Fusion rules and modular transformations in 2d conformal field theory. *Nucl. Phys.*, B300 (3): 360–376, 1988.

92. G. Moore and N. Seiberg. Classical and quantum conformal field theory. *Commun. Math. Phys.*, 123 (2): 177–254, 1989.

93. A. Einstein, B. Podolsky, and N. Rosen. Can quantum-mechanical description of physical reality be considered complete? *Phys. Rev.*, 47 (1): 777–780, 1935.

94. L. Carroll. *Alice's Adventures in Wonderland*. Macmillan, London, 1865.

95. E. Verlinde. A note on braid statistics and the non-Abelian Aharonov-Bohm effect. In S. Das et al., eds., *Modern Quantum Field Theory*, (Bombay, 1990), 1991. World Scientific, River Edge, NJ, pages 450–461.

96. C. C. Adams. *The Knot Book: An Elementary Introduction to the Mathematical Theory of Knots*. Freeman, New York, 1994.

97. L. H. Kauffman. *Knots and Physics*. World Scientific, Singapore, 1991.

98. M. Peshkin and A. Tonomura. *The Aharonov-Bohm Effect*, volume 340 of *Lecture Notes in Physics*. Springer-Verlag, Berlin-New York, 1989.

99. A. D. Thomas and G. V. Wood. *Group Tables*, volume 2 of *Shiva Mathematics Series*. Shiva Publishing Ltd., Nantwich, 1980.

100. F. A. Bais, A. Morozov, and M. de Wild Propitius. In preparation.

101. J. March-Russell, J. Preskill, and F. Wilczek. Internal frame dragging and a global analog of the Aharonov-Bohm effect. *Phys. Rev. Lett.*, 68 (17): 2567–2571, 1992.

102. M. V. Khazan. Analog of the Aharonov-Bohm effect in superfluid He^3-A. *Pis'ma Zh. Eksp. Teor. Fiz.*, 41 (9): 396–398, 1985; M. V. Khazan. Analog of the Aharonov-Bohm effect in superfluid He^3-A. *JETP Lett.*, 41 (9): 486–488, 1985.

103. A. C. Davis and A. P. Martin. Global string and the Aharonov-Bohm effect. *Nucl. Phys.*, B419: 341–351, 1994.

104. S. Deser and R. Jackiw. Classical and quantum scattering on a cone. *Commun. Math. Phys.*, 118 (3): 495–509, 1988.

105. G. 't Hooft. Nonperturbative 2 particle scattering amplitudes in (2 + 1)-dimensional quantum gravity. *Commun. Math. Phys.*, 117 (4): 685–700, 1988.

106. E. Witten. (2 + 1)-dimensional gravity as an exactly soluble system. *Nucl. Phys.*, B311 (1): 46–78, 1988/89.

9

Quantum Hall Fluids as $W_{1+\infty}$ Minimal Models

Andrea Cappelli, Carlo A. Trugenberger, and Guillermo R. Zemba

ABSTRACT We review our recent work on the algebraic characterization of quantum Hall fluids. Specifically, we explain how the incompressible quantum fluid ground states can be classified by effective edge field theories with the $W_{1+\infty}$ dynamical symmetry of "quantum area-preserving diffeomorphisms." Using the representation theory of $W_{1+\infty}$, we show how all fluids with filling factors $\nu = m/(pm+1)$ and $\nu = m/(pm-1)$ with m and p positive integers and p even, correspond exactly to the $W_{1+\infty}$ minimal models.

1 Introduction

The quantum Hall effect provides fascinating examples of *quantum fluids*; for a review see Ref. 1. At low temperatures, interacting planar electrons in high magnetic fields B have strong quantum correlations which lead to collective motion and macroscopic quantum effects. These find their experimental evidence in a discrete series of plateaus at rational values of the Hall conductivity:

$$\sigma_{xy} = \frac{e^2}{h}\nu, \quad \nu = 1, \frac{1}{3}, \frac{1}{5}, \frac{2}{7}, \ldots, 2, \ldots . \tag{1.1}$$

Corresponding to these plateaus the longitudinal conductivity σ_{xx} vanishes. The same plateaus are observed in several materials, signalling *universality*. Another experimental result is the remarkable *exactness* of these rational values of ν; the experimental error is $\Delta\nu = 10^{-8}$ for integer ν.

The current understanding of the quantum Hall effect is based on the seminal work of Laughlin [2]; for a review see Ref. 3. The main idea is the existence of *incompressible quantum fluids* at specific rational values $\bar{\rho} = \nu B/2\pi$ ($\hbar = 1$, $c = 1$) of the electron density. These are very stable, macroscopical quantum states with uniform density and an energy gap. Incompressibility accounts for the lack of low-lying conduction modes, which causes σ_{xx} to vanish, while the Hall conduction is realized as an overall rigid motion of the uniform droplet, which gives eq. (1.1).

While Laughlin's theory is very successful, the observed exactness and universality calls for a *fundamental principle* underlying it. Indeed, the *universality* observed in experiments calls for an effective theory approach at long distances, while the extreme precision of the rational values of ν suggests that dynamics is constrained by *symmetry*. Both facts suggest an analogy with two-dimensional critical phenomena, which are classified by conformal field theories [4].

The effective field theory approach, developed by Landau, Ginsburg, Wilson, and others—see, for example Ref. 5—does not attempt to solve the microscopic many-body dynamics, but rather it guesses the macroscopic physics generated by this dynamics. The variables of the effective field theory are the relevant low-energy (long-distance) degrees of freedom, which are characterized by a specific symmetry. They describe *universal* properties, which are independent of the microscopic details. This approach is well suited for the quantum Hall effect, given the *very precise* and *universal* values of the Hall conductivity.

In the following, we shall present an overview of our [6–12] and related [13, 14] recent work on the effective field theory approach to the quantum Hall effect. This is based on an *algebraic characterization of incompressible fluids*.

2   Dynamical Symmetry and Kinematics of Incompressible Fluids

In this section, we shall review the dynamical symmetry characterizing *chiral*, two-dimensional, incompressible fluids and indicate how this leads uniquely to the construction of the Hilbert spaces of low-energy excitations.

2.1   Classical Fluids

A classical incompressible fluid is defined by its distribution function

$$\rho(z, \bar{z}, t) = \rho_0 \chi_{S_A(t)}, \quad \rho_0 \equiv \frac{N}{A},$$

where $\chi_{S_A(t)}$ is the characteristic function for a surface $S_A(t)$ of area A, and $z = x + iy$, $\bar{z} = x - iy$ are complex coordinates on the plane. Since the particle number N and the average density ρ_0 are constant, the area A is also *constant*. The only possible change in response to external forces is in the shape of the surface. The shape changes at constant area can be generated by *area-preserving diffeomorphisms* of the two-dimensional plane. Thus, the configuration space of a classical incompressible fluid can be generated by applying these transformations to a reference droplet.

Next we recall the Liouville theorem, which states that canonical transformations preserve the phase-space volume. Area-preserving diffeomorphisms are, therefore, canonical transformations of a two-dimensional phase space. In order to use the formalism of canonical transformations, we treat the original coordinate plane as a *phase space*, by postulating nonvanishing Poisson brackets between z and $\bar{z}$. We do this by defining the dimensionless Poisson brackets

$$\{f, g\} \equiv \frac{i}{\rho_0}(\partial f \bar{\partial} g - \bar{\partial} f \partial g), \tag{2.1}$$

where $\partial \equiv \partial/\partial z$ and $\bar{\partial} \equiv \partial/\partial \bar{z}$, so that

$$\{z, \bar{z}\} = \frac{i}{\rho_0}.$$

Note that the Poisson brackets select a preferred *chirality*, because they are not invariant under the two-dimensional parity transformation $z \to \bar{z}$, $\bar{z} \to z$; in the quantum Hall effect, the parity breaking is due to the external magnetic field.

Area-preserving diffeomorphisms, i.e., canonical transformations, are usually defined in terms of a generating function $\mathcal{L}(z, \bar{z})$ of both "coordinate" and "momentum," as follows:

$$\delta z = \{\mathcal{L}, z\}, \quad \delta \bar{z} = \{\mathcal{L}, \bar{z}\}.$$

A basis of (dimensionless) generators is given by

$$\mathcal{L}_{n,m}^{(cl)} \equiv \rho_0^{(n+m)/2} z^n \bar{z}^m.$$

These satisfy the classical w_∞ algebra [15]

$$\{\mathcal{L}_{n,m}^{(cl)}, \mathcal{L}_{k,l}^{(cl)}\} = -i(mk - nl)\mathcal{L}_{n+k-1,m+l-1}^{(cl)}. \tag{2.2}$$

Let us now discuss how w_∞ transformations can be used to generate the configuration space of classical excitations above the ground state. These configurations have a classical energy due to the interparticle interaction and the external confining potential, whose specific form is not needed here. Let us assume a generic convex and rotation-invariant energy function, such that the minimal energy configuration ρ_{GS} has the shape of a disk of radius R:

$$\rho_{GS}(z, \bar{z}) = \rho_0 \Theta(R^2 - z\bar{z}),$$

where Θ is the Heaviside step function. The classical "small excitations" around this ground-state configuration are given by the infinitesimal deformations of ρ_{GS} under area-preserving diffeomorphisms,

$$\delta\rho_{n,m} \equiv \{\mathcal{L}_{n,m}^{(cl)}, \rho_{GS}\}. \tag{2.3}$$

Using the Poisson brackets (2.1), we obtain

$$\delta\rho_{n,m} = i(\rho_0 R^2)^{(n+m)/2}(m-n)e^{i(n-m)\theta}\delta(R^2 - z\bar{z}).$$

These correspond to density fluctuations localized on the sharp boundary (which is parametrized by the angle θ) of the classical droplet. Due to the dynamics provided by the energy function, they will propagate on the boundary with a frequency ω_k dependent on the angular momentum $k \equiv (n-m)$, thereby turning into *edge waves*. These are the eigenoscillations of the classical incompressible fluid.

Another type of excitations are classical vortices in the bulk of the droplet, which correspond to localized holes or dips in the density. The absence of density waves, due to incompressibility, implies that any localized density excess or defect is transmitted completely to the boundary, where it is seen as a further edge deformation. For each given vorticity in the bulk, we can then construct the corresponding basis of edge waves in a fashion analogous to (2.3). Thus, the configuration space of the excitations of a classical incompressible fluid (of a given vorticity) is spanned by infinitesimal w_∞ transformations. This is the *dynamical symmetry* of classical incompressible fluids.

2.2   Quantum Fluids and Their Edge Excitations

The quantum[1] version of the chiral, incompressible fluids is given by the Laughlin theory of the plateaus of the quantum Hall effect [2, 3]. The simplest example of such a macroscopic quantum state is a fully filled Landau level (filling fraction $\nu = 1$). Generically, it possesses three types of excitations. First, there are *gapless edge excitations*, which are the quantum descendants of the classical edge waves described before; for a review see Ref. 16. These are particle-hole excitations across the Fermi surface represented by the edge of the droplet of radius R [17]; therefore, they are called *neutral*. They are *gapless* because their energy, of $O(1/R)$, vanishes for $R \to \infty$. Second, there are localized quasi-particle and quasi-hole excitations, which have a finite gap. These are the quantum analogs of the classical vortices and correspond to the anyon excitations with fractional charge, spin, and statistics [2, 3]. As in the classical case, they manifest themselves as charged excitations at the edge, owing to incompressibility. The third type of excitations are two-dimensional density waves in the bulk, the magnetoplasmons, and (for $\nu < 1$) the magnetophonons [18]. These have higher gaps and are not included in our effective field theory approach.

In the previous section, we explained the connection between the classical edge waves and the generators of the algebra w_∞ of area-preserving

---

[1]Throughout this paper we shall use units such that $c = 1$, $\hbar = 1$.

diffeomorphisms. In the quantum theory, there is a corresponding relation between edge excitations and the generators of the quantum version of w_∞, called $W_{1+\infty}$ [15]. This algebra is obtained by replacing the Poisson brackets (2.1) with quantum commutators: $i\{\ ,\ \} \to [\ ,\ ]$, and by taking the thermodynamic limit [7].

In this limit, the radius of the droplet grows as $R \propto \ell\sqrt{N}$, where $\ell = \sqrt{2/(eB)}$ is the magnetic length and B the magnetic field. Quantum edge excitations, instead, are confined to a boundary annulus of finite size $O(\ell)$. In the $N \to \infty$ limit, therefore, edge excitations become the particle-hole excitations of a relativistic theory describing a Weyl (chiral) fermion living on the one-dimensional edge of the droplet [7]. In this limit, the quantum incompressible fluid becomes the Dirac sea for this relativistic theory. Charged fermions represent instead quasi-particle excitations.

The field operator for the Weyl fermion[2] is given by

$$F_R(\theta) = \frac{1}{\sqrt{R}} \sum_{k=-\infty}^{\infty} e^{i(k-1/2)\theta} b_k \quad (|z| = R, t = 0),$$

where θ parametrizes the circular boundary, b_k and $b_k^\dagger$ are fermionic Fock space operators satisfying $\{b_l, b_k^\dagger\} = \delta_{l,k}$, and k is the angular momentum measured with respect to the ground-state value.

The generators of the quantum algebra $W_{1+\infty}$ are represented in this Fock space by the bilinears

$$V_n^j = \int_0^{2\pi} \frac{d\theta}{2\pi} :F^\dagger(\theta)e^{-in\theta}g_n^j(i\partial_\theta)F(\theta):$$

$$= \sum_{k=-\infty}^{\infty} p(k,n,j):b_{k-n}^\dagger b_k:, \quad j \geq 0. \qquad (2.4)$$

In this expression, $F(\theta) = F_R(\theta)e^{i\theta/2}\sqrt{R}$ is the canonical form of the Weyl field operator of conformal field theory. The factor $g_n^j(i\partial_\theta)$ is a jth-order polynomial in $i\partial_\theta$, whose form specifies the basis of operators and guarantees the hermiticity $(V_n^j)^\dagger = V_{-n}^j$. The coefficients $p(k,n,j)$ are also jth-order polynomials in k which we do not need to specify here (see Ref. 11). The $W_{1+\infty}$ algebra reads

$$[V_n^i, V_m^j] = (jn - im)V_{n+m}^{i+j-1} + q(i,j,m,n)V_{n+m}^{i+j-3}$$
$$+ \cdots + c^i(n)\delta^{i,j}\delta_{n+m,0}. \qquad (2.5)$$

Here, $i + 1 = h \geq 1$ represents the "conformal spin" of the generator V_n^i, while $-\infty < n < +\infty$ is the angular momentum (the Fourier mode on the

---

[2]Hereafter, we choose units such that $\ell = 1$.

circle). The first term on the right-hand-side of (2.5) reproduces the classical w_∞ algebra (2.2) by the correspondence $\mathcal{L}_{i-n,i}^{(cl)} \to V_n^i$ and identifies $W_{1+\infty}$ as the algebra of "quantum area-preserving diffeomorphisms." The additional terms are quantum operator corrections with polynomial coefficients $q(i,j,n,m)$, due to the algebra of higher derivatives [15]. Moreover, the c-number term $c^i(n)$ is the quantum *anomaly*, a relativistic effect due to the renormalization of operators acting on the infinite Dirac sea. It is diagonal in the spin indices for our choice of basis for the g_k^i (see Ref. 11). Finally, the normal ordering (: :) of the Fock operators takes care of the renormalization [4].

Let us analyze the generators V_n^0 and V_n^1 of lowest conformal spin. From (2.4) we see that the V_n^0 are Fourier modes of the fermion density evaluated at the edge $|z| = R$; thus, V_0^0 measures the edge charge. Instead, the V_n^1 are vector fields which generate angular momentum transformations on the edge, such that V_0^1 measures the angular momentum of edge excitations. Their algebra is given by

$$[V_n^0, V_m^0] = cn\delta_{n+m,0}, \qquad (2.6)$$

and

$$
\begin{aligned}
&[V_n^1, V_m^0] = -mV_{n+m}^0, \\
&[V_n^1, V_m^1] = (n-m)V_{n+m}^1 + \frac{c}{12}(n^3 - n)\delta_{n+m,0},
\end{aligned}
\qquad (2.7)
$$

with $c = 1$. These equations show that the V_n^0 and V_n^1 operators satisfy an Abelian Kac-Moody algebra and a Virasoro algebra, respectively [4]. For unitary $W_{1+\infty}$ theories, the central charge c can be any *positive integer* [19, 20].

Following the standard procedure of two-dimensional conformal field theory, we define a $W_{1+\infty}$ theory as the Hilbert space given by a set of irreducible, highest-weight representations of the $W_{1+\infty}$ algebra, closed under the *fusion rules* for making composite excitations [4]. Any representation contains an infinite number of states, corresponding to all the neutral excitations above a bottom state, the so-called *highest weight* state. This can be, for example, the ground-state $|\Omega\rangle$ corresponding to the incompressible quantum fluid. The excitations can be written as

$$
|k, \{n_1, n_2, \ldots, n_s\}\rangle = V_{-n_1}^0 V_{-n_2}^0 \cdots V_{-n_s}^0 |\Omega\rangle,
$$
$$
n_1 \geq n_2 \geq \cdots \geq n_s > 0, \quad (2.8)
$$

while the positive modes ($n_i < 0$) annihilate $|\Omega\rangle$. Here $k = \sum_j n_j$ is the total angular momentum of the edge excitation.

Moreover, any charged edge excitation, together with its tower of neutral excitations, also forms an irreducible, highest-weight representation of $W_{1+\infty}$. The states in this representation have the same form of (2.8), but the bottom state $|Q\rangle$ now represents a quasi-particle inside the droplet. The charge and spin of the quasi-particle are given by the eigenvalues of the operators[3] $(-V_0^0)$ and V_0^1, respectively:

$$V_0^0|Q\rangle = Q|Q\rangle, \quad V_0^1|Q\rangle = J|Q\rangle.$$

Actually, all the operators V_0^i are simultaneously diagonal and assign other quantum numbers to the quasi-particle, $V_0^i|Q\rangle = m_i(Q)|Q\rangle$, $i \geq 2$, which are known polynomials in the charge Q (see Ref. 11). These quantum numbers measure the radial moments of the charge distribution of a quasi-particle; their fixed functional form indicates the rigidity of density modulations of the quantum incompressible fluid.

Besides the explicit example for $\nu = 1$, leading to a theory with $c = 1$, it has been shown in general that the algebra (2.5) is the unique quantization of the w_∞ algebra in the $(1+1)$-dimensional field theory on the circle [21]. We shall therefore characterize quantum incompressible fluids as $W_{1+\infty}$ theories.

2.3 Classification of QHE Universality Classes

This characterization provides a powerful classification scheme for quantum Hall universality classes. These can in fact be classified by using the recently developed representation theory [19, 20] of $W_{1+\infty}$. We shall classify quantum Hall universality classes by the following *kinematical data*:

(i) the quantum numbers V_0^i of quasi-particle excitations;

(ii) the number of neutral edge excitations of given angular momentum, i.e., the *degeneracies* of the states (2.8) at fixed k;

(iii) the filling fraction of the ground state.

The filling fraction can be computed by identifying the *chiral anomaly* of the (1+1)-dimensional $W_{1+\infty}$ theory with the Hall current produced by an electric field [7].

3 Existing Theories of Edge Excitations and Experiments

Before developing this classification program, we would like to briefly review the existing theories of the quantum Hall effect and discuss their description of the experimental data.

---

[3]The minus sign is due to the fact that V_0^0 measures the charge on the edge. Due to overall charge conservation, the charge of a quasi-particle in the bulk has the opposite sign of its edge counterpart.

3.1 Hierarchical Trial Wave Functions

The Laughlin theory of the incompressible fluid [2, 3] was originally developed for the Hall conductivities $\sigma_{xy} = (e^2/h)\nu$, where $\nu = 1, \frac{1}{3}, \frac{1}{5}, \frac{1}{7}, \ldots$ are the filling fractions. Afterward, a hierarchical generalization of these trial wave functions was introduced by Haldane [22] and by Halperin [23], in order to describe other observed filling fractions. Therefore, by the *hierarchy* problem we usually mean the classification of stable ground states (and their excitations) corresponding to all observed plateaus. Naturally, the stability is related to the order of iteration of the hierarchical construction, starting from the integer fillings, then the Laughlin fillings, and so forth.

The Haldane-Halperin hierarchy is not completely satisfactory, because it produces ground states for too many filling fractions, already at low order of iteration. On the contrary, the experiments show only some stable ground states (see Figure 1). Although numerical experiments show that the hierarchical wave functions are rather accurate, their construction lacks a good control of stability.

Another hierarchical construction of wave functions, which match most of the experimental plateaus to lowest order of the hierarchy, has been proposed by Jain; for a review see Ref. 25. Jain abstracted from Laughlin's work the concept of *composite fermion*, a local bound state of the electron and an even number of flux quanta. Due to yet unknown dynamical reasons, the composite fermions are stable quasi-particles, which interact weakly among themselves. Moreover, the strongly interacting electrons at fractional filling can be mapped into composite fermions at effective integer filling. Therefore, the stability of the observed ground states with fractional filling can be related with the stability of completely filled Landau levels. The composite fermion picture was successfully applied [26] to the independent dynamics of the compressible fluid at $\nu = \frac{1}{2}$. This strongly interacting, gapless ground state can be described as a Fermi liquid of composite fermions with vanishing effective magnetic field. Experiments [27] have confirmed this theory by observing the free motion of the composite fermions.

3.2 The Chiral Boson Theory of the Edge Excitations

After the original works of Halperin [28] and Stone [17], a general theory of edge excitations, corresponding to the hierarchical constructions of wave functions, has been formulated [16, 29, 30]. This is the $(1 + 1)$-dimensional theory of the chiral boson [31]. An equivalent description is given by Abelian Chern-Simons theories on $(2 + 1)$-dimensional open domains [30]. The edge excitations of the Laughlin fluid are described by a one-component chiral boson, while the hierarchical fluids require many components. Every boson describes an independent edge current, and thus the incompressible fluids

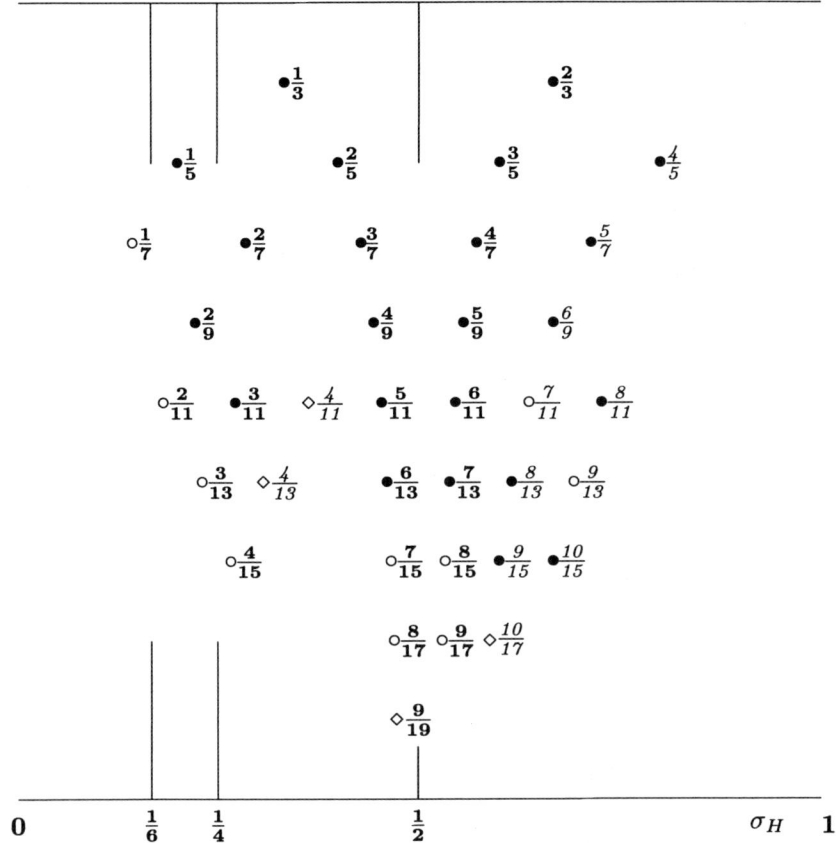

FIGURE 1. Experimentally observed plateaus in the range $0 < \sigma_H < 1$: their Hall conductivity $\sigma_H = (e^2/h)\nu$ is displayed in units of (e^2/h). The points denote stability: (•) very stable, (o) stable, and (⋄) less stable plateaus. Theoretically understood plateaus are in **bold**, unexplained ones are in *italic*. Observed cases of coexisting fluids are displayed as $\nu = 2/3$, $6/9$, $10/15$, $\nu = 3/5$, $9/15$, and $\nu = 5/7$, $15/21$ (but $15/21$ is not displayed). (Adapted from Ref. 24)

have generically a composite edge structure. Each current gives rise to the Abelian current algebra in $(1 + 1)$ dimensions, denoted by $\widehat{U(1)}$, which implies the Virasoro algebra with central charge $c = 1$ [4].

On an annulus geometry, with edge circles $|\mathbf{x}| = R_1$ and $|\mathbf{x}| = R_2$, one introduces m independent one-dimensional *chiral* currents

$$J^i(R_1\theta - v_i t) = -\frac{1}{2\pi R_1}\frac{\partial}{\partial\theta}\phi^i \quad (|\mathbf{x}| = R_1),$$

and corresponding ones with opposite chirality $J^i(R_2\theta + v_i t)$ at the other edge $|\mathbf{x}| = R_2$. The dynamics of these currents on the edge circle $|\mathbf{x}| = R_1$

is governed by the action,

$$S = -\frac{1}{\pi} \int dt\, dx \sum_{i=1}^{m} \kappa_i (\partial_t \phi^i + v_i \partial_x \phi^i) \partial_x \phi^i, \quad \text{for } x \equiv R_1 \theta, \qquad (3.1)$$

for the m $(1+1)$-dimensional *chiral boson* fields ϕ^i [31]. The corresponding action for the other circle $x \equiv R_2 \theta$ is obtained by replacing $v_i \rightarrow (-v_i)$. The dynamics on the two edges are identical and independent, only constrained by the conservation of the total charge: thus we describe one of them only. We can change the normalization of the fields, and reduce each coupling constant to a sign, $\kappa_i \rightarrow \pm 1$. The equations of motion imply that the fields are chiral, $\phi^i = \phi^i(x - v_i t)$, and canonical quantization implies the following commutation relations for the currents,

$$[J^i(x_1), J^k(x_2)] = \frac{1}{2\pi\kappa_i} \delta^{ik} \delta'(x_1 - x_2) \quad (t_1 = t_2), \qquad (3.2)$$

which are those of the multicomponent Abelian current algebra $\widehat{U(1)}^{\otimes m}$ [4]. The positive definiteness of the Hamiltonian requires the signs of the velocities v_i and the couplings κ_i to be related: $v_i \kappa_i > 0$, $i = 1, \ldots, m$.

Let us discuss one particular chiral current, $v_i > 0$ (i.e., $\kappa_i = 1$). The quantization of the chiral boson is equivalent to the construction of the representations of the current algebra (3.2)). Actually, all the states in the Hilbert space of the theory can be fitted into a set of representations [4]. To this end, we introduce the Fourier modes of the currents,

$$J^i(R\theta - v_i t) = \frac{1}{2\pi} \sum_{n=-\infty}^{\infty} \alpha_n^i e^{in(\theta - v_i t)},$$

which satisfy,

$$[\alpha_n^i, \alpha_m^j] = \delta^{ij} \frac{n}{\kappa_i} \delta_{n+m,0}. \qquad (3.3)$$

The positivity of the ground-state expectation value $\langle \Omega | \alpha_n^i \alpha_{-n}^i | \Omega \rangle \equiv \| \alpha_n^i | \Omega \rangle \|^2 \geq 0$, and the commutation relations (3.3) with $\kappa_i = 1$ imply the conditions

$$\alpha_n^i | \Omega \rangle = 0, \quad n > 0 \quad (v_i > 0).$$

An irreducible highest-weight representation of the $\widehat{U(1)}$ current algebra is made by the highest-weight state $|\Omega\rangle$ and by all states obtained by applying any number of α_n^i, $n < 0$, to it. The weight of the representation is given by the eigenvalue of α_0^i, which is the single-edge charge, in units to be specified later. For the ground state, we have

$$\alpha_0^i | \Omega \rangle = 0.$$

Other unitary highest-weight representations can be similarly built on top of other highest-weight states $|r\rangle$, $r \in \mathrm{R}$, which satisfy

$$\alpha_n^i|r\rangle = 0 \quad n > 0, \qquad \alpha_0^i|r\rangle = r|r\rangle,$$

and are built by applying the *vertex operators* to the ground state [4]. These representations correspond to the quasi-particle excitations of this edge theory. The Virasoro generators are defined by the Sugawara construction [4],

$$L_n^i = \frac{\kappa_i}{2} \sum_{l=-\infty}^{\infty} :\alpha_{n-l}^i \alpha_l^i:.$$

They give rise to the Virasoro algebra (2.7) with $c = 1$, for each current component i.

The m-edge theory has $\widehat{\mathrm{U}(1)}^{\otimes m}$ symmetry, $c = m$, and is parametrized by an integer, symmetric $(m \times m)$ matrix, with odd diagonal elements, the K matrix [30]. This determines the Hall conductivity and the fractional charge, spin and statistics [32] of the edge excitations, which correspond to the anyon quasi-particles of the incompressible fluid [2, 3].

In the chiral boson theory, the fusion rules are the addition of weight vectors $\vec{r}$; the set of representations which is closed under these rules is the *lattice* Γ,

$$\Gamma = \left\{ \vec{r} \,\middle|\, \vec{r} = \sum_{i=1}^{m} n_i \vec{v}_i, n_i \in \mathbb{Z} \right\}. \tag{3.4}$$

The basis vectors $\vec{v}_i$ represent a physical elementary excitation in the ith edge component, which may not correspond to the previous basis of propagating modes. The *physical charge* of an excitation with labels $n_i \in \mathbb{Z}$ is thus given by the sum of the components in the physical basis [12],

$$Q = \sum_{i,j=1}^{m} K_{ij}^{-1} n_j, \tag{3.5}$$

where

$$K_{ij}^{-1} = \sum_{l=1}^{m} \Lambda_{il} \frac{1}{\kappa_l} \Lambda_{lj}^T = (\vec{v}_i \cdot \eta \cdot \vec{v}_j).$$

Similarly, the fractional spin and statistics of this excitation is given by the sum of Virasoro eigenvalues L_0^i,

$$\frac{\theta}{\pi} = \sum_{i,j=1}^{m} n_i K_{ij}^{-1} n_j, \quad n_i \in \mathbb{Z}. \tag{3.6}$$

In general, the metric K^{-1} of the lattice Γ in the basis $\vec{v}_i$ is pseudo-Euclidean with signature $\eta_{ij} = \delta_{ij}\kappa_i$, due to the possible presence of excitations with different chiralities.

The Hall conductivity in the annulus geometry can be measured by applying a uniform electric field along all the edges, $E^i = E$. The chiral anomaly of the edge theory actually corresponds to a radial flow of particles in the annulus, which move from the inner edge to the outer edge. The Hall conductivity can be thus found to be [7],

$$\sigma_H = \frac{e^2}{h}\nu, \quad \nu = \sum_{i,j=1}^{m} K_{ij}^{-1}. \tag{3.7}$$

Equations (3.5)–(3.7) for the Hall conductivity and the spectrum of the charge and fractional statistics of edge excitations are the basic data of the quantum incompressible fluid described by the chiral boson theories [16, 30]. The existence of m electron excitations with unit charge and integer statistic relative to all excitations, requires that K has integer entries with odd integers on the diagonal [30].

3.3   The Jain Hierarchy

The Jain fluids have been described by the subset of the chiral boson theories characterized by the following K matrices [30],

$$K_{ij} = \pm\delta_{ij} + p\, C_{ij}, \quad C_{ij} = 1\,\forall i,j = 1,\ldots,m, \ p > 0 \text{ even}, \tag{3.8}$$

and the following spectra of edge excitations (eqs. (3.6), (3.7)),

$$\nu = \frac{m}{mp \pm 1}, \quad p > 0 \text{ even}, (c = m),$$

$$Q = \frac{1}{pm \pm 1}\sum_{i=1}^{m} n_i, \quad \frac{\theta}{\pi} = \pm\left(\sum_{i=1}^{m} n_i^2 - \frac{p}{mp \pm 1}\left(\sum_{i=1}^{m} n_i\right)^2\right). \tag{3.9}$$

Note that K has $(m-1)$ degenerate eigenvalues $\lambda_i = 1$, $i = 1, \ldots, m-1$ (resp. $\lambda_i = -1$), and a single-value $\lambda_m = \pm 1 + mp$. If the sign $\pm$ is negative, one edge has opposite chirality to the others. There is one basic charged quasi-particle excitation with label $n_i = (1,\ldots,1)$ and $m(m-1)/2$ *neutral* excitations for $n_i = (\delta_{ik} - \delta_{il})$, $1 \le k < l \le m$, with identical integer statistics.

The corresponding trial wave functions for the ground state have been constructed by Jain as [25]

$$\Psi_\nu = D^{p/2}L^m \mathbf{1}, \quad p \text{ even}, \tag{3.10}$$

where $L^m\mathbf{1}$ represents schematically the wave function of m filled Landau levels and $D^{p/2}$ multiplies the wave function by the pth power of the Vandermonde determinant, which "attaches p flux tubes to each electron," and transforms them into "composite fermions." This construction has been implemented in the multicomponent chiral boson theory (3.1) in Refs. 29 and 30.

The Jain hierarchy covers most of the experimentally observed plateaus, as we discuss in the next section. However, within the chiral boson approach, there is no clear motivation for selecting the special K matrices (3.8). The size of the gap for bulk density waves is usually invoked for solving this puzzle: the observed fluids are supposed to have the largest gaps, while the general K fluids have small gaps and are destroyed by thermal fluctuations and other effects. It is also true that edge theories give kinematically possible incompressible fluids and their universal properties, but cannot describe the size of the gaps, which is determined by the microscopic bulk dynamics.

Nevertheless, we have found a natural way to select the Jain hierarchy within the $W_{1+\infty}$ edge theory approach [12]. Indeed, the Jain fluids correspond to the $W_{1+\infty}$ minimal models, which are characterized by possessing less states than their chiral boson counterparts. We propose this reduction of available states as a natural stability principle.

3.4 Experiments

We first discuss the spectrum of fractional Hall conductivities in eq. (3.9). According to Jain, the stability of the ground states (3.10) should be approximately independent of m, which counts the number of Landau levels filled by the composite fermion. This is in analogy with the integer Hall plateaus, which are all equally stable. On the other hand, the stability decreases by increasing $|p|$, as observed for the Laughlin fluids ($m = 1$). Therefore, the most stable family of plateaus is,

$$\nu = \frac{m}{2m \pm 1} \quad (p = 2), \tag{3.11}$$

which accumulate at $\nu = \frac{1}{2}$. The next less stable family is

$$\nu = \frac{m}{4m \pm 1} \quad (p = 4), \tag{3.12}$$

which accumulate at $\nu = 1/4$. This behavior is clearly seen in the experimental data of Figure 1. For these filling fractions, the Jain wave functions for the ground state and the simplest excited states have a good overlap with those obtained numerically by diagonalizing the microscopic Hamiltonian with a small number of electrons.

A closer look into the experimental values of the filling fractions in Figure 1 shows other points (in italic), like $\nu = 4/5$, $5/7$, $8/11$, which fall outside the Jain main series ((3.11), (3.12)) (in bold). These points were originally interpreted as "charge conjugates" of these series [25],

$$\nu = 1 - \frac{m}{2m \pm 1}, \quad \nu = 1 - \frac{m}{4m \pm 1}. \tag{3.13}$$

A charge-conjugated fluid is a fluid of holes in a ($\nu = 1$) electron fluid. The corresponding K matrix is easily obtained as the $((m+1) \times (m+1))$-dimensional matrix [30],

$$\overline{K} = \begin{pmatrix} 1 & 0 \\ 0 & -K \end{pmatrix}.$$

These charge-conjugate models actually belong to the second iteration of the Jain hierarchy [25]. Unfortunately, the charge conjugate states do not fit well the data in Figure 1. The $\nu = 1/2$ family would be self-conjugate; thus there should be two fluids per filling fraction, which are not observed, apart from two cases. Actually, coexisting fluids can be detected by experiments where the magnetic field is tilted from the orthogonal direction to the plane [33].

Furthermore, the conjugate of the observed fractions in the $\nu = \frac{1}{4}$ family are not observed in half of the cases. Finally, there are fractions which do not belong to any previous group: $\nu = 4/11, 7/11, 4/13, 8/13, 9/13, 10/17$.

In conclusion, all the fractions outside the main Jain families (3.11), (3.12) are not well understood at present (and will not be explained here). Any known extension of the previous theory which explains these few extra fractions also introduces many more unobserved fractions, with an unclear pattern of stability. Besides the second iteration of the Jain hierarchy [25], we also quote the approach proposed by Fröhlich and collaborators [24]. They analyzed all lattices Γ (3.4), with positive-definite, integer (inverse) metric K, for small values of $\det(K)$, whose classification is known in the mathematical literature. These lattices can be related to the SU(m), SO(k), and exceptional Lie algebras. The stability of the corresponding fluids does not follow a clear pattern related to these algebras, besides the case of the chiral Jain fluids (3.8), whose SU(m) symmetry will be explained in the next section. Moreover, in this approach, the K matrices for the Jain filling fractions $\nu = m/(mp-1)$, $p > 0$, are different from the Jain proposal (3.8) which is not positive definite.

In Figure 2, we study the limitations of phenomenological descriptions of the stability of the fluids. Besides all the observed (bold) fractions of Figure 1, we report the unobserved (italic) ones $\nu = p/q$, which satisfy the conservative cuts of the "phase space"

$$\frac{2}{11} < \nu = \frac{p}{q} < \frac{4}{5}, \text{ and } p \le 10, q \le 17.$$

Namely, we display all fractions that would be observed if the gap were a smooth function of the parameters (ν, p, q) interpolating the data, a typical phenomenological hypothesis. Figure 2 shows that, besides the families

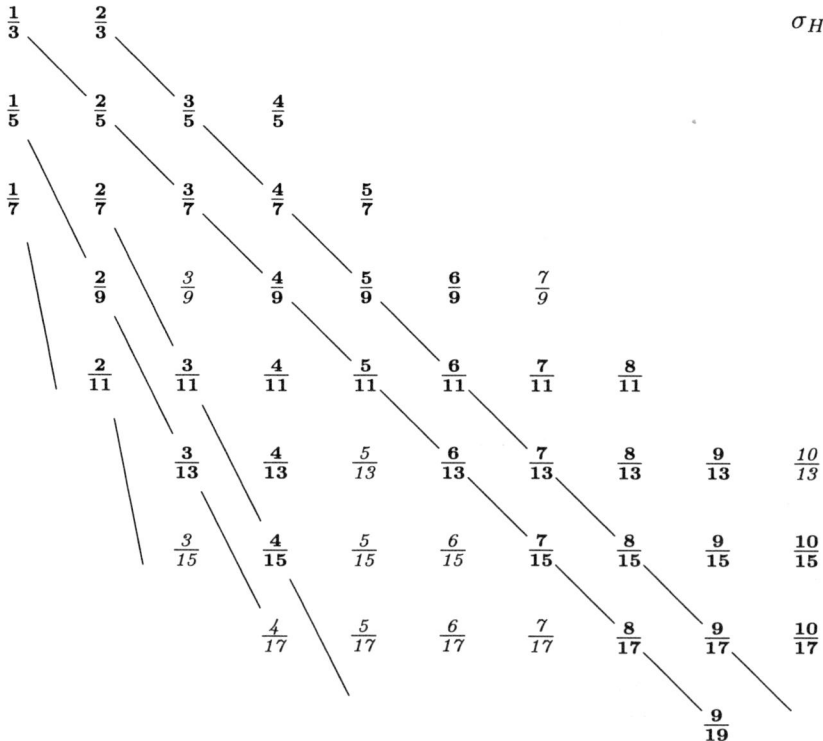

FIGURE 2. List of all fractions $\nu = p/q$, with $2/11 < \nu < 4/5$, $1 < p \leq 10$ and $3 \leq q \leq 17$, q odd. The fractions corresponding to experimental values of the Hall conductivity $\sigma_H = (e^2/h)\nu$ are in **bold**; the unobserved fractions are in *italic*. Observed fractions joined by lines are explained by the Jain hierarchy (3.11), (3.12).

(3.11), (3.12), about half of the fractions are unexplained observed fillings and half are unobserved but phenomenologically possible. This implies that the gap is not a smooth function of simple parameters like (ν, p, q)—deeper theories are needed to explain stability.

Actually, a major virtue of the Jain hierarchy is that of representing one-parameter families of Hall states, within which the gap *is* a smooth function of the above parameters. We think of these families as the set of *kinematically allowed* quantum incompressible fluids (at first level of the hierarchy).

More specific confirmations of the edge theory (3.8) come from the experimental tests of the spectrum of excitations (3.9). An experiment with high time resolution [34] has measured the propagation of a single chiral charge excitation for $\nu = \frac{1}{3}$ ($m = 1$, $p = 2$), and $\nu = \frac{2}{3}$ ($m = 2$, $p = 2$); this is in agreement with the Jain theory, although the neutral excitations have not been seen yet. The resonant tunnelling experiment [35] has verified the conformal dimensions (3.9) for the simplest Laughlin fluid $\nu = \frac{1}{3}$ [36]. Extensions of this experiment to $(m > 1)$ fluids have been suggested, as well as tests of the neutral edge spectrum [37]. We shall discuss more these experiments in Section 4.

4 $W_{1+\infty}$ Minimal Models

4.1 The Theory of $W_{1+\infty}$ Representations

We now develop the classification program of incompressible quantum Hall fluids outlined in Section 2. The basic piece of information we need is the mathematical theory of $W_{1+\infty}$ representations. Luckily enough, all unitary, irreducible $W_{1+\infty}$ representations were obtained in the fundamental work by Kac and collaborators [19, 20]: they exist for integer central charge $(c = m)$ and can be regular, i.e., *generic*, or *degenerate*. In Ref. 10, we used the generic representations to build the *generic* $W_{1+\infty}$ theories, which were shown to correspond to the previously described m-component chiral boson theories parametrized by generic $(m \times m)$ K matrices. In the algebraic approach, this is proven by identifying the generic $W_{1+\infty}$ representations with $\widehat{U(1)}^{\otimes m}$ representations—this qualifies the word "generic." Both representations are labelled by the same weight vectors $\vec{r}$. A complete equivalence requires also a one-to-one map between the states built on top of the respective highest-weight states.

The general theory of unitary, irreducible (quasi-finite) $W_{1+\infty}$ representations, developed in Refs. 19 and 20, leads to the following relations between irreducible representations of the two algebras,

$$M(W_{1+\infty}, 1, r) \sim M(\widehat{U(1)}, 1, r),$$

$$M(W_{1+\infty}, m > 1, \vec{r}) \sim M(\widehat{U(1)}^{\otimes m}, m, \vec{r}), \text{ for } (r_i - r_j) \notin \mathbb{Z}, \forall i \neq j, \quad (4.1)$$

$$M(W_{1+\infty}, m > 1, \vec{r}) \subset M(\widehat{U(1)}^{\otimes m}, m, \vec{r}), \text{ if } \exists (r_i - r_j) \in \mathbb{Z}.$$

Generically, $W_{1+\infty}$ and $\widehat{U(1)}^{\otimes m}$ representations are one-to-one equivalent. The exceptions appear for $c > 1$, when the weight has some integer components $(r_i - r_j)$. In these cases, the relation is many-to-one, i.e., an irreducible $\widehat{U(1)}^{\otimes m}$ representation is *reducible* with respect to the $W_{1+\infty}$ algebra. We call *generic* the $W_{1+\infty}$ representations that are one-to-one equivalent to $\widehat{U(1)}^{\otimes m}$ ones $((r_i - r_j) \notin \mathbb{Z}, \forall i \neq j)$, and *degenerate* the remaining $W_{1+\infty}$ representations $(\exists (r_i - r_j) \in \mathbb{Z})$.

The results (4.1) allow the construction of several types of $W_{1+\infty}$ symmetric theories. The generic $W_{1+\infty}$ theories [10] are defined by lattices Γ (3.4) which contain generic $W_{1+\infty}$ representations only: for these, the basis vectors satisfy $\left((\vec{v}_\alpha)_i - (\vec{v}_\alpha)_j\right) \notin \mathbb{Q}$, $\forall \alpha$, $i \neq j = 1, \ldots, m$. These theories are thus equivalent to chiral boson theories.[4] Other $W_{1+\infty}$ theories, containing only degenerate representations, are instead different. These are the *minimal models*, which we shall describe below. They are the basic new $W_{1+\infty}$ theories, and are actually very important because they will be shown to correspond to the experimentally observed Jain fluids. The mathematical rules for building the degenerate representations have a hierarchical structure similar to the Jain construction: in Ref. 12, we fully explained this correspondence to the lowest order of the hierarchies.

On the other hand, the chiral boson theories of the Jain hierarchy have been widely used in the literature and partially confirmed by the experiments, as we discuss hereafter. These are also $W_{1+\infty}$ symmetric, but are not the simplest realizations of this symmetry, because their $\widehat{U(1)}^{\otimes m}$ representations are reducible. Reducible and irreducible degenerate representations have the same quantum numbers of fractional charge, spin, and statistics. The existing experiments at hierarchical filling fractions were sensible to these data only; therefore their successful interpretation within the chiral-boson theory is consistent with our theory. More refined experiments are needed to test the difference.

4.2   The $W_{1+\infty}$ Minimal Models

Degenerate representations are common in conformal field theory: if the central charge and the weight of a given representation satisfy certain algebraic relations, some of its states decouple, and should be projected out to obtain an irreducible representation. A general fact is that the fusion of degenerate representations only gives degenerate representations of the same type; thus it is possible to find sets of degenerate representations that are closed under the fusion rules [4]. These build the minimal models [4]. There are specific minimal models for any symmetry algebra: the well-known ones are the $c < 1$ Virasoro minimal models [4]; larger symmetry algebras, like $W_{1+\infty}$, have $c > 1$ minimal models. The minimal models have less states than the generic theories with the same symmetry, due to the projection; for the same reason, they have richer dynamics.

The $W_{1+\infty}$ minimal models are *not* realized by the multicomponent chiral boson theories with $\widehat{U(1)}^{\otimes m}$ symmetry, because the latter do not incorporate the projection for making irreducible the $W_{1+\infty}$ representations of degenerate type. They are instead realized by the $\widehat{U(1)} \otimes \mathcal{W}_m(p = \infty)$ conformal theories [20], where the $\mathcal{W}_m(p)$ are the Zamolodchikov-Fateev-Lykyanov models with $c = (m-1)[1 - m(m+1)]/[p(p+1)]$ [38]. We have

---

[4]The ground-state representation ($\vec{r} = 0$) must also be a $\widehat{U(1)}^{\otimes m}$ representation for the closure of the fusion rules.

found the minimal set of representations which are closed under the fusion rules of these models, which is again a lattice (3.4)) satisfying special conditions, which makes it similar to the weight lattice of the $SU(m) \otimes U(1)$ Lie algebra [12]. We have also identified the physical charge of the excitations and the Hall conductivity with arguments analogous to the one described before in the chiral boson theory. We have obtained the spectrum

$$\nu = \frac{m}{mp \pm 1}, \qquad\qquad p > 0 \text{ even}, c = m,$$

$$Q = \frac{1}{pm \pm 1} \sum_{i=1}^{m} n_i, \qquad\qquad n_1 \geq n_2 \geq \cdots \geq n_m, \qquad (4.2)$$

$$\frac{\theta}{\pi} = \pm \left(\sum_{i=1}^{m} n_i^2 - \frac{p}{mp \pm 1} \left(\sum_{i=1}^{m} n_i \right)^2 \right).$$

These spectra agree with the experimental data and match the results of the lowest-order Jain hierarchy (3.9)[5] discussed in Section 2.

This result has far-reaching consequences, both theoretical and experimental. The physical mechanism which stabilizes the observed quantum Hall fluids has both short- and long-distance manifestations. At the microscopic level, it can be described by the Jain composite-electron picture and by the size of the gaps; in the scaling limit, by the minimality of the $W_{1+\infty}$ edge theory. Actually, we find it rather natural that the theories with a minimal set of excitations are also dynamically more stable. This long-distance stability principle leads to a logically self-contained edge theory of the fractional Hall effect: a thorough derivation of experimental results is obtained from the principle of $W_{1+\infty}$ symmetry, which is the basic property of the Laughlin incompressible fluid. This independent hierarchical construction is the main result of our approach.

Furthermore, the detailed predictions of the $W_{1+\infty}$ minimal theories are different from those of the chiral boson theories. The main differences are as follows:

(i) There is a *single* Abelian current, instead of m independent ones, and therefore a single elementary (fractionally) charged excitation; there are neutral excitations, but they cannot be associated to $(m - 1)$ independent edges.

(ii) The dynamics of these neutral excitations is new: they have an $SU(m)$ (not $\widehat{SU(m)}_1$) "isospin" quantum number, because their fusion rules are given by the branching rules of this group. Therefore, they are quarklike and their fractional statistics is non-Abelian [39]. For example, the edge excitation corresponding to the electron, for the filling

---

[5] Note, however, the reduced multiplicities of eq. (4.2).

fractions $\nu = m/(mp \pm 1)$, is a composite made of (mp) anyons and one "quark," and carries both the additive electric charge and the $SU(m)$ isospin.

(iii) The degeneracy of particle-hole excitations at fixed angular momentum is modified by the projection of the minimal models. This counting of states is provided by the characters of degenerate $W_{1+\infty}$ representations, which are known [20]. If the neutral $SU(m)$ excitations have a bulk gap, the particle-hole degeneracy of the ground state (the Wen topological order on the disk [40]) is different from the corresponding one of $\widehat{U(1)}^{\otimes m}$ excitations. This can be tested in numerical diagonalizations of few electron systems; existing data are not accurate enough [16].

4.3 Non-Abelian Fusion Rules and Non-Abelian Statistics

In the previous section, we have identified the physical electron as the minimal set of $W_{1+\infty}$ representations with unit charge and integer statistics relative to all excitations. These conditions are fulfilled by a composite edge excitation $n_i = (1 + p, p, \ldots, p)$, which is made of (mp) elementary charged *anyons* and the *quark* elementary neutral excitation, i.e., the fundamental $SU(m)$ isospin representation, due to $(n_i - n_{i+1}) = \delta_{i,1}$.

A conduction experiment that could show the composite nature of the electron has been proposed [37]. It is a modification of the "time domain" experiment [34] in which a very fast electric pulse was injected at the boundary of a disk sample and a chiral wave was detected at another boundary point. The proposed experiment will also detect the neutral excitation in the electron, which propagates at a different speed.

The compositeness of the electron also plays a role in the resonant tunnelling experiment [35], in which two edges of the sample are pinched at one point, such that the corresponding edge excitations, having opposite chiralities, can interact. At $\nu = 1/3$, the point interaction of two elementary anyons is relevant and determines the *scaling law* $T^{2/3}$ for the conductance [36]. This scaling of the tunnelling resonance peaks is verified experimentally. On the other hand, off-resonance and at low temperature, the conductance is given by the tunnelling of the whole electron, with a different scaling law in temperature [16].

These experiments involve processes with one or two electrons: their quark compositeness can be seen in four-electron processes, like scattering. Indeed, the expansion of the four-point function of the electrons in intermediate channels is determined by the fusion ($SU(2)$ isospin for $m = 2$) of an electron pair. This is, schematically,

$$\langle \Omega | \Psi^\dagger(1) \Psi^\dagger(2) \Psi(3) \Psi(4) | \Omega \rangle = \sum_{s=0,1} \langle \Omega | \Psi^\dagger(1) \Psi^\dagger(2) | \{s\} \rangle \langle \{s\} | \Psi(3) \Psi(4) | \Omega \rangle,$$

where the two channels follow from the addition of two one-half isospin values. More than one intermediate channel are also created in the adiabatic transport of two electrons around each other, in presence of two other excitations, because the amplitude for this process is again a four-point function. For generic excitations, the monodromy phases form a matrix, which gives a non-Abelian representation of the braid group [32]. This is precisely the notion of non-Abelian statistics.[6] These monodromy properties also determine the degeneracy of the ground state on a torus geometry, the so-called topological order [16]. This depends on the type of the representations carried by the edge excitations [4], and should be computed for the $\widehat{U(1)} \otimes \mathcal{W}_m$ ones. We hope to develop these issues in a separate work.

4.4 The Degeneracy of Excitations Above the Ground State

In order to discuss this point, we must rewrite the spectrum (4.2). Let us consider the $m = 2$ chiral theories, relevant for $\nu = \frac{2}{5}, \ldots$; the extension of the analysis to any m and mixed chiralities is straightforward. Recall that the $\mathcal{W}_{1+\infty}$ minimal model in this case is constructed from degenerate representations of the type $\widehat{U(1)} \otimes \mathcal{W}_2$ only, where the $\mathcal{W}_2$ algebra is the $c = 1$ Virasoro algebra. As explained in [12], these degenerate Virasoro representations carry an SU(2) isospin quantum number, as required by the fusion rules [4]. Consider any excitation (n_1, n_2) associated to a $\widehat{U(1)} \otimes$ Vir representation, labelled by the $\widehat{U(1)}$ charge $Q \propto (n_1 + n_2)$ and the SU(2) isospin $s = |n_1 - n_2|/2$. Divide the square lattice (n_1, n_2) into charged excitations and their neutral daughter excitations by introducing the change of integer variables $(n_1, n_2) \to (l, n)$:

$$\text{I} : \begin{cases} 2l = n_1 + n_2 \\ 2n = n_1 - n_2 > 0 \quad (n_1 + n_2 \text{ even}), \end{cases}$$

$$\text{II} : \begin{cases} 2l + 1 = n_1 + n_2 \\ 2n + 1 = n_1 - n_2 > 0 \quad (n_1 + n_2 \text{ odd}). \end{cases}$$

The spectrum (4.2) can be rewritten, for $\nu = 2/(2p + 1)$,

$$\text{I} : \begin{cases} Q = 2l/(2p + 1), \\ \frac{1}{2}\frac{\theta}{\pi} = (1/2p + 1)l^2 + n^2, \end{cases}$$

$$\text{II} : \begin{cases} Q = 2/(2p + 1)(l + \frac{1}{2}) \\ \frac{1}{2}\frac{\theta}{\pi} = 1/(2p + 1)(l + \frac{1}{2})^2 + (2n + 1)^2/4. \end{cases} \tag{4.3}$$

---

[6]For a general discussion of non-Abelian statistics in the quantum Hall effect, see Ref. 39.

The $\widehat{U(1)}$ charged excitations have the same spectrum $Q = \nu k$, $\theta/\pi = \nu k^2$, of the simpler Laughlin fluids ($m = 1$). Moreover, the infinite tower of neutral daughters ($n > 0$) are characterized by the conformal dimensions $h = (2n)^2/4$ (resp. $h = (2n+1)^2/4$).

The number of excitations above the ground state depends on whether the neutral excitations have a bulk gap or not. This affects also the thermodynamic quantities like the specific heat.

As remarked before, the charged edge excitations correspond to Laughlin quasi-particles vortices in the bulk of the incompressible fluid, which spill their density excess or defect to the boundary. They have a (nonuniversal) gap proportional to the electrostatic energy of the vortex core, which is not accounted for by the edge theory [2, 3]. On the other hand, the bulk excitations corresponding to *neutral* edge excitations are not well understood yet. If they have a gap, they could exhibit the internal structure of the quasi-particle vortex, or be bound states of a quasi-particle and a quasi-hole; these would be localized two-dimensional excitations. Neutral and charged gapful excitations can be thought of as analogs of the *breathers* and *solitons* of one-dimensional integrable models, respectively; see, for example Ref. 41. On the other hand, gapless neutral excitations would be pure effects of the structured edge.

In the gapful case, the excitations above the ground state are particle-hole excitations described by the $W_{1+\infty}$ representation of the ground state ($n = l = 0$) in (4.3). In the gapless case, there are also contributions from the neutral daughter Virasoro representations ($n > 0$, $l = 0$), because they have integer spin (Virasoro dimension) and are indistinguishable. Actually, the infinite tower of Virasoro representations ($n > 0$, l fixed) of each charged parent state (l, $n = 0$) can be summed (with their multiplicity one) into a single $\widehat{U(1)}$ representation by using the relations between characters of the corresponding representations [12]. In this case, the $m = 2$ $W_{1+\infty}$ square-lattice spectrum (4.3) reduces to a one-dimensional array of $\widehat{U(1)}^{\otimes 2}$ representations, with spectrum

$$
\begin{aligned}
\text{I} &: \frac{1}{2}\frac{\theta}{\pi} = \frac{1}{2p+1}l^2, \\[2mm]
\text{II} &: \frac{1}{2}\frac{\theta}{\pi} = \frac{1}{2p+1}\left(l+\frac{1}{2}\right)^2 + \frac{1}{4},
\end{aligned}
\tag{4.4}
$$

where the second $\widehat{U(1)}$ eigenvalue is not observable.

Let us repeat this analysis for the corresponding chiral boson theory of the Jain fluid. The spectrum of charge and fractional statistics is again given by (4.3), with multiplicities given by $n \in \mathbb{Z}$: each (l,n) value corresponds to a $\widehat{U(1)} \otimes \widehat{U(1)}$ representation now. If they are gapless, the neutral daughter $\widehat{U(1)} \otimes \widehat{U(1)}$ representations $((l,n), n \neq 0 \in \mathbb{Z})$ of each charged representation $(l,0)$ can be similarly summed up into one representation of

TABLE 1.

ΔJ	0	1	2	3	4	5
$\widehat{U(1)} \otimes \mathrm{Vir}$	1	1	3	5	10	16
$\widehat{U(1)} \otimes \widehat{U(1)}$	1	2	5	10	20	36
$\widehat{U(1)} \otimes \widehat{SU(2)}_1$	1	4	9	20	42	80

the larger algebra $\widehat{U(1)} \otimes \widehat{SU(2)}_1$, the non-Abelian current algebra of level one [4]. The spectrum of $\widehat{U(1)} \otimes \widehat{SU(2)}_1$ representations is again given by (4.4).

We can now compare the predictions of the $W_{1+\infty}$ minimal models and the chiral boson theories for the degeneracy of the excitations above the ground state. This degeneracy can be measured in numerical simulations of a few electron system in the disk geometry, by charting the eigenstates of the Hamiltonian below the bulk gap [16, 40]. Consider, for example, the $\nu = \frac{2}{5}$ ($m = p = 2$) ground state (($l = 0, n = 0$) in (4.3)). In the following table, we report the degeneracies encoded in the $\widehat{U(1)} \otimes \mathrm{Vir}$ character and the $\widehat{U(1)}^{\otimes 2}$ character, for $\vec{r} = 0$, as well as those of the $\widehat{U(1)} \otimes \widehat{SU(2)}_1$ one for $r = s = 0$ [12]:

If neutral daughter excitations have a gap, they should not be counted, and the degeneracy is only given by the particle-hole excitations encoded in the ground-state character of the theory. On the other hand, gapless neutral excitations contribute and the total degeneracy is given by the resummed characters [12]. We conclude that:

(i) The observation of $\widehat{U(1)} \otimes \mathrm{Vir}$ degeneracies confirms the $W_{1+\infty}$ minimal theory with gapful neutral excitations;

(ii) The $\widehat{U(1)} \otimes \widehat{U(1)}$ degeneracies are found both in the $W_{1+\infty}$ minimal theory with gapless neutral excitations and in the chiral boson theory with gapful ones;

(iii) The $\widehat{U(1)} \otimes \widehat{SU(2)}_1$ degeneracies support the chiral boson theory with gapless neutral excitations.

Numerical results known to us [16] are not accurate enough to see the differences in Table 1. Note the characteristic reduction of states of $W_{1+\infty}$ minimal models.

These remarks on the gap for neutral excitations do not affect the previous discussion of the conduction experiments, where excitations move along one edge or are transferred between two edges at the same Fermi energy, such that bulk excitations are never produced. Although the resummation of the neutral daughter $\mathcal{W}_m$ excitations gives Abelian excitations, these are not $W_{1+\infty}$ irreducible, and thus unlikely to be produced experimentally. We think that only irreducible $W_{1+\infty}$ excitations, i.e., the elementary ones, can be naturally produced in a real system by an external probe, for example by injecting an electron at the edge.

4.5   Remarks on the $\mathrm{SU}(m)$ and $\widehat{\mathrm{SU}(m)}_1$ Symmetries

We would like to explain the type of non-Abelian symmetry of the $W_{1+\infty}$ minimal models and clarify the differences with the chiral boson theories of the Jain hierarchy, which have been also assigned the $\mathrm{SU}(m)$ and $\widehat{\mathrm{SU}(m)}_1$ symmetries [24, 30, 37, 42].

Due to the $\widehat{\mathrm{U}(1)} \otimes \mathcal{W}_m$ construction of the $W_{1+\infty}$ models, their excitations carry a quantum number which adds up as a $\mathrm{SU}(m)$ isospin. This *does not* imply that these models have the full $\mathrm{SU}(m)$ symmetry, in the usual sense of, say, the quark model of strong interactions, because the states in each $\mathcal{W}_m$ representation do not form $\mathrm{SU}(m)$ multiplets. As shown by the $m = 2$ case, the quantum number $s = n/2$ of Virasoro representations is like the total isospin $S^2 = s(s + 1)$, but the S_z component is missing. In some sense, the effects of the $\mathcal{W}_m$ non-Abelian fusion rules can be thought of as a *hidden* $\mathrm{SU}(m)$ symmetry.

On the other hand, it has been claimed that the chiral boson theories of the Jain hierarchy have a $\mathrm{SU}(m)$ symmetry. The correct statement is, however, that they possess $\widehat{\mathrm{U}(1)} \otimes \widehat{\mathrm{SU}(m)}_1$ symmetry. This means that their $\widehat{\mathrm{U}(1)}^{\otimes m}$ representations can be rearranged into representations of the $\widehat{\mathrm{U}(1)} \otimes \widehat{\mathrm{SU}(m)}_1$ current algebra. In the $\widehat{\mathrm{SU}(m)}_k$ current algebra, the weights cannot be arbitrary, but are cut off by the *level* k (e.g., for $m = 2$, the spin s can be $0 \leq s \leq k/2$) [4]. The level-one non-Abelian current algebra has very elementary representations and their fusion rules are made Abelian by this cut-off.

Therefore, the $\widehat{\mathrm{SU}(m)}_1$ symmetry has no non-Abelian physical effect, it is only a convenient reorganization of the Abelian current algebra. The non-Abelian character of the excitations is a characteristic feature of the $W_{1+\infty}$ minimal models.

5   Further Developments

We have reviewed the $W_{1+\infty}$ theory of the edge excitations in the quantum Hall effect. In particular, we considered the *simplest* $W_{1+\infty}$ minimal models, which are made of one-congruence-class degenerate representations only. It would be interesting to generalize this construction in view of describing the experimentally observed filling fractions 4/11, 7/11, 4/13, 8/13, 9/13, 10/17, ... , not explained here. The $W_{1+\infty}$ minimal models can be generalized by considering two (or more) congruence classes [12]. There are analogies between this mathematical construction and the Jain hierarchical construction of wave functions, which read

$$\Psi_\nu = D^{q/2} L^l D^{p/2} L^m 1, \quad p, q \text{ even}, \qquad (5.1)$$

to second-order of iteration [25]. The number of fluids in any $W_{1+\infty}$ congruence class corresponds to the number of Landau levels in (5.1); in both constructions, there are two independent elementary anyons, each one accompanied by neutral excitations. However, we have not yet proven a complete equivalence of the two second-order hierarchies: the Jain construction assigns a definite filling fraction to each wave function (5.1), while we have a large modular degeneracy (of the group $SL(2, \mathbb{Z})$) in the definition of the physical charge of the two independent anyons, leading to many values of the filling fraction for each minimal model. On the contrary, we would like to find *more* constraints than in the Jain construction, because most of its second-order filling fractions are not observed experimentally. We guess that our algebraic construction of $W_{1+\infty}$ minimal model Hilbert spaces should be supplemented by the construction of other physical quantities, like the partition function [43], which could impose further conditions on the physical theories.

Acknowledgments: We would like to thank the organizers of the Banff summer school for the opportunity to present our results. Throughout our project, we have greatly benefited from discussions with V. Kac. We would also like to acknowledge the continuing support of L. Alvarez-Gaumé and S. Fubini. This work was supported in part by the CERN Theory Division, the MIT Center for Theoretical Physics, and the European Community program "Human Capital and Mobility." C.A.T. is supported by a Profil 2 fellowship of the Swiss National Science Foundation.

6   References

1. R. A. Prange and S. M. Girvin, eds. *The Quantum Hall Effect.* Springer-Verlag, New York, 1990.

2. R. B. Laughlin. Anomalous quantum Hall effect: An incompressible quantum fluid with fractionally charged excitations. *Phys. Rev. Lett.*, 50 (18): 1395–1398, 1983.

3. R. B. Laughlin. Elementary theory: The incompressible quantum fluid. In Prange and Girvin [1].

4. A. A. Belavin, A. M. Polyakov, and A. B. Zamolodchikov. Infinite conformal symmetry in two-dimensional quantum field. *Nucl.*

Phys., B241 (2): 333–380, 1984; P. Ginsparg. Applied conformal field theory. In É. Brézin and J. Zinn-Justin, eds., *Champs, cordes et phénomènes critiques*, (Les Houches, 1988), 1990. North-Holland, Amsterdam, pages 1–168.

5. J. Polchinski. Effective field theory and the Fermi surface. Technical Report NSF-ITP-92-132, ITP, UCSB, 1992.

6. A. Cappelli, C. A. Trugenberger, and G. R. Zemba. Infinite symmerty in the quantum Hall effect. *Nucl. Phys.*, B396 (2-3): 465–490, 1993.

7. A. Cappelli, G. V. Dunne, C. A. Trugenberger, and G. R. Zemba. Conformal symmetry and universal properties of quantum Hall states. *Nucl. Phys.*, B398 (3): 531–567, 1993.

8. A. Cappelli, G. V. Dunne, C. A. Trugenberger, and G. R. Zemba. Symmetry aspects and finite-size scaling of quantum Hall fluids. In L. Alvarez-Gaumé et al., eds., *Common Trends in Condensed Matter and High-Energy Physics*, (Chia, Sardinia, 1992), volume 33C of *Nuclear Phys. B. Proc. Suppl.*, 1993. pages 21–34.

9. A. Cappelli, C. A. Trugenberger, and G. R. Zemba. Large N limit in the quantum Hall effect. *Phys. Lett.*, B306 (1-2): 100–107, 1993.

10. A. Cappelli, C. A. Trugenberger, and G. R. Zemba. Classification of quantum Hall universality classes by $W_{1+\infty}$ symmetry. *Phys. Rev. Lett.*, 72 (12): 1902–1905, 1994.

11. A. Cappelli, C. A. Trugenberger, and G. R. Zemba. $W_{1+\infty}$ dynamics of edge excitations in the quantum Hall effect. *Ann. Phys.*, 246 (1): 86–120, 1996.

12. A. Cappelli, C. A. Trugenberger, and G. R. Zemba. Stable hierarchical quantum Hall fluids as $W_{1+\infty}$ minimal models. *Nucl. Phys.*, B448 (3): 470–504, 1995.

13. S. Iso, D. Karabali, and B. Sakita. One-dimensional fermoins as two-dimensional droplets via Chern-Simons theory. *Nucl. Phys.*, B388 (3): 700–714, 1992; S. Iso, D. Karabali, and B. Sakita. fermions in the lowest Landau level: Bosonization, W_∞ algebra, droplets, chiral bosons. *Phys. Lett.*, B296 (1-2): 143–150, 1992.

14. M. Flohr and R. Varnhagen. Infinite symmetry in the fractional quantum Hall effect. *J. Phys A: Math. Gen.*, 27 (11): 3999–4010, 1994; D. Karabali. Algebraic aspects of the fractional quantum Hall effect. *Nucl. Phys.*, B419 (3): 437–454, 1994; D. Karabali. W_∞ algebras in the quantum Hall effect. *Nucl. Phys.*, B428 (3): 531–544, 1994.

15. I. Bakas. The large-N limit of extended conformal symmetries. *Phys. Lett.*, B228 (1): 57–63, 1989; C. N. Pope, X. Shen, and L. J. Romans. W_∞ and the Racah-Wigner algebra. *Nucl. Phys.*, B339 (1): 191–221, 1990; X. Shen. W infinity and string theory. *Int. J. Mod. Phys.*, A7 (28): 6953–6993, 1992.

16. X. G. Wen. Theory of the edge states in fractional quantum Hall effects. *Int. J. Mod. Phys.*, B6 (10): 1711–1762, 1992.

17. M. Stone. Edge waves in the quantum Hall effect. *Ann. Phys.*, 207 (1): 38–52, 1991; M. Stone. Schur functions, chiral bosons and the quantum-Hall-effect edges states. *Phys. Rev.*, 42B (13): 8399–8404, 1990; M. Stone. Vertex operators in quantum Hall effect. *Int. J. Mod. Phys.*, B5 (3): 509–527, 1991.

18. S. M. Girvin, A. H. MacDonald, and P. M. Platzman. Magneto-roton theory of collective excitations in the fractional quantum Hall effect. *Phys. Rev.*, B33 (4): 2481–2494, 1986; S. M. Girvin. Collective excitations. In Prange and Girvin [1].

19. V. Kac and A. Radul. Quasi-finite highest-weight modules over the Lie algebra of differential operators on the circle. *Commun. Math. Phys.*, 157 (3): 429–457, 1993; H. Awata, M. Fukuma, Y. Matsuo, and S. Odake. Representation theory of the $W_{1+\infty}$ algebra. In *Quantum Field Theory, Integrable Models, and Beyond*, (Kyoto, 1994), number 118 in Prog. Theor. Phys. Suppl, 1995. pages 343–373.

20. E. Frenkel, V. Kac, A. Radul, and W. Wang. $\mathcal{W}_{1+\infty}$ and $\mathcal{W}(\mathfrak{gl}_N)$ with central charge N. *Commun. Math. Phys.*, 170 (2): 337–257, 1995.

21. I. Vaysburd and A. Radul. Differential operators and W-algebra. *Phys. Lett.*, B274 (3-4): 317–322, 1992.

22. F. D. M. Haldane. Fractional quantization of the Hall effect: A hierarchy of incompressible quantum fluid states. *Phys. Rev. Lett.*, 51 (7): 605–608, 1983.

23. B. I. Halpern. Statistics of quasi-particles and the hierarchy of fractional quantized Hall states. *Phys. Rev. Lett.*, 52 (18): 1583–1586, 1984.

24. J. Fröhlich, U. M. Studer, and E. Thiran. An ADE-O classification of minimal incompressible quantum Hall fluids. cond-mat/9406009.

25. J. K. Jain. Microscopic theory of the fractional quantum Hall effect. *Adv. Phys.*, 41 (2): 105–146, 1992.

26. B. I. Halperin, P. A. Lee, and N. Read. Theory of the half-filled Landau level. *Phys. Rev.*, B47 (12): 7312–7343, 1993.

27. R. R. Du, H. Stormer, D. C. Tsui, L. N. Pfeiffer, and K. W. West. Experimental evidence for new particles in the fractional quantum Hall effect. *Phys. Rev. Lett.*, 70 (19): 2944–2947, 1993; W. Kang, H. L. Stormer, L. N. Pfeiffer, K. W. Baldwin, and K. W. West. How real are composite fermions. *Phys. Rev. Lett.*, 71 (23): 3850–3853, 1993.

28. B. I. Halperin. Quantized Hall conductance, current-carrying edges states, and the existence of extended states in two-dimensional disordered potential. *Phys. Rev.*, B25 (4): 2185–2190, 1982.

29. X.-G. Wen. *Mod. Phys. Lett.*, B5: 39, 1991.

30. J. Fröhlich and A. Zee. Large-scale physics of the quantum Hall fluids. *Nucl. Phys.*, B364 (3): 517–540, 1991; X.-G. Wen and A. Zee. Classification of Abelian quantum Hall states and matrix formulation of topological fluids. *Phys. Rev.*, 46B (4): 2290–2301, 1993.

31. R. Floreanini and R. Jackiw. Self-dual fields as charge-density solitons. *Phys. Rev. Lett.*, 59 (17): 1873–1876, 1987.

32. F. Wilczek, ed. *Fractional Statistics and Anyon Superconductivity*. World Scientific, Teaneck, NJ, 1990.

33. L. W. Engel, S. W. Hwuang, T. Sajoto, D. C. Tsui, and M. Shayegan. Fractional quantum Hall effect at $\nu = 2/3$ and $3/5$ in tilted magnetic fields. *Phys. Rev.*, B45 (7): 3418–3425, 1992; J. Frölich et al. The fractional quantum Hall effect, Chern-Simons theory, and integral lattices. Technical Report ETH-TH/94-18, ETH-Zentrum, 1994.

34. R. C. Ashoori, H. L. Stormer, L. N. Pfeiffer, K. W. Baldwin, and K. West. Edge magnetoplasmons in the time domain. *Phys. Rev.*, B45 (7): 3894–3897, 1992.

35. F. P. Milliken, C. P. Umbach, and R. A. Webb. Evidence for a Luttinger liquid in the fractional quantum Hall effect. Technical report, IBM, 1994.

36. K. Moon, H. Yi, C. L. Kane, S. M. Girvin, and M. P. A. Fisher. Resonant tunneling between quantum Hall edge states. *Phys. Rev. Lett.*, 71 (26): 4381–4383, 1993; P. Fendley, A. W. W. Ludwig, and H. Saleur. Exact conductance through point contacts in the $\nu = \frac{1}{3}$ fractional quantum Hall effect. *Phys. Rev. Lett.*, 74: 3005–3008, 1995.

37. C. L. Kane and M. P. A. Fisher. Impurity scattering and transport of fractional quantum hall edge states. cond-mat/9409028.

38. V. A. Fateev and A. B. Zamolodchikov. Conformal quantum field theory models in two dimensions having Z_3 symmetry. *Nucl. Phys.*, B280 (4): 644–600, 1987; V. A. Fateev and S. L. Lykyanov. The models of two-dimensional conformal quantum field theory with Z_n symmetry. *Int. J. Mod. Phys.*, A3 (2): 507–520, 1988.

39. G. Moore and N. Read. Nonabelions in the fractional quantum Hall effect. *Nucl. Phys.*, B360 (2-3): 362–396, 1991.

40. X.-G. Wen. Topological order and edge structure of $\nu = \frac{1}{2}$ quantum Hall state. *Phys. Rev. Lett.*, 70: 355, 1993.

41. R. Rajaraman. *Solitons and Instantons. An Introduction to Solitons and Instantons in Quantum Field Theory*. North-Holland, Amsterdam, 1982.

42. N. Read. Excitation structure of the hierarchy scheme in the fractional quantum Hall effect. *Phys. Rev. Lett.*, 65 (12): 1502–1505, 1990.

43. A. Cappelli, C. Itzykson, and J.-B. Zuber. Modular invariant partition functions in two dimensions. *Nucl. Phys.*, B280 (3): 445–465, 1987.

10

On the Spectral Theory of Quantum Vertex Operators

Pavel I. Etingof

ABSTRACT We prove a conjecture from Ref.1 on the asymptotics of the composition of n quantum vertex operators for the quantum affine algebra $U_q(\widehat{\mathfrak{sl}_2})$, as n goes to ∞. For this purpose we define and study the leading eigenvalue and eigenvector of the product of two components of the quantum vertex operator. This eigenvector and the corresponding eigenvalue were recently computed by M. Jimbo. The results of his computation are given in Section 4.

1 Basic Definitions

1.1 Quantum Groups

Let $U_q(\widehat{\mathfrak{sl}_2})$ be the quantum group generated over $\mathbb{C}(q)$ by the elements e, f, $t^{\pm 1}$, satisfying the standard relations:

$$tet^{-1} = q^2 e, \quad tft^{-1} = q^{-2}f, \quad [e, f] = \frac{t - t^{-1}}{q - q^{-1}}. \tag{1.1}$$

For an integer n, set $[n] = (q^n - q^{-n})/(q - q^{-1})$.

Let $U_q(\widehat{\mathfrak{sl}_2})$ be the quantum affine algebra generated over $\mathbb{C}(q)$ by the elements e_i, f_i, $t_i^{\pm 1}$, $i = 0, 1$, satisfying the standard relations:

$$t_i e_i t_i^{-1} = q^2 e_i, \quad t_i f_i t_i^{-1} = q^{-2} f_i, \quad [e_i, f_i] = \frac{t_i - t_i^{-1}}{q - q^{-1}}, \quad i = 0, 1;$$

$$[e_i, f_j] = 0, \quad e_i^3 e_j - [3]e_i^2 e_j e_i + [3]e_i e_j e_i^2 - e_j e_i^3 = 0, \tag{1.2}$$

$$f_i^3 f_j - [3]f_i^2 f_j f_i + [3]f_i f_j f_i^2 - f_j f_i^3 = 0, \quad i \neq j.$$

We define the coproduct by $\Delta(t_i) = t_i \otimes t_i$, $\Delta(e_i) = e_i \otimes 1 + t_i \otimes e_i$, $\Delta(f_i) = f_i \otimes t_i^{-1} + 1 \otimes f_i$. Tensor product of representations of $U_q(\widehat{\mathfrak{sl}_2})$ is defined with the help of this coproduct.

For $z \in \mathbb{C}^*$, let $p_z \colon U_q(\widehat{\mathfrak{sl}_2}) \to U_q(\mathfrak{sl}_2)$ be the evaluation homomorphism defined by $e_0 \to zf$, $f_0 \to z^{-1}e$, $t_0 \to t^{-1}$, $e_1 \to e$, $f_1 \to f$, $t_1 \to t$.

1.2   Representations

Let Λ_0, Λ_1 be the fundamental weights for the $U_q(\widehat{\mathfrak{sl}_2})$. Let $L_0 = V(\Lambda_0)$, $L_1 = V(\Lambda_1)$ denote the irreducible integrable highest weight representations of the quantum affine algebra $U_q(\widehat{\mathfrak{sl}_2})$ with highest weights Λ_0, Λ_1, respectively. Let v_0, v_1 be their highest weight vectors. Let $L = L_0 \oplus L_1$. Let $\hat{L}$, $\hat{L}_i$ be the completions of the modules L, L_i with respect to the homogeneous grading.

Let V be the two-dimensional irreducible representation of $U_q(\mathfrak{sl}_2)$ in which the spectrum of t is q, q^{-1}. Let v_+, v_- be a basis of this representation such that $tv_\pm = q^{\pm 1}v_\pm$, and $v_- = fv_+$. Let $V(z) = p_z^* V$ be the representation of $U_q(\widehat{\mathfrak{sl}_2})$ obtained by pullback of V by p_z.

1.3   Vertex Operators

Quantum vertex operators were introduced by I. Frenkel and N. Reshetikhin. It is known [2] that for any $z \in \mathbb{C}^*$ there exist unique intertwining operators

$$\Phi^0(z)\colon L_0 \to \hat{L}_1 \otimes V(z), \quad \Phi^1(z)\colon L_1 \to \hat{L}_0 \otimes V(z), \qquad (1.3)$$

such that $\Phi^0(z)v_0 = v_1 \otimes v_- +$ lower weight terms, $\Phi^1(z)v_1 = v_0 \otimes v_+ +$ lower weight terms (by "weight" we mean the weight of the first component). These operators are called quantum vertex operators. Let $\Phi(z)\colon L \to \hat{L} \otimes V$ be defined by $\Phi = \Phi^0 \oplus \Phi^1$. We define the operators $\Phi_\pm(z)\colon L \to \hat{L}$ by

$$\Phi(z) = \Phi_+(z) \otimes v_+ + \Phi_-(z) \otimes v_-. \qquad (1.4)$$

It is easy to see that $t\Phi_\pm t^{-1} = q^{\mp 1}\Phi_\pm$.

1.4   The Fock Space

We would like to study the dependence of vertex operators on the parameter q. For this purpose we will

1. realize the $\mathbb{C}(q)$-vector space L as $\mathbb{C}(q) \otimes_\mathbb{C} H$, where H is a complex vector space called the Fock space, and

2. write down the action of the quantum group and vertex operators in $\mathbb{C}(q) \otimes H$ as series in q whose coefficients are operators on H.

This construction is called bosonization and comes from Refs.3 and 4.

Let us now define the Fock space H. Let $\mathfrak{h}$ be the Heisenberg Lie algebra with the basis $\{b_i, i \in \mathbb{Z} \setminus \{0\}; Z\}$, and relations

$$[b_m, b_n] = m\delta_{m+n,0}Z; \quad [X, Z] = 0, \quad X \in \mathfrak{h}. \qquad (1.5)$$

Let $H_0 = \mathbb{C}[b_{-1}, b_{-2}, \dots]$. Then H_0 is naturally a representation of $\mathfrak{h}$, in which $Z = 1$, and b_n acts by multiplication by itself for $n < 0$, and by differentiation $n\partial/\partial b_{-n}$ for $n > 0$. Let $H = H_0 \otimes \mathbb{C}[\mathbb{Z}]$.

We denote the element of $\mathbb{C}[\mathbb{Z}]$ corresponding to the integer n by ε^n. We introduce the homogeneous gradation in H in a standard way: the degree of b_{-n} is $-n$, $n > 0$, and the degree of ε^n is $(i - n^2)/4$, where $i = 1$ if n is odd and 0 if n is even.

1.5   Bosonization of $U_q(\widehat{\mathfrak{sl}_2})$

Now let us define the action of $U_q(\widehat{\mathfrak{sl}_2})$ in H. Set

$$a_n = q^{-n/2} \frac{[n]}{n} b_n, \quad a_{-n} = q^{n/2} \frac{[2n]}{n} b_{-n}, \qquad n > 0. \tag{1.6}$$

Then we have

$$[a_m, a_n] = \delta_{m+n,0} \frac{[m][2m]}{m} Z. \tag{1.7}$$

Let

$$X^{\pm}(z) = \sum_{n \in \mathbb{Z}} X_n^{\pm} z^{-n-1}$$

$$= \exp\left(\pm \sum_{n=1}^{\infty} \frac{a_{-n}}{[n]} q^{\mp n/2} z^n\right) \exp\left(\mp \sum_{n=1}^{\infty} \frac{a_n}{[n]} q^{\mp n/2} z^{-n}\right) \varepsilon^{\pm 2} z^{\pm \partial_\varepsilon}, \quad (1.8)$$

where the first component acts in H_0, the second component acts in $\mathbb{C}[\mathbb{Z}]$, and ∂_ε is defined by $\partial_\varepsilon \varepsilon^n = n \varepsilon^n$. Then all Fourier coefficients of this series define linear operators on the space $\mathbb{C}(q^{1/2}) \otimes H$.

Theorem 1.1 (I. Frenkel-N. Jing, [3]). *There exists a unique representation of $U_q(\widehat{\mathfrak{sl}_2})$ in $\mathbb{C}(q^{1/2}) \otimes H$ such that*

$$t_1 \to 1 \otimes q^{\partial_\varepsilon}, \quad t_0 \to 1 \otimes q^{1-\partial_\varepsilon}, \quad e_1 \to X_0^+, \quad f_1 \to X_0^-,$$
$$e_0 \to X_1^-(1 \otimes q^{-\partial_\varepsilon}), \quad f_0 \to (1 \otimes q^{1-\partial_\varepsilon})X_{-1}^+. \tag{1.9}$$

This representation is isomorphic to $\mathbb{C}(q^{1/2}) \otimes_{\mathbb{C}(q)} L$. The gradation in $\mathbb{C}(q^{1/2}) \otimes H$ introduced above coincides with the homogeneous gradation in L.

Let us rewrite (1.8) in terms of $\{b_n\}$:

$$X^+(z) = \exp\left(\sum_{n=1}^{\infty} \frac{b_{-n}}{n}(q^n + q^{-n})z^n\right)\exp\left(-\sum_{n=1}^{\infty}\frac{b_n}{n}q^{-n}z^{-n}\right)$$
$$\otimes \varepsilon^2 z^{\partial_\varepsilon}, \quad (1.10)$$

$$X^-(z) = \exp\left(-\sum_{n=1}^{\infty}\frac{b_{-n}}{n}(q^{2n}+1)z^n\right)\exp\left(\sum_{n=1}^{\infty}\frac{b_n}{n}z^{-n}\right)$$
$$\otimes \varepsilon^{-2}z^{-\partial_\varepsilon}.$$

It is seen from this equation that in fact, the representation of $U_q(\widehat{\mathfrak{sl}_2})$ defined by (1.9) is well defined over $\mathbb{C}(q)$ if considered in the basis of polynomials of b_{-n} (it is not necessary to take the square root of q). We can also see from (1.10) that X_n^- are actually defined over polynomials in q. This fact will be used later.

From now on we identify $\mathbb{C}(q)\otimes H$ and L by the $U_q(\widehat{\mathfrak{sl}_2})$-isomorphism $\mathbb{C}(q)\otimes H \to L$ fixed by the conditions $1\otimes\varepsilon^0 \to v_0, 1\otimes\varepsilon^1 \to v_1$.

1.6   Bosonization of Vertex Operators

Let $I\colon \mathbb{C}[\mathbb{Z}]\to\mathbb{C}[\mathbb{Z}]$ be defined by $I\varepsilon^n = \frac{1}{2}(1-(-1)^n)\varepsilon^n$.

Theorem 1.2 ([4]). *The vertex operators $\Phi_\pm(z)\colon L\to\hat{L}$ are given by the formulas*

$$\Phi_-(z) = \exp\left(\sum_{n=1}^{\infty}\frac{a_{-n}}{[2n]}q^{7n/2}z^n\right)\exp\left(-\sum_{n=1}^{\infty}\frac{a_n}{[2n]}q^{-5n/2}z^{-n}\right)$$
$$\otimes \varepsilon^1(-q^3z)^{(\partial_\varepsilon+I)/2}, \quad (1.11)$$

$$\Phi_+(z) = \Phi_-(z)X_0^- - qX_0^-\Phi_-(z).$$

Let us write down the expression of the vertex operators in terms of $\{b_n\}$. We have

$$\Phi_-(-q^{-3}z) = \exp\left(\sum_{n=1}^{\infty}\frac{(-1)^n q^n b_{-n}}{n}z^n\right)\exp\left(-\sum_{n=1}^{\infty}\frac{(-1)^n q^n b_n}{n(1+q^{2n})}z^{-n}\right)$$
$$\otimes \varepsilon^1 z^{(\partial_\varepsilon+I)/2}. \quad (1.12)$$

This shows, in particular, that we do not in fact need $q^{1/2}$, i.e., everything is defined over $\mathbb{C}(q)$.

1.7   Boson-Fermion Correspondence

Boson-fermion correspondence was first discussed in physics literature [5]. A representation-theoretic description of this correspondence is given in Ref. 6.

Consider the following formal series in z:

$$\psi(z) = \sum_{n \in \mathbb{Z}} \psi_n z^{-n}$$

$$= \exp\left(\sum_{n=1}^{\infty} \frac{b_{-n}}{n} z^n\right) \exp\left(-\sum_{n=1}^{\infty} \frac{b_n}{n} z^{-n}\right) \otimes \varepsilon^1 z^{\partial_\varepsilon + 1},$$

$$\psi^*(z) = \sum_{n \in \mathbb{Z}} \psi_n^* z^{-n} \tag{1.13}$$

$$= \exp\left(-\sum_{n=1}^{\infty} \frac{b_{-n}}{n} z^n\right) \exp\left(\sum_{n=1}^{\infty} \frac{b_n}{n} z^{-n}\right) \otimes \varepsilon^{-1} z^{-\partial_\varepsilon}.$$

Fourier components of these series define linear operators on H.

Theorem 1.3 (Boson-fermion correspondence; [6]). *The series ψ, ψ^* satisfy the fermionic commutation relations*

$$\psi(z)\psi(w) + \psi(w)\psi(z) = \psi^*(z)\psi^*(w) + \psi^*(w)\psi^*(z) = 0,$$

$$\psi^*(z)\psi(w) + \psi(w)\psi^*(z) = \delta(z - w) = \sum_{n \in \mathbb{Z}} z^n w^{-n}. \tag{1.14}$$

In particular, the operators ψ_n, ψ_n^* satisfy the relations of the Clifford algebra, i.e.,

$$\psi_m \psi_n + \psi_n \psi_m = \psi_m^* \psi_n^* + \psi_n^* \psi_m^* = 0,$$

$$\psi_n^* \psi_m + \psi_m \psi_n^* = \delta_{m+n,0}. \tag{1.15}$$

Furthermore, we have an inverse formula to (1.13):

$$b_n = \sum_{m \in \mathbb{Z}} \psi_m \psi_{n-m}^* \tag{1.16}$$

(as operators in H).

2   Spectral Properties of Vertex Operators

2.1   Vertex Operators as Power Series in q

Let $\mathbb{C}(q)_0$ be the ring of all rational functions of q smooth at the point $q = 0$. This ring is naturally a subring of the ring of formal power series $\mathbb{C}[\![q]\!]$, so we have a natural topology on $\mathbb{C}(q)_0$ which defines the notion of convergence of a Taylor series to a rational function.

Theorems 1.1 and 1.2 imply the following important proposition.

Proposition 2.1. *The Fourier components (with respect to z) of the operators* $\Phi_{\pm}(-q^{-3}z)\colon L \to \hat{L}$ *define* $\mathbb{C}(q)_0$*-linear endomorphisms of* $\mathbb{C}(q)_0 \otimes H$. *More precisely, the operators* $\Phi_{\pm}(-q^{-3}z)$ *can be written in the form*

$$\Phi_{\pm}(-q^{-3}z) = \sum_{n=0}^{\infty} \Psi_{\pm}^n(z) q^n, \tag{2.1}$$

where $\Psi_{\pm}^n(z)$ *are Laurent polynomials in* z *with coefficients in* $\mathrm{End}(H)$. *Furthermore, if* $v \in H$ *then every homogeneous component of the series* $\Phi_{\pm}(z)v$ *is convergent* q*-adically (as a series with values in a finite rank free* $\mathbb{C}(q)_0$*-module).*

Let $H[\![q]\!]$ denote the $\mathbb{C}[\![q]\!]$-module consisting of all formal series $w = \sum_{n \geq 0} w_n q^n, w_n \in H$. Then we have

Corollary 2.1. *For any complex number* $z \in \mathbb{C}^*$, *the operators* $\Phi_{\pm}(-q^{-3}z)$ *define* $\mathbb{C}[\![q]\!]$*-endomorphisms of* $H[\![q]\!]$.

From now on vertex operators will be regarded as such endomorphisms.

2.2   Composition of Vertex Operators

Theorem 1.3 implies that we can define composition of any number of vertex operators, as a formal series in q. In particular, we can define

$$F_{\varepsilon_1 \varepsilon_2 \cdots \varepsilon_n}(q) = \Phi_{\varepsilon_n}(-q^{-3}) \cdots \Phi_{\varepsilon_2}(-q^{-3}) \Phi_{\varepsilon_1}(-q^{-3}), \quad \varepsilon_i \in \{+, -\} \tag{2.2}$$

(this was first shown in Ref. 2). We will be especially interested in the operators $F_{-+}(q) = \Phi_+(-q^{-3})\Phi_-(-q^{-3})$ and $F_{+-}(q) = \Phi_-(-q^{-3})\Phi_+(-q^{-3})$, in particular, their leading eigenvectors and eigenvalues.

Remark. The operator F is defined over $\mathbb{C}[\![q]\!]$ but, in general, not over $\mathbb{C}(q)_0$ (if it contains two factors or more). Indeed, according to Ref. 7, the diagonal matrix element of F_{-+} corresponding to the vacuum vector in L_0 equals $(q^6; q^4)_\infty / (q^4; q^4)_\infty$, where $(a, p)_\infty$ denotes $\prod_{n=0}^{\infty}(1 - ap^n)$. This function is obviously not rational, but it is defined as an element of $\mathbb{C}[\![q]\!]$.

2.3   The Operators $F_{+-}(0)$ and $F_{-+}(0)$

Let us denote by H_n the subspace of H spanned by all the vectors $P \otimes \varepsilon^n$, $P \in \mathbb{C}[b_{-1}, b_{-2}, \dots]$. Clearly, the operators $F_{+-}(q)$, $F_{-+}(q)$ preserve the space H_n for all $n \in \mathbb{Z}$.

Proposition 2.2. (i) *The operator* $F_{-+}(0)$ *preserves degree in* H_0. *It satisfies the equation* $F_{-+}(0)v_0 = v_0$ *and is nilpotent in homogeneous subspaces of strictly negative degree in* H_0. *In* H_n, *the operator* $F_{-+}(0)$ *lowers the degree by* n.

(ii) *The operator $F_{+-}(0)$ preserves degree in H_1. It satisfies the equation $F_{-+}(0)v_1 = v_1$ and is nilpotent in homogeneous subspaces of strictly negative degree in H_1. In H_n, the operator $F_{+-}(0)$ lowers the degree by $n-1$.*

The rest of Section 2.3 is the proof of this proposition. Since (ii) is analogous to (i), we prove only (i).

Substituting $q = 0$ in (1.10)–(1.12), we get

$$F_{-+}(0) = \phi_0^*,$$

where

$$\sum_{n \in \mathbb{Z}} \phi_n^* z^{-n} = \exp\left(-\sum_{n=1}^{\infty} \frac{b_{-n}}{n} z^n\right) \exp\left(\sum_{n=1}^{\infty} \frac{b_n}{n} z^{-n}\right)(1 \otimes z^{-\partial_\varepsilon}), \qquad (2.3)$$

From this formula, it is obvious that $F_{-+}(0)$ lowers degree by n in H_n—in particular, it preserves degree in H_0—and that it fixes the vector v_0. It remains to prove the nilpotency of this operator on vectors of negative degree.

Lemma 2.1. *The operators ϕ_n^* in H satisfy the quadratic relations $\phi_n^* \phi_{m-1}^* + \phi_m^* \phi_{n-1}^* = 0$.*

Proof. This lemma follows from the boson-fermion correspondence (Theorem 1.3). Indeed, we see that $\phi^*(z) = \psi^*(z)(1 \otimes z^{\partial_\varepsilon} \varepsilon^1)$. This means that

$$\phi^*(z)\phi^*(w)|_{H_0} = w(1 \otimes \varepsilon)\psi^*(z)\psi^*(w)(1 \otimes \varepsilon)|_{H_0}, \qquad (2.4)$$

which implies $w^{-1}\phi^*(z)\phi^*(w) + z^{-1}\phi^*(w)\phi^*(z) = 0$ in H_0. This is equivalent to the identities $\phi_n^* \phi_{m-1}^* + \phi_m^* \phi_{n-1}^* = 0$. $\qquad \square$

In particular, Lemma 2.1 implies that $\phi_0^* \phi_{-1}^* = 0$ in H_0. Similarly, $(\phi_0^*)^2 \phi_{-2}^* = -\phi_0^*(\phi_{-1}^*)^2 = 0$. Continuing this, by induction we obtain $(\phi_0^*)^k \phi_{-k} = 0$. Therefore, the nilpotency in Proposition 2.2 follows from the following lemma.

Lemma 2.2. *The vectors $(\phi_{-k}^*)^{n_k} \cdots (\phi_{-2}^*)^{n_2}(\phi_{-1}^*)^{n_1} v_0$, where $k, n_1, \ldots, n_k$ are any nonnegative integers, form a basis in H_0.*

Proof. Let H_0' be the space spanned by the vectors from Lemma 2.2. Note that there are exactly as many vectors of each degree among them as the dimension of the corresponding homogeneous subspace in H_0. So, in order to prove the Lemma, it suffices to show that $H_0' = H_0$.

In order to establish this, let us first show that H_0' is invariant under the operators ϕ_n^*, $n \in \mathbb{Z}$. Indeed, using relations from Lemma 2.1, we can rearrange factors in any monomial of ϕ_n^*-s so that the subscripts increase from left to right. But such a monomial reduces to a monomial with only negative indices, since $\phi_n^* v_0 = 0$, $n > 0$, and $\phi_0^* v_0 = v_0$. This implies that ϕ_n^* maps H_0' to itself for any n.

Now let us introduce a new series

$$\phi(z) = \psi(z)(1 \otimes z^{-\partial_\varepsilon - 1} \varepsilon^{-1})$$

$$= \exp\left(\sum_{n=1}^{\infty} \frac{b_{-n}}{n} z^n\right) \exp\left(-\sum_{n=1}^{\infty} \frac{b_n}{n} z^{-n}\right). \tag{2.5}$$

Similarly to Lemma 2.1, we can prove the relations

$$\phi_n \phi_{m+1} + \phi_m \phi_{n+1} = 0, \quad \phi_m \phi_n^* + \phi_n^* \phi_m = \delta_{m+n,0}. \tag{2.6}$$

We also have $\phi_0 v_0 = v_0$, as follows from (2.5).

Let us show that the operators ϕ_n leave H_0' invariant. For $n \geq 0$, this is obvious because of (2.6). In the case $n < 0$, it is enough to prove that $\phi_n v_0 \in H_0'$.

Consider the series

$$u(s_1, \ldots, s_m, z) = \phi^*(s_1 z)\phi^*(s_2 z) \ldots \phi^*(s_m z)v_0$$

$$= \prod_{i<j}\left(1 - \frac{s_j}{s_i}\right) \exp\left(-\sum_{n=1}^{\infty} \frac{b_{-n}}{n}\left(\sum_j s_j^n\right) z^n\right)v_0. \tag{2.7}$$

Let $u(s_1, \ldots, s_m, z) = \sum_{n \geq 0} u_{-n}(s_1, \ldots, s_m)z^n$. It is clear that $u_{-n} \in H_0$ for any numbers $s_1, \ldots, s_m$ such that $|s_1| > |s_2| > \cdots > |s_m|$ (since it is a sum of a convergent series of homogeneous vectors in H_0). By analytic continuation $u_{-n} \in H_0$ for any nonzero values of $s_1, \ldots, s_m$. In particular, setting $s_k = e^{2\pi i(k-1)/m}$, we get $u_0 = v_0$, $u_{-1} = u_{-2} = \cdots = u_{-m+1} = 0$, $u_{-m} = Cb_{-m}v_0$, where C is a nonzero constant. We conclude that $b_{-m}v_0 \in H_0'$. But due to (1.16) we have $b_{-m}v_0 = \phi_{-m}v_0 + \phi_{-m+1}\psi_{-1}^* v_0 + \cdots + \phi_0 \phi_{-m}^* v_0$. By induction in m, we get that $\phi_{-m}v_0 \in H_0'$, i.e H_0' is invariant under ϕ_{-m}.

Because of (1.16), this implies that H_0' is invariant under b_{-m}, $m > 0$, i.e., $H_0' = H_0$. $\qquad\square$

Proposition 2.2 is proved.

Remark. The connection between the $q \to 0$ limit of the vertex operator construction of level one $U_q(\widehat{\mathfrak{sl}}_2)$-modules and the boson-fermion correspondence which was utilized in our proof was found by I. Frenkel and N. Jing (private communication).

2.4   The Highest Eigenvalue of $F_{-+}(q)$, $F_{+-}(q)$

Proposition 2.3. (i) *There exists a unique vector $u_0(q) = v_0 + u_0^1 q + \cdots \in H[\![q]\!]$ such that its zero degree component is v_0, and a unique formal series $\lambda(q) = 1 + \lambda_1 q + \cdots \in \mathbb{C}[\![q]\!]$ such that $F_{-+}(q)u_0(q) = \lambda(q)u_0(q)$.*

(ii) *There exists a unique $F_{-+}(q)$-invariant $\mathbb{C}[\![q]\!]$-submodule U_0 in $H[\![q]\!]$ such that $H[\![q]\!] = \mathbb{C}[\![q]\!]u_0(q) \oplus U_0$.*

(iii) *There exists a unique vector $u_1(q) = v_1 + u_1^0 q + \ldots \in H[\![q]\!]$ such that its zero degree component is v_1, and a unique formal series $\lambda^*(q) = 1 + \lambda_1^* q + \cdots \in \mathbb{C}[\![q]\!]$ such that $F_{+-}(q)u_1(q) = \lambda^*(q)u_1(q)$. The series λ^* coincides with λ.*

(iv) *There exists a unique $F_{+-}(q)$-invariant $\mathbb{C}[\![q]\!]$-submodule U_1 in $H[\![q]\!]$ such that $H[\![q]\!] = \mathbb{C}[\![q]\!]u_1(q) \oplus U_1$.*

Proof. Since (iii), (iv) are analogous to (i), (ii), we prove (i), (ii) only.

(i) Let

$$F_{-+}(q) = \sum_{n \geq 0} F_n q^n. \tag{2.8}$$

Let us look for u_0, λ in the form

$$u_0(q) = \sum_{n \geq 0} u_0^n q^n, \quad \lambda(q) = \sum_{n \geq 0} \lambda_n q^n, \quad \lambda_0 = 1, \quad u_0^0 = v_0. \tag{2.9}$$

Then from $F_{-+}u_0 = \lambda u_0$ we get

$$\sum_{m=0}^{n} F_m u_0^{n-m} = \sum_{m=0}^{n} \lambda_m u_0^{n-m}, \quad n \geq 0. \tag{2.10}$$

This can be rewritten as a recursive relation

$$(F_0 - 1)u_0^n = \lambda_n v_0 - F_n v_0 + \sum_{m=1}^{n-1} (\lambda_{n-m} - F_{n-m})u_0^m. \tag{2.11}$$

This implies, in particular, that all vectors u_0^n must belong to H_0. The operator $F_0 - 1$ is not invertible (it kills v_0), but it is invertible on vectors of negative degree in H_0, by virtue of Proposition 2.2. Therefore, we must choose λ_n in such a way that the right hand side of (2.11) does not have a zero degree term. This can be done in a unique way. After λ_n is chosen, u_0^n is determined uniquely by

$$u_0^n = (F_0 - 1)^{-1}(\lambda_n v_0 - F_n v_0 + \sum_{m=1}^{n-1} (\lambda_{n-m} - F_{n-m})u_0^m). \tag{2.12}$$

(because u_0^n has to have a trivial zero degree component).

(ii) To define an invariant complement U_0 to the eigenvector u_0 is the same as to define a $\mathbb{C}[\![q]\!]$-linear function $\theta: H[\![q]\!] \to \mathbb{C}[\![q]\!]$ such that $F_{-+}^* \theta = \lambda \theta$ and $\theta(u_0) = 1$ (θ is the projection along U_0, U_0 is the kernel of θ). It is shown in the same way as in the proof of (i) that such a function is unique. $\square$

3   The Semi-Infinite Tensor Product Construction

3.1   The Kyoto Conjecture

Consider the matrix elements

$$G_n^0(q) = \langle v_0^*, F_{-+}(q)^n v_0 \rangle, \quad G_n^1(q) = \langle v_1^*, F_{+-}(q)^n v_1 \rangle, \quad (3.1)$$

where v_i^* are the lowest-weight vectors in L_i^* such that $\langle v_i, v_i^* \rangle = 1$.

Clearly, $G_n^i(q) \in \mathbb{C}[\![q]\!]$.

The following statement was conjectured in Ref. 1 (we call it "the Kyoto conjecture").

Theorem 3.1. *The sequence $G_n^i(q)^{1/n}$ for $i = 0$ or 1 is q-adically convergent, and its limit equals $\lambda(q)$.*

Proof. We give the proof in the case $i = 0$. The case $i = 1$ is analogous.

Let us write v_0 in the form $v_0 = \xi(q)u_0(q) + w(q)$, where $\xi \in \mathbb{C}[\![q]\!]$, $w \in U_0$. This can be done in a unique way. Then by Proposition 2.6 we have

$$F_{-+}(q)^n v_0 = \xi(q)\lambda(q)^n u_0(q) + F_{-+}(q)^n w(q). \quad (3.2)$$

So, it is enough to show that for any $N > 0$ $F_{-+}(q)^n w(q)$ is zero in $U_0/q^N U_0$ for a sufficently large n. That is, to show that $F_{-+}(q)$ is locally nilpotent in $U_0/q^N U_0$.

Let W be the subspace of H_0 spanned by all vectors of strictly negative degree. Let $P : U_0 \to W[\![q]\!]$ be the projection parallel to v_0. Let $M(q) = PF_{-+}(q)P^{-1}: W[\![q]\!] \to W[\![q]\!]$. Then $M(q) = \sum_{n \geq 0} M_n q^n$, $M_n \in \text{End } W$, and $M_0 = F_0|_W$. It is enough to prove local nilpotency of $M(q)$ in $W[\![q]\!]/q^N W[\![q]\!]$.

Fix N. We have $M(q) = \sum_{n=0}^{N-1} M_n q^n$ in $W[\![q]\!]/q^N W[\![q]\!]$. Let $w \in W$ be a homogeneous vector of degree m. Let d_n be the smallest degree of a nontrivial homogeneous component of M_n (remember that this degree is nonpositive). Let $d^* = \min_n d_n$. Let r be a positive integer such that $F_0^{r+1} = 0$ on vectors in H_0 of degree $\geq m + (N - 1)d^*$. Such r exists because of Proposition 2.3.

Then $M(q)^{Nr+N} w = 0$ in $W[\![q]\!]/q^N W[\![q]\!]$. Indeed, let us expand the power of $M(q)$. Then any term contributing to the coefficient to q^k, $k \leq N - 1$, will look like $F_0^{r_1} M_{s_1} F_0^{r_2} M_{s_2} \cdots F_0^{r_l} M_{s_l} F_0^{r_{l+1}} w$, where $l \leq k$. Since $l + \sum_{j=1}^{l+1} r_j = Nr + N$, we have that at least one r_j is $\geq r + 1$. Since the degree of any homogeneous component of the vector to which $F_0^{r_j}$ is applied in our term is clearly $\geq m + (N-1)d^*$ (remember that F_0 preserves degree), it follows from the choice of r that the whole term is zero.   □

Remark. In Ref. 1, the authors use the operators $\Phi(1)$ rather than $\Phi(-q^{-3})$. However, this variation does not affect quantity (3.1), so all our arguments remain valid.

Actually, our method of proof of Theorem 3.1 allows to prove a more general statement, also conjectured in Ref. 1.

Theorem 3.2. (i) *Let $w \in H$. Then there exist formal limits*

$$\eta_0(w) = \lim_{n \to \infty} \lambda(q)^{-n} \langle v_0^*, F_{-+}(q)^n w \rangle,$$
$$\eta_1(w) = \lim_{n \to \infty} \lambda(q)^{-n} \langle v_1^*, F_{+-}(q)^n w \rangle. \tag{3.3}$$

(ii) $\eta_i(w) = \theta_i(w) \langle v_i^*, u_i \rangle$, *where u_0, u_1 are the eigenvectors of the operators F_{-+}, F_{+-}, and θ_0, θ_1 are the linear functionals defined by $w \in \theta_i(w) u_i + U_i$, $i = 0, 1$.*

Proof. Analogous to Theorem 3.1.   □

3.2   The Kyoto Homomorphism

Let S be the set of sequences $\{p_n, n \geq 1\}$, $p_n \in \{+, -\}$, such that there exists $N = N(p)$ such that for $n > N$ $p_n = -p_{n-1}$. An element $p \in S$ is called a path. A path p is called odd if $p_n = (-1)^{n-1}$ for sufficiently large n, and even if $p_n = (-1)^n$ for sufficiently large n. The set of odd (even) paths is denoted by S_1 (respectively S_0), so $S = S_0 \cup S_1$. Let $T_i = \mathbb{C}[S_i]$, $i = 0, 1$, and $T = \mathbb{C}[S] = T_0 \oplus T_1$ be the spaces of functions on S_i, S which vanish almost everywhere. One can interpret T as a semiinfinite tensor product $\cdots \otimes V \otimes V$, where $V = \mathbb{C}v_+ \oplus \mathbb{C}v_-$ is a two-dimensional representation of $U_q(\widehat{\mathfrak{sl}}_2)$. Let $T^*[\![q]\!]$, $T^*(\!(q)\!)$ be the sets of all linear maps from T to $\mathbb{C}[\![q]\!]$, $\mathbb{C}(\!(q)\!)$ (here $\mathbb{C}(\!(q)\!)$ is the field of formal Laurent series). Following Ref. 1, let us define a $\mathbb{C}(q)$-linear map $K \colon L \to T^*(\!(q)\!)$, as follows.

Definition. The Kyoto homomorphism is the linear map $K \colon L \to T^*(\!(q)\!)$ defined by

$$(Kw)(p) = \lambda(q)^{-n} \eta_i(\Phi_{p_{2n}}(1) \cdots \Phi_{p_1}(1) w)$$
$$= \lambda(q)^{-n} \eta_i(F_{p_1 p_2 \cdots p_{2n}}(-q)^{3d} w), \quad w \in L, p \in S_i, \tag{3.4}$$

where n is any positive integer for which $p_{N+1} = -p_N$, $N \geq 2n$, and d is the operator of homogeneous degree.

Lemma 3.1. *The map K is well defined, i.e., does not depend on the choice of n.*

Proof. Let n, m be two positive integers satisfying the conditions of the definition, and let $n < m$. Then they give the same value of K because of the identity $\eta_0(F_{-+}w) = \lambda(q)\eta_0(w)$, $\eta_1(F_{+-}w) = \lambda(q)\eta_1(w)$.   □

It is clear that the map K sends L_i into $T_i^*(\!(q)\!)$, $i = 0, 1$.

In particular, we can define the "half-vacuum state" $s_0(q) = Kv_0$. Clearly, $s_0 \in T_0^*[\![q]\!]$.

Proposition 3.1. $(s_0(0))(p)$ *equals* 1 *if* $p_n = (-1)^n$, $n \geq 1$, *and* 0 *otherwise.*

Proof. It is easy to check that at $q = 0$ the product $\Phi_{\varepsilon_{2n}}(-q^{-3})\cdots\Phi_{\varepsilon_1}(-q^{-3})$ vanishes whenever $\varepsilon_j = \varepsilon_{j+1} = +$ for some j. This is easy to show using the relation $\phi_m^* \phi_{m-1}^* = 0$. Therefore, if the product $\Phi_{\varepsilon_{2n}}(-q^{-3})\cdots\Phi_{\varepsilon_1}(-q^{-3})v_0$ does not vanish at $q = 0$, then $\varepsilon_1 = -$ and two pluses cannot stand beside each other. So, if in addition the total numbers of pluses and minuses are the same (i.e the considered vector is in $H_0[\![q]\!]$) then the only possibility is $\varepsilon_j = (-1)^j$. This implies the proposition. $\qquad\square$

Remark. The main part of the conjecture in Ref. 1 is to show that when the map K (whose very existense was so far conjectural) is applied to the vector $G(p)$ of Kashiwara's upper global base of L corresponding to the path p, then the obtained functional in $T^*(\!(q)\!)$ $(KG(p))$ is actially in $T^*[\![q]\!]$, and tends to the characteristic function of p at $q \to 0$. The above arguments do not settle this question, at least without some additional work; one needs a certain technique of keeping track of leading degrees of q. We will discuss it in a later paper.

4   Computation of the Leading Eigenvalue and Eigenvector

It turns out that the eigenvalue $\lambda(q)$ and the eigenvectors $u_0(q)$, $u_1(q)$ of $F_{-+}(q)$, $F_{+-}(q)$ can be computed explicitly. The following theorem was recently proved by M. Jimbo.

Theorem 4.1. *The following identities hold:*

$$\lambda(q) = \frac{(q^6; q^8)_\infty^2}{(q^4; q^8)_\infty^2}, \tag{4.1}$$

where $(a, p)_\infty$ *denotes* $\prod_{n=0}^\infty (1 - ap^n)$;

$$u_0(q) = e^{\mathcal{F}_0} v_0, \quad u_1(q) = e^{\mathcal{F}_1} v_1, \tag{4.2}$$

where

$$\mathcal{F}_0 = -\frac{1}{2}\sum_{n=1}^\infty \frac{1+q^{2n}}{n}q^{2n}b_{-n}^2 - \sum_{n=1}^\infty \frac{1}{n}(-q)^{3n}b_{-n} - \sum_{n=1}^\infty \frac{1-q^{2n}}{2n}q^{2n}b_{-2n},$$

$$\mathcal{F}_1 = -\frac{1}{2}\sum_{n=1}^\infty \frac{1+q^{2n}}{n}q^{2n}b_{-n}^2 + \sum_{n=1}^\infty \frac{(-q)^n}{n}b_{-n} - \sum_{n=1}^\infty \frac{1-q^{2n}}{2n}q^{2n}b_{-2n} \tag{4.3}$$

In particular, the series $\lambda(q)$ defines a nonvanishing analytic function in the region $|q| < 1$.

Remark. We see that $\lambda(q)^{1/2} = 1 + q^4 - q^6 + q^8 \mod q^{10}$, which was found in Ref. 1.

Proof. Let us first prove the formula for u_0. In the sequel we assume that q is a complex number with $|q| < 1$. For brevity we will write Φ_- instead of $\Phi_-(-q^{-3})$, and $\sum$ for $\sum_{n=1}^{\infty}$. The index n will always take positive integer values.

According to Section 1, we have

$$F_{-+} = \frac{1}{2\pi i} \int_{|z|=1} Y(z) \frac{dz}{z},$$
$$Y(z) = z(\Phi_- X^-(z) - qX^-(z)\Phi_-)\Phi_-, \quad (4.4)$$

(the contour is oriented anticlockwise). Substituting (1.10) and (1.12) into (4.4), after normal ordering (i.e., putting terms with b_{-n} to the right, and with b_n to the left) we obtain

$$Y(z)\big|_{H_0} = \frac{1 - q^2}{(1 + qz^{-1})^2(1 + qz)} \frac{(q^2; q^4)_\infty}{(q^4; q^4)_\infty}$$
$$\times \exp\left(\sum (2(-q)^n - (1 + q^{2n})z^n) \frac{b_{-n}}{n}\right)$$
$$\times \exp\left(-\sum (\frac{2(-q)^n}{1 + q^{2n}} - z^{-n}) \frac{b_n}{n}\right). \quad (4.5)$$

Let us look for an eigenvector of F_{-+} in H_0 in the form of an exponential function of a quadratic polynomial:

$$u_0 = \exp\left(\sum (\beta_n b_{-n}^2 + \gamma_n b_{-n})\right) v_0, \quad (4.6)$$

where β_n, γ_n are undetermined coefficients depending on q. Applying F_{-+} to (4.6) and using (4.4), (4.5), after normal ordering we get

$$F_{-+} u_0 = \frac{1}{2\pi i} \exp\left(\sum (\beta_n b_{-n}^2 + \gamma_n b_{-n})\right)$$
$$\times \exp\left(\sum \frac{2(-q)^n}{n} \left(1 - \frac{2\beta_n n}{1 + q^{2n}}\right) b_{-n}\right)$$
$$\times \int_{|z|=1} \exp\left(\sum \left(2\beta_n z^{-n} - \frac{(1 + q^{2n})z^n}{n}\right) b_{-n}\right) g(z) \frac{dz}{z}, \quad (4.7)$$

where

$$g(z) = \frac{1 - q^2}{(1 + qz^{-1})^2(1 + qz)} \frac{(q^2; q^4)_\infty}{(q^4; q^4)_\infty} h(z)$$

$$h(z) = \exp\left(\sum \beta_n \left(z^{-n} - \frac{2(-q)^n}{1 + q^{2n}}\right)^2 + \sum \gamma_n \left(z^{-n} - \frac{2(-q)^n}{1 + q^{2n}}\right)\right). \quad (4.8)$$

Therefore, the identity $F_{-+}u_0 = \lambda u_0$ that we would like to satisfy can be rewritten in the form

$$\frac{1}{2\pi i} \int_{|z|=1} \exp\left(\sum\left(2\beta_n z^{-n} - \frac{(1+q^{2n})z^n}{n}\right)b_{-n}\right)g(z)\frac{dz}{z}$$

$$= \lambda(q)\exp\left(-\sum \frac{2(-q)^n}{n}\left(1 - \frac{2\beta_n n}{1+q^{2n}}\right)b_{-n}\right). \quad (4.9)$$

Let us compute the integral on the l.h.s. of (4.9). Assume that $h(z)$, as a power series in z^{-1}, defines a function holomorphic in the region $|z| \geq |q|\delta$ for some $\delta < 1$ (including $z = \infty$), and $h(-q) = 0$, $h'(-q) \neq 0$. (We will later check that these conditions are satisfied for the undetermined coefficients we are going to choose). Then the function $g(z)$ has a simple pole at $z = -q$. Therefore, by the residue formula, the l.h.s. of (4.9) equals

$$\frac{1}{2\pi i} \int_{|z|=|q|\delta} \exp\left(\sum\left(2\beta_n z^{-n} - \frac{(1+q^{2n})z^n}{n}\right)b_{-n}\right)g(z)\frac{dz}{z}$$

$$+ \exp\left(\sum\left(2\beta_n(-q)^{-n} - \frac{(1+q^{2n})(-q)^n}{n}\right)b_{-n}\right)$$

$$\times \lim_{z\to-q} g(z)(1+qz^{-1}) \quad (4.10)$$

(we have moved the contour of integration through the pole).

We would like the integral term in (4.10) to be proportional to the integral on the left-hand side of (4.9); then we can express the integral explicitly, via the nonintegral term in (4.10). To see if we can do this, let us consider the change of variable $z \to q^2 z^{-1}$ in the integral term in (4.10) (this will bring us to the contour $|z| = |q|\delta^{-1}$, which can then be deformed to $|z| = 1$, since there is no singularities between these two contours). We obtain

$$\frac{1}{2\pi i} \int_{|z|=1} \exp\left(\sum\left(2\beta_n q^{-2n} z^n - \frac{(1+q^{2n})q^{2n}z^{-n}}{n}\right)b_{-n}\right)$$

$$\times g(q^2 z^{-1})\frac{dz}{z}. \quad (4.11)$$

Clearly, this integral is proportional to the l.h.s. of (4.9) if two conditions are satisfied:

1. $2\beta_n q^{-2n} = -(1+q^{2n})/n$, and

2. $g(q^2 z^{-1}) = -g(z)$ in the neighborhood of $|z| = |q|$.

So we choose $\beta_n = -\frac{1}{2}(1+q^{2n})q^{2n}/n$ to satisfy the first condition, and assume that the second condition holds (we will later choose the undetermined coefficients γ_n in such a way that it does). Then we get

$$
\frac{1}{2\pi i} \int_{|z|=|q|\delta} \exp\left(\sum\left(2\beta_n z^{-n} - \frac{(1+q^{2n})z^n}{n}\right)b_{-n}\right)g(z)\frac{dz}{z}
$$
$$
= \frac{1}{2}\exp\left(\sum\left(2\beta_n(-q)^{-n} - \frac{(1+q^{2n})(-q)^n}{n}\right)b_{-n}\right)\lim_{z\to-q}g(z)(1+qz^{-1})
$$
$$
= \frac{1}{2}\exp\left(-\sum\left(\frac{2(1+q^{2n})(-q)^n}{n}\right)b_{-n}\right)\lim_{z\to-q}g(z)(1+qz^{-1}).
$$
$$(4.12)$$

Substituting this into (4.9), we get

$$
\lambda(q) = \frac{\lim_{z\to-q}g(z)(1+qz^{-1})}{2}.
$$
$$(4.13)$$

Let us now find $h(z)$, i.e., the sequence γ_n. We have

$$
\frac{g(z)}{g(q^2 z^{-1})} = z^2 q^{-2}\frac{1+q^3 z^{-1}}{1+qz}\frac{h(z)}{h(q^2 z^{-1})}.
$$
$$(4.14)$$

To satisfy property 2, we want this ratio to be equal to -1. This implies that $h(z)$ vanishes at $z=q$. Therefore, it is natural to look for $h(z)$ in the form

$$
h(z) = (1 - q^2 z^{-2})\exp(f(z)),
$$
$$(4.15)$$

where f is regular in the region $|z| \geq |q|\delta$. Then from $g(z)/g(q^2 z^{-1}) = -1$ and (4.14) we get

$$
\exp(f(z) - f(q^2 z^{-1})) = \frac{1+qz}{1+q^3 z^{-1}},
$$
$$(4.16)$$

or

$$
f(z) = c - \ln(1+q^3 z^{-1}),
$$
$$
\ln h(z) = c + \ln(1 - q^2 z^{-2}) - \ln(1+q^3 z^{-1}),
$$
$$(4.17)$$

where c depends on q. From this equation and (4.8) we get

$$
-\frac{1}{2}\sum\frac{1+q^{2n}}{n}q^{2n}\left(z^{-n} - \frac{2(-q)^n}{1+q^{2n}}\right)^2 + \sum\gamma_n\left(z^{-n} - \frac{2(-q)^n}{1+q^{2n}}\right)
$$
$$
= -\sum\frac{q^{2n}z^{-2n}}{n} + \sum\frac{(-q)^{3n}z^{-n}}{n}. \quad (4.18)
$$

From this it is easy to obtain equations for γ_n:

$$
\gamma_n = -\frac{(-q)^{3n}}{n} - \frac{1+(-1)^n}{4}\frac{(1-q^n)q^n}{n}.
$$
$$(4.19)$$

Thus, we have obtained the first formula in (4.2).

It remains to compute $\lambda(q)$. From (4.13) we get

$$
\begin{aligned}
\lambda(q) &= -\frac{q}{2}\frac{(q^2;q^4)_\infty}{(q^4;q^4)_\infty}h'(-q) = \frac{(q^2;q^4)_\infty}{(q^4;q^4)_\infty}\exp(f(-q))\\
&= \frac{(q^2;q^4)_\infty}{(q^4;q^4)_\infty}\frac{e^c}{1-q^2} = \frac{(q^6;q^4)_\infty}{(q^4;q^4)_\infty}e^c,
\end{aligned}
\tag{4.20}
$$

where c is defined by (4.17) So, we need to compute e^c. From (4.17) it is seen that c is the free term in $\ln h(z)$, so from (4.8) we get

$$
\begin{aligned}
c &= \sum \beta_n \frac{4q^{2n}}{(1+q^{2n})^2} - \sum \gamma_n \frac{2(-q)^n}{1+q^{2n}}\\
&= \sum_n \frac{q^{4n}-q^{6n}}{1+q^{4n}} = \ln\frac{(q^6;q^8)_\infty(q^8;q^8)_\infty}{(q^4;q^8)_\infty(q^{10};q^8)_\infty}.
\end{aligned}
\tag{4.21}
$$

Thus, from (4.20) we finally obtain

$$
\lambda(q) = \frac{(q^6;q^4)_\infty(q^6;q^8)_\infty(q^8;q^8)_\infty}{(q^4;q^4)_\infty(q^4;q^8)_\infty(q^{10};q^8)_\infty} = \frac{(q^6;q^8)_\infty^2}{(q^4;q^8)_\infty^2},
\tag{4.22}
$$

which proves (4.1).

It is easy to check that our regularity assumptions on the function $h(z)$ hold true, so the proofs of (4.1) and the first part of (4.2) are complete.

It remains to prove the second part of (4.2). Now it is immediate. Indeed, it is clear that u_1 is proportional to Φ_-u_0, so after normal ordering in this expression we get $\mathcal{F}_1 = \sum(\beta_n b_{-n}^2 + \tilde\gamma_n b_{-n})$ where

$$
\tilde\gamma_n = \gamma_n - 2\beta_n\frac{(-q)^n}{1+q^{2n}} + \frac{(-q)^n}{n},
\tag{4.23}
$$

which yields formula (4.3) for $\mathcal{F}_1$.   □

Acknowledgments: I would like to thank M. Jimbo for useful remarks and corrections to the first version of this paper, and for sharing with me the contents of Section 4. I am also grateful to E. Frenkel, I. Frenkel, D. Kazhdan, and T. Miwa for useful discussions.

5   REFERENCES

1. B. Davies, O. Foda, M. Jimbo, T. Miwa, and A. Nakayashiki. Diagonalization of the XXZ Hamiltonian by vertex operators. *Commun. Math. Phys.*, 151 (1): 89–153, 1993.

2. E. Date, M. Jimbo, and M. Okado. Crystal base and q-vertex operators. *Commun. Math. Phys.*, 155 (1): 47–69, 1993.

3. I. B. Frenkel and N. Jing. Vertex representations of quantum affine algebras. *Proc. Nat. Acad. Sci. USA*, 85 (24): 9373–9377, 1988.

4. M. Jimbo, K. Miki, T. Miwa, and A. Nakayashiki. Correlation functions of the XXZ model for $\Delta < -1$. *Phys. Lett. A*, 168 (4): 256–263, 1992.

5. K. Bardakci and M. B. Halpern. New dual quark models. *Phys. Rev.*, D3: 2493–2506, 1971.

6. I. B. Frenkel. Two constructions of affine Lie algebra representations and boson-fermion correspondence in quantum field theory. *J. Funct. Anal.*, 44 (3): 259–327, 1981.

7. M. Jimbo and T. Miwa. *Algebraic Analysis of Solvable Lattice Models*, volume 85 of *CBMS Regional Conference Series in Mathematics*. Amer. Math. Soc., Providence, RI, 1995.

Index